T-Labs Series in Telecommunication Services

Series editors

Sebastian Möller, Berlin, Germany
Axel Küpper, Berlin, Germany
Alexander Raake, Berlin, Germany

For further volumes:
http://www.springer.com/series/10013

Sebastian Möller · Alexander Raake
Editors

Quality of Experience

Advanced Concepts, Applications and Methods

Springer

Editors
Sebastian Möller
Quality and Usability Lab, Telekom Innovation Laboratories
TU Berlin
Berlin
Germany

Alexander Raake
Assessment of IP-based Applications, Telekom Innovation Laboratories
TU Berlin
Berlin
Germany

ISSN 2192-2810 ISSN 2192-2829 (electronic)
ISBN 978-3-319-37526-7 ISBN 978-3-319-02681-7 (eBook)
DOI 10.1007/978-3-319-02681-7
Springer Cham Heidelberg New York Dordrecht London

Printed on acid-free paper

Springer is part of Springer Science+Business Media (www.springer.com)

Preface

This book is meant as a handbook centered around Quality of Experience of information and communication systems and services, its underlying concepts, and its application examples. It is based on the work accomplished in roughly the last two decades by researchers and practitioners in many diverse fields, such as telecommunications engineering, speech, audio and video processing, psychophysics, human–computer interaction, psychology, ergonomics, and human-factors research, as well as innovation and economics.

Starting point for the book were the activities on the definition of the term "Quality of Experience" (QoE) and related concepts which have been initiated by an international group gathered in the "European Network on Quality of Experience in Multimedia Systems and Services," Qualinet (COST Action IC 1003). These activities resulted in a so-called "White Paper on Definitions of Quality of Experience," compiled in a first version in 2012, and updated in 2013, following an intense discussion among the Qualinet members and external experts. The Qualinet White Paper being limited to mere definitions, we felt the necessity to explain the concepts to a larger community in a more detailed way, paving the way for their application for different types of systems and services. The result is the present book, the motivation for which is outlined in more detail in Chap. 1.

The editors would like to thank all authors who have contributed to the book, as well as all authors and contributors of the Qualinet White Paper which formed its inspiring basis. We would like to extend our thanks to some external and anonymous reviewers who helped shaping the content of some of the chapters. The compilation and formatting of the book was largely supported by Marc Hanisch, M.A., whose work we are very grateful for. Finally, thanks to Christoph Baumann and his team at Springer for organizing the publication process.

Berlin, November 2013

Sebastian Möller
Alexander Raake

Contents

Contributors

Ercan Altinsoy Chair of Communication Acoustics, TU Dresden, Dresden, Germany, e-mail: ercan.altinsoy@tu-dresden.de

Jan-Niklas Antons Quality and Usability Lab, Telekom Innovation Laboratories, TU Berlin, Berlin, Germany, e-mail: jan-niklas.antons@tu-berlin.de

Savvas Argyropoulos Assessment of IP-based Applications, Telekom Innovation Laboratories, TU Berlin, Berlin, Germany, e-mail: savvas@ieee.org

Sebastian Arndt Quality and Usability Lab, Telekom Innovation Laboratories, TU Berlin, Berlin, Germany, e-mail: sebastian.arndt@telekom.de

Marcus Barkowsky University of Nantes and IRCCyN, Nantes, France, e-mail: marcus.barkowsky@univ-nantes.fr

Sergio Beker European Research Center, Huawei Technologies, Munich, Germany, e-mail: sergio.beker@huawei.com

Benjamin Belmudez Quality and Usability Lab, Telekom Innovation Laboratories, TU Berlin, Berlin, Germany, e-mail: ben.belmudez@gmail.com

Jens Berger SwissQual AG—A Rohde & Schwarz Company, Zuchwil, Switzerland, e-mail: jens.berger@swissqual.com

Gunilla Berndtsson Multimedia Technologies, Ericsson Research, Ericsson AB, Stockholm, Sweden, e-mail: gunilla.berndtsson@ericsson.com

Justus Beyer Quality and Usability Lab, Telekom Innovation Laboratories, TU Berlin, Berlin, Germany, e-mail: justus.beyer@qu.tu-berlin.de

Adam Borowiak Department of Electronics and Telecommunications, Norwegian University of Science and Technology, Trondheim, Norway, e-mail: adam.borowiak@iet.ntnu.no

Kjell Brunnström Acreo Swedish ICT AB, Stockholm and Mid Sweden University, Sundsvall, Sweden, e-mail: kjell.brunnstrom@acreo.se

Rahul Chaudhari Institute for Media Technology, TU Munich, Munich, Germany, e-mail: rahul.chaudhari@tum.de

Nicolas Côté ISEN Department, Institute of Electronics, Microelectronics and Nanotechnology, Lille, France, e-mail: nicolas.cote@isen.fr

Katrien De Moor Department of Telematics, Norwegian University of Science and Technology, Trondheim, Norway; iMinds-MICT, Ghent University, Ghent, Belgium, e-mail: katrien.demoor@item.ntnu.no

Touradj Ebrahimi École Polytechnique Fédérale de Lausanne, Lausanne, Switzerland, e-mail: touradj.ebrahimi@epfl.ch

Sebastian Egger Telecommunications Research Center Vienna (FTW), Vienna, Austria, e-mail: egger@ftw.at

Bernhard Feiten Telekom Innovation Laboratories, Deutsche Telekom, Berlin, Germany, e-mail: bernhard.feiten@telekom.de

Markus Fiedler Blekinge Institute of Technology (BTH), Karlskrona, Sweden, e-mail: mfi@bth.se

Matthias Frank Institute of Electronic Music and Acoustics, University of Music and Performing Arts Graz, Graz, Austria, e-mail: frank@iem.at

Marie-Neige Garcia Assessment of IP-based Applications, Telekom Innovation Laboratories, TU Berlin, Berlin, Germany, e-mail: marie-neige.garcia@telekom.de

Dennis Guse Quality and Usability Lab, Telekom Innovation Laboratories, TU Berlin, Berlin, Germany, e-mail: dennis.guse@telekom.de

Ulrich Heute Digital Signal Processing and System Theory, CAU Kiel, Kiel, Germany, e-mail: uh@tf.uni-kiel.de

Florian Hinterleitner Quality and Usability Lab, Telekom Innovation Laboratories, TU Berlin, Berlin, Germany, e-mail: florian.hinterleitner@tu-berlin.de

Tobias Hoßfeld Chair of Communication Networks, Institute of Computer Science, University of Würzburg, Würzburg, Germany, e-mail: hossfeld@informatik.uni-wuerzburg.de

Lucjan Janowski AGH University of Science and Technology, Kraków, Poland, e-mail: janowski@kt.agh.edu.pl

Christian Keimel Institute for Data Processing, TU Munich, Munich, Germany, e-mail: christian.keimel@tum.de

Mohamed-Chaker Larabi XLIM, Université de Poitiers, Poitiers, France, e-mail: chaker.larabi@univ-poitiers.fr

Patrick Le Callet University of Nantes and IRCCyN, Nantes, France, e-mail: patrick.lecallet@univ-nantes.fr

Pierre Lebreton Assessment of IP-based Applications, Telekom Innovation Laboratories, TU Berlin, Berlin, Germany, e-mail: pierre.lebreton@telekom.de

Mikołaj Leszczuk AGH University of Science and Technology, Kraków, Poland, e-mail: leszczuk@kt.agh.edu.pl

Blazej Lewcio Quality and Usability Lab, Telekom Innovation Laboratories, TU Berlin, Berlin, Germany, e-mail: blazej.lewcio@telekom.de

Sebastian Möller Quality and Usability Lab, Telekom Innovation Laboratories, TU Berlin, Berlin, Germany, e-mail: sebastian.moeller@telekom.de

Niall Murray Athlone Institute of Technology, Athlone, Ireland, e-mail: nmurray@research.ait.ie

Matteo Naccari BBC R&D, London, UK, e-mail: matteo.naccari@bbc.co.uk

Christoph Norrenbrock Digital Signal Processing and System Theory, CAU Kiel, Kiel, Germany, e-mail: cno@tf.uni-kiel.de

Manuela Pereira University of Beira Interior, Covilhã, Portugal, e-mail: mpereira@di.ubi.pt

Andrew Perkis Norwegian University of Science and Technology, Trondheim, Norway, e-mail: andrew@iet.ntnu.no

Antonio Pinheiro University of Beira Interior, Covilhã, Portugal, e-mail: pinheiro@ubi.pt

Alexander Raake Assessment of IP-based Applications, Telekom Innovation Laboratories, TU Berlin, Berlin, Germany, e-mail: alexander.raake@telekom.de

Benjamin Rainer Alpen-Adria-Universität Klagenfurt, Klagenfurt, Austria, e-mail: benjamin.rainer@itec.aau.at

Peter Reichl University of Vienna, Vienna, Austria; UEB/Télécom Bretagne Rennes, Rennes, France, e-mail: peter.reichl@univie.ac.at

Ralf Reichmuth Quality and Usability Lab, Telekom Innovation Laboratories, TU Berlin, Berlin, Germany, e-mail: ralf.reichmuth@telekom.de

Ulrich Reiter Department of Electronics and Telecommunications, Norwegian University of Science and Technology, Trondheim, Norway, e-mail: reiter@iet.ntnu.no

Miguel Rios-Quintero Assessment of IP-based Applications, Telekom Innovation Laboratories, TU Berlin, Berlin, Germany, e-mail: miguel.rios-quintero@telekom.de

Raimund Schatz Telecommunications Research Center Vienna (FTW), Vienna, Austria, e-mail: schatz@ftw.at

Robert Schleicher Quality and Usability Lab, Telekom Innovation Laboratories, TU Berlin, Berlin, Germany, e-mail: robert.schleicher@tu-berlin.de

Katrin Schoenenberg Assessment of IP-based Applications, Telekom Innovation Laboratories, TU Berlin, Berlin, Germany, e-mail: katrin.schoenenberg@telekom.de

Lea Skorin-Kapov Faculty of Electrical Engineering and Computing, University of Zagreb, Zagreb, Croatia, e-mail: lea.skorin-kapov@fer.hr

Janto Skowronek Assessment of IP-based Applications, Telekom Innovation Laboratories, TU Berlin, Berlin, Germany, e-mail: janto.skowronek@telekom.de

Sascha Spors Institute of Communications Engineering, University of Rostock, Rostock, Germany, e-mail: sascha.spors@uni-rostock.de

Nicolas Staelens Department of Information Technology, Internet Based Communication Networks and Services, Ghent University—iMinds, Ghent, Belgium, e-mail: nicolas.staelens@intec.ugent.be

Eckehard Steinbach Institute for Media Technology, TU Munich, Munich, Germany, e-mail: eckehard.steinbach@tum.de

Dominik Strohmeier Assessment of IP-based Applications, Telekom Innovation Laboratories, TU Berlin, Berlin, Germany, e-mail: dominik.strohmeier@gmail.com

Peter Svensson Department of Electronics and Telecommunications, Norwegian University of Science and Technology, Trondheim, Norway, e-mail: svensson@iet.ntnu.no

Christian Timmerer Alpen-Adria-Universität Klagenfurt, Klagenfurt, Austria, e-mail: christian.timmerer@itec.aau.at

Anna Ukhanova Technical University of Denmark, Kongens Lyngby, Denmark, e-mail: annuk@fotonik.dtu.dk

Markus Vaalgamaa Skype Labs/Microsoft, Tallinn, Estonia, e-mail: markus.vaalgamaa@skype.net

Martín Varela VTT Technical Research Centre of Finland, Oulu, Finland, e-mail: martin.varela@vtt.fi

Marcel Wältermann Quality and Usability Lab, Telekom Innovation Laboratories, TU Berlin, Berlin, Germany, e-mail: marcel.waeltermann@alumni.tu-berlin.de

Markus Waltl Alpen-Adria-Universität Klagenfurt, Klagenfurt, Austria, e-mail: markus.waltl@itec.aau.at

Ina Wechsung Quality and Usability Lab, Telekom Innovation Laboratories, TU Berlin, Berlin, Germany, e-mail: ina.wechsung@telekom.de

Benjamin Weiss Quality and Usability Lab, Telekom Innovation Laboratories, TU Berlin, Berlin, Germany, e-mail: bweiss@telekom.de

Tilo Westermann Quality and Usability Lab, Telekom Innovation Laboratories, TU Berlin, Berlin, Germany, e-mail: tilo.westermann@telekom.de

Hagen Wierstorf Assessment of IP-based Applications, Telekom Innovation Laboratories, TU Berlin, Berlin, Germany, e-mail: hagen.wierstorf@tu-berlin.de

Junyong You Christian Michelsen Research AS, Bergen, Norway, e-mail: junyong.you@cmr.no

Andrej Zgank University of Maribor, Maribor, Slovenia, e-mail: andrej.zgank@uni-mb.si

Franz Zotter Institute of Electronic Music and Acoustics, University of Music and Performing Arts Graz, Graz, Austria, e-mail: zotter@iem.at

Part I
Concepts

Chapter 1
Motivation and Introduction

Sebastian Möller and Alexander Raake

Abstract In this chapter, we provide a motivation for the upcoming chapters of the book. We discuss how the concept of Quality of Experience (QoE) has evolved during the last decades, resulting in a need for a common terminology, as well as the need for applying the identified concepts to new applications and services. The first issue was already addressed by the "Qualinet White Paper on Definitions of Quality of Experience", the history of which will be briefly reviewed, but due to its very nature that White Paper could not cover all concepts, applications and methods in sufficient depth in order to be helpful for scientists and practitioners alike. We hope to overcome this limitation with the present book, and present an outline of the contents and the relationships between individual chapters.

1.1 Quality of Information and Communication Technology

With the increasing development of information and communication technology (ICT) systems and services, the need for evaluating their quality becomes urgent. Systems for transmitting information from one user to another (e.g. a data link) need to be evaluated with respect to their performance, i.e. whether they transport the information effectively and efficiently. Systems for delivering media to human consumers (e.g. an IP-based television service) can be evaluated with respect to their transmission quality, i.e. whether the user experiences a high quality when consuming the media content. Systems enabling human-to-human communication

S. Möller (✉)
Quality and Usability Lab, Telekom Innovation Laboratories, TU Berlin, Berlin, Germany
e-mail: sebastian.moeller@telekom.de

A. Raake
Assessment of IP-based Applications, Telekom Innovation Laboratories, TU Berlin, Berlin, Germany
e-mail: alexander.raake@telekom.de

S. Möller and A. Raake (eds.), *Quality of Experience*,
T-Labs Series in Telecommunication Services, DOI: 10.1007/978-3-319-02681-7_1,

(e.g. a Voice-over-IP communication system) can be evaluated with respect to their conversation quality, i.e. whether they enable a good communication of information between the partners. And finally, systems for human-machine interaction (e.g. a web site) can be evaluated with respect to their usability, i.e. whether they enable an effective, efficient and satisfying interaction to their users.

In all these cases, it is commonly assumed that high performance, transmission quality, conversation quality and usability will in one way or another lead to a high acceptance of the respective services, i.e. the actual number of users would be high. This assumption is partly based on empirical evidence, namely that low-quality systems sometimes suffer from low acceptance, but there are a number of contrasting examples (such as the first SMS systems), which despite low quality and usability resulted in an enormous success. This shows that the relationship between performance, quality, usability and acceptance of a system or service is still poorly understood.

Even more profoundly, there seems to be no agreement of what the term "quality" actually means.[1] Until the turn of the century, the term quality has mostly been used by engineers to describe the "totality of characteristics of an entity [...] that bear on its ability to satisfy stated or implied needs" (EN ISO 9000, 2000, cited after [2]). This "totality of characteristics" is related to the Latin origin "qualitas" of the English word "quality", and is something which we nowadays would call the "character" of an entity. This understanding of the term "quality" is also reflected by its use in "Quality of Service" (QoS) which in the networking community was for years a fixed synonym for a set of guaranteed performance characteristics of a network connection. Around the turn of the century, researchers from different disciplines such as computer science, telecommunication engineering, psychophysics, psychology, sociology, and communication sciences started to discuss about the meaning of this old term, and tried to understand the processes which lead to its formation in a human user.

As a consequence, the quality definition has been improved in 2005, stating that quality is the "degree to which a set of inherent characteristics [...] fulfils requirements" [1]. Other scientists defined quality from a perceiving person's point-of-view as the "result of judgment of the perceived composition of an entity with respect to its desired composition" [2]. This definition involves a perception and a judgment process, during which the perceiving person compares the perceptual event with a (so-far unknown) reference. The character of the perceived composition is not necessarily a permanent characteristic of an entity; in fact, the reference may influence what is actually perceived. In any case, as the result of the comparison, quality is always relative and happens as a "quality event" in a particular spatial, temporal and functional context. Such a context apparently needs to be taken into account when quality is to be evaluated.

[1] More details on the quality and QoE terminology and its history and can be found in Chap. 2.

In parallel to the re-consideration of the term “quality”, the term “Quality of Experience” (QoE) has gained momentum and followers, mainly with respect to media transmission systems and services (see the discussion in [3], and the detailed overview in Chap. 2. This term was born to counter-balance the term Quality of Service with something which addresses the user’s perceptions and experiences, because those were considered to be more appropriate for designing systems and services with a high acceptance. Thus, a paradigm shift could be observed for service providers to deliver services not with a high QoS, but with a high QoE to their customers. This trend can also be observed for interactive human-machine interfaces, where it coincides with a focus shift from classical “usability” (in terms of effectiveness and efficiency) towards the design of experiences that people have through the use of these interfaces, the so-called “User Experience” (UX). Also for these concepts, evaluation methods were scarce, and so systems and services could not be shaped to provide maximum UX. The underlying reason, again, was a missing solid theoretical and practical framework for these concepts, and—above all—a missing well-accepted definition.

1.2 The “Qualinet White Paper on Definitions of Quality of Experience”

Focussing on QoE and multimedia services, the European Network on Quality of Experience in Multimedia Systems and Services, Qualinet (COST Action IC 1003[2]), started in 2011 to foster the scientific discussion about the definition of the term QoE and related concepts. This discussion resulted from the need to agree on a working definition for this term which facilitates the communication of ideas within a multidisciplinary group, where a joint interest around multimedia communication systems existed, but was approached from different perspectives. The idea was to extend the notion of network-centric QoS by defining a user-centric concept of QoE. The main scientific objective of the network was the development of methodologies for subjective and instrumental quality metrics taking into account current and new trends in multimedia communication systems, as witnessed by the appearance of new types of content and interactions.

As a result of this discussion, the “Qualinet White Paper on Definitions of Quality of Experience” [3] was compiled on the basis of a first open call for ideas which was launched for the February 2012 Qualinet Meeting held in Prague, Czech Republic. The ideas were presented as short statements during that meeting, reflecting the ideas of individuals of different background and working interests. During the Prague meeting, the ideas were further discussed and consolidated in the form of a structure for the White Paper. An open call for authors was issued at that meeting, and coordinating authors were assigned for individual sections which were defined

[2] www.qualinet.eu

in the joint group. The individual sections were prepared by the authors, integrated and aligned by an editing group, and the entire document was iterated with the entire group of authors. Furthermore, the draft text was discussed with the participants of the Dagstuhl Seminar 12181 "Quality of Experience: From User Perception to Instrumental Metrics" (Schloss Dagstuhl, Germany, May 1–4 2012), at the November 2012 Qualinet Meeting in Zagreb, Croatia, and during an online conference in January 2013. This resulted in Version 1.2 of the document which is available under www.qualinet.eu.

Although the Qualinet White Paper is considered as a main scientific basis also for this book, its purpose was quite different from the beginning. Due to its focus on definitions, the authors of each chapter of the White Paper were asked to write a maximum of two pages (excl. references) so to avoid an imbalance of topics covered in the paper, and to keep the focus of the paper on definitions which should be bold and easily extractable. Although a chapter on applications was included in the White Paper, it became clear very early that such a chapter could not cover the diversity of applications in the ICT domain for which QoE is an important topic, nor could it address the wide range of subjective evaluation methods and instrumental prediction models which are available or being developed for these applications. Thus, apart from the White Paper, we saw a substantial need for a broader discussion of the term, its underlying concepts, and the many application cases for which these concepts are relevant today.

1.3 Topics Addressed in this Book

As a result, we decided to start the present book project. The book is meant as a handbook centered around Quality of Experience of information and communication systems and services, its underlying concepts, and its application examples. While starting from the definitions and the ideas of the White Paper, it became clear that we would not be limited to these, as we tried to reach out to a larger and more diverse community of scientists to contribute to the book. As a matter of fact, each chapter was prepared by well-selected scientists or practicioneers of the respective matter, reflecting also their personal thoughts on and experiences with the topic, and not necessarily a group opinion. Nevertheless, the editors took care in aligning the individual chapters so that a more-or-less congruent picture of the multi-faceted concept of Quality of Experience arose.

We roughly divided the book into two parts. Whereas the first part relates to general concepts and theories which are relevant for almost all types of applications, the second part addresses these concepts and ideas in the light of individual applications.

The book starts with a chapter on definitions of quality and Quality of Experience in Chap. 2. Alexander Raake and Sebastian Egger provide an overview of quality, service quality and QoE definition work, discuss processes of human perception, experience and judgment, and come up with a new definition of QoE which elaborates the one of the White Paper. The notion of experience is also taken up by Ina Wechsung and Katrien De Moor which relate QoE to User Experience. Their Chap. 3 reviews the developments in Human-Computer Interaction (HCI), and relates QoE to emotions and needs.

Chapters 4, 5 and 6 address the dichotomy of quality: On the one hand the factors which are expected to influence QoE, and on the other hand the resulting perception of the user, in terms of quality features. Chapter 4 by Ulrich Reiter, Kjell Brunnström, Katrien De Moor, Mohamed-Chaker Larabi, Manuela Pereira, Antonio Pinheiro, Junyong You and Andrej Zgank discusses the influence factors and classifies them into factors related to the human user, the technical system, and the context of use. In turn, Chap. 5 by Sebastian Möller, Marcel Wältermann and Marie-Neige Garcia discusses perceived QoE as an event in a multi-dimensional perceptual space, and illustrates psychophysical methods as well as empirical results for extracting such dimensions. This dichotomy is finalized by Martìn Varela, Lea Skorin-Kapov and Touradj Ebrahimi who elaborate on the relationship between QoS and QoE, and illustrate the evolution from QoS to QoE (and back).

Chapter 7 opens a new perspective on QoE, namely the one of a service provider or operator who has to take into account the business impact of his service. Andrew Perkis, Peter Reichl and Sergio Beker develop an ecosystem of QoE which analyzes the relationships between the individual "players" (user, service provider, application, content, network) and puts them into relation to QoE.

The following Chaps. 8, 9 and 10 address low-level processes which are important for QoE formation. In Chap. 8, Jan-Niklas Antons, Sebastian Arndt, Robert Schleicher and Sebastian Möller show how quality degradations are reflected in the human brain activity, using electroencephalogram data for continuous frequency analyses or triggering event-related potentials. Robert Schleicher and Jan-Niklas Antons show in Chap. 9 how emotions are evoked by perceiving different types of media, and how the impact of evoked affect can be assessed in practice. Whereas most standard evaluation methods address only short media stimuli, Benjamin Weiss, Dennis Guse, Sebastian Möller, Alexander Raake, Adam Borowiak and Ulrich Reiter discuss long-term effects related to QoE in Chap. 10. Analyzing cognitive and memory-related processes, they differentiate between momentary, episodic and multi-episodic QoE, and illustrate these categories through empirical data, assessment methods and prediction models.

Chapter 11, which is the final one of the first part of the book, brings in the aspect of interactivity. Discussing definitions and assessment methods for interactivity, Sebastian Egger, Peter Reichl and Katrin Schoenenberg highlight the differences between QoE assessed in static versus interactive situations, and illustrate these with examples from speech communication and human-machine interaction applications.

The second part of the book addresses applications and application-specific methods which are based on the concepts of the first part of the book. This part starts with services related to speech as the communication media. In Chap. 12, Nicolas Côté and Jens Berger review influence factors and quality features of speech communication services, and provide examples of subjective evaluation methods and instrumental quality prediction models which estimate overall quality as well as individual quality features on the basis of signals or parameters. In a similar vein, Florian Hinterleitner,

Christoph Norrenbrock, Sebastian Möller and Ulrich Heute provide exemplary methods for subjective quality evaluation as well as instrumental quality prediction for services involving synthetic speech generated by Text-To-Speech systems. Their approach is based on quality features as well, showing that the initial concepts can be successfully applied in practice.

These concepts are transferred to audio-visual and interactive situations in Chaps. 14 and 15. In Chap. 14, Markus Vaalgamaa and Benjamin Belmudez summarize findings for audio-visual communication situations, including the audio-visual integration happening when viewing audio-visual content, related evaluation methods, and time-varying quality perception. Janto Skowronek, Katrin Schoenenberg and Gunilla Berndtsson extend the concepts to conferencing and telemeeting situations involving more than two partners in Chap. 15.

Beyond speech, audio services are addressed in Chaps. 16 and 17. Whereas Bernhard Feiten, Marie-Neige Garcia, Peter Svensson and Alexander Raake in Chap. 16 focus on the effects of audio coding and transmission, summarizing also the corresponding subjective assessment methods and instrumental prediction models, Matthias Frank, Franz Zotter, Hagen Wierstorf and Sascha Spors extend these considerations to spatial audio services where localization, spatial width and timbre play a role. In Chap. 18, Rahul Chaudhari, Ercan Altinsoy and Eckehard Steinbach address haptics as a relatively new interaction modality. They review the bases of haptic perception, provide performance parameters and QoE aspects, and show that some aspects of QoE can already be predicted by instrumental models.

Similar to audio services, Chaps. 19 and 20 address quality aspects and prediction models for services involving video. Chapter 19 by Marie-Neige Garcia, Savvas Argyropoulos, Nicolas Staelens, Matteo Naccari, Miguel Rios-Quintero and Alexander Raake focusses mostly on quality prediction models, showing how quality indices for video streaming services can be derived from information on different levels of the media stream. Complementary to this, Chap. 20 by Pierre Lebreton, Marcus Barkowsky, Alexander Raake and Patrick Le Callet focusses on perceptual quality features as well as technical influence factors of 3D video services.

Chapter 21 differs from the previous ones in that it focusses on a method rather than on an application, namely on the use of crowdsourcing in QoE evaluation. Tobias Hoßfeld and Christian Keimel review the background of the crowdsourcing concept and analyze usage scenarios for QoE evaluation. They particularly focus on the impact that the crowdsourcing experiment set-up might have on the obtained results.

Whereas speech, audio and audio-visual services are quite established and the respective methods for subjective quality evaluation and instrumental quality prediction are relatively well researched and (to a large extent) standardized, the subsequent applications addressed in the reminder of the book are relatively new; thus, the corresponding methods are less stable and subject to active research. Chapter 22 by Dominik Strohmeier, Sebastian Egger, Alexander Raake, Tobias Hoßfeld and Raimund Schatz addresses web browsing, where subjective evaluation methods are currently under discussion in the International Telecommunication Union (ITU-T). In Chap. 23 on mobile human-computer interaction by Robert Schleicher, Tilo

Westermann and Ralf Reichmuth, the authors explicitly highlight current research paradigms, goals and questions, as standardized methods are still largely missing. In Chap. 24 on the role of sensory experience beyond audio-visual (by Christian Timmerer, Markus Waltl, Benjamin Rainer and Niall Murray), it is discussed how sensory effects such as force feedback, background light, wind and odor need to be specified and assessed. Chapter 25 addresses Gaming QoE (by Justus Beyer and Sebastian Möller) by introducing a new taxonomy of influence factors, interaction performance and QoE aspects for gaming. In Chap. 26 (by Lucjan Janowski, Mikołaj Leszczuk, Mohamed-Chaker Larabi and Anna Ukhanova), recognition tasks are addressed, and it is shown how definitions, subjective evaluation methods and quality prediction models need to be adapted towards the specificities of these tasks.

The last two chapters provide an outlook into how knowledge on QoE can be used for optimizing ICT services. In Chap. 27, Blazej Lewcio and Sebastian Möller analyze the impact of time-varying quality in heterogeneous wireless networks, and provide guidelines on aspects that are helpful for service optimization. Similarly, in Chap. 28 Raimund Schatz, Markus Fiedler and Lea Skorin-Kapov show that QoE information can be used for network and application management depending on the streaming technique, and how these two approaches can be integrated to one unique approach, e.g. in a multi-operator setting.

The selection of concepts, applications and methods presented in this book is by far not complete. In fact, more and more applications arise, and with them also the need for evaluating and optimizing their quality. Still, we think that the approach presented and the exemplary applications illustrating the approach will form a solid basis for the scientist and practitioner alike. We hope that it will inspire ideas on how QoE can be addressed for new services, or be addressed in a better way for more traditional ones, and will help to identify which factors need to be taken into account in order to come to valid and reliable results.

References

1. EN ISO 9000 (2005) Quality management systems—fundamentals and vocabulary. International Organization for Standardization, Geneva
2. Jekosch U (2005) Voice and speech quality perception: assessment and evaluation. Springer series in signals and communication technology. Springer, Berlin
3. Le Callet P, Möller S, Perkis A (eds) (2013) Qualinet white paper on definitions of quality of experience. European network on quality of experience in multimedia systems and services (COST Action IC 1003), Lausanne, Version 1.2, Novi Sad, March 2013

Chapter 2
Quality and Quality of Experience

Alexander Raake and Sebastian Egger

Abstract The chapter discusses the processes of human perception and experiencing, and of quality formation. In this context, definitions of relevant terms are re-visited and adapted to the presented, updated view, and different aspects of research into quality at large and into Quality of Experience are summarized. Using a conceptual model, the quality formation process is analyzed in view of different contexts and tasks, such as taking part in a quality test under controlled conditions, experiencing a video presentation or concert, or exploring a system or device when considering a purchase in a shop. We provide a short overview of different quality assessment methods, and outline related trends in QoE research.

2.1 Introduction

The present chapter lays out the basis for the understanding of Quality of Experience (QoE) as it is followed by the book.[1] The terms *quality* and *Quality of Experience* are typically used with an engineering goal in mind, reflecting the fact that perceived quality is a key criterion for evaluating systems, services or applications during the design phase or during operation. As such, QoE research often takes a measurement-centered, reductionist's perspective, to assess known services and identify quality-relevant criteria. How to create certain (possibly new) types of "experiences" typically

[1] The authors of this chapter have been the corresponding authors for Chaps. 2 and 3 of the Qualinet White Paper on QoE. The present chapter is an updated and more in-depth consideration of the quality and QoE concepts.

A. Raake (✉)
Assessment of IP-based Applications, Telekom Innovation Labs, TU Berlin, Berlin, Germany
e-mail: alexander.raake@telekom.de

S. Egger
Telecommunications Research Center Vienna (FTW), Vienna, Austria
e-mail: egger@ftw.at

S. Möller and A. Raake (eds.), *Quality of Experience*,
T-Labs Series in Telecommunication Services, DOI: 10.1007/978-3-319-02681-7_2,

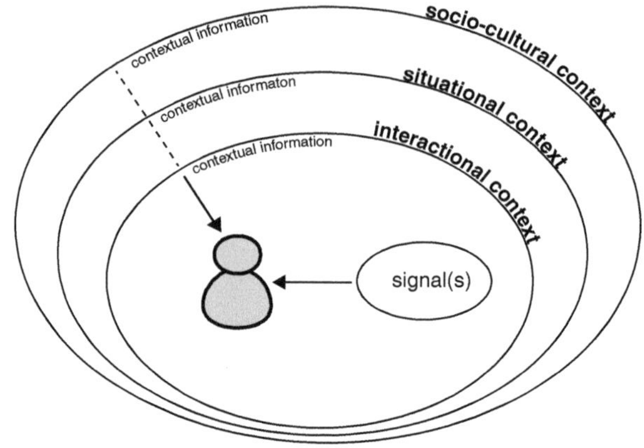

Fig. 2.1 Different contexts a person may be embedded in, inspired by De Moor and Geerts, cf. [14, 19]. Each context is associated with a specific ecosystem that several different stake-holders are involved in, and in which the person takes different roles (as a viewer, friend, customer etc.), as further discussed in Chap. 7

is the domain of User Experience (UX) research. An in-depth comparison of QoE and UX is given in Chap. 3. In the present chapter, a combined engineering- and perception-oriented view is used to discuss work on QoE from different fields.[2]

In the present chapter, quality and QoE are addressed from the perspective of a person whose experiencing in a given situation involves a technical application, service or system. Figure 2.1 depicts the multi-layered context that characterizes the person's situation. The signal(s) as well as the different contexts influence the perception and quality formation processes discussed in this chapter. The different contexts as well as the associated ecosystem of multimedia usage are discussed in more detail in Chap. 7. The contextual information is addressed in more detail in Chap. 4, in terms of factors influencing quality and QoE. In turn, the present chapter presents definitions and considerations in the context of QoE, focusing on the perceptual and cognitive processes underlying the quality formation in the perceptual world of the person.

This perspective can be illustrated using the following example: A person watches a soccer match on TV at home with friends. Here, the signals are of acoustic and visual form [*signal(s)* in Fig. 2.1]. The person interacts with the other persons, possibly with the TV set and the home environment (*interactional context*). Jointly watching the soccer match in the home environment sets the *situational context*. The socio-cultural background of the group of friends forms the *socio-cultural context*. How the

[2] It must be noted that the engineering, computer science and networking communities sometimes still use "QoE" in a misleading way in terms of technological aspects that are *likely to impact QoE as perceived by users*, without actually assessing or quantifying QoE or the QoE impact. For further discussion of QoE and service performance in terms of Quality of Service (QoS) see Sect. 2.2 and especially Chap. 6.

person under consideration *experiences* the soccer match and evaluates the quality of the (technically mediated) experience depends on the audiovisual signals and the contextual settings. As such, this information represents the inputs to the quality formation process discussed in this chapter.

The remainder of this chapter is structured as follows: Sect. 2.2 reviews the related work on quality and QoE in different fields, and provides an updated view of QoE, introducing complementary terms and concepts. In Sect. 2.3, a conceptual model of the quality formation process is presented. In Sect. 2.4, general considerations on quality assessment and evaluation are summarized, and Sect. 2.5 discusses open issues and trends.

2.2 QoE Foundations, Terms and Definitions

In this section, we discuss the terms and concepts of quality and Quality of Experience. In the first step, we introduce our view of the concepts 'perception' and 'experience'—or 'experiencing'—as used in this book.

Here, *perception* is the conscious processing of sensory information the human subject is exposed to. Perception is assumed to involve two subsequent processing stages before a percept finally appears in the perceivers world, namely,

1. Conversion of stimuli via the respective physiologically adequate sensory organs into neural signals.
2. Processing and transmission of these neural signals in the central nervous system up to the cortex, finally resulting in the appearance of specific percepts in the person's perceptual world.

Based on this view, we define *experiencing* as follows:

> **Experiencing** is the individual stream of perceptions (of feelings, sensory percepts and concepts) that occurs in a particular situation of reference.

Here, we follow the widely accepted understanding that experiencing[3] can have hedonic (feelings) and pragmatic (concepts) aspects (see Chap. 5). In terms of the application-domain of this book, *experiencing* may result, for example, from an encounter of a human being with a system, service or artifact. Experiencing in this definition does not include a quality judgement. Quality judgements are considered to be the result of additional cognitive processes on top of experiencing, as described in more detail in the remainder of the section. A conceptual model of the perception, experiencing and quality formation processes is presented in Sect. 2.3.

[3] It is noted that in case of *experiencing* as it may, for example, happen during dreams, or processes of thinking, conception or design, parts of the sensory information are replaced by sketches from memory. This type of experiencing is explicitly excluded here.

2.2.1 Quality and Quality of Experience: Related Work

In the following, we discuss different concepts of quality and important contributions from other authors, before we present an updated view of quality and quality formation.

Qualia

The concept of Quality of Experience can be related with the concept of *Qualia*. Based on the considerations by Jackson [24], *Qualia* can be seen as an inherent property to experiencing that cannot be shared by verbal description or technical means, that is, it can only be accessed via *individual* experiencing. The respective perceptual features may be referred to as *Quale*. Jackson writes [24]: "Tell me everything physical there is to tell about what is going on in a living brain, the kind of states, their functional role, their relation to what goes on at other times and in other brains, [...] you won't have told me [...] about the characteristic experience of tasting a lemon, smelling a rose, hearing a loud noise or seeing the sky." In the context of the present chapter, examples according to the Qualia concept are the listening to a spatial audio production, or the use of a smartphone with intuitive touch input, representing experiencing that cannot be explained verbally to a person who has never had a comparable encounter.

Qualitas and quality

Martens and Martens [35] discuss two extremes of existing approaches for understanding quality: (1) an "objective", rationalistic and product-oriented approach, and (2) a perceptual, "subjective" approach. The authors discuss these two approaches along four quality definitions, whereby (QD1) as *qualitas* and (QD2) as *EXCELLENCE / GOODNESS* are most useful in the context of this book. The approach of (QD1) focuses on generalizable characteristics and properties of the item under consideration in terms of quality as the description of the item's characteristics. In contrast, the perceptual approach (QD2) requires the human evaluator to actually experience the perceptual 'event' under consideration and evaluate the experience in terms of "evaluated excellence or goodness". This approach is strongly related with the degree of need fulfillment [35] or utility. Note that the two notions of quality (QD1 and QD2) are inline with Letowsky's work on sound quality [34].

Utility and "Quality of Experiencing"

Two connotations of the term *utility*[4] in the context of experiencing have been distinguished by Kahneman [29]:

> **Experienced Utility** ...as the judgment in terms of good/bad of a given experience, related with individually perceived "pleasure and pain", "point[ing] out what we ought to do, as well as determine what we shall do" (Kahneman [28], making reference to Bentham [4]). Experience(ing) in this context may refer to painful medical

[4] Note that "utility" and "utility function" also are central terms in micro-economics, however referring to the mapping of a resource to the value for a customer. Economic aspects related with QoE are further discussed in Chap. 7.

investigations such as colonoscopy as in [29], or pleasant phases of experiencing, for example during a concert, or to the quality of life at large [29].

Decision Utility is considered by (external) observation in terms of whether or not certain decisions have been taken, for example on whether or not a service is being used, a low-quality phone call is being ended or a web-item is being clicked.

In principle, both connotations of utility are of relevance for this book: *Experienced utility* is related with perception and experiencing from an individual perspective. In turn, *decision utility* is a useful concept when it comes to whether or not a service or application is actually being used, and thus relates to the concept of acceptance (see next section and Chap. 7).

The previous and following discussions mainly focus on *experienced utility*, and it is noted that Kahneman explicitly uses the term *quality of experience* in this regard. Note that Kahneman has illustrated his ideas referring to quite different domains than the ones addressed in this book, such as medical treatments like colonoscopy, or a person's own life (at large!). For assessment, Kahneman distinguishes a *moment-based* and a *memory-based* approach [28]: For the moment-based approach, momentary or instantaneous judgments of experience are asked for, and for remembered utility (memory-based), respective judgments refer to past or just ended phases (or episodes) of experience(ing). The so-called peak-end effect and temporal integration properties related with momentary or remembered quality (utility) are addressed in Chap. 9.

Standards' Views

One of the most comprehensive reviews of quality definitions has been given by Reeves and Bednar [54]. They identify the most pervasive definition to be "the extent to which a product or service meets and/or exceeds a customer's expectations", which they account to be a definition coming from the service marketing literature. According to their review, services were what was most difficult to include in previous quality definitions up to that date. Around 1990, it was acknowledged that "only customers judge quality" and "all other judgments are essentially irrelevant" (cited by [54] from [67]). It is noteworthy that this perspective is well reflected in standardized quality definitions, such as the one in ISO 9000:2000 [21]:

> Quality ..."is the ability of a set of inherent characteristics of a product, system or process to fulfill requirements of customers and other interested parties".

The current definition of Quality of Service (QoS) by the ITU-T is similar to the ISO-definition of quality given above, with an explicit view from a service operator's or manufacturer's perspective:

> **Quality of Service** [The] Totality of characteristics of a telecommunications service that bear on its ability to satisfy stated and implied needs of the user of the service.

The standardized QoE definition most frequently used in the QoE (and QoS) context is the one according to ITU-T Rec. P.10 (Amendment 2, 2008):

QoE(P.10) ..."The overall acceptability of an application or service, as perceived subjectively by the end user."

Note 1: Includes the complete end-to-end system effects.
Note 2: May be influenced by user expectations and context.

It was pointed out by Möller [38] and others that the inclusion of the term *acceptability* as the basis for a QoE definition is not ideal. As a consequence, during the Dagstuhl Seminar 09192, May 2009, acceptability has been newly defined [38]:

Acceptability ..."is the outcome of a decision [yes/no] which is partially based on the Quality of Experience."

It is noted that this definition is inline with Kahneman's *decision utility*.

Several authors such as Martens and Martens [35] and Jekosch [26] have made reference to quality as defined by earlier engineering-, service- or production-related standardization bodies.

Quality, Quality Elements and Quality Features

A definition of *quality* extending the standards' view is that by Jekosch [26]:

Quality results from the "judgment of the perceived composition of an entity with respect to its desired composition".

Here, the *desired composition* refers to the set of internal references and expectations against which the *perceived composition* is being compared.

To reflect the design process in typical quality management or engineering concepts, in [26] Jekosch takes up the definition of *quality element* from the Deutsches Institut für Normung (DIN):

Quality element ...is the "contribution to the quality of a material or immaterial product... in one of the planning, execution or usage phases." [26]

In simple terms, quality elements can be seen as the material or immaterial knobs and screws that may affect perceived quality. In contrast, a *quality feature* can be described as [26]:

Quality feature ...is the the perceived characteristics of an entity "that is relevant to the entity's quality".

Factors affecting quality perception (that is, *quality elements*) are summarized in Chap. 4, and *quality features* for different multimedia services are outlined in Chap. 5. An in-depth discussion of the relation between QoS and QoE is given in Chap. 6.

2.2.2 *Quality of Experience: Updated Terminology*

From the previous discussions, it is obvious that the term *quality* has different connotations, depending on the context it is used in (see work by Parasuraman et al. [45], Reeves and Bednar [54], Blauert and Jekosch [7], Jekosch [26] and Martens and Marten [35]). In this subsection, we present our synthesis of the different views on *quality* and present new or updated definitions of relevant terms.

For the following considerations, we apply Jekosch's definition of *quality* [26] so as to exclusively address perception that involves sensory processing of external stimuli:

Quality (based on experiencing) results from the "judgment of the perceived composition of an entity with respect to its desired composition".

This way, we explicitly distinguish it from *assumed quality*:

Assumed quality corresponds to the quality and quality features that users, developers, manufacturers or service providers *assume* regarding a system, service or product that they intend to be using, or will be producing, without however grounding these assumptions on an explicit assessment of *quality based on experiencing*.

Here, it is noted that the underlying *assumptions* or expectations are positioned at a different level of the perceptual/cognitive system than actual sensory and emotional references,[5] namely, at the level of *concepts*. Assumed quality as introduced here comprises the traditional views of quality as it was used up to the 1990s in the context of quality management, for example in the production cycle in terms of *excellence* and *conformance to specifications* (cf. Reefes and Bednar [54]). Yet, to a certain extent, it also includes the view of quality in terms of "meeting and/or exceeding customer's expectations" [54], which is more inline with the definition of *quality* (*based on experiencing*) as given above. However, assumed quality excludes explicit experiencing involving sensory processing of external stimuli.

Another term used in the following is *quality of experiencing*. This term is equivalent to Kahneman's use of "quality of experience" [29] and the related concept of *experienced utility* outlined in more detail in Sect. 2.2.1. We here define this concept as follows:

[5] Of course quality-related assumptions may be associated with sensory or emotional references, too.

Quality of experiencing is the degree of delight or annoyance of a person during the process of experiencing.[6] It results from the person's evaluation of the fulfillment of his or her expectations and needs with respect to the utility (pragmatic and hedonic) in the light of the person's context, personality and current state.

In the above definition, *context* refers to the multi-layered view discussed in Sect. 2.1, see Fig. 2.1. *Personality* refers to "...those characteristics of a person that account for consistent patterns of feeling, thinking and behaving", following Pervin and John [48], and *current state* is used in terms of "situational or temporal changes in the feeling, thinking or behavior of a person" (translated from German from Amelang [1]). Note that the current state is both an influencing factor of experiencing (see also Chap. 4), and a consequence of the experiencing.

In this chapter, *quality of experiencing* refers to judgments during or after experiencing (cf. *momentary utility/experience* versus *remembered utility/experience* as in [28, 29], see previous subsection). In the following, for the applications addressed in this book, let us consider that the *experiencing* explicitly involves some kind of technology that impacts the signals presented to the person. For example, this may be a person's overall judgment on the quality of experiencing a concert show, or a soccer match on television together with friends. Note that we use the term *experience* here referring to an evaluation of the *experiencing* at a given moment in time, or in retrospect, considering a certain period of experiencing (cf. remembered utility or experience, [28], discussed in detail in Chap. 10).

For the special case of *quality of experiencing* addressing the context of using multimedia services and applications, in the Qualinet White paper [40] we had proposed the following definition:

> **Quality of Experience** (QoE): "is the degree of delight or annoyance of the user of an application or service. It results from the fulfillment of his or her expectations with respect to the utility and/or enjoyment of the application or service in the light of the user's personality and current state."
> Here, an application is defined as:
> **Application:** "A software and/or hardware that enables usage and interaction by a user for a given purpose. Such purpose may include entertainment or information retrieval, or other." [40]
> **Service:** "An episode in which an entity takes the responsibility that something desirable happens on the behalf of another entity." (Dagstuhl Seminar 09192, May 2009, cited after [60])

[6] Note that in our view presented here, *experiencing* is the process, which however is evaluated in terms of the features associated with the perceptual events happening during that process. Here, it is interesting to note that the German translation of *quality of experiencing* or "quality of experience" as used by Kahneman is *Qualität des Erlebens*, which explicitly reflects that experiencing is a process.

For the definition of QoE, a number of specifications are added to the definition of *quality of experiencing* by the context of *applications or services*: A snapshot is taken, resulting in the exchange of *experiencing* by *experience*. Further, the person takes the role of a *user* [14, 19]. The experiencing happens in the context of using the *application or service*. In our definition of *quality of experiencing*, *utility* is considered to have both pragmatic and hedonic connotations, where *enjoyment* is implicitly considered in terms of a (perceived) need.

However, we identify a major limitation of the above QoE definition in the fact that it addresses the *explicit* experiencing of an application or service. Instead, we believe that a more global view should be taken that also comprises the evaluation of the contribution of a given application, system or service implementation to the *quality of experiencing* as defined above in a more global sense. Further, the delight or annoyance related with the experiencing needs to be *evaluated* to come to QoE, which appears less clear from the above definition. As a result, the following updated definition of QoE is proposed:

Quality of Experience (new) (QoE) is the degree of delight or annoyance of a person whose experiencing involves an application, service, or system. It results from the person's evaluation of the fulfillment of his or her expectations and needs with respect to the utility and/or enjoyment in the light of the person's context, personality and current state.

With the inclusion of the term *system*, even the use of, for example, concert halls, public address systems or television sets can be included in the QoE definition. We here acknowledge the fact that a person that uses an ICT (Information and Communication Technology) product actually takes the role of a *user*, see De Moor's and Geerts work [14, 19]. However, it appears less evident that a person attending a concert and possibly judging upon the quality of experiencing the concert including the employed PA system is an actual *user*. As a consequence, we have re-introduced the *person* instead of the *user*. It is clear that, if the interaction with the application, service or system is at the core of the consideration, the person mainly takes the role of a user.

2.3 Experiencing and Quality Formation

In the following, we present a conceptual model of the quality formation process, taking the perspective of the experiencing person. The quality formation process comprises a perception-component at its basis, as well as the higher-level reference-based quality formation, which we consider as parallel and interactive processes.

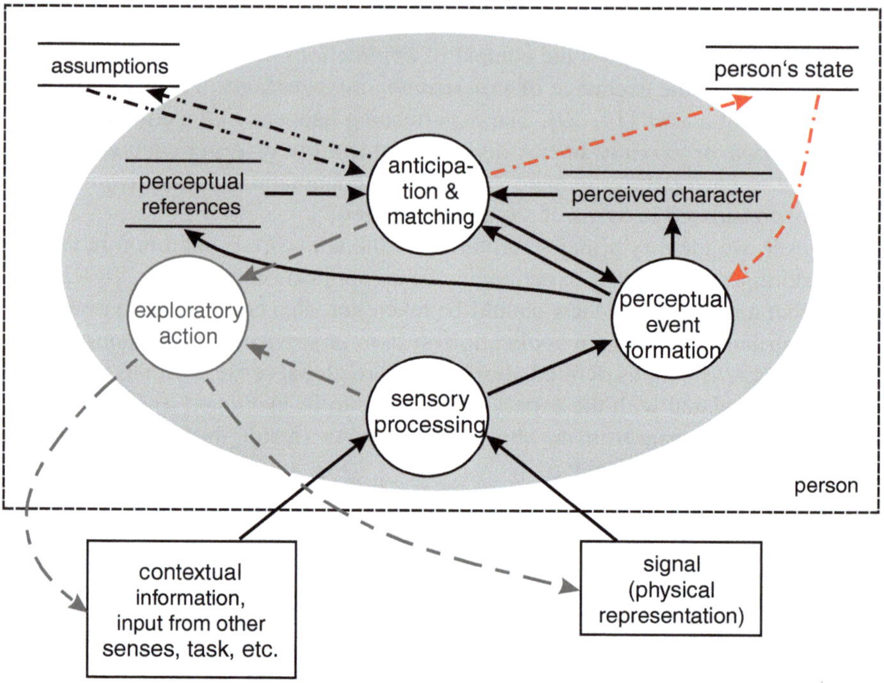

Fig. 2.2 Schematic illustration of the authors' concept of the perception process. *Circles* represent perceptual processes, *two parallel horizontal lines* represent storages for different types of representations, and *boxes outside of the person* represent input information. Note that *continuous lines* represent direct input to the perception process. Here, for simplification, contextual information is assumed to be processed, too, but by parallel perceptual processes not shown in the picture. The *person's state* refers to both the cognitive as well as the physiological, current state of the person. In turn, *assumptions* here refer to the person's attitude and concepts. See text for further details

2.3.1 Perception and Experiencing Process

The basis for quality and QoE as addressed in this book is perception. Figure 2.2 schematically depicts a current concept of how the neural signal processing during perception takes place in an iterative way. The process of perception starts by the incidence of respective stimuli to one or multiple of the human sensory organs. In the sensory organ(s), the physical representations of the stimuli are converted into neural representations that include characteristic electric signals. This representation is conveyed to the brain through neural transmission for further processing. Throughout the transmission to the respective brain region, these representations are transformed from initial representations of stimuli into more abstract, symbolic representations. For details on assessing the neurophysiological basis of these processes refer to, for example, Chap. 8.

Current physiological knowledge supports the following model assumptions with regard to different levels of neural processing. In the first, *sensory processing* step, neural processing by the sensory periphery results in a multidimensional neural topologically organized representation, covering aspects of time, space, frequency and activity (see e.g. Raake and Blauert [51] for a model framework for spatial audio perception and quality, and complementary considerations in the work by Blauert et al. [8]). The neural representation precedes the actual formation of perceptual objects (*perceptual event formation*). Further steps towards this goal are conducted at higher levels of the brain, where multiple parallel processors perform the bottom-up pre-segmentation of the multidimensional feature representation, leading to a Gestalt-analysis[7] of features for object- and event identification. Subsequent processing steps analyze the pre-segmented features in terms of objects in the specific modalities, such as visual objects or aural scene objects, or words in an utterance. Already at these levels, perception is influenced by remembered perceptual events and subsequent feedback-based adaptation of the processing, such as, for example, noise suppression once a human voice is sensed. As a consequence, the neural features likely to belong to the same object are associated. The pre-segmentation and object-formation can be subsumed under the process *perceptual event formation*. At this stage, information from other modalities is already integrated, via respective sensory processing.

Based on internal references and rules, hypotheses are created in a top-down manner that are verified against the bottom-up perceptual evidence [8]. In Fig. 2.2, this process is denoted as *anticipation and matching*. The result of the iterative processes of *perceptual event formation* and *anticipation and matching* are recognized objects of perception, that have a specific *perceived character*. The presence of certain stimuli may lead to *exploratory action*, such as the so-called turn-to-reflex in audio-visual perception, where a low-level representation of an impulsive sound from a given direction typically causes a reflexive turning of the head towards the sound source [11]. Similarly, in a top-down manner, actions such as exploratory head-movements [5], tactile exploration [33] or overt attention type eye-movements may be carried out due to salient properties of the stimuli and/or contextual information, or may be governed by higher-level cognitive processes that direct visual attention [16, 22]. This, in turn, alters the sensory and subsequent neural input information (see also [42]).

It is noted that contextual and/or task-related information given to persons are processed via their sensory organs and the subsequent neural processing, too, possibly in other modalities. Such information either directly affects the perceptual process, or does so via information made available in terms of higher-level concepts, here referred to as *assumptions* (see Fig. 2.2). Further, in principle, perception is largely co-determined by the *person's (current) state*. It reflects the "situational or temporal changes in the feeling, thinking or behavior of a person" (translated from German from Amelang et al. [1]).

[7] Initial works on Gestalt-theory are those by its founders Wertheimer [61, 62] and Koffka [31]. Its use in, for example, auditory scene analysis has been discussed in detail by Bregman [10].

Memory and Perceptual References

In Fig. 2.2, different stores (storages) are depicted by parallel lines. According to authors such as Cowan [13], Coltheart [12] and Baddeley [2], different levels of memory have been identified, with respective roles in the perception process, and respective storage durations. Such memory levels are:

Sensory memory: Is a peripheral memory that stores sensory stimulus representations for short durations between 150 ms and 2 s so as to be retrieved by higher processing stages [2, 12, 13, 36].

Working memory: Stores re-coded information at symbolic level for longer durations from a few up to tens of seconds [3].

Long-term memory: Covers longer time spans up to years or even a full lifetime. It involves multiple stages of encodings in terms of symbolic and perceptual representations [2]. Current theories assume that a central executive component controls the linking between long-term memory and working memory via an episodic buffer at working memory level that integrates information into episodes, and that this central component is associated with attention [3].

Perceptual references as depicted in Fig. 2.2 can be present at different levels of memory: Working memory for the perceptual integration of a scene and respective scene analysis, as well as information retrieved from long-term memory, for example for the identification of objects in a scene or words in an utterance. Similarly, the *perceived character* or the respective perceptual event or flow of events can be situated in working memory, or be stored in long-term memory, for example after verbal or episodic re-coding has occurred. (cf. Chap. 9).

In this context, learning of perceptual or conceptual references is directly associated with expertise and know-how. In Fig. 2.2, learning is considered as being implicitly integrated into the processes that are involved in perception, which enables more fine-grained performance with learning, as well as the increasing availability of respective detailed references in long-term memory. For example, a person that is fluent in a learned language can relate utterances with respective references, and a skilled musician or sound engineer will be able to associate a given auditory percept with respective actions—while an unskilled person is usually not able to do so. Considerations on categories of references can be found in Neisser's cognitive system theory, cf. [42].

2.3.2 *Quality-Formation Process*

A conceptual model of the quality formation process has been proposed by Jekosch [26], further adapted in [50]. This view is extended in the following, see Fig. 2.3. The quality-formation process can be seen as a parallel but higher-level cognitive process related with the process of experiencing (cf. Sect. 2.3.1). Here, it is assumed that experiencing itself may be subject to *quality of experiencing* evaluation, in case

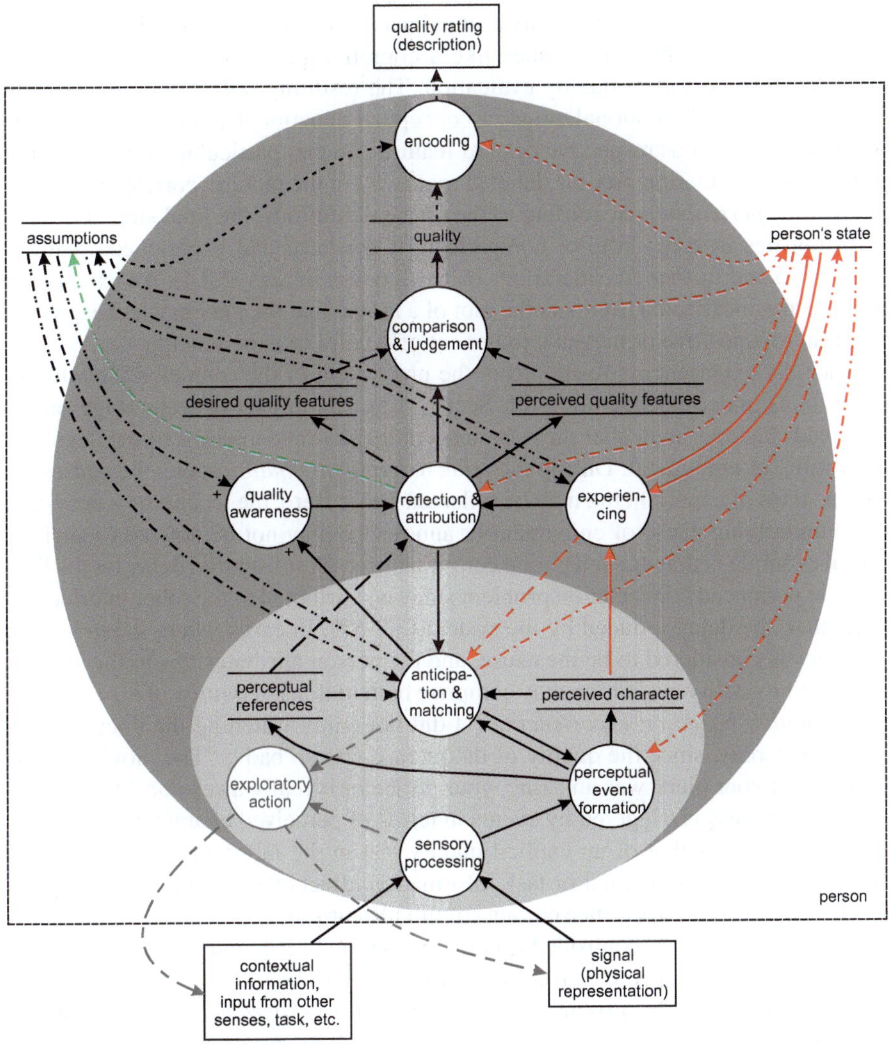

Fig. 2.3 Quality formation process during experiencing (includes Fig. 2.2). The picture extends the initial ideas from [26, 50]. See text for details. Note that this picture does not include the interaction components that need to be added for services such as human-to-human communication or human-computer interaction (for further discussion see Chap. 11)

that the person reflects upon it (*reflection & attribution* in Fig. 2.3). This reflection can be triggered by an external task to evaluate what has been experienced (for example in a quality test), during or after the process of experiencing [28, 29]. Here, the task is contained in the *assumptions* as the abstract conceptual expectations and attitude of the person (Fig. 2.3). Or, the reflection may be triggered by unexpected events, where the experiencing deviates from assumptions.

The triggering of an actual quality evaluation is represented in Fig. 2.3 by a *quality-awareness* component that operates like a cognitive gate, focussing the person's attention on some sort of quality evaluation. The resulting *reflection* is linked with the identification of emotional, sensory, conceptual or actional quality features of the experience, as well as respective desired features. In this particular case, the output of the quality formation process, labeled as *quality* in the picture, corresponds to the *quality of experiencing*. According to the above definitions, the final step of quality formation lies in some kind of comparison of expected and experienced features (cf. Chap. 5 and further considerations on expectation in Sect. 2.2.1). An example of this case are unexpected events in the plot of a movie that the person watches, which may lead to a positive or negative judgment of quality of experiencing.

Since the experiencing results from the processing of the (*perceived*) *character* of the items under consideration (cf. Sect. 2.3.1), any impact of technology on the perceived character may alter the experienced, for example in terms of the degree of immersion, or enjoyment. During the *reflection and attribution* stage, the causes for certain states of experiencing may be reflected upon. Here, the technology or system as the underlying cause of enjoyment or annoyance may not be noticed as such. A typical example is that of a telephone conversation with substantial delay on the line, where experienced conversation problems may be attributed to the other interlocutor rather than the delay induced by the system [53, 57]. In cases where the *perceived character* is considered to be the cause, and the person attributes this to the system, the resulting quality evaluation may comprise both notions of *quality of experiencing* and of *quality based on experiencing* ("I did not enjoy watching the documentary on TV yesterday, since the quality of the picture was so bad"; "The movie session yesterday at your place was amazing, your projector is really awesome!"). Here, the quality awareness is triggered by events in terms of perceived character.

Another case is that of an explicit quality test in the laboratory or in the field. Here, test-specific contextual or task information affects the assumptions based on which a person may experience certain stimuli. According to Jekosch's terminology, the person conducts a "controlled quality evaluation" [26], and *quality awareness* is triggered by the respective task- or context-based assumptions.[8] Here, the resulting quality typically corresponds to *quality based on experiencing*, of course also depending on the employed test method (see Sect. 2.4). A similar case is that of a person who has, for example, the intention to buy a new multimedia device. Then, too, respective assumptions may trigger an evaluation of different systems in the shop, leading to a judgment of *quality based on experiencing*.

In all of the cases discussed up to here, the quality evaluation of the service, application or system involves an actual *experiencing* including the respective perception process. However, as mentioned earlier, often times users or system designers develop a notion of *assumed quality*, before or without an actual process of experiencing taking place. This notion may strongly be influenced, for example, by what people read

[8] According to Jekosch, "controlled" perception can be distinguished from "random" perception/quality assessment. Here, natural experience(ing) without a dedicated quality judgment task, for example, corresponds to "random", and task-driven quality assessment to "controlled".

or hear about a product, or what they think about the brand. Here, the perception- and experiencing-path shown on the right side of Fig. 2.3 does not carry information from direct sensory processing that can be exploited during quality formation. It is currently under debate within the QoE community, which criteria must be fulfilled by respective non-perceptual sources of information, where no ground truth data in terms of explicit *quality based on experiencing* is available. Such sources of information can include system specifications, quality metrics such as PSNR (Peak Signal to Noise Ratio) or SSIM (Structural Similarity index, cf. [64] and Chap. 19), quality prediction algorithms, or even models of the human quality formation process (Sect. 2.4). It is generally accepted that key performance indicators (KPIs) such as packet loss or stalling rate alone cannot be used as a direct measure of quality in the sense laid out here (e.g. [17, 20, 49]).

For all the cases discussed above, the person's state as well as his/her personality play a key role, and impact on multiple of the presented processes. Further, as outlined in the description of the perception and experiencing process in Sect. 2.3.1, personality is contained in the processes themselves, as well as the value system established by references. Due to their involvement of memory, and the respective access to this memory during quality formation, the underlying references are influenced by contextual factors and undergo temporal changes. In the process, perceptions and knowledge about the service or system are turned into references that belong to the domain of the person's expectations. The reference formation and assessment of features in terms of their plausibility are performed in a top-down manner, where attentional processes at different levels of the perceptual-cognitive system of the human person steer the information provided by the bottom-up components (cf. Raake and Blauert [51]).

References and Semiotics

Let us now take a closer look at internal references and their use during quality perception as discussed above. A reference-related concept initially suggested by Piaget is the one of *schemata* and the respective formation or adaptation processes in terms of *accommodation* and *assimilation* (see e.g. Neisser's [41] and Jekosch's [26] works). This concept is useful for the (qualitative) understanding of perceptual and conceptual references and their formation: In case of unknown perceptual/cognitive information, a disequilibrium with available references (schemata) and thus the anticipated event may result. In this context, *assimilation* refers to the adaptation of the stimulus-related representation, so as to fit to an existing schema. In turn, *accommodation* refers to the case that not the representation of the perceived or experienced, but the (reference) schema is adjusted. If a person encounters multiple similar phases of experiencing over time, the initially flexible schema may be crystalizing into a new schema during a learning process. These considerations help to understand the formation of references for example when using new types of technology such as spatial audio or 3D video.

It is useful to further consider that perception and cognition as well as communication can be discussed in terms of the underlying "signs". Jekosch [25, 26] has introduced semiotics, that is, the science of signs, into quality assessment research.

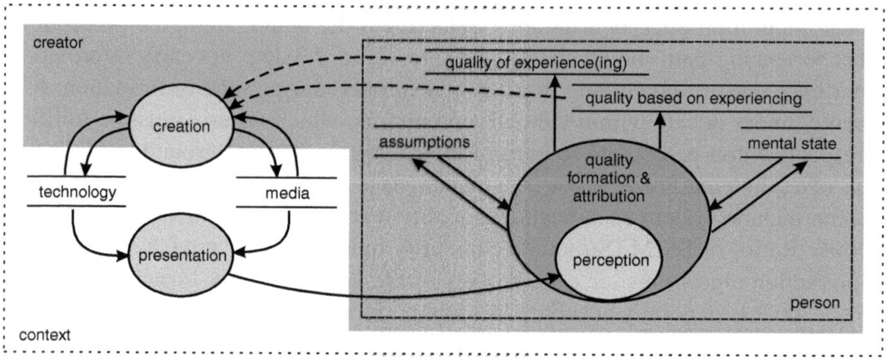

Fig. 2.4 Quality perception in the context of creation/production. The person and creator may be identical. *Quality of experiencing* or *quality based on experiencing* (for the respective definitions see Sect. 2.2.2) will be used by the creator as target for optimization. Obviously, for the creator, the creation process is comprised in the *experiencing*, too. Note that the creator's experiencing during creation is a different one from the experiencing of the created, cf. [26]

Semiotics addresses the relation between the sign carrier and the associated meaning, and the perspectives of different persons that may interact with a sign such as a picture, video sequence or speech message. To this aim, different sign models have been introduced [26, 43, 44, 47]. The classical triadic form is composed of a sign carrier, the referent, and the meaning. In our context, the sign carrier may be the physical form of the sign as in case of a transmitted video sequence or a word in a phone call. The referent is the item the sign stands for, and may be abstract or concrete (for example, the specific chair shown in a video sequence). The meaning results from the interpretation of the sign by the interpreting person. The dynamic process during which the effect of a sign (or rather of an interconnected set of signs) is created is referred to as *semiosis* [43]. Here, semiosis can be any kind of interpretation of a sign by a cognitive system. Obviously, a different meaning may be assigned by different interpreting persons, who can have the role of the creator or the receiver of the sign.

Semiotics is a very useful concept to discuss, for example, the criteria based on which quality is being judged by a given person, that is, whether the sign carrier, the referent or the meaning for the person have been addressed. For the example of a photography, the carrier could be judged upon (in terms of the camera, lighting conditions, framing, coding, resolution, paper used for printing, etc.), the referent (what objects are shown in the picture), or the meaning (what does the picture tell, what is its impact on me, etc.?).

The creation and *experiencing* processes of media, involving technology during its creation, is illustrated in Fig. 2.4. It should be noted that technology comes into play at different stages here, namely at the creation stage, the (post-)processing stage, and the presentation stage (includes possible transmission and display). Further note that the presentation may also apply to the viewing by the photographer during post-production. Artists or content producers create entities (carriers, signs) that can be experienced, and thereby may attempt to deliberately provoke or achieve specific experiencing. As discussed by the authors of this chapter also in the Qualinet QoE White Paper [40], in terms of semiosis, "meaning" is associated with the creator's intentions ("sender"), while at the "receiving" end, "meaning" results from interpreting the content during experiencing.

Expectations and Service Context

Let us now take a more service-oriented viewpoint, discussing that quality is based on the comparison of perception with expectations. The aspect of expectation has been addressed in a more global manner in the context of marketing research, considering the person's role as a customer (cf. Fig. 2.1). *Service Quality* is used in the respective works by Parasuraman et al. [46], Boulding et al. [9], and Zeithaml et al. [66] in terms of perception vs expectations. Here, *perception* may refer to both the perception during encounter with a service, and the conceptual impression of the Service Quality related with a given company after a number of encounters, namely in terms of *Customer Satisfaction and Dissatisfaction* (CS/D, [9, 46, 66]).

A model of expectations vs perceptions-based Service Quality has been proposed by Boulding et al. [9], introducing different types of expectations. Here, the impact of external information on expectations is explicitly considered. To this aim, two types of expectation are distinguished, namely *will expectations* in terms of what users expect *will* be happening for their next interaction with a company's service, and *should expectations* in terms of what *should* be happening for that next encounter, based also on what they may know about the performance of competitors' services. Both types of expectations are assumed to be time-varying and dependant on what has been perceived during previous service encounters. Boulding et al. further contrast *should expectations* from *ideal expectations* in terms of what the customer wants "in an ideal sense" for the respective type of a service.

Zeithaml et al. [66] distinguish two levels of expectations in relation to the acceptance of a certain service configuration in a given context: (1) The "desired service" corresponds to what the user wishes to have, in terms of a construct in-between the *should* and *ideal* expectations as of Boulding et al. [9]; (2) the "adequate service" reflects what the user may still perceive as acceptable under given contextual and situational constraints, for example related with the current weather or the given location she is in (and, for example, respective degradations, as they may be encountered during mobile service usage). Hence, the "adequate service" expectation-level is what determines the *acceptability* for the customer.[9] The zone in-between the two expectation-levels (1) and (2) is referred to as the "zone of tolerance" for what is being perceived [66]. This concept of expectation has been adopted in recent work by Sackl and Schatz [56], who have applied it for explaining different quality tests that varied in terms of the considered user-types (affecting the ideal or "desired services" expectation level), and the context-specific influences (assumed to be affecting mainly the level of "adequate services"). It is noted that this Service-quality perspective bears several similarities to the quality taxonomies developed by Möller for different types of telecommunications-related services, see [37, 38] and Chap. 5.

[9] This also includes whether perceived quality is currently relevant for the customer for acceptability.

Another noteworthy expectation-related perspective addresses the (product) features that underly customer satisfaction. According to Kano's model [30], features can be subsumed in terms of three types of requirements: (1) Must-be requirements (sometimes referred to as hygiene-factors)—their under-fulfillment leads to dissatisfaction, while their fulfillment does not lead to satisfaction (example: today's touch-control in smartphones); (2) one-dimensional requirements (i.e. performance-factors)—their fulfillment is linearly related with satisfaction (example: bandwidth of customer's home internet connection); (3) attractive requirements—unexpected features that, if fulfilled, lead to delight (example: high resolutions of smartphone displays when first introduced some time ago). It is obvious that with time, features that have initially been of type (3) will ultimately end up to be features of type (1), that is, are generally expected to be fulfilled. We will not further detail the Kano-model and surrounding work in marketing research here. It is obvious that it is a useful tool for describing why certain service innovations such as color TV or later high definition video eventually become must-be requirements.

2.4 Quality Assessment

The central question for quality assessment is how to operationalize the concept of QoE in terms of performing reliable and valid measurements. The respective *quality of quality assessment methods* [37] is of cardinal importance, since the respective results can easily be misused. The overarching question is: *How can we quantify quality, and how can we measure it?* This question is of course not unique to media-related quality (of experiencing) as mainly addressed in this book, but also extends to numerous other disciplines, for example food quality (cf. Lawless and Heymann [32]) or service quality in a broader sense (cf. Parasuraman et al. [45], Reeves and Bednar [54]). In this context, according to Jekosch [25], assessment is the "measurement of system performance with respect to one or more criteria. Typically used to compare like with like, whether two alternative implementations of a technology, or successive generations of the same implementation", with the criterion being *quality based on experiencing* or *quality of experiencing*. Ideally, quality assessment methodologies should act as a translator between the *quality elements* (see above and Chap. 4), and QoE, or the underlying *quality features* (see Chap. 5). Quality assessment methods can be classified into perception-based[10] and instrumental[11] ones, depending on whether human subjects are involved in the assessment process or not. A brief discussion of these two assessment approaches is given in the following. More details can be found, for example, in [37, 39, 50] for speech and audio quality, in [52, 63, 65] for video quality, and in [32] for food quality.

[10] Often referred to as "subjective", a somewhat misleading term avoided here.

[11] Often referred to as "objective", which is even less appropriate than "subjective", since it implies that instrumental measurements bear objectivity, which they only do in case that they can be generalized.

Perception-based methods are the most valid way to assess quality, and typically provide the ground-truth data for the development of instrumental methods. Perception-based methods are used in tests with human evaluators to gather quality-related information for a certain test condition or set of stimuli. To this aim, test subjects are presented with one or several simultaneously or subsequently available stimuli, or are involved in an interaction with a system or another person via the system. The test participants are asked for (quantitative) ratings of momentary or remembered quality on a set of scales, or of qualitative descriptions of the features of the stimuli. In a subsequent statistical analysis of their judgments, a QoE value for each of the test conditions is determined. This and more complex statistical analysis of the test data can provide information about the underlying structure and dependencies on the applied test conditions, that is, the quality elements.

Instrumental methods provide estimates of quality using an appropriate algorithm or instrument. These estimates are based on quality metrics such as the Peak-Signal to Noise Ratio [64], estimation algorithms such as the so-called E-model for speech [23], or explicit quality models that implement certain portions of the human perception and quality-formation process (peripheral signal processing, cognitive processing). The different algorithms are fed by a set of input features acquired from the technical system, or with signals as they would be presented to human assessors in a respective test. The type of model input can be utilized to classify different instrumental methods: (1) Signal-based models that employ the signal (as processed by the system) as single input (No-Reference methods, NR), or plus some reduced or explicit version of the reference (reduced- or full-reference models, RR, FR, respectively). (2) Parametric algorithms that predict QoE based on certain system or signal parameters. The latter can further be subdivided into (a) parametric planning models fed with a-priori known system parameters and (b) packet-level or bitstream models that extract parametric information at the packet level. (3) Hybrid algorithms, which apply a mix between signal-based and parametric information.

In addition to the above differentiation, assessment methods can be distinguished as *utilitarian* and *analytical*, depending on what type of output information they provide. Here, the term *utilitarian* makes direct reference to *utility*, and represents a typically single-valued index based on which systems or services can be ordered with regard to their quality. In turn, *analytic* means that the perceptual features relevant for quality are being assessed.

Utilitarian Quality Assessment The purpose of utilitarian measurements is to objectively quantify an "overall" or "general" impression of quality. This assumes that the subject is in some form of integrative state of mind, where the influence of the impression for the individual attributes, the context, the mood, the expectations, the previous experience, traditions and so on, are all combined into one single-valued rating (providing a ranking "worse-to-better") that establishes the basis for some form of action of the person.

Analytic Quality Assessment The main aim of analytic assessment methods is to decompose and measure certain quality features related with a given stimulus or system (Chap. 5). They result in a multi-dimensional description inherent to the character of the experience. These different features can then be used either for diagnostic purposes, that is, when systems are analyzed, or for analyzing the relation between utilitarian quality and underlying stimulus characteristics.

2.5 Discussion

In this chapter, we have introduced a procedural model of the quality formation process, and have linked it with related quality and QoE concepts and research. The goal was to take a perceptionist's view (cf. Blauert [6]) by treating QoE from an individual's perspective. There are still crucial issues to be addressed in the context of quality and QoE research, and the application of respective methods. For the time being, the majority of research efforts has been focused on *quality based on experiencing*. Only little work has been devoted to assessing actual QoE in terms of *quality of experiencing*. One of the key challenges here is the handling of the respective, let us call it, *Schrödinger's cat problem of QoE research*, namely, how can QoE be assessed without interfering with the experiencing, that is, how can random experiencing [26] be probed? This question is particularly important in the context of applications or systems that trigger new types of schemata or references, as in case of 3D Video or spatial audio (cf. Chaps. 17 and 20). Some approaches along these lines have been proposed by, for example, Staelens et al. [59] and Jumisko-Pykköo et al. [27]. Another approach is the assessment of an *inferred* quality of experiencing, for example in terms of the persons' acceptance: If users are dissatisfied with a given usage session, they may abandon it, which may be observed in measures such as call durations (see Skype's blog [55]), durations of watching individual videos (see Dobrian et al. [15]), or cancelation rates in web-browsing (see e.g. Shaikh et al. [58]). Another approach that is more instructive in terms of the quality-formation process, is not to ask persons for actual quality ratings, but rather try and understand what actually characterizes the experiencing, and what role the underlying quality elements play for it: Along these lines, physiological correlates of experiencing will be discussed in Chap. 8, the role of emotions in QoE will be addressed in Chap. 9 and Chap. 11 discusses the role of interaction performance for the QoE of interactive services or applications. Further work in this direction is related with the understanding of appeal of media such as pictures or movies, and the understanding of the role of quality elements and features in this context. These approaches will be supported by the explicit inclusion of exploratory and attentional processes in quality assessment and respective instrumental models, which is expected to gain further importance in future research [16, 18, 51].

Acknowledgments In the course of writing Chaps. 2 and 3 of the Qualinet QoE White Paper [40], the authors had different fruitful discussions and exchanges with the respective chapter co-authors, which are gratefully acknowledged. Further, we are grateful to Jens Blauert for his in-depth additional review of our chapter and the constructive proposals for improvement.

References

1. Amelang M, Bartussek DGS, Hagemann D (2006) Differentielle Psychologie und Persönlichkeitsforschung. W. Kohlhammer Verlag
2. Baddeley A (1997) Human memory—theory and practice. Taylor & Francis: Psychology Press, East Sussex
3. Baddeley A (2003) Working memory: looking back and looking forward. Nat Rev Neurosci 4:829–839
4. Bentham J (1948) An introduction to the principle of morals and legislations. Blackwell, reprinted. Oxford, UK (1789)
5. Blauert J (1997) Spatial hearing: the psychophysics of human sound localization. The MIT Press, Cambridge
6. Blauert J (2012) A perceptionists view on psychoacoustics. Arch Acoust 37(3):365–371. doi:10.2478/v10168-012-0046-z
7. Blauert J, Jekosch U (2003) Concepts behind sound quality: some basic considerations. In: Proceedings of Internoise 2003, pp 72–79
8. Blauert J, Kolossa D, Obermayer K, Adiloglu K (2013) Further challenges and the road ahead. In: Blauert J (ed) The technology of binaural listening, Chap. 18. Springer, Berlin
9. Boulding W, Karla A, Staelin R, Zeithaml V (1993) A dynamic process model of service quality: from expectations to behavioral intentions. J Mark Res 30(1):7–27
10. Bregman AS (1990) Auditory scene analysis. The MIT Press, Cambridge
11. Clifton R, Morongiello B, Kulig J, Dowd J (1981) Newborn's orientation towards sounds: possible implication for cortical development. Child Dev 52(3):833–838
12. Coltheart M (1980) Iconic memory and visible persistence. Percept Psychophysics 27(3):183–228. doi:10.3758/BF03204258
13. Cowan N (1984) On short and long auditory stores. Psychol Bull 96(2):341–370
14. De Moor K (2012) Are engineers from Mars and users from Venus? Bridging the gaps in quality of experience research: reflections on and experiences from an interdisciplinary journey. Ph.D. thesis, Universiteit Gent
15. Dobrian F, Awan A, Joseph D, Ganjam A, Zhan J, Sekar V, Stioca I, Zhang H (2013) Understanding the impact of video quality on user engagement. Commun ACM 56(3):91–99
16. Engelke U, Kaprykowsky H, Zepernik HJ, Ndjiki-Nya P (2011) Visual attention in quality assessment. IEEE Signal Process Mag (November):50–59
17. Garcia MN, Raake A (2011) Frame-layer packet-based parametric video quality model for encrypted video in IPTV services. In: Proceedings of International Workshop on Quality of Multimedia Experience (QoMEX)
18. Garcia MN, Schleicher R, Raake A (2011) Impairment-factor-based audiovisual quality model for IPTV: influence of video resolution, degradation type, and content type. EURASIP J Image Video Process 2011:1–14
19. Geerts D, Moor KD, Ketyko I, Jacobs A, den Bergh JV, Joseph W, Martens L, Marez LD (2010) Linking an integrated framework with appropriate methods for measuring QoE. In: Proceedings of International Workshop on Quality of Multimedia Experience (QoMEX)
20. Hoßfeld T, Strohmeier D, Raake A, Schatz R (2013) Pippi-Longstocking calculus for temporal stimuli pattern on YouTube QoE. In: Proceedings of the 5th workshop on mobile video, MoVid '13. ACM, New York, pp 37–42
21. International Organization for Standardization: ISO 9000:2000 (2000) Quality management systems: fundamentals and vocabulary
22. Itti L, Koch C (2001) Computational modelling of visual attention. Nat Rev Neurosci 2(3):194–203
23. ITU-T Rec (2009) G.107 The E-model, a computational model for use in transmission planning. International Telecommunication Union, Geneva
24. Jackson F (1982) Epiphenomenal qualia. Philos Q 32(127):127–136
25. Jekosch U (2005) Assigning meaning to sounds: semiotics in the context of product-sound design. In: Blauert J (ed) Communication acoustics. Springer, Heidelberg

26. Jekosch U (2005) Voice and speech quality perception—assessment and evaluation. Springer, Berlin
27. Jumisko-Pyykko S, Hakkinen J, Nyman G (2007) Experienced quality factors—qualitative evaluation approach to audiovisual quality. Proc SPIE 6507:65070M
28. Kahneman D (2003) Experienced utility and objective happiness: a moment-based approach. In: Brocas I, Carrillo JD (eds) The psychology of economic decisions. Oxford University Press, Oxford, pp 187–208
29. Kahneman D (2003) Objective happiness. In: Kahneman D, Diener E, Schwarz N (eds) Well-being: the foundations of hedonic psychology. Russell Sage Foundation, pp 3–25
30. Kano N, Seraku N, Takahashi F, Tsuji S (1984) Attractive quality and must-be quality. J Jpn Soc Qual Control 14(2):39–48
31. Koffka K (1922) Perception: an introduction to the Gestalt-Theorie. Psychol Bull 19(10):531–585
32. Lawless HT, Heymann H (2010) Sensory evaluation of food: principles and practices, vol 5999. Springer, Berlin
33. Lederman SJ, Klatzky RL (2009) Haptic perception: a tutorial. Attention, Percept Psychophysics 71(7):1439–1459
34. Letowski T (1989) Sound quality assessment: concepts and criteria. In: 87th Convention Audio Engineering Society
35. Martens H, Martens M (2001) Multivariate analysis of quality. Wiley, Chichester
36. Massaro DW (1975) Backward recognition masking. J Acoust Soc Am 58(5):1059–1065
37. Möller S (2000) Assessment and prediction of speech quality in telecommunications. Kluwer Academic Publishers, Boston
38. Möller S (2010) Quality engineering: Qualität kommunikationstechnischer Systeme. Springer, London
39. Möller S, Chan WY, Côté N, Falk T, Raake A, Wältermann M (2011) Speech quality estimation: models and trends. IEEE Signal Process Mag 28(6):18–28
40. Möller S, Le Callet P, Perkis A (eds) (2012) Qualinet white paper on definitions of Quality of Experience—output version of the Dagstuhl seminar 12181: European network on Quality of Experience in multimedia systems and services (COST Action IC 1003). Lausanne, 1.1 edn.
41. Neisser U (1978) Perceiving, anticipating and imagining. Minn Stud Philos Sci 9:89–106
42. Neisser U (1994) Multiple systems: a new approach to cognitive theory. Eur J Cogn Psychol 6(3):225–241
43. Nöth W (2000) Handbuch der Semiotik. Metzler, Stuttgart
44. Ogden CK, Richards IA (1960) The meaning of meaning, 10th edn. Routledge & Kegan Paul Ltd., London
45. Parasuraman A, Zeithaml V, Berry L (1985) A conceptual model of service quality and its implications for future research. J Mark 49(Fall 1985):41–50
46. Parasuraman A, Zeithaml V, Berry LL (1988) Servqual. J Retail 64(1):12–40
47. Peirce CS (1986) Semiotische Schriften, vol 1. H. Suhrkamp, Frankfurt/Main
48. Pervin LA, John OP (eds) (2001) Handbook of personality theory and research. The Guilford Press, New York, pp 102–138. Cited from Carducci B (2009) The psychology of personality. Wiley, New York
49. Raake A (2006) Short- and long-term packet loss behavior: towards speech quality prediction for arbitrary loss distributions. IEEE Trans Audio Speech Lang 14(6):1957–1968
50. Raake A (2006) Speech quality of VoIP: assessment and prediction. Wiley, Chichester
51. Raake A, Blauert J (2013) Comprehensive modeling of the formation process of sound-quality. In: Proceedings of Institute of Electrical and Electronics Engineers (QoMEX). Klagenfurt, Austria
52. Raake A, Gustafsson J, Argyropoulos S, Garcia MN, Lindegren D, Heikkila G, Pettersson M, List P, Feiten B (2011) IP-based mobile and fixed network audiovisual media services. IEEE Signal Process Mag 28(6):68–79
53. Raake A, Schoenenberg K, Skowronek J, Egger S (2013) Predicting speech quality based on interactivity and delay. In: Proceedings of INTERSPEECH

54. Reeves CA, Bednar DA (1994) Defining quality: alternatives and implications. Acad Manage Rev 19(3):419–445
55. Rosen JD (2010) The power of the SILK codec—skype blogs. http://blogs.skype.com/2010/09/28/the-power-of-silk
56. Sackl A, Schatz R (2013) Evaluating the impact of expectations on end-user quality perception. In: Proceedings of International Workshop Perceptual Quality of Systems (PQS), pp 122–128
57. Schoenenberg K, Raake A, Koeppe J (2013) Are you just a bit slow? About misattribution of transmission delay to attributes of the person at the far-end. Int J Human-Comput Stud (to appear)
58. Shaikh J, Fiedler M, Arlos P, Collange D (2008) On the use of TCP interruptions to assess user experience on web. In: Third Euro-NF workshop on socio-economic issues in networks of the future
59. Staelens N, Moens S, Van den Broeck W, Mariën I, Vermeulen B, Lambert P, Van de Walle R, Demeester P (2010) Assessing Quality of Experience of IPTV and video on demand services in real-life environments. IEEE Trans Broadcast 56(4):458–466
60. Weiss B, Möller S, Wechsung I, Kühnel C (2011) Quality of experiencing multi-modal interaction. In: Minker W, Lee GG, Nakamura S, Mariani J (eds) Spoken dialogue systems technology and design. Springer, New York, pp 213–230
61. Wertheimer M (1923) Untersuchungen zur Lehre von der Gestalt. II. Psychologische Forschung 4(1):301–350
62. Wertheimer M (1938) Gestalt theory. Hayes Barton Press. http://books.google.de/books?id=PhlN945ORCYC
63. Winkler S (2005) Digital video quality: vision models and metrics. Wiley, New York
64. Winkler S, Mohandas P (2008) The evolution of video quality measurement: from PSNR to hybrid metrics. IEEE Trans Broadcast 54(3):660–668
65. Wu HR, Rao KR (2006) Digital video image quality and perceptual coding. CRC/Taylor & Francis, Boca Raton
66. Zeithaml VA, Berry LL, Parasuraman A (1993) The nature and determants of customer expectations of service. J Acad Mark Sci 21(1):1–12
67. Zeithaml VA, Parasuraman A, Berry L (1990) Delivering quality service. The Free Press, New York

Chapter 3
Quality of Experience Versus User Experience

Ina Wechsung and Katrien De Moor

Abstract The current chapter discusses the concepts Quality of Experience and User Experience. As Quality of Experience is introduced in the previous chapter, this chapter starts with an introduction to the User Experience concept at the level of theory and practice. First its origins, definitions, and key attributes are discussed. This is followed by an overview of methods and approaches to evaluate User Experience in practice. Thereupon, we discuss both concepts in comparison. While a number of similarities are identified, these are exceeded by the number of differences, which are situated at the theoretical-conceptual level and the methodological-practical level. It is concluded that User Experience is the more mature concept, both at the level of theory and practice. Thus the literature within the User Experience domain can be of great value for the Quality of Experience-community, especially if the latter intends to really put the recently proposed more holistic definition of Quality of Experience into practice.

3.1 Introduction

Quality of Experience (QoE) and User Experience (UX) are relatively new concepts, which became increasingly popular during the last decade. Both can be situated in a broader paradigm shift towards the demand-side in general, and technology users

---

I. Wechsung (✉)
Quality and Usability Lab, Telekom Innovation Laboratories, TU, Berlin, Germany
e-mail: ina.wechsung@telekom.de

K. De Moor
Department of Telematics, Norwegian University of Science and Technology, Trondheim, Norway
e-mail: katrien.demoor@item.ntnu.no

K. De Moor
iMinds-MICT, Ghent University, Ghent, Belgium

S. Möller and A. Raake (eds.), *Quality of Experience*,
T-Labs Series in Telecommunication Services, DOI: 10.1007/978-3-319-02681-7_3,

in particular. They can also be framed within the shift from products and services to 'experiences' as sources of value and differentiation, which has manifested itself in several domains from the 1990s on [1].

In the course of their developments, QoE and UX have been labeled as both 'buzzwords' [2, 3] and central concepts for the design and evaluation of products, systems and services [4, 5]. Since both concepts refer to 'users' and their 'experience' with technology, QoE and UX are sometimes wrongly used as synonyms. A possible explanation for this confusion, beyond the semantic similarity, is that the relation between both concepts has—so far—only been discussed to a limited extent in the literature, see e.g. [6, 7].

At first sight, several similarities can be identified. Both terms are relatively young, but they were not entirely new before they started to gain importance. Both QoE and UX have their origins in other, related concepts: until the last decade, research and industry were predominantly focusing on Quality of Service (QoS) and Usability instead. Whereas the former stems from the field of Telecommunications, the latter originates from the field of Human-Computer Interaction (HCI).[1] Although QoS and Usability are by definition concerned with respectively the "totality of characteristics of a telecommunications service that bear on its ability to satisfy stated and implied needs of the user of the service" [8] and "the extent to which a product can be used by specified users to achieve [...] satisfaction [...]" [9], both concepts were mainly or even purely operationalized in terms of system and service performance-related measures. Assessments of QoS predominantly include network performance parameters such as delay or packet loss [10] and "stated or implied user needs" were not taken into account. Similarly, in the first waves of Usability studies more attention was given to users' efficiency and effectiveness in completing a certain task, rather than to user satisfaction in a broader sense. In 2006, a meta-analysis studying practices in measuring Usability showed that only a small proportion of the reviewed studies (around 20 percent) did not include measures of efficiency and effectiveness. User satisfaction, however, was not (or not explicitly)assessed in around 40 percent of the investigated studies [11]. In contrast to QoS evaluations, which rely exclusively on performance parameters and metrics, Usability evaluations usually involve actual human users (naïve users or experts) and could thus be considered as more human-oriented than QoS. However, the evaluation practices for both, QoS and Usability, implicitly assume that a well performing network, service, or system will lead to higher efficiency, effectiveness and ultimately, to satisfied and happy users. To some degree, evidence for this assumption can be found in the literature. For instance, Sauro and Kindlund [12] observed the expected correlations between user satisfaction and measures of efficiency and effectiveness. Similar findings are reported in [13]. However, several studies could not confirm these findings: For instance, Möller [14] did not find a correlation between task duration and perceived efficiency of spoken

[1] The multidisciplinary field of HCI is around 30 years old and can be considered as 'an amalgam' of several fields, such as computer science, sociology, communication, human factors and ergonomics, engineering [17].

dialogue systems. Also the studies by Frøkjær et al. [15] and Hornbæk and Law [16] suggest that the user's ratings of the interaction, and performance measures, such as task duration and error rates, show little concordance, or even negative correlations. Studies such as the ones above thus illustrate that it is indispensable to measure how the interaction or the service is experienced by the user, and that performance measures are not necessarily strongly related to the user's experiences.

The experience of a human user is obviously central for both QoE and UX, the concepts that are rooted in and to some degree even replaced QoS and Usability. These shifts have to some extent been characterized by a similar evolution from a rather narrow, utilitarian, and instrumental perspective, to a broader perspective that also acknowledges the importance of hedonic and affective aspects when considering human experiences.

Moreover, both fields and related research communities were characterized by similar growing pains, by initiatives to stimulate the evolution towards a more mature and coherent research field and by efforts to prove that they represent more than just 'old wine in new bottles' [18]. Both communities are very diverse and during the last decade, both the UX and the QoE community were taking initiatives to address the lack of a broadly supported, common understanding of what respectively UX and QoE cover, what goes beyond their scope and what distinguishes them from other, related concepts.[2]

Considering the above outlined similarities between QoE and UX, the question arises: what are the differences between the two? A short and easy answer would be that both terms refer to the same underlying concept, but that they simply originated from different research communities. Although the relation between both concepts is sometimes understood in this way, the answer is much more complex. Although it is not possible in the frame of this chapter to present an exhaustive overview of the history and of the abovementioned shifts from QoS to QoE[3] and from Usability to UX, a closer look at the literature shows that QoE and UX are essentially different, rooted in different research traditions, and based on fundamentally different assumptions.

Since QoE—unlike UX—has already been extensively discussed in Chap. 2, the first aim of this chapter is to provide an introduction to the UX concept, at the level of theory and practice. Thereupon, apparent and less apparent differences between QoE and UX are discussed. The rest of this chapter is organized as follows: the next section focuses on UX from a theoretical perspective and briefly discusses its origins, definitions, and key attributes. Next, Sect. 3.3 gives an overview of methods and approaches to evaluate UX in practice. Thereupon, in Sect. 3.4, we discuss UX in comparison to QoE. Finally, Sect. 3.5 concludes this chapter.

---

[2] Not only from QoS and Usability, but also from other concepts such as Customer Experience and User Acceptance.

[3] See also Chap. 6, in which the evolution from QoS to QoE is discussed in detail and in which a comparison of both concepts is made.

3.2 User Experience in Theory

As briefly mentioned in the introduction, HCI research in general (and Usability research in particular), was for a long time focusing on enhancing the efficiency and effectiveness of the system [19]. Major themes in the early years were ergonomics and the functional aspects of human–machine interaction. This is plausible, as the initial work within HCI—prior to the advent of personal computing—was mainly concerned with machines used for research and military purposes. However, Shackel's [20] extensive review on the history of HCI shows that while the technologies changed rapidly, the main research focus remained the same. Although Information and Communication Technologies (ICT) became increasingly ubiquitous, research was still predominantly focusing on work settings and task performance [2]. Moreover, a major concern was to prevent that negative emotions (e.g., frustration) would arise in/due to human-machine interaction [2]. According to Hassenzahl [18], this traditional perspective implies that technology usage is mainly motivated to enhance productivity and to gain time for doing other (non-technology related) things—thus using technology was not being considered as a pleasurable experience in itself. Even though Hassenzahl and Tractinsky [2] point out that hedonic aspects like fun, enjoyment and the relevance of the experience-concept were already suggested during the late 1980s, they also show that it took a number of years until these ideas were adopted by the majority of the HCI community. Nowadays, the aim is not only to prevent rage attacks as a result of a crashing computer, but to facilitate positive emotions while using an interactive system [2] and even to design systems in such a way that they lead to a specific emotional experience (e.g., pleasure). To do so, concepts of Positive or Hedonic Psychology have been embedded and adopted in HCI. Hedonic Psychology, as proposed by Kahneman [21], is focusing on concepts like enjoyment, pleasantness as well as unpleasantness, rather than on attention and memory, two key topics that have been the focus of traditional psychological research. Analogous to this development, HCI research also moved away from the classical cognitive information processing paradigm [22] towards concepts like Affective Computing [23] and Emotional Design [24]. At some point, the term User Experience, which probably originated from the work of Donald Norman at Apple Computers [25], became omnipresent.

3.2.1 Definitions and Attributes of User Experience

Despite its popularity, the concept UX itself was—as was already briefly mentioned above—neither well defined nor well understood [26]. The lack of a shared view on UX (and the subsequent need for one) became obvious, when many companies just exchanged the label Usability with the label User Experience, but kept on doing the same task-centered Usability testing and engineering they did before [18].[4] In academia on the other hand, several contributions have been made over the years to define UX and its properties. As a result, the literature on UX and its related concepts is very rich and diverse; see e.g. [24, 27–33]. Despite the high number of definitions and frameworks proposed in the literature however, a 'common' definition of UX was still missing. Finally in 2010, the International Organization for Standardization (ISO) presented its new ISO 9241-210 standard [34], which included the following definition of the term User Experience.

> A person's perceptions and responses that result from the use or anticipated use of a product, system or service.

This definition is very broad, which was pointed out by Bevan [35],[5] who offered different interpretations for the definition. For example UX may be understood as a counter-concept to Usability or as,

> an umbrella term for all the user's perceptions and responses [...].

Accordingly, even with a standard definition available, the attributes and characteristics of UX were still not clarified. This is illustrated by the work of Law et al. [26], who conducted a survey among researchers and practitioners regarding their conceptions of UX. Their study showed how heterogeneous the views on UX are; however, the survey's authors were able to deduce the following shared understanding: UX can be described as

> dynamic, context-dependent and subjective, stemming from a broad range of potential benefits users may derive from a product.

Whereas this definition does not contradict the definition provided in the ISO 9241-210 [34] it highlights four key characteristics of UX: (1) The first aspect is the (temporal) dynamic of UX, which means UX is changing over time. (2) Secondly, UX is context-dependent; this means each experience is influenced by the situational characteristics of its occurrence, this also means each experience is unique. However, different unique experiences may be similar [36]. The uniqueness of UX is also emphasized by Roto et al. [37]. (3) Furthermore UX is considered as something inherently subjective and individual. Hence, even when confronted with the same system in the same situation, two different persons will experience the system differently and give a different meaning to it. (4) The last aspect of the above definition implies that UX is something positive emphasizing the pleasantness and joy of interacting with technology rather than "the trouble with computers" highlighted in early years of HCI [38].

In 2010, a Dagstuhl seminar was organized to further clarify what UX is and what it is not. This work resulted in a white paper on UX [37], which also discusses the

---

[4] Note that in the years after the introduction of the QoE concept in the literature, similarly, a lot of research presented under the 'QoE flag', was actually much closer to traditional QoS research.

[5] Note that Bevan referred to a draft version of the ISO 9241-210, which was already available in 2008.

characteristics of UX[6] and factors that may influence UX. The latter are grouped into three categories, which correspond to the first three of the above key characteristics. The categories are 'the context around the user and system', 'the user's state', and 'system properties' [37]. In the white paper, the notion of 'use' is also further clarified: UX is about 'encounters' with systems, which can be active (actual use) or passive (for instance, seeing someone else use a system; [37]).

A meta-analysis on UX studies performed by Bargas-Avila and Hornbæk [39] clearly indicates that the research community seems to agree on those four aspects as the central attributes of UX. Similarly, Hassenzahl [36] understands the first three of the above characteristics as 'key properties' of experience in general. Regarding the focus on 'positive' experiences he suggests the terms 'worthwhile' or 'valuable' instead of 'positive', since an experience can be valuable although it is negative if it allows "for a higher, valuable end" [36]. However, he also points out that a "deliberately designed experience should be positive" as such experiences are "per se worthwhile" [36]. Hence, while experiences can be negative, a 'good' user experience is very likely characterized by being positive. A further aspect, which has been identified as a central characteristic of (user) experiences by both, Bargas-Avila and Hornbæk [39] and Hassenzahl [36], is the holistic approach taken by UX. While this is not explicitly addressed in the definition by Law et al. [26], it is implied by User Experience as being understood as something dynamic, situated, and subjective. Hence, experience "comprises of perception, action, motivation and cognition" [36].

3.2.2 *Emotions and Needs*

The previous section described characteristics, which are essential within the current understanding of UX. Although these attributes provide some guidance to unravel factors that may influence UX, an important question remains: What determines whether or not an experience is valuable or not? What are drivers in this respect?

According to McCarthy and Wright [31] any experience is closely related to emotions and affect. In their framework, they refer to the 'emotional thread' of experience in this respect. This position is strongly influenced by the work of the American philosopher Dewey [40], who describes emotions as "the moving and cementing force" which "selects what is congruous and dyes what is selected with its colour" and "provides unity in and through the varied parts of experience". Accordingly, positive[7] experiences are linked to positive emotions.

Based on psychological research, Hassenzahl and colleagues [41] argue that positive (or valuable) experiences are also related to the fulfillment of basic psychological

[6] Note that in [37], the positive nature of UX is not explicitly discussed, but rather the first three key characteristics mentioned above. However, it is implied to some extent through the emphasis on emotions and affect.

[7] The valence of valuable experiences can be positive or negative (cf. Sect. 2.1) hence valuable experiences can be linked to both, positive and/or negative emotions.

needs of humans,[8] which are assumed to be largely invariant across human beings [42]. It is difficult to think of situations where the 'technical quality' of a product, system, or network alone—without any underlying need that is being addressed—can actually lead to a worthwhile, valuable experience. For example, watching a movie that one is not interested in may be experienced as completely meaningless, even though it is presented using the latest 3D technology, in perfect quality. In contrast, watching an old degraded home video together with the family may be a meaningful experience due to the feeling of relatedness with the family members. Moreover, according to Hassenzahl et al. [41], the striving for the fulfillment of needs is the underlying motivation to use interactive technologies. They found that positive experiences with technology are related to need fulfillment and moreover, that need fulfillment is linked to specific product qualities. This means that products provide positive experiences if they are able to fulfill basic psychological needs. However, only certain product qualities were shown to be related to positive experience. Those qualities are labeled as 'hedonic qualities' and they differ from 'pragmatic product qualities' in Hassenzahl's terminology [43], which we will now briefly discuss.

3.2.3 Hedonic and Pragmatic Qualities and Be-goals and Do-goals

Hedonic qualities cover the system's non-functional aspects [44] while pragmatic qualities refer to the task-related aspects of a system and are closely related to the classical concept of Usability.[9] Hassenzahl et al. [41] found pragmatic qualities to be a factor that removes barriers to the fulfillment of the user's needs. A system's pragmatic qualities just enable need fulfillment, but are themselves not a source of a positive experience (as was also argued in the movie example). Hedonic qualities on the other hand are associated with a system's ability to evoke pleasure and to promote the psychological well-being of the user [44]; they are motivators and reflect the product's capability to create a positive experience [41].

Hassenzahl himself describes the relationship in terms of different types of goals—'do-goals' and 'be-goals' [46] (cf. Chap. 4). Do-goals are derived from higher-level be-goals. For example, missing somebody may lead to the desire to communicate with this person. Making a phone call is the do-goal then, the feeling of being related

[8] Note, that psychological needs do not match biological-physiological needs such as hunger or thirst. The most salient needs in the context of human-computer-interaction have been identified as the needs for stimulation, relatedness, competence, and popularity [41]. Examples of other psychological needs are e.g., autonomy and security.

[9] Hassenzahl et al. [41] adopted the terminology of Herzberg's Two Factor Theory of Job Satisfaction [45] to describe the different characters of pragmatic and hedonic qualities. In this theory, 'hygiene factors' and 'motivators' are distinguished: Hygiene factors (i.e., job context factors such as the environmental conditions) can in the best case just prevent dissatisfaction with the job, but cannot lead to satisfaction. However, their absence will results in dissatisfaction. The absence of motivators (job content factors, such as acknowledgment) on the other hand does not result in dissatisfaction, but their presence will facilitate satisfaction and motivation.

to this person is the be-goal. Do-goals are, according to Hassenzahl and Roto [46], related to pragmatic qualities, while be-goals are linked to hedonic qualities and need fulfillment. Although the framework of Hassenzahl is not the only one, it has been very influential within the UX community and beyond.

After this introduction to UX from the theoretical perspective, the next section briefly discusses UX in practice. Both are of course closely tied and the theoretical debates mentioned above, raised many issues and challenges for UX in practice, especially regarding the measurement of UX.

3.3 User Experience in Practice

It has often been claimed that Usability methods are not sufficient to assess UX, and consequently the need for new measurement methods has been postulated [47]. And indeed aspects like hedonic qualities or affect and emotions were not in the scope of many traditional HCI methods and approaches, such as Goals, Operators, Methods, and Selection rules (GOMS) [48], or questionnaires like the popular System Usability Scale (SUS) [49]. In addition, since experience is inherently a subjective construct, it can only be assessed through 'subjective' methods. Therefore, interaction parameters such as task duration or error rates do not allow to assess UX, although they can be related to the subjective experience.

Another critique often expressed is that Usability studies were often conducted in the domain of work-related systems employing task-centered study designs [47]. Such designs cannot simply be extrapolated to the use of e.g., entertainment-oriented systems. As a result, new methods were developed, and methods from a wide range of disciplines[10] were explored, adapted, and adopted.[11] For instance, methods such as the Self-Assessment-Manikin (SAM) [50] (cf. Chap. 9) or the Repertory Grid Technique [51, 52] were adopted from psychology. The latter illustrates one of the main differences between these new or newly adopted methods and traditional HCI methods: Whereas many UX methods aim to gather qualitative feedback, Usability was measured using quantitative approaches [17, 47]. Another important difference is the focus on the inclusion of non-functional aspects, such as emotions and other affective states. Several of the newer tools solely address these emotional aspects. Examples are LemTool [53], PrEmo [54] or the Joy-Of-Use-Button [55]. Furthermore, a recent review by Bargas-Avila and Hornbæk [39, 47] indicates that apart from 'generic UX', emotions and affect are the most assessed dimensions in UX research; in this context the SAM was found to be the most widely used instrument. Hence, two of the key attributes of UX, its subjective and positive character, are reflected in the evaluation methods.

[10] Including amongst others psychology, product design, social sciences, and anthropology.

[11] On the UX community-driven platform http://www.allaboutux.org, an interesting repository of UX evaluation methods (with a short description) can be found.

For the other key attributes mentioned above, namely temporal dynamics and context-dependency, Bargas-Avila and Hornbæk [47] come to a different conclusion: they found that UX, like Usability, is mainly measured after interacting with the system. Moreover, only very few of the studies included in their meta-analysis explicitly describe the context. These findings need to be nuanced however, since the debate on methods and the exploration of new methods were in full explosion at the time when the review study was conducted. Whereas the SAM for instance, has been available since the eighties, new methods for measuring the momentary UX (i.e., during the unfolding of the experience) or considering longer usage periods (i.e., the 'cumulative UX') are just emerging. An example is the Valence Method [56]. In this two-phase measure, which is based on the need-based approach by Hassenzahl et al. [41], users are asked to set positive and negative valence markers while exploring the system. This first phase is videotaped. In the second phase, the marked situations are presented to the participants again while the interviewer is asking which design aspect was the reason for setting the marker. At this point, the question why a certain attribute was mentioned is repeated until the underlying need is identified. Another new method, the UX Curve [57], is used to retrospectively assess UX over time and intends to measure the long-term experience and the influencing factors. Participants are asked to draw curves describing their experiences with the system. In addition, they are asked to explain major changes in the curves. Although 'real' long-term studies, ranging over months or years, are rather rare, the temporal dynamics of UX are far from being unaddressed by the UX community. Karapanos et al. [58] for example studied perceived product quality of a TV set-top box over a four week-period. The main finding was that in the beginning, pragmatic aspects determine satisfaction with the product, while after four weeks hedonic qualities dominate the perceived goodness of a product. Kujala et al. [59] assessed remembered experiences of Facebook and mobile phone usage. It was shown that changes in long-term UX are related to the hedonic qualities of a product, rather than to its pragmatic qualities. Further studies investigating UX over time were presented in [60, 61]. The study with the longest time-frame is probably the longitudinal study conducted by Minge [62] in the frame of his doctoral research. He assessed expectations before the purchase of a mobile phone. One week later first usage experiences were assessed. Afterwards UX measures were collected on a monthly basis for roughly 10 months after purchase.

While the above studies may not be representative for UX studies in general, they indicate that temporal dynamics have been considered. Similarly, also the complex context concept and its properties may have not received the necessary attention but it is more a light grey than a white spot in UX research as also here relevant studies can be found, e.g. in [63]. Most methods that can be used in this respect have their origins in other disciplines. Examples include the Experience Sampling Method (ESM) [64], Context-aware ESM [65], the diary method [66], or the combination of observations with interviewing, as is the case in Contextual Inquiry [67]. Moreover, many research efforts are nowadays 'put into context' as field studies are becoming increasing popular, especially in the context of mobile HCI; see e.g. [68, 69] and Chap. 23. However, this does not mean that more about the context is known, unless such situational variables are assessed.

In summary, the techniques which are employed to measure UX are quite diverse. Still, the review by Bargas-Avila and Hornbæk [39] revealed some favorites like the AttrakDiff and the SAM. Moreover, they report a shift from the quantitative to the qualitative and a lack of studies employing both methods. They criticize that many of the newer methods lack validation and that some dimensions of UX (emotions, affect, and aesthetics) receive much more attention compared to other aspects such as the temporal dynamics or context factors. However, their study demonstrates how the UX community is monitoring itself and offers the chance to improve the current measurement practice.

3.4 Discussion: User Experience Versus Quality of Experience

After this introduction to the literature on UX from both a theoretical and practical perspective, this section discusses UX versus QoE in terms of a number of key characteristics (we also refer the interested reader to the summarizing overview provided in the Appendix at the end of this chapter).

Origins. As was already mentioned above, a first clear and important difference is that QoE originates from Telecommunications and UX from HCI. Both the QoE and the UX research field can be typified as 'multi-disciplinary': a range of different disciplines are involved. Especially the UX field has been very multi-disciplinary from the very beginning, both in theory and practice and with representation of people from so-called soft and hard sciences. In the field of QoE, it has also been acknowledged from the beginning that a genuine multi-disciplinary perspective (going beyond the engineering domain) is needed. In practice, however, the establishment of links to disciplines that are closer to the user perspective[12] is still very much an ongoing process. Moreover, the origins and evolution of QoE have been strongly industry- and push-driven. The goal of delighting users and avoiding user frustration is not (only) a noble goal in itself, it is strongly driven by economical motivations of different players along the associated ecosystem,[13] for example to increase revenue or to enhance the loyalty of their customers. For instance, from a network perspective, a better understanding of end-user QoE is needed to optimally manage trade-offs and facilitate the cost-efficient allocation of limited (technical) resources. Although this economical dimension is less prominently discussed in the UX literature, the survey of Law et al. [26] indicated that the goal of 'making people happy' is not the only one and that UX is strongly linked to the objective of 'designing better products'. From an industry and more business-oriented perspective, this also implies a focus on increasing the possible market success (by putting users and the UX central). Still, economic aspects are a peripheral matter within the (academic) discussions of

[12] For instance, rooted in social sciences (sociology, anthropology, economics, etc.) and behavioral sciences (psychology, cognitive sciences, etc.).

[13] See e.g., the work of Kilkki [71] and Reichl et al. [72].

UX. In this line of thinking, it can be argued that QoE is actually much closer to the concept of Customer Experience[14] than it is to UX.

Driving force. The different origins of UX and QoE are linked to another fundamental difference between UX and QoE: One of the general, but essential properties of UX is that it is human-centered. Roto et al. [37] explicitly emphasize that UX is not driven by technology, and this is reflected both at the level of theory and practice. QoE on the other hand, is considered to be "in a large part of instances, highly dependent on QoS" [70], and research on QoE is still primarily system- and technology-centered.

Theoretical basis. In terms of the theoretical basis, the overview above illustrated that the theoretical basis of UX is rooted in different disciplines,[15] with strong influences from the field of hedonic psychology. These theoretical underpinnings have been thoroughly discussed and documented in the literature. The work of Law et al. [26] indicated that several heterogeneous views on UX exist, but that they share a number of important, common denominators: UX has a dynamic nature, it is context-dependent, inherently subjective, and individual (also implying that each experience is unique) and it is concerned with positive or valuable experiences. Moreover, UX is about encounters with systems, and more concretely, it includes both actual use (active or passive) and anticipated use.

In contrast to the field of UX and keeping in mind its origins, QoE seems to have evolved in an inverse manner, in an application- and practice-driven way. As a result, it is strongly tied to traditional measurement and instrumentation approaches and lacking a strong theoretical basis. The debate on the definition of QoE has been omnipresent during the last decade, and many different definitions have been introduced in the literature. However, these discussions were not that fundamental in the sense that the theoretical roots of the concept and the critical deconstruction of its elements have received considerably less attention compared to the discussion in the UX community. During the recent years, several more holistic conceptual frameworks which reach out to other fields and literature streams (including UX and HCI), have been presented in the literature. Examples include the QoE-taxonomy [75], the Gr@sp-framework of QoE [76], and the User-Centered Quality of Experience approach [6]. These and other efforts are very important in view of the establishment of a well-grounded, and commonly accepted understanding of QoE, both in theory and practice. However—as is also acknowledged by Le Callet et al. [70]—the work is not finished yet: a robust, theoretical foundation of QoE and its translation into a solid practical framework are still missing. As is discussed more intensively in Chap. 2 of this book, based on the recently published Qualinet White Paper [70] a new, more holistic definition of QoE has been introduced as

[14] For a thorough introduction to the literature on Customer Experience, see e.g. Palmer [73].

[15] Obrist et al. [74] conducted a survey on the theoretical roots of UX, which indicated 56 different theoretical perspectives stemming from nine disciplines. These activities illustrate not only the inherent multi-disciplinarity of the research field, but also its ongoing efforts to get a deeper understanding of UX beyond the somewhat agreed key attributes explained before.

> the degree of delight or annoyance of a person whose experiencing involves an application, service, or system. It results from the person's evaluation of the fulfillment of his or her expectations and needs with respect to the utility and / or enjoyment in the light of the person's context, personality and current state.

This new definition indicates that the understanding of QoE is increasingly going beyond the instrumental and expanding towards the non-instrumental and hedonic, which has a central role in the domain of UX. In contrast to earlier definitions, QoE is no longer expressed in terms of satisfaction, but as a degree of delight or annoyance, thus acknowledging its dynamic and transient character (delight in a certain context might be frustration in another context). In addition, the inherent subjective and individual character and the context-dependency of QoE—which are also seen as key characteristics of UX—are also implied by this definition. Both in the UX literature and in the approach to QoE discussed in Le Callet et al. [70], possible influencing factors are roughly grouped into three categories, namely user-related, system-related and context-related characteristics and factors.

Measurement and Evaluation. Regarding evaluation methods, the focus within the UX domain has increasingly shifted towards the development, use, and adaptation of a wide range of evaluation methods and towards a debate on the measurement and modeling of UX in order to appropriately address the practical implications of the theoretical understanding of UX [77]. Moreover, as the (non-exhaustive) overview of UX evaluation methods above indicated, a number of methods and tools originating from a wide range of disciplines are used. Although UX research draws on both quantitative and qualitative research methods, the UX community adopted a wide range of qualitative approaches from its related disciplines, which have a strong qualitative tradition, such as sociology or ethnology. As a result, UX has been quite heavily influenced by the interpretative and constructivism-based research paradigm, which seeks to understand, focuses on meaning and interpretation, and aims to gain a richer understanding of phenomena. As was discussed earlier in this chapter, the strong emphasis on non-instrumental aspects is reflected in the used methods and empirical work. As a result, human affect and emotions in particular have been very prominent in many UX studies. Section 3.3 also pointed to the focus on the temporal dynamics of UX and contextual factors in UX evaluations, albeit to a lesser degree. Moreover, meta-analyses such as the one by Bargas-Avila and Hornbæk [39] receive a lot of attention in the UX community and show that there is an on-going debate whether or not the current practice is sufficient to assess all dimensions of UX, and whether or not the methods match the constructs which are intended to be measured.

For QoE, in contrast, the implications of its recent more holistic definition are not (yet) reflected in the dominant measurement approaches, even though the number of QoE studies that try to go beyond the dominant framework is growing (see further). Traditional QoE measurement is strongly determined by a series of recommendations issued by the International Telecommunications Union (ITU). These recommendations stipulate in a detailed way which scales should be used and which procedures should be followed in the context of subjective quality assessment, and these are widely used for QoE experiments. Such experiments are typically conducted in an artificial, controlled research environment with the aim of isolating and investigating the impact of specific factors. This type of empirical-positivist research, which is predominantly based on quantitative approaches, undoubtedly has its value: it allows to quantify the impact and weight of investigated factors while keeping other possible influences under control. When well documented and rigorously conducted, such experiments can be easily replicated and data can be exchanged and compared [78]. Moreover, this type of research allows the development of models and use of statistical analysis techniques. However, it also has crucial limitations, which need to be addressed. The settings in which the users experience an application or service and are asked to evaluate the quality of their experience, has very little resemblance with the real, natural environment. Acknowledging that human experiences do not take place in isolation (as both the UX and QoE community do) also implies that they cannot be studied only in isolation. Moreover, given their highly subjective nature (implied in both the QoE and UX definitions), the measures that are used to evaluate their quality need to be carefully (re-)considered.

Unlike in UX evaluation, QoE measurement is still predominantly based on quantitative quality evaluations and numerical expressions of the inherently subjective QoE construct. A highly complex, subjective and multi-dimensional construct is thus reduced into one or a couple of numbers. More specifically, QoE is often operationalized in terms of a Mean Opinion Score (MOS) value. While such a widely used scale is helpful when comparing different studies, it illustrates, on the other hand, that while the paradigm changed from QoS to QoE, the dominant measurement scale remained for a large part the same. MOS ratings were used long before QoE became a trending topic, e.g. as early as 1969 it was part of the 'IEEE Recommended Practice for Speech Quality Measurements' [79]. While this may indicate the universal applicability of MOS, it can also indicate that the recent efforts of parts of the QoE community, to move from pure quantitative measurement to evaluation in the sense of gaining insights on how 'experiencing' takes place and how people give meaning to it, have not been adopted yet by the majority of the QoE community. Measures beyond MOS and ITU recommendations are still the exception and not the rule. Nevertheless, although no measurement revolution has taken place (yet), several initiatives and research efforts over the past years have initiated an important evolution. The methodologies to operationalize 'experience' are one of the major topics in current QoE research. Since a number of years, the dominant quantitative approach has been the subject of a lot of debate and other measures, methods and approaches are currently being explored. Several chapters in this book address these new approaches. For instance, Chap. 21 presents methods that enable larger scale measurement outside of the lab, specifically the recent focus on using crowdsourcing for QoE, see also [80, 81]. Other examples of QoE research in living lab and more realistic environments can be found in [82–84]. Chapter 9 illustrates the growing interest in emotions and affective states in relation to QoE and the use of physiological and behavioral measures; related studies include [85–87]. The recent work on temporal dimensions of QoE is described in Chap. 10; additional studies are e.g. [82, 88–93]. These examples indicate that the temporality of (quality of) 'experiences' is also reflected in recent QoE research and that new methods are being explored in this respect. Still the time frames are usually shorter compared to UX research. It is debatable whether or not the temporal resolutions of UX and QoE are equal, and whether or not QoE involves per se shorter time frames than UX. However, the recent QoE definition does not limit the time frame of QoE. Accordingly, also long-term studies ranging over months (as in a number of recent UX studies) can be considered as 'in scope' for QoE research and as an interesting topic for future research. Lastly, the recently introduced more qualitative, descriptive approaches such as sensory profiling and mixed-method approaches such as open profiling of quality (OPQ) are briefly addressed in Chaps. 5, 19 and 26; for more detailed information see e.g. [94, 95].

Experience Versus Perceptions. Finally, as was indicated earlier in this chapter, the 'experience' concept, its meanings and implications, have been thoroughly and critically deconstructed by scholars in the UX community. As a result, it has attained a central role and is well-embedded in UX evaluation and practice. In the domain of QoE however, the focus is more explicitly on the quality formation process and features that contribute to the perception of quality (i.e., 'quality features'). As a result, the focus is on quality assessment and much less on the evaluation of experiences and characteristics of the experience that impact their quality.

3.5 Conclusion

The current chapter first of all introduced the concept of UX both from a theoretical and practical perspective. Thereupon, it discussed the relationship between QoE and UX. Although a number of similarities were identified, this chapter mainly indicated that the number of differences exceeds the number of similarities. As was shown, these differences are profound, and they are situated both at the theoretical-conceptual level and the methodological-practical level. Within the UX domain, the complex, dynamic, inherently subjective and situated, context-dependent character of human experiences is fully acknowledged, both in theory and in practice. Moreover, attention has gone out to the temporal dimensions (e.g., evolution of UX over time), to human affect in general and emotions in particular, to the role of non-instrumental, hedonic product qualities and attributes, and how they relate to human needs and goals.

Despite the large distance between QoE and UX, the literature within the UX-domain can be of great value for the QoE-community, especially if the latter intends to really put the recently proposed more holistic definition of QoE into practice. This would however imply a fundamental reorientation of the dominant QoE measurement paradigm, which is based on a range of rigid recommendations and standardized approaches which describe in detail how specific factors should be isolated, and how QoE should be measured.

Notwithstanding the notable evolution of the field of QoE during the last years, it could be argued that the UX concept is in a more mature state, both at the level of theory and practice. Thus, a major challenge ahead for the QoE community lies in the operationalization of its new definition and the identified influencing factors into adjusted measurement approaches that allow to capture QoE as it has been defined.

Acknowledgments Katrien De Moor's work was carried out during the tenure of an ERCIM "Alain Bensoussan" Fellowship Programme and received funding from the European Union Seventh Framework Programme (FP7/2007-2013) under grant agreement no 246016.

Appendix to Chapter 3

	Quality of Experience	User Experience
Origins	Telecommunications	HCI
Driving force	Primarily technology-driven, technology-centered	Primarily human-driven, human-centered
Theoretical basis	Limited (more emphasis on practical applications)	Strong and diverse theoretical basis
Disciplinary nature of the research field	Multidisciplinary, increasingly also in practice	Multidisciplinary from the beginning, in theory and practice
Main focus	Evaluate (technical) quality perception, gather input to guide optimization of technical parameters at different layers	Evaluate and understand the User Experience / process of experiencing, gather input for designing and creating products and services that enable users to have valuable, pleasurable experiences, enable the fulfillment of be-goals
Main research 'objects'	Multimedia communication systems	Products, services, and artifacts that a person can interact with through a user interface
Perspective on use	Use of application or service	Encounter with a system (active or passive), anticipated use
Measurement and instrumentation	Standardized measurement and relatively rigid instrumentation (recommendations), predominantly operationalised in terms of MOS ratings	Not translated into standards and official recommendations, large range of methods and tools originating from wide range of disciplines
Research designs	Predominantly quantitative, increasingly also mixed-methods approaches	Both quanti- and qualitative, with strong emphasis on qualitative research
Research environment	Mostly controlled, laboratory research, but growing interest in field and online studies	Laboratory, field and online studies
Research aims	Quantifying, modeling	Understanding, modeling
Main focus	By definition: both pragmatic, utilitarian and hedonic aspects, in (measurement) practice: emphasis on the former	In theory and practice: both instrumental and non-instrumental aspect, strong emphasis on the latter (hedonic dimensions)

	Quality of experience	User experience
Research approach	Isolation of specific factors	Holistic approach
Temporal perspective	Growing emphasis on temporal QoE features and influencing factors, very little empirical work on how QoE changes over longer time	Different time spans of UX are considered, in theory and practice
Business perspective	Importance of and interest in monetary dimension (user as customer), willingness to pay	Little direct attention to monetary dimension

References

1. Pine JB, Gilmore JH (1999) The experience economy: work is theatre and every business a stage. Havard, Boston
2. Hassenzahl M, Tractinsky N (2006) User experience: a research agenda. Behav Inf Technol 25(2):91–97
3. Schatz R, Reichl P (2011) Quality of experience: just another Buzzword? In: Proceedings of Euroview 2011.
4. Agboma F, Liotta A (2008) QoE-aware QoS management. In: Proceedings of the 6th international conference on advances in mobile computing and multimedia (MoMM '08), pp 111–116
5. Wilson GM, Sasse MA (2004) From doing to being: getting closer to the user experience. Interact Comput 16(4):697–705
6. Jumisko-Pyykkö S (2011) User-centered quality of experience and its evaluation methods for mobile television. Doctoral thesis, Tampere University of Technology, Tampere
7. De Moor K (2012) Are engineers from Mars and users from Venus? Bridging gaps in quality of experience research: experiences from and reflections on an interdisciplinary journey. Unpublished doctoral thesis. Ghent University, Gent
8. ITU-T Recommendation E.800 (2008) Terms and Definitions Related to Quality of Service and Network Performance Including Dependability. International Telecommunication Union, Geneva
9. ISO 9241–11 (1998) Ergonomic requirements for office work with visual display terminals (VDTs). Part 11: guidance on usability. International Organization for Standardization (ISO), Geneva
10. Fiedler M, Kilkki K, Reichl P (2009) From quality of service to quality of experience. Executive summary of dagstuhl seminar 09192
11. Hornbæk K (2006) Current practice in measuring usability: challenges to usability studies and research. Int J Hum Comput Stud 64(2):79–102
12. Sauro J, Kindlund E (2005) A method to standardize usability metrics into a single score. In: Proceedings of the SIGCHI conference on human factors in computing systems (CHI '05), pp 401–409
13. Nielsen J, Levy J (1994) Measuring usability: preference versus performance. Commun ACM 37(4):66–75
14. Möller S (2006) Messung und Vorhersage der Effizienz bei der Interaktion mit Sprachdialogdiensten [Measurement and prediction of interaction efficiency with spoken dialogue systems]. In: Langer S, Scholl W (eds) Fortschritte der Akustik-DAGA 2006, pp 463–464

15. Frøkjær E, Hertzum M, Hornbæk K (2000) Measuring usability: are effectiveness, efficiency, and satisfaction really correlated?. In: Proceedings of the SIGCHI conference on human factors in computing systems (CHI '00), pp 345–352
16. Hornbæk K, Law EL (2007) Meta-analysis of correlations among usability measures. In: Proceedings of the SIGCHI conference on human factors in computing systems (CHI '07), pp 617–626
17. Lazar J, Feng JH, Hochheiser H (2010) Research methods in human-computer interaction. Wiley, Chichester
18. Hassenzahl M (2008) User experience (UX): towards an experiential perspective on product quality. In: Proceedings of the 20th international conference of the association francophone d'interaction homme-machine (IHM '08), pp 11–15
19. Preece J, Rogers Y, Sharp H, Benyon D, Holland S, Carey T (1994) Human-computer interaction: concepts and design. Addison-Wesley, Wokingham
20. Shackel B (2009) Human-computer interaction–Whence and whither? Interact Comput 21(5–6):353–366
21. Kahneman D (1999) Objective happiness. In: Kahneman D, Diener E, Schwarz N (eds) Well-being: foundations of hedonic psychology. Russell Sage Foundation Press, New York, pp 3–25
22. Kaptelinin V, Nardi B, Bødker S, Carroll J, Hollan J, Hutchins E, Winograd T (2003) Post-cognitivist HCI: second-wave theories. In: Proceedings of the conference on human factors in computing systems (CHI '03), pp 692–693
23. Picard RW (1997) Affective computing. MIT Press, Cambridge
24. Norman DA (2004) Emotional design: why we love (or hate) everyday things. Basic Books, New York
25. Norman DA, Miller J, Henderson A (1995) What you see, some of what's in the future, and how we go about doing it: HI at apple computer. In: Proceedings of the conference companion on human factors in computing systems (CHI 1995), p 155
26. Law EL, Roto V, Hassenzahl M, Vermeeren APOS, Kort J (2009) Understanding, scoping and defining user experience: a survey approach. In: Proceedings of the 27th international conference on human factors in computing systems (CHI 2009), pp 719–728
27. Hassenzahl M, Platz A, Burmester M, Lehner K (2000) Hedonic and ergonomic quality aspects determine a software's appeal. In: Proceedings of the conference on human factors in computing (CHI 2000), pp 201–208
28. Forlizzi J, Ford S (2000) The building blocks of experience: an early framework for interaction designers. In: Proceedings of the 3rd conference on designing interactive systems: processes, practices, methods, and techniques (DIS 2000), pp 419–423
29. Arhippainen L (2003) Capturing user experience for product design. Paper presented at the 26th information systems research seminar (IRIS26), Porvo, Finland
30. Desmet P, Hekkert P (2007) Framework of product experience. Int J Des 1(1):57–66
31. McCarthy J, Wright P (2004) Technology as experience. Interactions 11(5):42–43
32. Jordan p (2000) Designing pleasurable products. An introduction to the new human factors. Taylor & Francis, London
33. Forlizzi J, Battarbee K (2004) Understanding experience in interactive systems. In: Proceedings of the 5th conference on designing interactive systems: processes, practices, methods, and techniques (DIS '04), pp 261–268
34. ISO 9241–210 (2010) Ergonomics of human system interaction-part 210: human-centred design for interactive systems (formerly known as 13407). International organization for standardization (ISO), Geneva
35. Bevan N (2009) What is the difference between the purpose of usability and user experience evaluation methods. Paper presented at the workshop user experience evaluation methods in product development (UXEM'09). Retrieved from http://www.nigelbevan.com/papers/What_is_the_difference_between_usability_and_user_experience_evaluation_methods.pdf
36. Hassenzahl M (2010) Experience design: technology for all the right reasons. Princeton, Morgan & Claypool

37. Roto V, Law EL, Vermeeren A, Hoonhout J (eds) (2011) User experience white paper: bringing clarity to the concept of user experience, result of dagstuhl seminar 10373
38. Landauer TK (1995) The trouble with computers. MIT Press, Cambridge
39. Bargas-Avila JA, Hornbæk K (2011) Old wine in new bottles or novel challenges: a critical analysis of empirical studies of user experience. In Proceedings of the SIGCHI conference on human factors in computing systems (CHI 2011), pp 2689–2698
40. Dewey J (1980) Art as experience (first printed in 1934). Perigee Books, New York
41. Hassenzahl M, Diefenbach S, Göritz AS (2010) Needs, affect, and interactive products–Facets of user experience. Interact Comput 22(5):353–362
42. Sheldon KM, Elliot AJ, Kim Y, Kasser T (2001) What is satisfying about satisfying events? Testing 10 candidate psychological needs. J Pers Soc Psychol 89:325–339
43. Hassenzahl M, Burmester M, Koller F (2003) AttrakDiff: Ein Fragebogen zur Messung wahrgenommener hedonischer und pragmatischer Qualität [AttracDiff: a questionnaire to measure perceived hedonic and pragmatic quality]. In: Ziegler J, Szwillus G (eds) Mensch & Computer 2003. Interaktion in Bewegung. Teubner, Stuttgart, pp 187–196
44. Hassenzahl M (2003) The thing and I: understanding the relationship between user and product. In: Blythe M, Overbeeke C, Monk AF, Wright PC (eds) Funology: from Usability to Enjoyment. Kluwer Academic Publishers, Dordrecht, pp 287–302
45. Herzberg F (1968) One more time: how do you motivate employees? Harvard Bus Rev 46(1):53–62
46. Hassenzahl M, Roto V (2007) Being and doing: a perspective on User Experience and its measurement. Interfaces 72:10–12
47. Bargas-Avila J, Hornbæk K (2012) Foci and blind spots in user experience research. ACM interact 19(6):24–27
48. Card S, Moran T, Newell A (1983) The psychology of human-computer interaction. Lawrence Erlbaum Associates, Hillsdale
49. Brooke J (1996) SUS-A "quick and dirty" usability scale. In: Jordan P, Thomas B, Weerdmeester B, McClelland I (eds) Usability evaluation in industry. Taylor & Francis, London, pp 189–194
50. Bradley MM, Lang PJ (1994) Measuring emotion: the self-assessment Manikin and the semantic differential. J Behav Ther Exp Psychiatry 25(1):49–59
51. Hassenzahl M, Trautmann T (2001) Analysis of web sites with the repertory grid technique. In: Proceedings of conference on human factors in computing systems. Extended abstracts (CHI 2001), pp 167–168
52. Kelly GA (1955) The psychology of personal constructs. Norton, New York
53. Huisman G, Van hout M (2008) The development of a graphical emotion measurement instrument using caricatured expressions: the LEMtool. In: Proceedings of the workshop emotion in HCI–Designing for people, pp 5–7
54. Desmet P (2004) Measuring emotions: development and application of an instrument to measure emotional responses to products. In: Blythe M, Overbeeke C, Monk AF, Wright PC (eds) Funology: from usability to enjoyment. Kluwer Academic Publishers, Dordrecht, pp 111–123
55. Schleicher R, Trösterer S (2009) The 'joy-of-use'-button: recording pleasant moments while using a PC. In: Proceedings of the 12th IFIP TC 13 international conference on human-computer interaction: Part II (INTERACT '09), pp 630–633
56. Burmester M, Mast M, Jäger K, Homans H (2010) Valence method for formative evaluation of user experience. In: Proceedings of designing interactive systems conference (DIS '10), pp 364–367
57. Kujala S, Roto V, Vaananen-Vainio-Mattila K, Karapanos E, Sinnela A (2011) UX curve: a method for evaluating long-term user experience. Interact Comput 23(5):473–483
58. Karapanos E, Hassenzahl M, Martens J-B (2008) User experience over time. In: Proceedings of the 26th international conference on human factors in computing systems (CHI' 2008), pp 3561–3566
59. Kujala S, Roto V, Väänänen-Vainio-Mattila K, Sinnelä A (2011) Identifying hedonic factors in long-term user experience. In Proceedings of DPPI-11: designing pleasurable products and interfaces, pp 137–144

60. von Wilamowitz-Moellendorff M, Hassenzahl M, Platz A (2006) Dynamics of user experience: how the perceived quality of mobile phones changes over time. In: NordiCHI workshop. User experience–Towards a unified View, pp 74–78
61. Karapanos E, Zimmerman J, Forlizzi J, Martens J-B (2009) User experience over time: an initial frame-work, In: Proceedings of the 27th international conference on human factors in computing systems (CHI 2009), pp 729–738
62. Minge M (2011) Dynamische Aspekte des Nutzungserlebens der Interaktion mit technischen Systemen [Dynamic aspects of user experience of interaction with technical systems]. Doctoral thesis, Berlin Institute of Technology, Berlin. Retrieved from http://opus.kobv.de/tuberlin/volltexte/2011/3290/pdf/minge_michael.pdf
63. Partala T, Kallinen A (2012) Understanding the most satisfying and unsatisfying user experiences: emotions, psychological needs, and context. Interact Comput 24(1):25–34
64. Csikszentmihalyi M, Larson R (1987) Validity and reliability of the experience-sampling method. J Nerv Ment Dis 175(9):526
65. Intille SS, Rondoni J, Kukla C, Anacona I, Bao L (2003) A context-aware experience sampling tool. In: Proceedings of CHI '03 extended abstracts on human factors in computing systems (CHI EA '03), pp 972–973
66. Bolger N, Davis A, Rafaeli E (2003) Diary methods: capturing life as it is lived. Annu Rev Psychol 54:579–616
67. Beyer H, Holtzblatt K (1998) Contextual design: defining customer-centered systems. Morgan Kaufmann, San Francisco
68. Wechsung I, Jepsen K, Burkhardt F, Köhler A, Schleicher R (2012) View from a distance: comparing online and retrospective UX-evaluations. In: Proceedings of the 14th international conference on human-computer interaction with mobile devices and services companion (MobileHCI '12), pp 113–118
69. Schleicher R, Sahami A, Rohs M, Kratz S, Schmidt A (2011) WorldCupinion: experiences with an android app for real-time opinion sharing during Soccer World Cup Games. Int J Mob Hum Comput Interact 3(4):18–35
70. Le Callet P, Möller S, Perkis A (eds) (2012) Qualinet white paper on definitions of quality of experience–output version of the dagstuhl seminar 12181. European network on quality of experience in multimedia systems and services (COST Action IC 1003), Lausanne, Version 1.1
71. Kilkki K (2008) Quality of experience in communications ecosystems. J Univers Comput Sci 14(5):615–624
72. Reichl P, Tuffin B, Maillé P (2012) Economics of quality of experience. In: Hadjiantonis AM, Stiller B (eds) Telecommunication Economics. Springer, Berlin, pp 158–166
73. Palmer A (2010) Customer experience management: a critical review of an emerging idea. J Serv Mark 24(3):196–208
74. Obrist M, Roto V, Vermeeren A, Väänänen-Vainio-Mattila K, Law EL-C, Kuutti K (2012) Search of theoretical foundations for UX research and practice. In: Proceedings of the 2012 ACM annual conference extended abstracts on human factors in computing systems extended abstracts (CHI EA '12), pp 1979–1984
75. Möller S, Engelbrecht K-P, Kühnel C, Wechsung I, Weiss B (2009) A taxonomy of quality of service and quality of experience of multimodal human-machine interaction. In: Proceedings of the first international workshop on quality of multimedia experience (QoMEX'09), pp 7–12
76. Geerts D, De Moor K, Ketyko I, Jacobs A, Van den Bergh J, Joseph W, Martens L, De Marez L (2010) Linking an integrated framework with appropriate methods for measuring QoE. In: Proceedings of quality of multimedia experience (QoMEX), 2010 second international workshop on quality of multimedia experience, pp 158–163
77. Law E, van Schaik P (2010) Modelling user experience: an agenda for research and practice (editorial). Interact Comput 22:313–322
78. Brooks P, Hestnes B (2010) User measures of quality of experience: why being objective and quantitative is important. IEEE Netw 24(2):8–13

79. Rothauser EH, Chapman WD, Guttman N, Nordby KS, Silbiger HR, Urbanek GE, Weinstock M (1969) IEEE recommended practice for speech quality measurements. IEEE Trans Audio Electroacoust 17(3):225–246
80. Gardlo B (2012) Quality of experience evaluation methodology via crowdsourcing. Unpublished doctoral dissertation, University of Zilina, Zilina
81. Hoßfeld T, Seufert M, Hirth M, Zinner T, Phuoc T-G, Schatz R (2011) Quantification of YouTube QoE via crowdsourcing. In: Proceedings of the IEEE international symposium on multimedia (ISM)
82. Staelens N, Moens S, Van den Broeck W, Mariën I, Vermeulen B, Lambert P, Van de Walle R, Demeester P (2011) Assessing quality of experience of IPTV and video on demand services in real-life environments. IEEE Trans Broadcasting 56(4):458–466
83. De Moor K, Ketyko I, Joseph W, Deryckere T, De Marez L, Martens L, Verleye G (2010) Proposed framework for evaluating quality of experience in a mobile, testbed-oriented living lab setting. Mob Netw Appl 15(3):378–391
84. Schatz R, Egger S (2011) Vienna surfing: assessing mobile broadband quality in the field. In: Proceedings of the first ACM SIGCOMM workshop on measurements up the stack, pp 19–24
85. Laghari K, Gupta R, Arndt S, Antons J-N, Schleicher R, Möller S, Falk TH (2013) Neurophysiological experimental facility for quality of experience (QoE) assessment. In: Proceedings of the first IFIP/IEEE international workshop on quality of experience centric management (QCMan)
86. Arndt S, Antons JN, Schleicher R, Moller S, Curio G (2012) Perception of low-quality videos analyzed by means of electroencephalography. In: Proceedings of 4th international IEEE workshop on quality of multimedia experience (QoMEX), pp 284–289
87. Reiter U, De Moor K (2012) Content categorization based on implicit and explicit user feedback: combining self-reports with EEG emotional state analysis. In: Proceedings of 4th international workshop on quality of multimedia experience (QoMEX), pp 266–271
88. Hoßfeld T, Strohmeier D, Raake A, Schatz R (2013) Pippi Longstocking calculus for temporal stimuli pattern on YouTube QoE: $1 + 1 = 3 1 \bullet 4 \neq 4 \bullet 1$. In: Proceedings of the 5th workshop on mobile video, pp 37–42
89. Guse D, Möller S (2012) Long-term impact of varying multimedia service performance on quality ratings in a multiservice scenario. In: Proceedings of fortschritte der Akustik–DAGA 2013: Plenarvortrag und Fachbeitrag d. 39. Dtsch. Jahrestg. f. Akust. DEGA
90. Borowiak A, Reiter U, Svensson UP (2012) Evaluation of audio quality require-ments over extended periods of time using long duration audiovisual content. In: Proceedings of the 13th Pacific-Rim conference on advances in multimedia information processing (PCM'12), 10–20
91. Fröhlich P, Egger S, Schatz R, Mühlegger M, Masuch K, Gardlo B (2012) QoE in 10 seconds: are short video clip lengths sufficient for quality of experience assessment? In: Proceedings of 4th international IEEE workshop on quality of multimedia experience (QoMEX), pp 242–247
92. Möller S, Bang C, Tamme T, Vaalgamaa M, Weiss B (2011) From single-call to multi-call quality: a study on long-term quality integration in audio-visual speech communication. Proc Interspeech 2011:1485–1488
93. Hosfeld T, Biedermann S, Schatz R, Platzer A, Egger S, Fiedler M (2011) The memory effect and its implications on Web QoE modeling. In: Proceedings of the 23rd international teletraffic congress (ITC), pp 103–110
94. Strohmeier D, Jumisko-Pyykkö S, Kunze K (2010) Open profiling of quality: a mixed method approach to understanding multimodal quality perception. Advances in multimedia, 2010
95. Strohmeier D, Jumisko-Pyykkö S, Kunze K, Bici MO (2011) The extended-OPQ method for user-centered quality of experience evaluation: a study for mobile 3D video broadcasting over DVB-H. EURASIP Journal on Image and Video Processing, 2011

Chapter 4
Factors Influencing Quality of Experience

Ulrich Reiter, Kjell Brunnström, Katrien De Moor, Mohamed-Chaker Larabi, Manuela Pereira, Antonio Pinheiro, Junyong You and Andrej Zgank

Abstract In this chapter different factors that may influence Quality of Experience (QoE) in the context of media consumption, networked services, and other electronic communication services and applications, are discussed. QoE can be subject to a range of complex and strongly interrelated factors, falling into three categories:

U. Reiter (✉)
Department of Electronics and Telecommunications, Norwegian University of Science and Technology, Trondheim, Norway
e-mail: ulrich.reiter@ntnu.no

K. Brunnström
Visual Media Quality, Acreo Swedish ICT AB and Mid Sweden University, Stockholm, Sweden
e-mail: kjell.brunnstrom@acreo.se

K. De Moor
Department of Telematics, Norwegian University of Science and Technology, Trondheim, Norway
e-mail: katrien.demoor@item.ntnu.no

K. De Moor
iMinds-MICT, Ghent University, Ghent, Belgium

M.-C. Larabi
Université de Poitiers, Poitiers, France
e-mail: chaker.larabi@univ-poitiers.fr

M. Pereira and A. Pinheiro
University of Beira Interior, Covilhã, Portugal
e-mail: mpereira@di.ubi.pt

A. Pinheiro
e-mail: pinheiro@ubi.pt

J. You
Christian Michelsen Research AS, Bergen, Norway
e-mail: junyong.you@cmr.no

A. Zgank
University of Maribor, Maribor, Slovenia
e-mail: andrej.zgank@uni-mb.si

S. Möller and A. Raake (eds.), *Quality of Experience*,
T-Labs Series in Telecommunication Services, DOI: 10.1007/978-3-319-02681-7_4,

human, *system* and *context* influence factors (IFs). With respect to Human IFs, we discuss variant and stable factors that may potentially bear an influence on QoE, either for low-level (bottom-up) or higher-level (top-down) cognitive processing. System IFs are classified into four distinct categories, namely content-, media-, network- and device-related IFs. Finally, the broad category of possible Context IFs is decomposed into factors linked to the physical, temporal, social, economic, task and technical information context. The overview given here illustrates the complexity of QoE and the broad range of aspects that potentially have a major influence on it.

4.1 Introduction

In the context of media consumption, networked services, and other electronic communication services and applications, the human experience may be influenced by various and numerous factors that impact QoE. Some of these are more straightforward and their impacts have been thoroughly described and quantified. However, others are situation-dependent, are more difficult to describe, or are effective only under certain circumstances, e.g. in combination with or in absence of others. The Qualinet White Paper on Definitions of Quality of Experience defines these factors influencing QoE as follows:

Influence Factor (IF): Any characteristic of a user, system, service, application, or context whose actual state or setting may have influence on the Quality of Experience for the user [1].

In this sense, the Influence Factors discussed here are the independent variables, whereas the resulting QoE as perceived by the end user is the dependent variable. A certain set of Influence Factors may be described by users in terms of their impact on QoE. This means that users are not necessarily aware of the underlying IFs, but they are usually—to a certain extent—able to describe what they like or dislike about the experience.

In the following, we will group and discuss Influence Factors into three categories, namely Human IFs (HIFs), System IFs (SIFs), and Context IFs (CIFs), and we will give examples and in-depth explanations. However, the IFs must not be regarded as isolated, since they frequently interrelate, see Fig. 4.1. For example, HIFs and CIFs might determine in which way and how much the set of SIFs actually impacts on QoE: the same video clip might leave a totally different quality impression when watched on a mobile phone while riding on the bus than when watched on a TV screen in the user's home.

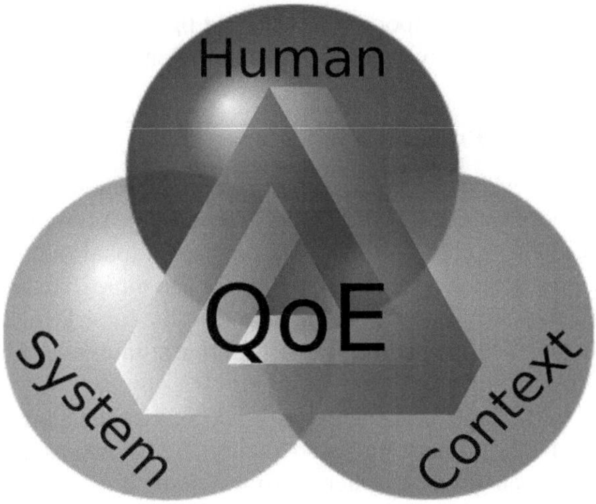

Fig. 4.1 Factors influencing quality of experience might be grouped into human, system, and context influence factors (IFs). These groups of factors frequently overlap, and together have a mutual impact on QoE

4.2 Human Influence Factors

A Human Influence Factor (HIF) is any variant or invariant property or characteristic of a human user. The characteristic can describe the demographic and socio-economic back-ground, the physical and mental constitution, or the user's emotional state [1].

HIFs may bear an influence on a given experience and how it unfolds, as well as on its quality. They are highly complex because of their subjectivity and relation to internal states and processes. This makes them rather intangible and therefore much more difficult to grasp. In addition, HIFs are strongly interrelated and may also strongly interplay with the other IFs described in this chapter. As a result, the influence of human factors on QoE cannot only be considered at a generic level.

At the theoretical and more conceptual level, the importance of human factors and their possible influence on QoE is often emphasized [2–6]. Moreover, at a more specific level, some studies have investigated the influence of specific human factors on perceived quality [7] and QoE [8]. In most empirical studies however, human factors are only taken into account to a limited extent. Common examples of HIFs usually include gender, age, expertise level (expert vs. naïve). As a result, due to their inherent complexity and the lack of empirical evidence, it is still poorly understood how human factors influence QoE.

In this section, we give examples of human factors that may influence the perceptual and quality formation process. More concretely, we consider relevant factors at both low- and higher-level processing [4]. Following the definition of HIFs, we distinguish between (relatively) stable and variant characteristics of human users.

It is, however, important to note that the overview presented here cannot be considered as exhaustive and that the distinction between stable and variant factors should not be seen as a black versus white one.

Low-level Processing and Human IFs

At the level of early sensory—or so-called low-level—processing, properties related to the physical, emotional and mental constitution of the user may play a major role. These characteristics can be dispositional (e.g., the user's visual and auditory acuity, gender, age) as well as variant and more dynamic (e.g., lower-order emotions, user's mood, motivation, attention). At the same level, characteristics that are closely related to human perception of external stimuli might bear the strongest influence on QoE.

In the human visual system (HVS), visual sensitivity might be the most important factor influencing visual quality. Traditional psychophysical studies assume that visual sensitivity to external stimuli is determined by the spatial and temporal frequencies of the stimuli [9]. Additionally, due to the non-uniform distribution of photo receptors (i.e., cones and rods) on the retina, the HVS has the highest sensitivity around the fixation point of the eyes (fovea) and drastically decreases away from this point. As the visual sensitivity mechanism always plays an essential role in the perceptual viewing experience, QoE of visual content can significantly be improved by taking it into account. For example, visual sensitivity models have been widely applied in many advanced video/image compression algorithms and quality assessment methods [10]. Similarly to the HVS, auditory quality and QoE depend on the sensory processing by the periphery of the human auditory system (HAS) [11]. Here, too, auditory processing models are also widely applied in audio coding and even signal-based quality prediction models.

Higher-level Processing and Human IFs

Top-down—or so-called higher-level—cognitive processing relates to the understanding of stimuli and the associated interpretative and evaluative processes. It is based on knowledge, i.e. "any information that the perceiver brings to a situation" [12]. As a result, a wide range of additional HIFs are important at this level. Some of them have an invariant or relatively stable nature. Examples in this respect include first of all the socio-cultural and educational background, life stage and socio-economic position of a human user.[1] Especially in the context of studies investigating the monetary dimension of QoE (e.g., willingness to pay [13], see also Chap. 7), the latter is of crucial importance. The above mentioned HIFs are strongly connected to a set of other human characteristics, which can also be considered as relatively stable. These include, for instance, the norms and beliefs that one has, which are often determined at a higher level and therefore strongly linked to the wider social and cultural context. Another higher-level characteristic that is often related to the

[1] Note that the socio-economic aspects are also considered to be part of the CIFs, demonstrating that some factors are very hard to disentangle and categorize. This is reflected in the overlapping areas of IFs in Fig. 4.1.

viewing or hearing behaviors when consuming multimedia services, is guided by the attention mechanism. Attention is a cognitive process of selectively concentrating on certain external objects (e.g., visual or auditory) while paying less or no attention to others [14]. Objects might be salient not only because of their characteristics but also because surrounding objects are not.

Other relatively stable HIFs that we will now shortly discuss include individual values, needs and goals, motivations, preferences and sentiments, attitudes and personality traits. QoE in general and the relative importance of specific QoE features in particular, may be strongly impacted by a human user's goals and corresponding values and needs. Several classifications have been proposed in the literature: in [15] a distinction is made between terminal and instrumental values. The former are linked to ultimate life goals (e.g., happiness, pleasure, comfortable life) and the latter correspond to modes of behavior and more pragmatic goals (e.g., cheerfulness, ambition). Hassenzahl [16] distinguishes between "be-goals" and "do-goals" that people want to fulfill in this respect (see Chap. 3 for a more extensive discussion). Such goals are underlying drivers of human behavior and orient people's motivations. In the literature, it is argued that motivation is very personal and subjective and may vary in terms of level and orientation (i.e., nature and focus) [17]. A common distinction that is made in motivational research, is the one between intrinsic and extrinsic motivation. Whereas the former implies that something is done because it is "inherently interesting or enjoyable", the latter refers to "doing something because it leads to a separable outcome" [17]. Chapter 25 of this book briefly discusses the importance of intrinsic motivation in relation to quality of gaming. In general, however, and although previous research on human motivation has shown that the type of motivation may strongly influence performance and QoE [17], the influence of motivation on QoE is still a largely unexplored territory.

Preferences and attitudes can also be considered as rather stable factors that may influence QoE at a higher level. Scherer [18] defines preferences as "relatively stable evaluative judgments in the sense of liking or disliking a stimulus, or preferring it or not over other objects or stimuli". Desmet [19] refers to such intentional and dispositional (dis)likes oriented towards a specific object or event as "sentiments". Preferences differ from attitudes (i.e., "relatively enduring beliefs and predispositions towards specific objects or persons" [18]), which have a cognitive (i.e., beliefs), affective (i.e., associated feelings) and motivational/behavioral (i.e., action tendency) component. Attitudes, the external and internal variables that influence them and their translation into behavioral intentions, have been extensively studied in research on technology adoption and acceptance. However, only a limited number of studies so far have explicitly investigated the influence of specific attitudes on QoE. In [7], it was shown that attitudes and perceived quality are related. In the same study, the possible influence of personality traits was also investigated. Personality traits have been defined as "consistent patterns of thoughts, feelings, or actions that distinguish people from one another" [20]. In the literature on human affective states, the concept of 'emotional traits' is also used to address the characteristics of someone's personality that are dispositional and enduring [19]. In the study of Wechsung et al. [7], no direct link between personality traits and perceived quality was found. Another study [21] investigated the impact of users' cognitive styles—which are linked to personality aspects—on perceived multimedia quality (and more specifically, the level of understanding and enjoyment). However, no strong correlation was found.

It can be argued that another set of influencing factors at the human level have a more dynamic and even acute character. At the level of human affective states, the influence of moods and emotions on QoE (and vice versa) has increasingly gained research interest [7, 22–24] (see also Chaps. 3, 8 and 9 of this book). Although both are characterized by their relatively short duration, moods usually last longer (ranging from hours to days) than emotions (ranging from seconds to minutes). Moreover, moods are neither triggered by one particular object nor oriented towards it [25]. Emotions in turn are momentarily reactions, that are oriented towards a specific object or event. Previous research has pointed to the influence of different affective states on perception (for instance, on the time spent on processing mood-consistent details and on evaluative judgments [26], on the motivation to process information and attention to details [27], and on perception of time [28]). Next to these affective characteristics of a human user, several other factors that have a variant and unstable character may bear a significant influence on QoE. These include for instance previous experiences, (prior) knowledge, skills and capabilities, and expectations. Previous experiences can relate to lived, previous experiences, and memories based upon those experiences (see also Chap. 2, in which different levels of memory are discussed in relation to the quality perception process), but also to indirect previous experiences (e.g., through stories from others) and these—in addition to other sources—contribute to the relevant knowledge that a human user has. Similarly, expectations may also be based on a range of sources. In [29], expectations are defined as "pre-trial beliefs about a product or service and its performance at some future time". A difference is made between different types of expectations. In [8], the influence of expectations (related to the type of access network used) on QoE was investigated and shown. However, only a limited number of studies so far have investigated the influence of expectations on QoE (see e.g. [30]), or explored how the test setup may influence expectations [31]. Prior knowledge and skills may also influence QoE and the related quality formation process. As was mentioned above, in subjective testing, a distinction is often made between expert test subjects (due to their specific prior knowledge and experiences) and so-called naïve users. Whereas the former tend to be more critical and answering in a more consistent way, it has been shown in some studies that the latter are less focused on impairments and tend to give higher ratings [32, 33]. In a recent study [34] in the context of HD telephony services, participants were categorized into six user segments with different characteristics in terms of their prior knowledge, but also their attitudes towards adoption of new technologies and socio-demographic and -economic position. The results pointed to significantly different quality ratings between these segments and call for a combined approach to take HIFs into account. Next to knowledge, skills may also bear a strong influence on QoE, for instance in the context of gaming: a lack of skills to master the controls of a game may lead to frustration and prevent the player to make progress. This was one of the findings from a field study on QoE in the context of a location-based real-time mobile Massively Multiplayer Online Role-Playing Game (MMORPG) [35].

The above mentioned aspects may, but do not necessarily, have a direct impact on QoE. They can also bear an indirect influence on QoE through affective factors, attitudes and preferences, etc. In addition to the aforementioned criteria and factors, Human Influence Factors are intimately linked to technical characteristics of a system. These are the focus of the next section.

4.3 System Influence Factors

System Influence Factors (SIFs) refer to properties and characteristics that determine the technically produced quality of an application or service [1].

Whereas Chap. 6 describes the difference between technically produced quality, perceptual quality, and QoE, here we will discuss in more detail the classification of SIFs into content-related, media-related, network-related and device-related SIFs.

Content-related System IFs

The content itself and its type is highly influential to the overall QoE of the system, as different content characteristics might require different system properties. For auditory information, the audio bandwidth and dynamic range are the two major SIFs, and their requirements vary with the content itself, e.g. for voice/spoken content versus musical content.

When it comes to visual information, the amount of detail as well as the amount of motion in the scene is important. To a large extent this has to do with HIFs such as contrast sensitivity and visual masking, but also with the fact that current compression techniques are affected by these. Furthermore, it is also influenced by the content itself [36], as well as influenced by the higher-level processing as described above in Sect. 4.2. In 3D image and video content, the amount of depth is an aspect that also influences the quality and especially the viewing comfort [37]. Aspects of 3D video are discussed in more detail in Chap. 20 of this book.

Media-related System IFs

The media-related SIFs refer to media configuration factors, such as encoding, resolution, sampling rate, frame rate, media synchronization [38]. They are interrelated with the content-related SIFs. Media-related SIFs can change during the transmission due to variation in network-related SIFs [39].

In most cases the resources for distributing media are limited. There are both economical as well as hardware-related reasons for limiting the size of media. This is usually accomplished by applying compression, which can be either lossless or lossy. Lossy compression gives higher compression rates at the cost of quality. However, the influence depends on the principle the lossy coding is built upon. For instance for image and video, block-based compression techniques as in JPEG, and MPEG4/AVC a.k.a. H.264, are the most common. For stronger compression, these will usually give visible blocking (rectangular shaped) distortions and blurring, whereas wavelet based techniques mostly give blurring distortions as in JPEG 2000 (cf. Chap. 19 for more details).

For audio, the coding also depends on the content type and service/application scenario. Telephone codecs (such as G.711, G.729) are used for voice-only scenarios (e.g. VoIP). Better QoE can usually be achieved if wideband codecs (e.g. AMR-WB) are supported over the complete transmission chain. Several lossy compression codecs are used for audio media (MP3, AC-3, and Vorbis). For lossy compression, perceptual coding based on psychoacoustic principles is a widely used method. The sampling rates and resolutions vary between codecs and their usage scenarios, and are compromises between codecs' rates and achieved quality. Delays are highly undesirable in conversational communication services (see Chap. 11). The media synchronization can have an important influence if the media (e.g. movie) contains audio and video [40].

Network-related System IFs

Network-related SIFs refer to data transmission over a network. The main network characteristics are bandwidth, delay, jitter, loss and error rates and distributions, and throughput [41, 42]. The network-related SIFs may change over time or as a user changes his location, and are tightly related to the network Quality of Service (QoS).

Network-related SIFs are impacted by errors occurring during the transmission over a network. Especially in case of delay, the impact of SIFs also depends on whether the service is interactive or more passively consumed (see Chap. 11), as for instance in telephony versus radio broadcast, or video conferencing versus streaming video. In an interactive, e.g., conversational service, delay may have a negative impact on QoE. The delay can present a major limitation if older mobile network technologies are used for real-time audio applications such as VoIP. Streaming video and IPTV are examples of services with more passive consumption, but depending on how they are distributed over the network, they will be very differently affected. Most often the video is deliberately delayed by using strategically placed buffers in order to be more resilient towards network capacity variations and errors.

For User Datagram Protocol (UDP) and Real-time Transport Protocol (RTP) based transmission, the most severe errors are packet losses [43]. The visibility of these mostly depend on the applied concealment at the receiving end, and on the content and the coding scheme itself: larger parts of the image might disappear in a blocky fashion for some time (see Chap. 19). Speech is often presented in bursts during the VoIP service. Therefore, the packet loss distribution plays an important role. The same level of packet loss can result in a more severe impact if audio is used for some additional processing, as in the case of spoken dialog telephone systems [44].

Recently, the popularity of over-the-top (OTT) streaming video, e.g. Youtube or Netflix, has increased very rapidly. The distribution method is TCP- and http-based (Transmission Control Protocol and Hypertext Transfer Protocol, respectively), and here the influence of packet loss and bandwidth limitations is quite different. Network problems will result in freezes without loss of content in the video. Freezing also has a bad influence on the experienced quality [45], but can be avoided by using adaptive or scalable codecs in conjunction with OTT video services [46].

Device-related System IFs

Device-related SIFs refer to the end systems or devices of the communication path. The visual interface to the user is the display. Its capacity will have a tremendous impact on the end-user experience, but the content and signal quality will interact with it. For instance, if a high-quality, high-resolution (here meaning in terms of number of pixels) image is shown on a low resolution display with few colors, most of the original intent of the image might be lost. However, if a low resolution image is shown on a large high resolution display, most likely a very blocky and blurry image will be displayed, but the end result will be highly dependent on the final image scaling procedure. For an in-depth treatment of the influence of scaling and display rendering, as well as the influence of the dynamic capabilities of the screen for reproducing motion, see e.g. [47].

In recent years the technical development of displays has been progressing very fast, both on the TV side and the mobile side. One important trend, especially in the smartphone market, is the increase in display resolution. Also, the colors and brightness have improved. On the TV side, the development is taking place in larger steps over several years or even decades, e.g. the transition from standard definition TV to high definition TV. The most influential trend in recent years, with a substantial influence on experience, are stereoscopic 3D devices, see Chap. 20 of this book. The basic principle is to present two views of the same scene. Depending on how this is done technically, many device-related IFs will be present [48–50]. For instance, leakage of one view into the other a.k.a. crosstalk will lead to visible ghosting [51, 52].

The 1.75 billion mobile devices (e.g. smartphones, tablets) sold throughout the world in 2012 [53] greatly outperformed the numbers of any other terminal equipment types in usage. In regard to devices' form-factor dimensions, the built-in loudspeakers represent only an average possibility for playing audio. The main progress in the area of input devices is the increased usage of touchscreens, which are addressing the human tactile modality. The touchscreen as an input device can present a limitation, if the user needs to input a larger amount of information. The state-of-the-art mobile devices with multi-core processors and advanced Graphics Processing Unit (GPU) can deliver a substantial amount of computational power, but at the cost of autonomy. Mobility, which is a Context IF, thus strongly influences various characteristics of devices.

4.4 Context Influence Factors

Context Influence Factors (CIFs) are factors that embrace any situational property to describe the user's environment [1].

CIFs have been considered in different multimedia applications and services [54–59]. In most of these works, context factors appear mixed with *human* and *system* factors, thus without any structure or categorization. However, different literature places a strong emphasis on multimedia quality progress, resulting in a properly structured categorization of the different kinds of influence factors. In the case of CIFs the latest and most complete categorization was proposed in [60].

Following previous work described in [61–65] and mobile work contexts, the CIFs were broken down in [60] in terms of physical, temporal, social, economic, task, and technical characteristics. These factors can occur on different levels of magnitude (micro vs. macro), behavior (static vs. dynamic), and patterns of occurrence (rhythmic vs. random), either separately or as typical combinations of all three levels. Furthermore, in [66] another context categorization is presented. Six different context categories are defined: personal context, social context, event-based context, application based context, historic context, and intra-user context difference. However, according to the present factors' categorization, the application-based context data shall be considered as a *system* factor. The variability of categorization is confirmed in [67] where the following CIFs are considered: those capturing the physical environment (e.g. home, office, mobile, or public usage; space, acoustic, and lighting conditions; transmission channels involved; potential parallel activities of the user; privacy and security issues) as well as the service factors (e.g. access restrictions, availability of the system, resulting costs).

Modelling CIFs might provide a selection of appropriate quality levels for the given experience, improving efficiency and reliability of the application/system, or adapting the content characteristics. The importance of CIFs knowledge on the provided Quality of Experience can be understood with the following examples: long duration content is not interesting at lunch time on a weekday, music with a fast beat is better than slow music in a gym, and advertisements in a social network shall typically consider the user profile. Moreover, different contexts might change the user profile (e.g. using a service at home or at work).

Following the idea of CIFs, context-aware multimedia services/infrastructures have attracted considerable research activity in recent years [59, 68]. For instance, an infrastructure for context-aware multimedia services in a smart home environment is proposed in [59]. Such a system is supposed to be adaptive to typical preferences of the multimedia system user, like for example, record the favorite TV programs of the family members, show suitable content based on the user's social activity (e.g. holding a birthday party), and show content in an appropriate form according to the technical capabilities. The multi-layered system is based on the triptych for context: aggregation, reasoning and learning.

The description given in the remainder of this section is based on the context factors categorization proposed in [60], whereas links to the categories of [66] are also given.

Physical context

The physical context describes the characteristics of location and space, including movements within and transitions between locations; spatial location (e.g. outdoor or indoor, in a personal, professional or social place), functional place and space; sensed environmental attributes (e.g. peaceful place vs. noisy place, lights and temperature); movements and mobility (e.g. sitting, standing, walking or jogging); artifacts. The personal context described in [66] can be partially included here, namely at the user location, user activity[2] and user physiological information level. Several works use the physical context to model the application quality. User's preferences can vary in different contexts, such as location, time, movement state and temperature. For example, someone jogging might prefer hip-hop over classical music. A survey showed that activity significantly affects the listener's mood [56]. Authors in [54] use this finding and conclude that context information is an important element for a music selection recommender that suits the listener's mood. They propose to group the users under similar context conditions to find implicit and more applicable perceptual patterns. Through mining integration of both context information and musical content, appropriate ubiquitous music recommendations are provided. Hence, physical factors like heartbeat, body temperature, air temperature, noise volume, humidity, lighting conditions, motion and spatial location are used to get similar user clusters. These physical context factors also allow for context-specific processing to increase QoE, e.g. the adjustment of screen brightness on a mobile, depending on lighting conditions. Moreover, the use of spatial context is proposed to provide a better visualization and tracking in multi-camera video surveillance systems in [69, 70].

Temporal context

The temporal context is related with temporal aspects of a given experience, e.g. time of day (morning, afternoon or evening), week, month, season (spring, summer, fall or winter) and year; duration (see e.g. [71] or Chap. 10 for aspects of content duration, and Chap. 2 for memory effects), and frequency of use (of the service/system); before/during/after the experience; actions in relation to time; synchronism. It is quite common in literature to include physical and temporal contexts in the same category. For instance, the categorization in [66] includes the temporal context in the personal context, namely the time of the system access and the task list influence. In fact, these two context categories' influences are typically highly correlated. Moreover, a historic context is considered, that uses the subject's past context information stored in a database similar to a user profile or a resource profile (e.g. Twitter offers a

[2] User activity context may be strongly related to task context, for instance when the user tries to achieve a certain goal.

rich source of user context in terms of current and past activities; the last 10-min physiological or one's ambient data stored in a smartphone). Authors in [66] also define a sixth category that can be considered inside the temporal context, defined as the intra-user context difference. This sub-category results from the change in one particular user's context throughout a day. This separation is considered because every user might access different services or communicate with different categories of people during different periods of a day. Returning to the music recommender example [54], factors like time of day and season were also considered.

Social context

The social context is defined by the inter-personal relations existing during the experience. Hence, it is important to consider if the application/system user is alone or with other persons, and even how different persons are involved in the experience, namely including inter-personal actions. Moreover, the cultural, educational, professional levels (namely hierarchical dependencies, internal vs. external), and entertainment (dependent of random or rhythmic use) also need to be considered. In [66] also the contact list, social ties through social nets and interactions, and types of shared information are considered. Furthermore, in [66] another category defined as event-based context (e.g. appointments, or meetings) can also be considered as a sub-category belonging to the social context.

In [58], the analysis of the user's social context permits to infer interesting data about the user's interests via information provided spontaneously by the user himself, and analyzing behavior and habits of his friends' network. Along the same lines, several research efforts intend to understand and to automatically extract from the social information deposit the users' relationships, interests, and even their mood. More recently the new Google "Search, plus your world,"[3] makes intrinsic integration of the user's social environment for the searching mechanisms. Some contemporary context-aware recommenders attempt to enhance recommendations with more considerations of environmental metadata [72, 73].

A combination of physical and social context is proposed in [74] to foster a more efficient delivery of mobile services. That model exploits the fact that a very lightweight component such as the mobile nodes, can be deployed to monitor socio-technical information in three main areas: user physical location and activity (running, driving, …), user social context (friends, common interests, …), and service usage (frequency of use, last login, …). A solution for IPTV services personalization based on context-awareness relying on physical, temporal and social categories, is introduced in [55–57] by a real-time gathering of context information on the user, his environment (devices and network) and the service.

As the previous examples have shown, the social context becomes very important at the recommendation level. Content recommendation based on the gathered context information allows guaranteeing better users' experience. Collaborative recommendation, where the user recommends items that are consumed by other users with similar preferences, can also be made possible.

[3] http://www.google.com/insidesearch/plus.html

Economic context

Costs, subscription type, or brand of the application/system are part of the economic context. Chap. 7 in this book focuses on QoE from a business perspective and discusses more details of its influence. Network cost information (e.g. relative distances between the peers) is used in [75], jointly with some physical and social factors, to enable network optimization strategies for media delivery.

Task context

The task context is determined by the nature of the experience. Depending on this, three situations may arise: multitasking (potential parallel activities of the user [67]), interruptions, or task type. For example, a recent paper by Sackl et al. investigates the impact of additional tasks on perceived quality in a QoE evaluation experiment in which the effect of video stalling is explored [45]. The authors conclude that an additional task does not have an influence on the perceived quality, independently of the difficulty (hard or easy) of that task, as stalling did affect the perceived quality to a similar extent under both task conditions. However, the relationship between QoE and task may not be this simple *per se*: Reiter et al. have previously shown in a series of experiments that a challenging task can indeed have an effect on perceived quality in an interactive scenario, especially when both the main varying (or salient) quality attribute and the task are located in the same modality [76–78]. According to these studies, inner-modal task influence (or distraction) is significantly greater than cross-modal task influence. This is also suggested by the common theories of capacity limits in human attention [79].

Technical and information context

Finally, the technical and information context describes the relationship between the system of interest and other relevant systems and services including: devices (e.g. existing interconnectivity of devices over Bluetooth or Near Field Communication, NFC), applications (e.g. availability of an application instead of the currently used browser-based solution of a service), networks (e.g. availability of other networks than the one currently used), or additional informational artifacts (e.g. additional use of pen and paper for better information assimilation from the service used). Characteristics like interoperability, informational artifacts and access, or mixed reality also need to be considered.

4.5 Conclusions

The above discussion of factors influencing the user's individual Quality of Experience of a device or service demonstrates that QoE can be influenced by wide a range of factors, which are complex and strongly interrelated. It is currently still poorly understood which factors influence QoE under which circumstances, how exactly they influence QoE, and what their possible influence implies for the field of QoE research.

Table 4.1 Overview and examples of potential IFs

IF	Type	Examples
HIF	Low-level: physical, emotional, mental constitution	Visual / auditory acuity and sensitivity; gender, age; lower-order emotions; mood; attention level
	High-level: understanding, interpretation, evaluation	Socio-cultural background; socio-economic position; values; goals; motivation; affective states; previous experiences; prior knowledge; skills
SIF	Content-related	Audio bandwidth, dynamic range; video motion and detail
	Media-related	Encoding, resolution, sampling rate, frame rate; synchronization
	Network-related	Bandwidth, delay, jitter, loss, error rate, throughput; transmission protocol
	Device-related	Display resolution, colors, brightness; audio channel count
CIF	Physical context	Location and space; environmental attributes; motion
	Temporal context	Time, duration and frequency of use
	Social context	Inter-personal relations
	Economic context	Costs, subscription type, brand
	Task context	Nature of experience; task type, interruptions, parallelism
	Technical / informational context	Compatibility, interoperability; additional informational artifacts

We classified IFs into *human, system* and *context* influencing factors. With respect to HIFs, we have discussed both, variant and relatively stable, factors that may potentially bear an influence on QoE, both in the context of low-level or bottom-up processing and top-down, higher-level cognitive processing. SIFs were classified into four distinct categories, namely content-, media-, network- and device-related IFs. Finally, the broad category of possible CIFs was further decomposed into factors related to the physical, temporal, social, economic, task and technical and information context. Table 4.1 provides a checklist containing the most important IF examples for the practitioner to cross-check when designing QoE experiments and reporting.

Although the overview given in this chapter should not be considered as exhaustive, it illustrates the complexity of QoE and the broad range of aspects that potentially have a major influence on it. The amount of factors with influence on QoE results in a very difficult modeling and in a high level of subjectivity. However, the knowledge of these factors and an appropriate categorization might provide patterns and tools that allow to predict or even to improve the level of QoE. A challenge for future research is to develop adequate methodological approaches to take into account relevant influencing factors and to better understand their interrelations.

Acknowledgments Katrien De Moor's work was carried out during the tenure of an ERCIM "Alain Bensoussan" Fellowship Programme and received funding from the European Union Seventh Framework Programme (FP7/2007-2013) under grant agreement no 246016.

References

1. Qualinet White Paper on Definitions of Quality of Experience — Output version of the Dagstuhl seminar 12181 (2012) In: Le Callet P, Möller S, Perkis A (eds) European network on quality of experience in multimedia systems and services (COST Action IC 1003), Version 1.1, Lausanne
2. Geerts D, De Moor K, Ketykó I, Jacobs A, Van den Bergh J, Joseph W, Martens L, De Marez L (2010) Linking an integrated framework with appropriate methods for measuring QoE. In: 2010 second international workshop on quality of multimedia experience, pp 158–163
3. Wechsung I, Engelbrecht K-P, Kühnel C, Möller S, Weiss B (2012) Measuring the quality of service and quality of experience of multimodal human-machine interaction. J Multimodal User Interfaces 6(1–2):73–85. doi:10.1007/s12193-011-0088-y
4. Jumisko-Pyykkö S (2011) User-centered quality of experience and its evaluation methods for mobile television. Doctoral thesis, Tampere University of Technology, Tampere
5. Quintero MR, Raake A (2011) Towards assigning value to multimedia QoE. In: Third international, workshop on quality of multimedia experience (QoMEX), pp 1–6
6. Laghari KUR, Crespi N, Connelly K (2012) Toward total quality of experience: a QoE model in a communication ecosystem. Commun Mag IEEE 50(4):58–65
7. Wechsung I, Schulz M, Engelbrecht K-P, Niemann J, Möller S (2011) All users are (Not) equal-the influence of user characteristics on perceived quality, modality choice and performance. In: Delgado RL-C, Kobayashi T (eds) Proceedings of the paralinguistic information and its integration in spoken dialogue systems workshop. Springer, New York, pp 175–186
8. Sackl A, Masuch K, Egger S, Schatz R (2012) Wireless vs. wireline shootout: how user expectations influence quality of experience. In: Fourth international workshop on quality of multimedia experience (QoMEX), 5–7 July 2012, pp 148–149
9. Burbeck CA, Kelly DH (1980) Spatiotemporal characteristics of visual mechanisms: excitatory-inhibitory model. JOSA 70(9):1121–1126
10. You J, Xing L, Perkis A, Ebrahimi T (2012) Visual contrast sensitivity guided video quality assessment. In: 2012 IEEE international conference on multimedia and expo (ICME). IEEE, pp 824–829
11. Greenberg S, Ainsworth WA (2004) Speech processing in the auditory system: an overview. In: Speech processing in the auditory system. Springer, pp 1–62
12. Goldstein EB (2009) Sensation and perception, 8th edn. Cengage Learning, Wadsworth
13. Sackl A, Egger S, Zwickl P, Reichl P (2012) The QoE alchemy: turning quality into money. Experiences with a refined methodology for the evaluation of willingness-to-pay for service quality. In: Fourth international workshop on quality of multimedia experience (QoMEX), 5–7 July 2012, pp 170–175
14. Reiter U (2010) Perceived quality in game audio. In: Grimshaw M (ed) Game sound technology and player interaction: concepts and developments. IGI Global, New York
15. Rokeach M (1973) The nature of human values. The Free Press, New York
16. Hassenzahl M (2008) User experience (UX): towards an experiential perspective on product quality. In: Proceedings of the 20th international conference of the association francophone d'interaction homme-machine, ACM Press, New York, pp 11–15
17. Ryan RM, Deci EL (2000) Intrinsic and extrinsic motivations: classic definitions and new directions. Contemp Educ Psychol 25(1):54–67
18. Scherer KR (2005) What are emotions? And how can they be measured? Soc Sci Inf 44(4):695–729
19. Desmet PMA (2002) Designing emotions. Unpublished doctoral dissertation, TU Delft, Delft, The Netherlands
20. Robert S, John J, Hogan B (1997) Handbook of personality psychology. Academic Press, San Diego
21. Ghinea G, Chen SY (2003) The impact of cognitive styles on perceptual distributed multimedia quality. Br J Educ Technol 34(4):393–406

22. Rainer B, Waltl M, Cheng E, Shujau M, Timmerer C, Davis S, Burnett I, Ritz C, Hellwagner H (2012) Investigating the impact of sensory effects on the quality of experience and emotional response in web videos. In: Fourth international workshop on quality of multimedia experience (QoMEX). IEEE, pp 278–283
23. Arndt S, Antons J-N, Schleicher R, Möller S, Curio G (2012) Perception of low-quality videos analyzed by means of electroencephalography. In: 2012 fourth international workshop on quality of multimedia experience (QoMEX). IEEE, pp 284–289
24. Reiter U, De Moor K (2012) Content categorization based on implicit and explicit user feedback: combining self-reports with EEG emotional state analysis. In: 2012 fourth international workshop on quality of multimedia experience (QoMEX). IEEE, pp 266–271
25. Frijda NH (1994) Varieties of affect: emotions and episodes, moods, and sentiments. In: Ekman P, Davidson R (eds) The nature of emotions: fundamental questions. Oxford University Press, New York, pp 59–67
26. Forgas JP, Bower GH (1987) Mood effects on person-perception judgments. J Pers Soc Psychol 53(1):53
27. Bless H, Clore GL, Schwarz N, Golisano V, Rabe C, Wölk M (1996) Mood and the use of scripts: does a happy mood really lead to mindlessness? J Pers Soc Psychol 71(4):665
28. Angrilli A, Cherubini P, Pavese A, Manfredini S (1997) The influence of affective factors on time perception. Percept Psychophysics 59(6):972–982
29. Higgs B, Polonsky MJ, Hollick M (2005) Measuring expectations: forecast vs. ideal expectations. Does it really matter? J Retail Consum Serv 12(1):49–64
30. Sackl A et al (2013) Evaluating the impact of expectations on end-user quality perception. PQS workshop 2013, Vienna, Austria
31. Staelens N, Van den Broeck W, Pitrey Y, Vermeulen B, Demeester P (2012) Lessons learned during real-life QoE assessment. In: 10th European conference on interactive TV and video (Euro ITV-2012). Ghent University, Department of information technology, pp 1–4
32. Rumsey F, Zielinski S, Kassier R, Bech S (2005) Relationships between experienced listener ratings of multichannel audio quality and naive listener preferences. J Acoust Soc Am 117(6):3832
33. Speranza F, Poulin F, Renaud R, Caron M, Dupras J (2010) Objective and subjective quality assessment with experts and non-experts viewers. In: Proceedings of the second international workshop on quality of multimedia experience, Trondheim, Norway, pp 46–51
34. Quintero MR, Raake A (2012) Is taking into account the subjects degree of knowledge and expertise enough when rating quality? In: 2012 fourth international workshop on quality of multimedia experience (QoMEX), IEEE, pp 194–199
35. De Moor K (2012) Are engineers from Mars and users from Venus? Bridging gaps in quality of experience research: reflections on and experiences from an interdisciplinary journey. Unpublished doctoral dissertation. Ghent University
36. Radun J, Leisti T, Häkkinen JP, Ojanen HJ, Olives J, Vuori T, Nyman GS (2008) Content and quality: interpretation-based estimation of image quality. ACM Trans Appl Percept 4(4):21:1–21:15
37. Chen W, Fournier J, Barkowsky M, Le Callet P (2013) New stereoscopic video shooting rule based on stereoscopic distortion parameters and comfortable viewing zone. In: Stereoscopic displays and applications XXII, proceedings of SPIE-IS&T electronic imaging, vol: SPIE, vol. 7863
38. Zinner T, Hohlfeld O, Abboud O, Hoßfeld T (2010) Impact of frame rate and resolution on objective QoE metrics. In: Proceedings of second international workshop on quality of multimedia experience (QoMEX)
39. Jammeh E, Mkwawa I, Khan A, Goudarzi M, Sun L, Ifeachor E (2012) Quality of experience (QoE) driven adaptation scheme for voice/video over IP. Telecommun Syst 49(1):99–111
40. ITU BT.1359: Relative timing of sound and vision for broadcasting
41. Nahrstedt K, Steinmetz R (1995) Resource management in networked multimedia systems. IEEE Comput 1995:52–63

42. Fiedler M, Hoßfeld T, Tran-Gia P (2010) A generic quantitative relationship between quality of experience and quality of service. Netw IEEE 24(2):36–41
43. Brunnström K, Stålenbring D, Pettersson M, Gustafsson J (2010) The impact of transmission errors on progressive 720 lines HDTV coded with H.264. In: Rogowitz B, Pappas TN (eds) Proceedings of SPIE-IS&T human vision and electronic imaging XV, vol 7527, paper 56
44. Pratsolis D, Tsourakis N, Digalakis V (2007) Degradation of speech recognition performance over lossy data networks. In: Wmunep'07: Proceedings of the third ACM workshop on wireless multimedia networking and performance modeling, pp 88–91
45. Sackl A, Seufert M, Hoßfeld T (2013) Asking costs little? The impact of tasks in video QoE studies on user behavior and user ratings. In: PQS workshop 2013. Vienna, Austria
46. Tavakoli S, Gutiérrez J, García N (2013) Quality assessment of adaptive 3D video streaming. In: Three-dimensional image processing (3DIP) and applications. Burlingame, California, USA. 03 Feb 2013, vol Proc. SPIE 8650
47. Klompenhouwer MA (2006) Flat panel display signal processing: analysis and algorithms for improved static and dynamic resolution. PhD thesis, Technische Universiteit Eindhoven, The Netherlands
48. Kaptein R, Kuijsters A, Lambooij M, IJsselsteijn WA, Heynderickx I (2008) Performance evaluation of 3D-TV systems. In: Proceedings of SPIE Image quality and system performance V, vol SPIE 6808, p 680819
49. Lambooij M, IJsselsteijn W, Heynderickx I (2009) Visual discomfort and visual fatigue of stereoscopic displays: a review. J Imaging Sci Technol 53(3):030201-1–030201-14
50. Wang K, Barkowsky M, Brunnström K, Sjöström M, Cousseau R, Le Callet P (2012) Perceived 3D TV transmission quality assessment: multi-laboratory results using absolute category rating on quality of experience scale. IEEE Trans Broadcast 58(4):544–557
51. Woods AJ, Docherty T, Koch R (1993) Image distortions in stereoscopic video systems. In: Proceedings of SPIE volume 1915 stereoscopic displays and applications IV, pp 36–48
52. Patterson R (2009) Review paper: human factors of stereo displays: an update. J Soc Inf Display 17(12):987–996
53. Gartner: Gartner says worldwide mobile phone sales declined 1.7 percent in 2012
54. Su J-H, Yeh H-H, Yu PS, Tseng VS (2010) Music recommendation using content and context information mining. Intell Syst IEEE 25(1):16–26
55. Song S, Moustafa H, Afifi H (2012) Advanced IPTV services personalization through context-aware content recommendation. IEEE Trans Multimedia 14(6):1528–1537
56. Adomavicius G, Tuzhilin E (2005) Toward the next generation of recommender systems: a survey of the state-of-the-art and possible extensions. IEEE Trans Knowl Data Eng 17(6):734–749
57. Chen Y-C, Huang H-C, Huang Y-M (2009) Community-based program recommendation for the next generation electronic program guide. IEEE Trans Consum Electron 55(2):707–712
58. Qi X, Davison BD (2009) Web page classification: features and algorithms. ACM Comput Surv 41:1–31
59. Yu Z, Zhou X, Yu Z, Zhang D, Chin C-Y (2006) An OSGi-based infrastructure for context-aware multimedia services. Commun Mag IEEE 44(10):136–142
60. Jumisko-Pyykkö S, Vainio T (2010) Framing the context of Use for mobile HCI. Review paper about mobile contexts of use between 2000–2007. Int J Mob Hum Comput Interact (IJMHCI) 3(4):1–28
61. Bradley NA, Dunlop MD (2005) Toward a multidisciplinary model of context to support context-aware computing. Hum Comput Interact 20(4):403–446
62. Roto V (2006) Web browsing on mobile phones: characteristics of user experience. Doctoral dissertation, TKK Dissertations 49, Helsinki University of Technology, Helsinki, Finland
63. Väänänen-Vainio-Mattila K, Ruuska S (2000) Designing mobile phones and communicators for consumers' needs at nokia. In: Bergman E (ed) Information appliances and beyond: interaction design for consumer products, Morgan Kaufmann, Morgan Kaufmann
64. Wigelius H, Väätäjä H (2009) Dimensions of context affecting user experience in mobile work. In: Proceedings of INTERACT 2009, Aug 2009, Uppsala, Sweden

65. Korhonen H, Arrasvuori J, Väänänen-Vainio-Mattila K (2010) Analysing user experience of personal mobile products through contextual factors. In: International conference on mobile and ubiquitous multimedia. Limassol, Cyprus
66. Rahman MA, El-Saddik A, Gueaieb W (2011) Augmenting context awareness by combining body sensor networks and social networks. IEEE Trans Instrum Measur 60(2):345–353
67. Möller S, Engelbrecht K-P, Kühnel C, Wechsung I, Weiss B (2009) A taxonomy of quality of service and quality of experience of multimodal human-machine interaction. In: International workshop on quality of multimedia experience, pp 7,12, 29–31
68. Zhou L, Xiong N, Shu L, Vasilakos A, Yeo S-S (2010) Context-aware middleware for multimedia services in heterogeneous networks. Intell Syst IEEE 25(2):40–47
69. You S, Neumann U (2005) V-sentinel: a novel framework for situational awareness and surveillance. Proc SPIE 5778(713):713–724
70. Wang Y, Krum DM, Coelho EM, Bowman DA (2007) Contextualized videos: combining videos, with environment models to support situational understanding. IEEE Trans Vis Comput Graph 13(6):1568–1575
71. Borowiak A, Reiter U, Svensson UP (2013) Audio quality requirements and comparison of multimodal vs. unimodal perception of impairments for long duration content. J Sig Process Syst, May 2013. doi:10.1007/s11265-013-0777-8
72. Hong J, Suh E-H, Kim J, Kim SY (2009) Context-aware system for proactive personalized service based on context history. Expert Syst Appl 36(4):7448–7457
73. Reynolds G, Barry D, Burker T, Coyle E (2008) Interacting with large music collections: towards the use of environmental metadata. In: Proceedings of IEEE Int'l conference on multimedia and expo, pp 989–992
74. Cardone G, Corradi A, Foschini L, Montanari R (2012) Socio-technical awareness to support recommendation and efficient delivery of IMS-enabled mobile services. Commun Mag IEEE 50(6):82–90
75. Chakareski J, Frossard P (2010) Context-adaptive information flow allocation and media delivery in online social networks. IEEE J Sel Top Sig Process 4(4):732–745
76. Reiter U, Weitzel M, Cao S (2007) Influence of interaction on perceived quality in audio visual applications: subjective assessment with n-back working memory task. In: Proceedings of AES 30th international conference. Saariselkä, Finland
77. Reiter U, Weitzel M (2007) Influence of interaction on perceived quality in audio visual applications: subjective assessment with n-back working memory task, II. In: AES 122nd convention. Vienna, Austria. Preprint 7046
78. Reiter U, Weitzel M (2007) Influence of interaction on perceived quality in audiovisual applications: evaluation of cross-modal influence. In: Proceedings of 13th international conference on auditory displays (ICAD). Montreal, Canada
79. Pashler HE (1999) The psychology of attention. 1st paperback edition, The MIT Press. Cambridge, MA, USA. ISBN 0-262-66156-X

Chapter 5
Features of Quality of Experience

Sebastian Möller, Marcel Wältermann and Marie-Neige Garcia

Abstract In this chapter we describe how the factors of the user, system and context of use, which influence QoE, are perceived by the user. For this purpose, we use the notion of a *feature*, i.e., a perceivable, recognized and nameable characteristic of an experience. Such a feature can be considered as a dimension of the perceptual space, and we will analyze the nature and dimensionality of this space. We will then review features which have been extracted via empirical methods for several multimedia services and group them on several levels. For two exemplary services (speech transmission and video streaming/communication), we will describe the features and corresponding methods in more detail. We conclude by discussing the links between influence factors and quality features, and by identifying open issues of research.

5.1 Introduction

As described in Chap. 2, quality can be seen as the outcome of an individual's comparison and judgment process, requiring perception, reflection and description processes to take place. Unfortunately, little is known about the characteristics of these processes, even for well-delimited situations like when listening to a transmitted spoken utterance, or when viewing a video clip. Knowledge about these

S. Möller (✉) · M. Wältermann
Quality and Usability Lab, Telekom Innovation Laboratories, TU Berlin, Berlin, Germany
e-mail: sebastian.moeller@telekom.de

M. Wältermann
e-mail: marcel.waeltermann@alumni.tu-berlin.de

M.-N. Garcia
Assessment of IP-based Applications, Telekom Innovation Laboratories, TU Berlin, Berlin, Germany
e-mail: marie-neige.garcia@telekom.de

S. Möller and A. Raake (eds.), *Quality of Experience*,
T-Labs Series in Telecommunication Services, DOI: 10.1007/978-3-319-02681-7_5,

characteristics is however necessary when investigating *why* a specific experience is sub-optimum, i.e., not judged with the highest-possible rating, and what can be done in order to improve the situation (quality *diagnosis*). Thus, in this chapter, we will try to analyze the characteristics of the individual's experience by decomposing it into so-called *quality features*.

Using the terminology of Jekosch [8], a quality feature is

> "A perceivable, recognized and nameable characteristic of the individual's experience of a service which contributes to its quality."

Thus, features are characteristics of perceptual events. As stated in Chap. 2 and following the processes hypothesized in [12, 20], the perceptual event is triggered by a physical event, i.e., the physical signal reaching the individual's sensory organs in a specific situation. The physical event is first processed by the low-level sensation and perception processes, resulting in a *perceptual character* of the event. This perceptual character is then reflected by the individual during the quality judgment process, resulting in the *perceptual event* which is characterized by its decomposing *features*. Thus, a feature can be seen as a dimension of a multidimensional perceptual event, in a multidimensional perceptual space. As a perceptual event is always situation- and context-dependent, also a feature depends on the situation and the context it has been extracted in. Thus, an empirical analysis of features commonly reveals only those features which are perceivable and nameable in that respective context (the others cannot be discerned).

In order for a *feature* to become a *quality feature*, the feature has to be relevant for quality. One can argue that all features are always perceivable and nameable (if the context allows), but they are only under certain conditions relevant for quality, thus quality features. The reference in the quality formation process can thus be considered as the instance which decides whether a feature is relevant or not, i.e., whether it is considered in the following comparison process or not. This is illustrated by the arrows originating from the anticipation process which itself is influenced by the reference-building process, see Fig. 1 of [12].

In the following sections, we will first address the nature of the perceptual space by analyzing its dimensionality, and the relationships which can be built in this space (Sect. 5.2). We will then present empirical methods which have been used in the past to identify and to quantify features in this space (Sect. 5.3). Using these and other methods, a number of features have been extracted for different types of multimedia services. We will group these features on several levels in Sect. 5.4. We will then provide examples of features (and corresponding extraction methods) for speech transmission services (Sect. 5.5) and for video streaming and communication services (Sect. 5.6). We will conclude by discussing relationships between influence factors and quality features, and by identifying missing features and corresponding methods for identifying and quantifying them.

5.2 Feature Space

The perceptual event a physical stimulus provokes can be conceived as being located in a multidimensional feature space. In this feature space, each of the underlying axes corresponds to one feature of the perceptual event. The perceptual event can mathematically be described by a position vector, where its coordinates correspond to specific *feature values* of the perceptual event. If the underlying axes of this space, and thus the features, are orthogonal, the features can also be referred to as *perceptual dimensions*. The number of the dimensions, i.e., the nameable perceptual features that are orthogonal to each other, corresponds to the dimensionality of the feature space. Typically, 2 . . . 5 dimensions can be identified in one multidimensional experiment, because the number of stimuli which can be presented is limited, and also because the human reasoning capabilities seem to be limited; more features can commonly only be discerned by comparing several such experiments.

The concept of a feature space can be helpful for explaining the relation between features and the quality of a perceptual event. The (integral) quality is a scalar value that can in general not directly be represented in the feature space. However, functional relations can be established by mapping the feature values of perceptual events onto corresponding quality scores. Depending on the form of this mapping, the nature of the features with respect to quality can be analyzed.

A simple mapping of the features onto quality is a linear one. In geometrical terms, a *quality vector* can be conceived to reside in the feature space, pointing towards optimum quality, see Fig. 5.1. Thus, quality is monotonically related to the projection of a point (i.e., a perceptual event) onto this vector. The cosines of the angles between the vector and the feature-space axes measure the importance of a feature with regard to quality (corresponding to coefficients in a linear combination of the features). A feature which is perpendicular to the quality vector is irrelevant for quality, and thus not a quality feature, see Sect. 5.1.

The model can be interpreted as relating the features towards quality in a "the more the better—the less the worse" way: Assuming that quality is negatively related to the vector, the lower the feature values (for example, the lower the "noisiness"), the better the quality. Or, if quality is positively related to the vector, the higher the feature values (e.g., the higher the signal-to-noise ratio), the better the quality. This so-called *vector model* is one case in the linear-quadratic hierarchy of models introduced by Carroll [1].

Another example of Carroll's model framework is the so-called unfolding model (*ideal-point model*), cf. the right panel of Fig. 5.1. Quality can here inversely and monotonically be related to the distance between the perceptual event and an ideal point with regard to one or more underlying features. Beyond this ideal point, quality decreases in each direction. One example is a feature describing the "brightness" of a sound: Both a too "dark" and a too "bright" sound can be detrimental for quality. Such a feature certainly exhibits an ideal point of "brightness", beyond which the quality decreases.

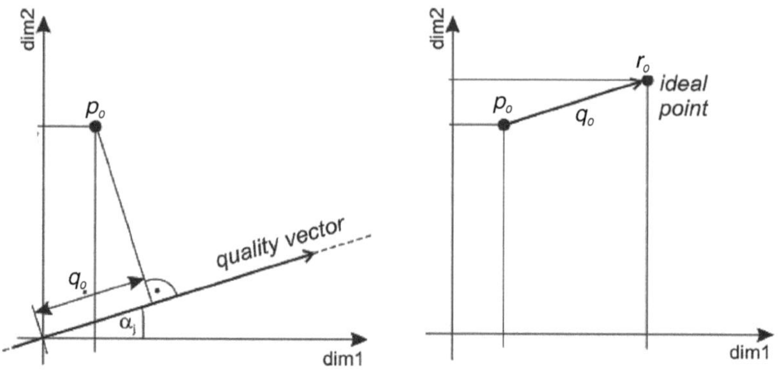

Fig. 5.1 Mapping of perceptual dimensions to quality, adapted from [20]. *Left panel* vector model; *right panel* ideal-point model. p_0 illustrates the location of the perceptual event in the perceptual space opened by dim_1 and dim_2; r_0 the position of the reference; and q_0 the quality value

In the literature, the vector approach has also been combined with a multiplicative approach for mapping features to quality [3, 6, 7]. For instance, in [3, 6, 7], the models are composed of both additive and interactive terms for mapping the impairments due to coding and transmission degradations to the perceived quality. This combined approach reflects the observation that the quality impact of one feature depends on the magnitude of the other features. Finally, Wältermann has shown in [25] that the perceived quality is most accurately estimated by computing the square root of the sum of squares of the quality dimensions.

5.3 Feature Extraction Methods

Subjective tests aiming at measuring perceived features of perceptual events belong to analytical-type of tests, see [19, 20] and Chap. 2. The feature scores can here be obtained in different ways. With *attribute scaling*, a direct way of judgment solicitation, attribute scales are presented to the participants of a subjective test, where the attributes verbally describe the features to be judged. A prominent example of such a scaling method is the Semantic Differential (SD) technique developed by Osgood et al. [18]. In SD, a set of bipolar scales with antonym labels is used, a technique which has been deployed in many fields of psychological research. Figure 5.2 shows examples of SD scales.

The Diagnostic Acceptability Measure (DAM), see [19, 24], was particularly developed for attribute scaling in the context of speech communication systems. In this method, 19 different attribute scales are used for rating the features of samples of transmitted speech. Separate scales are used for assessing the speech signal on one hand, and the background (e.g., ambient noise at sender side) on the other hand. This shows that the analysis can also extend to the situative (in this case background-noise)

noisy	○	○	○	○	◉	○	○	not noisy
interrupted	○	◉	○	○	○	○	○	continuous
dark	○	◉	○	○	○	○	○	bright
distant	○	○	○	◉	○	○	○	close
unintelligible	○	○	○	○	○	◉	○	intelligible

Fig. 5.2 Examples of semantic differential scales

context of the actually desired signal. For analysis, the scales are subsumed to 10 "parametric" scales, in addition to four "metametric" and "isometric" scales that are related to integral quality. For the DAM method, expert (i.e. not naïve) listeners are necessary.

Attribute scaling can also be adapted to an individual's own vocabulary and to a more realistic usage situation, For example, [21] adapted the Free Choice Profiling (FCP) method commonly used in the sensory evaluation of food [11] to the analysis of the perceptual dimensions underlying the quality for 3D audiovisual applications such as mobile 3DTV. This new method is referred to as Open Profiling of Quality (OPQ) and consists in three sessions. The first session is an Absolute Category Rating test in which test participants rate the integral quality of the stimuli. In the second session, participants are asked to think about quality features they have used to evaluate overall quality in the first session. Features which are not unique or cannot be defined are excluded. The resulting list of attributes is written on a scoring card, adjacent to a continuous rating scale for each attribute, with a "min" (minimum sensation) and a "max" (maximum sensation) label at the extreme ends of the scale. In the third session, each participant rates the stimuli using all scales on his/her scoring card. The method can also be combined with semi-structured interviews [9], mixing quantitative and qualitative data for discerning perceptual features.

As it is commonly very difficult to identify nameable characteristics of perceptual events, a different approach to their solicitation is to ask test participants to rate differences or similarities between stimuli presented in pairs or triads. The differences or similarities then also determine a perceptual space, but the dimensions of this space are not yet characterized by corresponding verbal attributes. This task remains for the second operation necessary for extracting features, namely the data analysis of the obtained judgments. Most commonly, the *perceptual dimensions* of the feature space, that is, the components that are orthogonal to each other, are of interest, as they describe the perceptual space with least redundancy, i.e., with a minimum set of features. They can be extracted by two different paradigms: (a) Principal Component Analysis (PCA) or other types of factor analysis of attribute scales, or (b) Multidimensional Scaling (MDS) of similarity scores of stimulus pairs.

With PCA, correlating attribute scales can be subsumed to principal components, reflecting the perceptual dimensions of the feature space. With MDS, in the contrary, proximity data of a set of stimuli is transformed into distances between points representing perceptual events in the feature space. Common to both the PCA and

MDS technique is that the experimental data is converted into a low-dimensional representation, providing a parsimonious picture of the data by a meaningful description of the underlying dimensions of the experimenter. The paradigm of similarity scaling has the advantage that no a-priori definition of attributes is required. In contrast to attribute scaling, where care must be taken to cover the complete feature scales with the employed attributes, it can be assumed that all features are taken into account when judging the perceptual dissimilarity of a stimulus pair. However, the interpretation of the dimensions provided by MDS might be difficult, whereas the correlating attributes may help interpreting the principal components of attribute-scaling data.

The experimental paradigms of attribute scaling and similarity scaling have been proven useful for the listening-only or viewing-only modality of tests, thus on the level of direct perception, see Sect. 5.4. In order to be useful in interactive services, these paradigms have to be elaborated, for example, for a realistic (speech and/or video) conversation situation. It might be difficult to apply the pairwise similarity paradigm or an SD experiment in a communication situation, as new and complex factors such as the communication behavior as well as the longer duration of conversations are highly influencing the scaling process. Moreover, time-varying degradations, possibly resulting in features that also vary with time, in turn might necessitate instantaneous assessment and/or temporal integration, see Sect. 9.5.

Once the perceptual features have been identified, they can be mapped to integral quality judgments obtained for the same physical events, and triggering hopefully the same perceptual events. This way, it becomes possible to identify the weighting or importance of each perceptual feature for the integral quality. This process is called *external preference mapping*, and the resulting weights can be used for modeling overall quality on the basis of features. Both simple linear regression models as well as k-nearest neighbor classifiers have been used for this purpose, see e.g., [2, 25].

5.4 Feature Levels

So far, we have addressed quality features only on the level of the perceptual event, and we have described methods which are able to extract features and feature values on this level. However, when referring to a service which is potentially interactive, where the individual usage situation spans over a delimited period of time, and/or which is being used regularly over a longer time, there are additional features which may play a role for the global quality, utility, and acceptability of the service. Several proposals have been made in the past to classify and relate these features, e.g. in [14, 16] for telephony services, in [15] for spoken dialogue services, and in [17, 26] for multimodal interactive services. A broader classification of features has been proposed in [12]. We mainly adopt this classification here, but extend and detail it with some ideas of [16, 17, 26] to illustrate the respective types of features, as it is shown in Fig. 5.3.

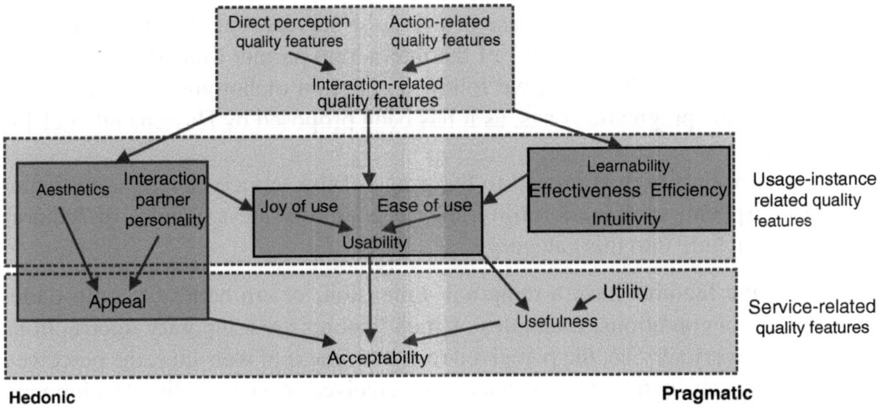

Fig. 5.3 Organization of quality features on different levels. Taxonomy adapted from [17, 26], taking into account some of the levels introduced in [12]

According to this new classification, features are grouped on five layers:

- *Level of direct perception*, On this layer, we summarize all quality features which are related to the perceptual event in a narrow sense, and which are created immediately and spontaneously during the experience. These features may be related to individual sensory channels, such as visual features, auditory features or tactile features, but may also be linked to the perception via multiple senses in parallel (e.g., audio-visual features). For the visual channel, examples include sharpness, darkness, brightness, contrast, flicker, distortion, and color perception, see Sect. 5.6. For the auditory channel, example features of audio-streaming services are localization and timbre, and example features of speech-transmission services include coloration, noisiness, loudness, or continuity, see Sect. 5.5. For services that address multiple sensory channels simultaneously, relevant features are e.g. balance and synchronism, see, e.g., Sect. 24.4.
- *Level of action*, i.e., the level which relates to the human perception of his/her own actions. In case of video services, this may include involvement and immersion, the perception of space (as far as this is supported by the user perceiving his/her own motions in the virtual space), as well as the perception of own motions. In the case of speech services, this may include talking-only features such as the perception of sidetone, echo, or double-talk degradations.
- *Level of interaction*, i.e., the level that includes the constant exchange of actions and reactions, be it between humans (human-to-human interaction) or between humans and systems (human-to-machine interaction), see also Chap. 11. Features on this level include responsiveness, naturalness of interaction, communication efficiency, and conversation effectiveness.
- *Level of the usage instance of the service*, which includes also the physical and social usage situation. Examples of such features are the learnability and intuitivity of the service, its effectiveness and efficiency for reaching a particular goal during

the current usage instance, the ease of using the service, but also non-functional features such as the "personality" of the interaction partner (human or machine), or its aesthetics. On this level, we follow a common dichotomy of features into "hedonic" and "pragmatic" ones, as it has been proposed by Hassenzahl et al. [4] for example.

- *Level of service*, which is related to the usage of the service beyond a particular instance. Appeal, usefulness, utility and acceptability are examples of features which we include into this category.

Some of the features have a temporal dimension, or are nameable only under certain temporal conditions. Examples of such features are temporary interruptions of a media delivery service, the perceived responsiveness of web sites, the perceived availability and set-up time for a service, the perceived service quality development and the perceived service reliability over longer periods of time. Thus, in some classifications, the temporal dimension is sometimes considered as a "feature" as well (e.g. in [12]). As this would contradict the organization of features on the mentioned levels, we prefer to consider the temporal aspect to be a part of the above-mentioned feature categories.

In the following two sections, we will provide examples for quality features for two popular use cases, namely speech and video transmission and communication services, see also Chaps. 12 and 19. The features identified so far are mostly on the level of direct perception, but some ideas for identifying features also on other levels are outlined in Sect. 5.7.

5.5 Case 1: Features of Speech Services

For narrowband (300–3,400 Hz audio transmission bandwidth) as well as for wideband (50–7,000 Hz) speech services, the feature space for the listening situation was explored in [25]. The whole end-to-end speech transmission chain was considered, including, for example, user terminals with different electro-acoustic properties, speech codecs and other signal processing algorithms (e.g., noise reduction), and both packet-based and public-switched networks. For different sets of stimuli reflecting different transmission set-ups, the paradigms of dissimilarity scaling as well as Semantic Differentials were applied in several auditory experiments. For the definition of the attributes in the SD experiments, the attributes that were used in related literature were reviewed and partially included in the test (e.g., [13, 24]). The resulting scores were analyzed with subsequent Multidimensional Scaling and Principal Component Analysis. As a result, the following quality-relevant perceptual dimensions were identified:

- *Discontinuity*,
- *noisiness*, and
- *coloration*.

The discontinuity dimension describes the perceptual effect of time-varying distortions (such as packet loss in VoIP), whereas noisiness reflects noise perception due to background or circuit noise, as well as to signal-correlated noise stemming from certain speech coding algorithms. Linear distortions causing deviations from an expected "timbre" on the perceptual level are subsumed under the label coloration. In scenarios where speech level differences can be expected, a *loudness* dimension can be added to the above set. These dimensions were shown to be mostly independent of whether narrowband or wideband speech has been assessed. The dimensions can be regarded to cover most aspects encountered in today's speech services.

Moreover, in [25], a test method was developed for the direct scaling of the three dimensions, using three rating scales. This measurement method is an efficient new tool for meaningful and reliable analytic feature assessment, as it allows a much larger number of stimuli to be assessed by non-expert listeners. At the time of writing, the method is considered as a part of a new subjective test methodology in the International Telecommunication Union, ITU-T, Study Group 12, Question 7 (work item P.MULTI) and as the basis for signal-based dimension *estimators* in Question 9 (work item P.AMD). Details as well as a comprehensive literature overview can be found in [25].

5.6 Case 2: Features of Video Services

In the case of video, Teunissen and Westerink [22] conducted a study using six TV sets (CRT) varying in spatial resolution, color reproduction, peak luminance, and luminance contrast. Video stimuli were presented to the subjects in two ambient illumination environments. In an attribute scaling table, eight quality features were found to be relevant for the application: Color naturalness, sharpness, darkness (of black areas), brightness, contrast, flicker, smear/geometrical distortion. The authors identified that *color naturalness* (which is affected by color rendering) was the most important factor, followed by *perceived sharpness*. They further conclude that the combined scores for sharpness and naturalness give a good prediction of overall perceived quality. They observed that the correlation between the sharpness cluster and quality scores is higher (0.83) than between the naturalness and the quality scores (0.69), but that a difference in color (RGB) balance (color naturalness) affects the perceived quality more than a difference in resolution (sharpness).

In a recent study carried out by Tucker at TU Berlin [23], video streaming of IPTV services was analyzed using the Semantic Differential technique with subsequent factor analysis. The degradations considered included coding and packet-loss induced degradations such as freezing and slicing (see Chap. 19). Three perceptual video quality dimensions were identified for this scenario:

- *Fragmentation*, describing impairments due to compression (yielding blocking artifacts), or combinations of compression and packet-loss induced slicing,
- *movement disturbance*, which describes the perceptual effect of freezing, and

- *spatial frequency content*, which depends on the video compression (this time yielding blurring artifacts).

The dimension *fragmentation* was found to be the main contributor, as it explained the largest part of the variance of the data.

Yamagishi and Hayashi in [27] addressed the audiovisual case for interactive multimedia service such as video-telephony. Using the Semantic Differential technique with subsequent factor analysis, they found that two perceptual dimensions are contributing to the perceived quality: *aesthetic feeling* and *feeling of activity*. The aesthetic feeling is linked to audio and video packet loss, and video bitrate. This dimension is related to attribute pairs such as quiet/clamorous, clear/cloudy or beautiful/dirty. Feeling of activity is related to one-way transmission delay and video frame rate. Respective attributes are, for instance, dynamic/static, slow/fast, or light/heavy.

5.7 Conclusions

In this chapter, we have analyzed the concept of *quality features*, i.e., perceivable, recognized and nameable characteristics of perceptual events which are relevant for Quality of Experience. Such features can be considered as axes in a multidimensional perceptual space. Their relevance for QoE can mathematically be formulated by the vector model or the ideal-point model. When referring to a pure perception level, quality features can be identified by Semantic Differential scaling and subsequent factor analysis, or by (dis-) similarity scaling and subsequent MDS. However, it has been pointed out that there are several other levels where quality features are of relevance, such as when acting and interacting with a human or machine partner, when considering individual service usage instances which span over a certain period of time, or when integrating quality over longer service usages and considering service usefulness, utility and acceptability.

For quality features on those layers, appropriate experimental paradigms identifying the relevant dimensions are not yet available. A first attempt has been made in [10] to extend the SD approach towards the level of action, but the results obtained are not yet conclusive and need to be analyzed in a real interactive context as well. The longer the usage intervals get, the more difficult it will be to make use of the MDS paradigm, as the length of the period which is necessary for formulating a judgment prevents direct comparisons to be made. In turn, the SD approach might be disputed as well, as validated collections of relevant attributes are mostly missing for quality features on those layers.

Once the relevant features have been identified and quantified, it will be very helpful to draw links between perceptual quality features and *related* quality factors, so that cause–effect relationships can be identified and used for diagnosing reasons of sub-optimal quality. As an example, perceptual quality features of speech transmission services quantified with the direct scaling method outlined in Sect. 5.5 can be put into a relationship with technical causes, which in turn might be identified through expert-listening procedures which are currently discussed in ITU-T Study Group 12, see [5]. This way, it becomes possible to not only gain insights into the perceptual and judgment-related processes underlying a QoE judgment, but also to use this knowledge for quality engineering.

Acknowledgments The ideas presented in this chapter are partially based on the concepts presented in Chap. 6 of [12]; the contributions of the co-authors of that chapter are gratefully acknowledged.

References

1. Carroll J (1972) Individual differences and multidimensional scaling. In: Shepard RN, Romney AK, Nerlove SB (eds) Multidimensional scaling—Theory and applications in the behavioral sciences, Vol I: theory, pp 105–155
2. Côté N (2011) Integral and diagnostic intrusive prediction of speech quality. Springer, Berlin
3. Garcia MN, List P, Argyropoulos S, Lindegren D, Pettersson M, Feiten B, Gustafsson J, Raake A (2013) Parametric model for audiovisual quality assessment in IPTV: ITU-T Rec. P.1201.2. In: Proceedings of IEEE MMSP, Sept 2013
4. Hassenzahl M, Platz A, Burmester M, Lehner K (2000) Hedonic and ergonomic quality aspects determine a software's appeal. In: Proceedings of CHI 2000, Den Haag, pp 201–208
5. ITU-T Contribution COM 12-14 (2013) Validation of the P.TCA annotation methodology and comparison to perceptual dimensions from P.AMD. Source: Deutsche Telekom AG; authors: Köster F, Möller S, Schiffner F, Skowronek J, ITU-T SG12 Meeting, Geneva, 19–28 Mar 2013
6. ITU-T Recommendation G.107 (2005) The E-model, a computational model for use in transmission planning. International Telecommunication Union, Geneva
7. ITU-T Recommendation P.1201.2 (2012) Parametric non-intrusive assessment of audiovisual media streaming quality—Higher resolution application area. International Telecommunication Union, Geneva
8. Jekosch U (2005) Voice and speech quality perception—Assessment and evaluation. Springer series in signals and communication technology. Springer, Berlin
9. Jumisko-Pyykkö S, Hakkinen J, Nyman G (2007) Experienced quality factors—Qualitative evaluation approach to audiovisual quality. In: Proceedings of SPIE/IS&T human vision and electronic imaging (HVEI)
10. Köster F, Möller S (2013) Towards a new test paradigm for the subjective quality assessment of conversational speech. In: Fortschritte der Akustik—DAGA 2013: Plenarvortr. u. Fachbeitr. d. 39. Dtsch. Jahrestg. f. Akust., Meran, Dtsch. Ges. Akust., Berlin
11. Lawless HT, Heymann H (1999) Sensory evaluation of food: principles and practices. Chapman & Hall, New York
12. Le Callet P, Möller S, Perkis A (eds) (2013) Qualinet white paper on definitions of quality of experience. In: European network on quality of experience in multimedia systems and services (COST Action IC 1003), Lausanne, Version 1.2, Novi Sad, March 2013
13. Mattila V-V (2001) Perceptual analysis of speech quality in mobile communications, vol 340. Doctoral dissertation, Tampere University of Technology, Tampere
14. Möller S (2000) Assessment and prediction of speech quality in telecommunications. Kluwer Academic Publishers, Boston
15. Möller S (2005) Quality of telephone-based spoken dialogue systems. Springer, New York
16. Möller S, Berger J, Raake A, Wältermann M, Weiss B (2011) A new dimension-based framework model for the quality of speech communication services. In: Proceedings of the third international workshop on quality of multimedia experience (QoMEX'11), Mechelen, 7–9 Sept 2011

17. Möller S, Engelbrecht K-P, Kühnel C, Wechsung I, Weiss B (2009) A taxonomy of quality of service and quality of experience of multimodal human-machine interaction. In: Proceedings of the first international workshop on quality of multimedia experience (QoMEX'09), San Diego CA, 29–31 July 2009
18. Osgood C, Suci G, Tannenbaum P (1957) The measurement of meaning. University of Illinois Press, Urbana
19. Quackenbush SR, Barnwell TP III, Clements MA (1988) Objective measures of speech quality. Prentice Hall, Englewood Cliffs
20. Raake A (2006) Speech quality of VoIP—Assessment and prediction. Wiley, Chichester
21. Strohmeier D, Jumisko-Pyykkö S, Reiter U (2010) Profiling experienced quality factors of audiovisual 3D perception. In: Proceedings of international workshop on quality of multimedia experience (QoMEx'10), Trondheim
22. Teunissen K, Westerink J (1995) A multidimensional evaluation of the perceptual quality of television sets. J SMPTE 105:31–38
23. Tucker I (2011) Perceptual video quality dimensions. Master thesis, Technische Universität Berlin, Berlin
24. Voiers WD (1977) Diagnostic acceptability measure for speech communication systems. In: Proceedings of international conference on acoustics, speech, and signal processing (ICASSP'77), Hartfort CT, pp 204–207
25. Wältermann M (2013) Dimension-based quality modeling of transmitted speech. Springer, Berlin
26. Wechsung I, Engelbrecht K-P, Kühnel C, Möller S, Weiss B (2012) Measuring the quality of service and quality of experience of multimodal human-machine interaction. J Multimodal User Interf 6:73–85
27. Yamagishi K, Hayashi T (2005) Analysis of psychological factors for quality assessment of interactive multimodal service. In: Proceedings of SPIE/IS&T human vision and electronic imaging (HVEI), pp 130–138

Chapter 6
Quality of Service Versus Quality of Experience

Martín Varela, Lea Skorin-Kapov and Touradj Ebrahimi

Abstract It is often the case that in the current literature, the term "QoE" is used in contexts where "QoS" would be more appropriate. This is likely due to several reasons, one of which being the current popularity of all things related with QoE, but more fundamentally it is due to the boundaries between QoS and QoE not being clearly defined—and indeed, sometimes hard to define clearly. QoE is an intrinsically multi-disciplinary field, and practitioners from different backgrounds see it, quite naturally, from different perspectives colored by their own expertise. For networking people, in particular, QoE is sometimes seen as a simple extension, or even a re-branding, of the well-established concept of QoS. In this chapter we will delve into the differences and commonalities between the two, with the goal of easing and encouraging their proper use.

6.1 What QoS Is, and How It is not QoE

Quality of Service, or QoS, is a well-established research domain that has seen an enormous amount of activity for over 20 years. According to the ITU (Rec. E.800) [18],[1] QoS is defined as

[1] This definition was also used more recently in the third amendment to ITU-T P.10/G.100 [21].

M. Varela (✉)
VTT Technical Research Centre of Finland, Oulu, Finland
e-mail: martin.varela@vtt.fi

L. Skorin-Kapov
Faculty of Electrical Engineering and Computing, University of Zagreb, Zagreb, Croatia
e-mail: lea.skorin-kapov@fer.hr

T. Ebrahimi
École Polytechnique Fédérale de Lausanne, Lausanne, Switzerland
e-mail: touradj.ebrahimi@epfl.ch

S. Möller and A. Raake (eds.), *Quality of Experience*,
T-Labs Series in Telecommunication Services, DOI: 10.1007/978-3-319-02681-7_6,

> The totality of characteristics of a telecommunications service that bear on its ability to satisfy stated and implied needs of the user of the service.

Contrasting this definition with the definition of QoE (cf. Chap. 2), we notice that the ITU-T definition of QoS is narrower in scope (clearly geared at telecommunications services, whereas the QoE definition is not limited to only such services) and in depth (it does not explicitly address things that are basic to the definition of QoE, such as the context of usage and the user's personality traits and current state of mind). The mention of the user's "*implied needs*" also indicates a rather utilitarian view of quality, and closer to the notion of *assumed quality* presented in Chap. 2, whereas QoE includes a hedonic component as well ("*...utility and/or enjoyment...*"), and corresponds to the notion of *quality based on experiencing*. More importantly, QoS is defined from a system's perspective —"*...characteristics of a telecommunications service...*", whereas QoE is defined entirely from the user's perspective —"*...the degree of delight or annoyance of a person...*", considering the system's aspects as subordinate, by their influence on the degree of fulfillment of the user's expectations. The ETSI takes a similar approach to the ITU's in their definition [8], which is in turn based on an old (*ca.* 1988) version of the E.800 recommendation. Gozdecki et al. provide a detailed overview of QoS-related terminology in [11]. The IETF has, even more than the ITU and ETSI, taken a network-centric view of QoS, giving the following definition for QoS [6]:

> A set of service requirements to be met by the network while transporting a flow.

In this definition, there is no mention whatsoever of users, and the QoS is defined in terms of "*service requirements*", which are not further specified. A summary and comparison of definitions of QoS put forth by different standards bodies is given in Table 6.1.

Naturally, definitions and common usage are not always aligned, and that is the case for both QoS and QoE. The term QoS is commonly used in the literature with two different meanings, none of which abides by the ITU definition. In its first acception,[2] QoS refers to concepts and measures of network performance, such as throughput, delay, jitter, etc. In its second acception, it refers to mechanisms such as Differentiated Services (*DiffServ*, where forwarding of packets is done according to their DSCP field in the IP header, allowing different so-called *per-hop behaviours* to be implemented), Integrated Services (*IntServ*, which is an approach based on resource reservation along a network path), or other forms of traffic engineering (also sometimes referred to as "*QoS architectures*" or "*mechanisms*"), which aim at improving said measures of performance. This latter meaning is particularly used by the IETF. In common usage, we might read or hear that "a network has bad QoS", meaning that it has poor performance; or that a service provider "has implemented QoS in their network", meaning that it has deployed some performance-improving mechanism on said network. It is worth noting that in most of these common usage

[2] The reader will note that this ordering is rather arbitrary. Researchers with a background in network performance evaluation will probably have a different view than those with a background in traffic engineering, etc.

Table 6.1 Summary of QoS definitions

Definition	Key terms	Acception	Differences with QoE	References
ITU-T	Characteristics of a telecommunications service; stated and implied needs of the user	Network/system performance	Focus on the system	[18]
ETSI	*Idem*	Network/system performance	Focus on the system	[8, 18]
IETF	Service requirements met by the network	Architectures	User not considered	[6]

Table 6.2 Summary of distinguishing factors between QoS and QoE. Expanded from [27]

	QoS	QoE (Qualinet white paper)	QoE (Chap. 2)
Stance	Utilitarian	Utilitarian/Hedonic	Utilitarian/Hedonic
Scope	Typically telecom services	Broader domain (not necessarily network-based)	Broader domain (not necessarily network-based)
Perspective	System's	User's	Person's
Focus	Performance aspects of telecom systems; mechanisms such as DiffServ	ICT service or application	ICT service, application or system
Methods	Technology-oriented; empirical or simulated measurements	Multi-disciplinary and multi-methodological approach	Multi-disciplinary and multi-methodological approach

patterns, the user of the service is not really taken into consideration, and if so, in an indirect way.

Based on the above discussion, we can stipulate that QoS and QoE are two different concepts, which in practice have intersections. While it would be incorrect to try and classify one as a sub- or super-set of the other, there is a large overlap between them, insofar as some dimensions of networked multimedia applications' QoE are heavily affected by network QoS, and QoE does in many cases provide a higher-level understanding of network performance (Table 6.2).

QoE, as a term, is also often used in ways that do not really follow its definition. For multi-media services in particular, it is common to find results purporting to "improve QoE", where the actual achievement is for example a reduction in transport delays. It can be argued that in many cases lower delays can indeed result in a better QoE, but in omitting to directly take into account aspects related to the service's users, their context of usage, and so on, the application of the term QoE in such cases is at the borderline of abusing the language.

Table 6.1, expanded from [27] with the aspects discussed in Chap. 2, provides a summary of the major conceptual differences between QoS and QoE.

6.2 From QoS to QoE...

For the remainder of this chapter we will, unless explicitly noted, consider QoS in the first acception given above (concepts and measures of network performance), as we are here interested in the conceptual understanding of quality as perceived by the users, and not necessarily on how networks can be instrumented to improve it (this will be further addressed in Chaps. 27 and 28).

6.2.1 QoS as Quality Evaluation

The study of QoS in networks spans a variety of sub-fields, ranging from analysis of queuing systems to metrology, traffic characterization, etc. Staples of these disciplines are performance metrics such as throughput, good-put, packet loss rate, delay and jitter, as well as dependability (availability, reliability, maintainability...) measures and models that define overall how well a network performs. QoS, as a concept, can also be extended beyond its original network-related aspects to other system- and operations-related aspects. For example, the latest version of the ITU Rec. E.800 [18] identifies QoS as comprising both network-related performance (e.g., bit error rate, latency) and non-network related performance (e.g., service provisioning time, different tariffs, complaint resolution time, etc.).

In the context of multi-media systems, there are well-known effects caused by the network performance (QoS) on some dimensions of the QoE that the user perceives. In particular, the perceptual dimensions of QoE, which in some cases might be quite dominant, can be strongly affected by impairments in the network such as losses and delays. While QoS, as a concept, does not explicitely take into account user perceived quality and degree of satisfaction, the combined notions of user perceived/experienced quality and QoS do appear in the literature [5] and standards going back over a decade. ITU Rec. G.1000 [12] defines four different viewpoints of QoS, going from customer QoS requirements, QoS offered by a service provider, QoS achieved by a service provider, to QoS perceived by the customer. Complementary to Rec. G.1000, ITU Rec. G.1010 [13] specifically addresses the "customer" viewpoint. By considering user expectations for a range of multi-media applications (involving various media such as voice, video, image and text), eight distinct categories are proposed based on tolerance to information loss and delay, intended to provide a basis for defining realistic QoS classes for underlying transport networks and associated QoS control mechanisms. Hence, key performance parameters and corresponding target values for a wide range of multi-media applications are outlined. Further ITU recommendations also define standard performance parameters for packet transfer in IP-based networks [19, 20].

Application-level quality considerations have been gaining importance in the QoS literature. Coupled with a simplistic understanding of QoE, which basically reduces it to its perceptual aspects, this has yielded a large number of results focused on network performance, but under a QoE banner. Such research is clearly important, but its limitations in scope require further exploration. In general, this type of work is based on models of human perception that have varying complexity, but that at best reach some physiological aspects of the end-users' perception (e.g., via models of the visual or auditory systems), and not other user-related aspects that are critical for QoE, such as the users' emotional state, socio-cultural background and environment, context of usage, etc. These other aspects of QoE are, incidentally, likely to be relevant for people traditionally concerned with pure network performance, such as Internet Service Providers (ISPs) and other service providers; these "higher layer" aspects of QoE are closely related to the users' expectations and their valuation of the service, and considering them is likely to yield a more useful estimation of QoE from a business perspective (churn, willingness to pay, etc.). Last but not least, personalization of quality to define performance metrics for a specific user or a cluster of users with specific profiles, as opposed to the mean opinion score measured on average users, can have an important impact on the estimation of actual QoE.

6.2.2 Perceptual Quality

The perceptual aspects of multimedia QoE have been studied for a long time. By perceptual, we refer to the physical characteristics of the media and their interaction with the users' physiology, and the resultant quality judgment: "*this conversation had very bad sound*!" or "*this video had excellent quality*". There are several dimensions of QoE that relate to perceptual quality, as has been discussed before in Chap. 5, and they often relate to specific properties of the media as observed by the user, such as the presence of artifacts (or lack there of), intelligibility, continuity, and so on, as well as other aspects that are related to the system's performance (e.g. interactivity, in the case of telephony systems, which is partly dependent on the delay), and user interface.

Studies dealing with the quality of media services date back to the early days of telephony speech and television system quality evaluation, to digital media services delivered via packet switched networks [33]. Today, numerous ITU standards recommend various quality models and assessment methodologies [23].

Subjective methods for assessing the quality of telephony and television systems have been around for a long time, and so have several so-called objective models of perceptual quality. The former are considered as the reference (or *ground truth*), for after all, it is the users alone who can judge the perceptual quality of a given service. The latter introduce mechanisms of varying complexity to produce estimates of those ground truth values, aiming at reducing the cost of the assessments and improving their reproducibility, as well as providing in-service quality monitoring and control mechanisms.

Many of these objective models were not designed with the transmission of media over IP networks in mind, and thus did not consider transmission impairments explicitly or at times even correctly, and they were not up to the task of accurately estimating their impact on the quality perceived by the users. Newer models tend to be designed around the idea of media transmitted over packet networks [22], and can significantly improve their estimates by doing so. A good example of this, in the context of speech quality, could be the better performance of POLQA [15] versus PESQ [14] when time-aligning the signals is problematic.

This shift from models concerned mostly with the effects of encoding on quality towards network-aware models is in a way a dual of the shift in the QoS domain from purely network performance-oriented metrics towards perceptual estimates as a measure of network performance.

6.2.3 Transitioning Towards QoE

While it is clear that QoE may be influenced by a broad range of factors, from a service provider or network operator's point of view there has been a need for an understanding of the fundamental relationships between QoE and measurable QoS parameters, paving the way for practical in-service QoE monitoring and management solutions [39]. Along these lines, a number of studies have focused on identifying the generic relationships between QoS and QoE, most frequently observing exponential or logarithmic relationships.

The IQX hypothesis [10] expresses the generic exponential dependency of QoE on QoS, and builds on the assumption that the higher the current level of QoE, the greater the impact of additional QoS disturbances (e.g., loss, jitter, throughput). On the other hand, logarithmic relationships [17, 34] consider the magnitude of change of QoE for a user as a function of the reciprocal QoS. Such relationships stem from the Weber-Fechner Law (WFL), which (based on human perceptive abilities) states that the just-noticeable difference between two levels of a certain stimulus is proportional to the magnitude of the stimuli (in this case referring to QoS level). It has been noted in [37] that while the WFL applies mostly in cases of the signal- or application-level stimulus directly being perceived by the user (e.g., latency), exponential relationships based on the IQX hypothesis provide accurate insight in cases of network-level QoS impairments not directly perceivable by the end-user (e.g., packet loss).

Observations of the relationship between QoE and QoS may be considered with respect to different types of QoS measures, namely failure- (e.g., loss rate), success- (e.g., packet success rate) and resource-related measures (e.g., throughput) [9]. Aside from objective perceptual quality models put forth by standards, numerous research efforts have attempted to model QoE in terms of application or network-level QoS measures. Among commonly-studied applications are real-time voice and video, usually based on UDP transport, for which numerous models of varying accuracy have been developed linking QoE to intrinsic network metrics such as packet loss rate and loss distribution or delay. On the other hand, in the case of media delivery relying on TCP-based transport (e.g., YouTube [40]), QoE is often modeled in terms of application-specific buffering metrics, which in turn depend indirectly on network-level QoS. In the case of interactive request/response type services such as Web browsing, page loading time has been identified as a key factor impacting QoE [7], indirectly related to offered bandwidth and network delays. In the context of networked games, studies have addressed user perceived quality as a function of network impairments (most commonly delay, jitter or loss) [32, 35].

Following the different viewpoints of QoS mentioned previously, the notion of "QoS experienced by the user" (QoSE) has been identified as being "influenced by the delivered QoS and the psychological factors influencing the perception of the user" [18]. It is stated that while the quantitative component of QoSE can be influenced by the network infrastructure, "the qualitative component can be influenced by user expectations, ambient conditions, psychological factors, application context, etc." While still focusing on the perception of the quality of the *service*, this description brings us closer to the current definition of QoE. Hence, going beyond ensuring that the technical performance requirements are met, QoE is based on adopting a user perspective in judging that the actual needs and expectations of the end-user are met.

Given the need to relate parameters expressed at the user/application level with parameters specifying network performance, both standards [16] and literature [25] have addressed QoS specification and mapping across different levels. More recently, layered approaches have been discussed relating network-level Key Performance Indicators (KPIs, e.g., delay, loss, throughput, etc.) with end-to-end user-level application specific Key Quality Indicators (KQIs, e.g., service availability, media quality, reliability, etc.), which then provide input for QoE estimation models [4, 34]. Taking as an example a video streaming service, transmission parameters such as loss or delay will result in video artifacts impacting the media quality, which may in turn be translated to end-user QoE.

As discussed by Reichl et al. [34] additional input to a QoE estimation model may then be provided by user and context influencing factors. Such knowledge regarding the mapping from KPIs to KQIs provides valuable input regarding the analysis of the root causes of QoE degradation (cf. Fig. 6.1, adapted from [4]).

6.3 . . .and Back

Transitioning from provider-centric QoS to user-centric QoE clearly provides deeper insight into the wide variety of influencing factors impacting the actual end-user experience, going beyond only technological parameters by also considering psychological and sociological factors. However, from a provider's point of view, the goal of reliable QoE models and estimators is to provide the necessary input for

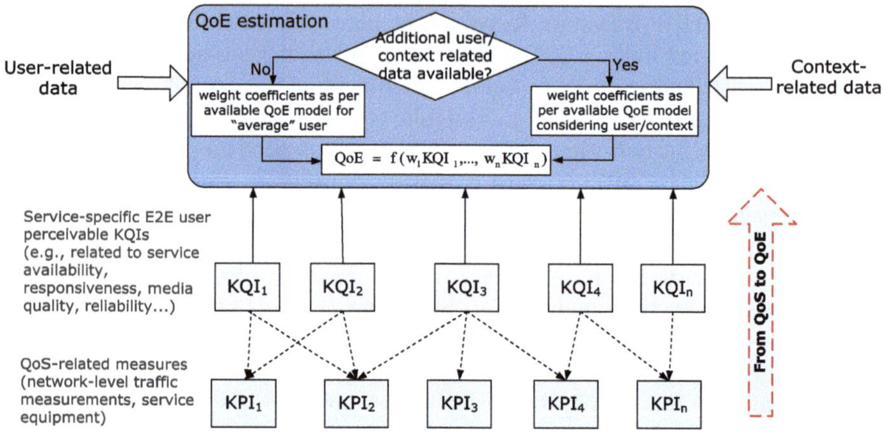

Fig. 6.1 QoE as a weighted function of user perceivable KQIs, further linked to QoS-related KPIs (adapted from [4])

effective QoE control and optimization mechanisms, mainly by means of network QoS management. In a network environment, different providers involved in the service delivery chain (e.g., service provider, network operator, content provider, device provider) will ultimately address QoE optimization strategies from their specific viewpoint. We increasingly see the transition from "QoS management" to "QoE management" [1, 29], whereby traditional QoS management mechanisms (e.g., QoS-based routing, resource allocation algorithms, policy control, service adaptation, etc.) are being reconsidered so as to incorporate the notion of end-user subjectivity. While the majority of approaches incorporate subjective quality perception models (e.g., in the context of QoE-driven resource allocation mechanisms [2, 41]), others are driven by explicitly provided end-user QoE-related feedback (e.g., in the context of radio resource management mechanisms [3]).

The idea of using application-level quality measurements or estimates to drive changes in the network has been around for some time, and a slew of cross-layer mechanisms for controlling some network aspects based on application-level performance exist in the literature. These mechanisms may act for example by performing application-level adaptations as reactions to changing network QoS, and also at the network level, both on the terminal and network sides. In many cases, the estimations are based on simplistic notions and models of quality, but more recently, perceptual quality models have become more common as optimization targets for these cross-layer mechanisms. Results such as those by Lewcio et al. [28] provide useful insight as to the impact of application- and network-level adaptations on perceptual quality. Understanding these relations between QoS and certain aspects of QoE (notably its perceptual dimensions, in the case of multi-media services) enables the development of smarter ways of controlling network performance, for example by performing mobility management [42], admission control [38], traffic shaping, bandwidth adaptation [24], or managing the priorities of different service types and

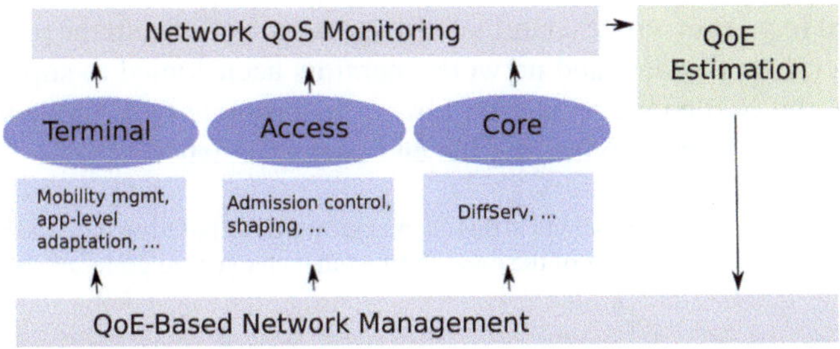

Fig. 6.2 Conceptual architecture for QoE-based network management, with commonly used management mechanisms

subscriber classes [38], for instance by using a QoS architecture such as DiffServ. Figure 6.2 presents a conceptual view of how such systems work in general, with some concrete examples of common QoS management mechanisms.

Other, more indirect ways of using QoE to manage network performance is via the input of QoE models into network planning (*à la* E-model, or using them together with classic network performance evaluation tools [36]), SLA creation (QoE-based SLAs), application-aware network elements [26], and plain customer experience management systems.

Beyond the network-oriented management mechanisms mentioned above, it is also possible for application developers and service providers to use QoE models to make dynamic adaptation at the application level (e.g. adapt error correction mechanisms, change encoding parameters, etc.) or make larger scope operational optimizations (e.g. addition of caching nodes in a CDN closer to a location where users are experiencing lower quality) by exploiting application-level QoS or QoE information that is available to them.

Challenges related to implementing "true" QoE in the application domains currently dominated by QoS/perceptual quality clearly lie in the collection and processing of data from the client/end-user, and in feeding back this data to relevant network or application management mechanisms. We leave a further in-depth discussion of QoE-based application and network management to Chaps. 27 and 28.

6.4 Conclusions

In this chapter, we addressed both the conceptual differences and the links between QoS and QoE, discussing the shift from purely technical network performance metrics to estimates of subjective user perceived quality. While clear relationships between the terms exist, true measures of QoE must ultimately take into account end-user subjectivity and the impact of additional contextual and user-related factors. Consequently, subjective and objective quality assessment methods have evolved over the years, aiming at modeling the impact of both technical (QoS-related) and non-technical (e.g., user, context) influence factors on QoE. From the practical point of view of service providers and network operators accustomed to supporting QoS mechanisms, the challenges remain on how to incorporate QoE models in driving such mechanisms towards optimizing the end-user experience.

Acknowledgments The ideas presented in this chapter are partially based on the concepts presented in Chap. 7 of [27]; the contributions of the co-authors of that chapter are gratefully acknowledged.

References

1. Agboma F, Liotta A (2008) QoE-aware QoS management. In: Proceedings of the 6th international conference on advances in mobile computing and multimedia, ACM, pp 111–116
2. Ameigeiras P, Ramos-Munoz JJ, Navarro-Ortiz J, Mogensen P, Lopez-Soler JM (2010) QoE oriented cross-layer design of a resource allocation algorithm in beyond-3G systems. Comput Commun 33(5):571–582
3. Aristomenopoulos G, Kastrinogiannis T, Kaldanis V, Karantonis G, Papavassiliou S (2010) A novel framework for dynamic utility-based QoE provisioning in wireless networks. In: Global telecommunications conference (GLOBECOM 2010), IEEE, pp 1–6
4. Batteram H, Damm G, Mukhopadhyay A, Philippart L, Odysseos R, Urrutia-Valdés C (2010) Delivering quality of experience in multimedia networks. Bell Labs Tech J 15(1):175–193
5. Bouch A, Sasse MA, DeMeer H (2000) Of packets and people: a user-centered approach to quality of service. In: Eighth international workshop on quality of service, 2000. IWQOS, IEEE, pp 189–197
6. Crawley E, Sandick H, Nair R, Rajagopalan B (1998) A framework for QoS-based routing in the internet. IETF RFC 2386
7. Egger S, Reichl P, Hoßfeld T, Schatz R (2012) Time is bandwidth? Narrowing the gap between subjective time perception and quality of experience. In: IEEE international conference on communications (ICC 2012), Ottawa, Canada
8. ETSI: ETR 003 (1994) Network aspects (NA); general aspects of quality of service (QoS) and network performance (NP)
9. Fiedler M, Hoßfeld T (2010) Quality of experience-related differential equations and provisioning-delivery hysteresis. In: 21st ITC specialist seminar on multimedia applications-traffic, performance and QoE, Miyazaki, Japan
10. Fiedler M, Hoßfeld T, Tran-Gia P (2010) A generic quantitative relationship between quality of experience and quality of service. IEEE Netw 24(2):36–41
11. Gozdecki J, Jajszczyk A, Stankiewicz R (2003) Quality of service terminology in IP networks. IEEE Commun Mag 41(3):153–159
12. ITU-T: Recommendation G.1000 (2001) Communications quality of service: a framework and definitions
13. ITU-T: Recommendation G.1010 (2001) End-user multimedia QoS categories
14. ITU-T: Recommendation P.862 (2001) Perceptual evaluation of speech quality (PESQ), an objective method for end-to-end speech quality assessment of narrowband telephone networks and speech codecs
15. ITU-T: Recommendation P.863 (2001) Perceptual objective listening quality assessment
16. ITU-T: Recommendation H.360 (2004) An architecture for end-to-end QoS control and signalling
17. ITU-T: Recommendation G.1030 (2005) Estimating end-to-end performance in IP networks for data applications
18. ITU-T: Recommendation E.800 (2008) Definitions of terms related to quality of service

19. ITU-T: Recommendation Y.1540 (2011) Internet protocol data communication service G IP packet transfer and availability performance parameters
20. ITU-T: Recommendation Y.1541 (2011) Network performance objectives for IP-based services
21. ITU-T: Recommendation P.10/G.100 Amendment 3 (2012) New definitions for inclusion in recommendation ITU-T P.10/G.100
22. ITU-T: Recommendation P.1201 (2012) Parametric non-intrusive assessment of audiovisual media streaming quality
23. ITU-T: Recommendation G.1011 (2013) Reference guide to quality of experience assessment methodologies
24. Jammeh E, Mkwawa I, Khan A, Goudarzi M, Sun L, Ifeachor E (2012) Quality of experience (QoE) driven adaptation scheme for voice/video over IP. Telecommun Syst 49(1):99–111
25. Jin J, Nahrstedt K (2004) QoS specification languages for distributed multimedia applications: a survey and taxonomy. IEEE Trans Multimedia 11(3):74–87
26. Lazzara S, Roella A, Moschetti L (2012) Deliverable 8.2: live test report. Technical report, FP7 Optiband project
27. Le Callet P, Möller S, Perkis A (eds) (2012) Qualinet white paper on definitions of quality of experience
28. Lewcio B, Belmudez B, Enghardt T, Möller S (2011) On the way to high-quality video calls in future mobile networks. In: Third international workshop on quality of multimedia experience (QoMEX), 2011, pp 43–48
29. Martini MG, Chen CW, Chen Z, Dagiuklas T, Sun L, Zhu X (2012) Guest editorial: QoE-aware wireless multimedia systems. IEEE J Sel Areas Commun 30(7):1153–1156
30. Möller S, Schmidt S, Beyer J (2013) Gaming taxonomy: an overview of concepts and evaluation methods for computer gaming QoE. In: International workshop on quality of multimedia experience, QoMEX, pp 1–6
31. Möller S et al (2009) A taxonomy of quality of service and quality of experience of multimodal human-machine interaction. In: International workshop on quality of multimedia experience, QoMEX, pp 7–12
32. Picovici D, Denieffe D, Kastrati Z (2013) Enhanced network based model for measuring online games quality of experience. Dev Appl Syst 13
33. Raake A, Gustafsson J, Argyropoulos S, Garcia M, Lindegren D, Heikkila G, Pettersson M, List P, Feiten B (2011) IP-based mobile and fixed network audiovisual media services. IEEE Signal Process Mag 28(6):68–79
34. Reichl P, Egger S, Schatz R, D'Alconzo A (2010) The logarithmic nature of QoE and the role of the Weber-Fechner law in QoE assessment. In: IEEE international conference on communications (ICC), 2010, IEEE, pp 1–5
35. Ries M, Svoboda P, Rupp M (2008) Empirical study of subjective quality for massive multiplayer games. In: 15th International conference on systems, signals and image processing. IWSSIP 2008, IEEE, pp 181–184
36. Rubino G, Varela M (2004) A new approach for the prediction of end-to-end performance of multimedia streams. In: First international conference on quantitative evaluation of systems (QEST'04)
37. Schatz R, Hoßfeld T, Janowski L, Egger S (2013) From packets to people: quality of experience as a new measurement challenge. In: Biersack E, Callegari C, Matijasevic M (eds) Data traffic monitoring and analysis. Lecture notes in computer science, vol 7754. Springer, Berlin, pp 219–263
38. Seppänen J, Varela M (2013) QoE-driven network management for real-time over-the-top multimedia services. In: IEEE wireless communications and networking conference 2013, Shanghai, China
39. Shaikh J, Fiedler M, Collange D (2010) Quality of Experience from User and Network Perspectives. Annales des télécommunications (Ann Telecommun) 65(1–2):47–57
40. Staehle B, Hirth M, Pries R, Wamser F, Staehle D (2011) Aquarema in action: improving the YouTube QoE in wireless mesh networks. In: Internet communications (BCFIC Riga), 2011 Baltic congress on future, IEEE, pp 33–40

41. Thakolsri S, Kellerer W, Steinbach E (2011) QoE-based cross-layer optimization of wireless video with unperceivable temporal video quality fluctuation. In: IEEE international conference on communications (ICC), 2011, IEEE, pp 1–6
42. Varela M, Laulajainen JP (2011) QoE-driven mobility management—integrating the users' quality perception into network-level decision making. In: Third international workshop on quality of multimedia experience (QoMEX), 2011, pp 19–24

Chapter 7
Business Perspectives on Quality of Experience

Andrew Perkis, Peter Reichl and Sergio Beker

Abstract The current paradigm change towards Quality of Experience (QoE) does not only have conceptual and methodological consequences, but at the same time exhibits a profound impact on corresponding economic and business models, especially in the telecommunications market. This chapter deals with related issues from several layers of abstraction. We consider the general ecosystem level, proceed to resulting Customer Experience Management (CEM) systems, discuss consequences for Service Level Agreements (SLA), and finally analyze the implications if it comes to charging for QoE. As a result, it should become clear that integrating the economic dimension into QoE research provides an indispensable step towards enabling the future commercial success of the telco industry as such.

7.1 Introduction

Integrating quality as a major component into a business model is neither new nor far-fetched, given that any business model relies on the basic idea of bringing added value to the intended end user. However, both quality and added value are often hard to quantify and define, and will of course differ immensely from sector to sector. This

A. Perkis (✉)
Norwegian University of Science and Technology, Trondheim, Norway
e-mail: andrew@iet.ntnu.no

P. Reichl
University of Vienna, Vienna, Austria
e-mail: peter.reichl@univie.ac.at

P. Reichl
UEB/Télécom Bretagne, Rennes, France

S. Beker
Huawei Technologies, European Research Center, Munich, Germany
e-mail: sergio.beker@huawei.com

S. Möller and A. Raake (eds.), *Quality of Experience*,
T-Labs Series in Telecommunication Services, DOI: 10.1007/978-3-319-02681-7_7,

is also the case for media experience which supports natural interactions between people and their environment. The media considered consists of audio and visual presentations, and their interactions as well as user interactions including traditional interactivity as well as novel methods through Natural User Interfaces creating real world presence. In order to measure the corresponding user's perceived quality we need to shift from using simple Quality of Service (QoS) as a quality measure to the broader concept of Quality of Experience [1] (see also Chap. 2).

For the converged media and telecommunication sector, where delivery and consumption of audiovisual content is crucial, the concept of Quality of Experience is maturing, especially since the Qualinet White Paper on the definition of Quality of Experience is gathering acceptance [2]. Once a definition of QoE is agreed upon, it will also be possible to measure it, and thus QoE becomes a major component of the business perspective for all the stakeholders of the value chain [3].

Among those, service providers are increasingly aware of the importance of their customers' experience to increase loyalty in a more and more competitive market, especially since service quality has started to replace tariffing as the key selling point, if it comes to guaranteeing sustainable economic success. Hence, solutions that would help them to gain a comprehensive view of the end-user perceived quality together with means and methods to improve it are key for their business. However, due to the elusive nature of QoE, which comprises many layers of interaction between the elements enabling the delivery of a service or product and the human being as its user, measuring and improving user experience is a challenging task, which must be tackled taking into account both technical and non-technical aspects [4]. More specifically, the composing elements of communication services range from technical elements such as network transport (e.g. response times or throughput), coding and compression techniques, screen resolution, etc., over aspects of user expectation and context to business-related elements such as charging and pricing, after-sales customer support, etc. In this context, the present chapter focuses on business aspects of QoE and discusses corresponding ecosystem frameworks, supporting tools (with a special emphasis on customer experience management systems) and successful business factors and strategies.

In order to be able to correctly assess the user experience in terms of the impact that those different aspects have on the user perception of the received service or purchased product, a framework in which the different aspects are identified and put in perspective of the user is needed. Therefore, Fig. 7.1 depicts such a framework, called Media Experience Model. Note that, for the users, once upon a time media started with storytelling and wall drawing around the fire in the caves of early men, while today multimedia is about sharing experiences (real or imaginary) with others. This can be described in a layered model where QoE is the overriding factor and is seen as a tool for monitoring and managing the users' experience at each of the interfaces between the model layers, providing cross-layer optimization.

The first layers in the model consider the physical representation and delivery of the content. Today's media content is evolving around optimal utilization of 2D media and has focused on HD (High Definition) issues of resolution, frame rates, dynamic range, colour space and formats. There are numerous advances in these

Fig. 7.1 Media experience model

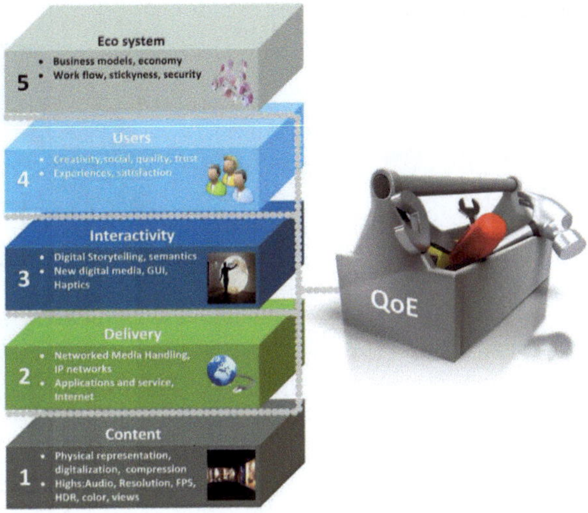

fields, among others Ultra High Definition TV (UHDTV), High Dynamic Range (HDR) and 3D. The future looks at increasing the user's experience by moving to multi-scopic, multi-view, free viewpoint and omnidirectional. Together with the advances in audio technology all the way to auralization and 3D audio we see the possibility of offering interactive holistic rendering of our real world to the user, with the ultimate goal being to digitally create real world presence where we can build business models and an economy based on the ecosystem at the top layer. As an example of a concrete cross layer optimisation we see the interaction between the content and delivery layer by efforts within Networked Media Handling.

7.2 Ecosystem: High Level and Generic Concepts

Customer experience management is central to the future business ecosystems, as the service providers are more and more subject to the market pressure for attaining increasing levels of provisioning efficiencies while at the same time facing shrinking revenues. In such a context, customer loyalty becomes the main enabler for customer experience management, where assessing the QoE of users constitutes the key element in any customer management system. An ecosystem is necessary to clearly identify the different actors in the value chain of producing and consuming a service, as well as their interactions. The ecosystem provides the interdependencies between these roles and identifies the interfaces where quality plays a major role. Taking this into account, the following ecosystem has been introduced in [2] (Fig. 7.2).

The introduced ecosystem illustrates the elusive nature of QoE as an intrinsic characteristic of the experience of the user with a given service, and as such it models the impossibility of being fully assessed but through its interactions with the context

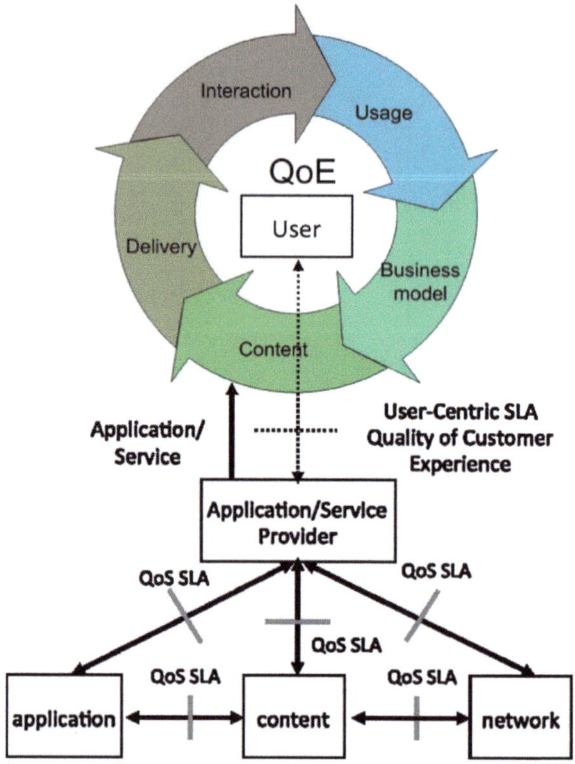

Fig. 7.2 Communications ecosystem

of use (for a detailed view on the concepts and definitions of the context and the QoE itself, please refer to Chap. 2). In this chapter, the accent is put on the business perspective of assessing the observable aspects of the experience and of mapping them at any possible extent to those aspects of the service which can be observed or measured.

All of the service aspects around the user, such as the interaction, the usage, the service delivery process, the nature of the content itself and the business model (especially the price to be paid, as we will analyse in detail in Sect. 7.5), they all impact the assessment that the user will do of her own experience with the service. Thus, the application is the instrument through which the user accesses the service, and as such it constitutes the main interface between the user and the service provider. As it is discussed in the next section, contractual obligations are increasingly being defined in terms of customer experience. A user-centric Service Level Agreement (SLA) is then defined at this interface, and must encode all the mentioned aspects of user experience. The service provisioning can be seen as a composition of service elements from one or more service providers. All these elements are closer to the technical implementation of the service, and their quality can be described in terms of performance metrics generally termed as Quality of Service (QoS). Translating the QoS into the QoE is

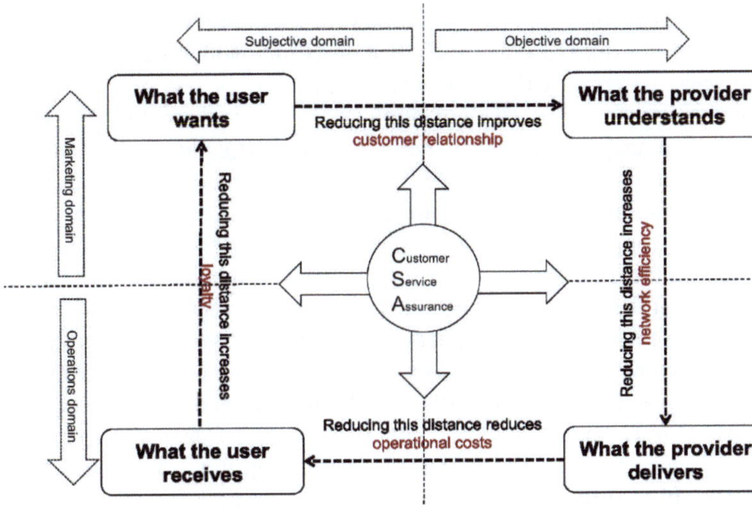

Fig. 7.3 CEM system components

nowadays one of the main research areas around customer experience management, as service providers are in demand for tools to help them understanding and managing their investments in CAPEX (capital expenditure, e.g. network infrastructure) and OPEX (operational expenditure, e.g. customer care, marketing) as a function of the impact on customer loyalty and the associated revenues. Research on the QoE domain around QoE assessment and the relationship between QoS and QoE requires a multidisciplinary approach in which subjective QoE assessment through field testing with actual users is driven by precise knowledge about the technical aspects of the service [5].

7.3 Management and System Aspects

The ecosystem introduced in the previous section helps in identifying the different aspects of the service impacting the user experience, the different actors and their roles, and the interfaces where contractual user experience or service performance metrics can be defined. A Customer Experience Management system (CEM) then consists of a set of tools which allow the management of the user experience and the associated business aspects of the provisioned service. Figure 7.3 depicts the main high level components of a CEM system. A detailed functional block description is out of the scope of this chapter, and can be found in the different CEM product descriptions existing in the market.

Most of existing CEM solutions follow the traditional Telemanagement Forum (TMF) service modelling. Data is collected from the different network and service elements and transformed into Key Performance Indicators (KPIs). Aggregations of KPIs are transformed into Key Quality Indicators (KQIs) which are in general equated to QoE metrics. Different CEM products present more or less refined aggregation models, with room for some customization. As such, traditional CEM systems do not truly allow managing the user experience, since KQIs are only obtained from performance metrics and hence can only reflect the technical performance aspects of the service, assuming they can be directly mapped to user perception.

However, this approach, although dominating the market today, has proven to be inefficient in tackling the customer loyalty. In fact, related research extensively demonstrates that the correlation between service performance and user experience cannot be represented by such aggregations [6]. Hence, some CEM systems have introduced different elements of usage context (i.e. user location, terminal type, application, etc.) into their dashboards in order to filter and segment the user base into groups of users having common contexts and apply data mining to discover common service performance problems affecting that group's experience. However, in order to be able to truly follow the user experience, a per-user–per-session follow up is required which records and interprets the user interaction with the service. The ecosystem is then a fundamental element, together with models on how to structure the different service aspects around the user found in works like [5].

To this end, Big Data techniques are increasingly being introduced as a way to make appear correlations between the performances related data collected from the service and the subjective data collected from the users. Big Data applications to QoE constitute a promising research domain. Also, service models which look into a per-user–per-service–per-session granularity and integrating the usage context into the QoE indicators as opposed to the KQI aggregations can be of help in driving those applications to ease the complexity of monitoring and troubleshooting [7].

Summarizing, most of the leading CEM tools on the market evolve towards a more user centric service management. It is interesting to note that CEM products claiming more customer experience oriented metrics and analytics are those mostly collecting data from user terminals, and the collected data is mostly oriented to detect user behaviours than to measure actual service performance. On the other side of the spectrum, systems which claim having a more end-to-end view are collecting network wide data about service performance, while higher layers of customer experience analysis on those data is less developed.

A broad end-to-end view requires data collection in a comprehensive way from various interfaces at the different service components. Observe that in most cases, end-to-end views do not integrate subjective data, hence an additional effort has to be made in correlating the subjective observations at device/user level with measurements in the network, and the amount and complexity of data that needs then to be processed becomes a real challenge.

Moreover, a CEM system should also provide tools to manage the key elements to the transitions between the subjective and objective domains and between the operations and market domains, as the user expectations progress to become service offerings and these service offerings are provisioned. As already discussed in Chap. 2, the difference between the user expectations and the received service constitutes the customer experience, which at the same time integrates the user experience into the ecosystem above. Hence, this QoE toolset is an integral part of the QoE framework introduced in the beginning of the chapter.

7.4 Contractual and Non-contractual Obligations of the User Experience

Having discussed CEMs to some extent, we will now turn towards Service Level Agreements (SLAs) which are increasingly becoming an important part of the supply chain, attaining now the end user as a customer of the service. Within the ecosystem, the QoE of the user needs to be instantiated into the manifestations of the received experience through actions of the customer, with the ultimate objective of being able to predict and manage those actions: loyalty, increased spending, service recommendations, etc.

The problem of SLA definition and management has been around for some time, particularly in the Enterprise Market, where service contracts include almost by default an SLA section. Those SLAs have traditionally been defined in terms of service performance parameters that need to be met individually in order to consider the service to be of acceptable quality. On the other hand, this quality was rarely related to the "utility" of the service as such,[1] but rather related to an objectively measurable target that could be used to decide whether the contract is being honoured or not.

At least in recent years the actual utility of the provided service for the customer has more and more made its way into what is termed as Next-Generation SLAs (NG-SLA). The quality of the service provider business processes that are components of the service provided to the end user impact directly their experience, and as such their behavior as customers or their efficiency in the context of enterprise market. The metrics for NG-SLA management systems are then more related to business process efficiency than to technical performance itself. Figure 7.4 shows the lifecycle of such NG-SLAs: the user satisfaction translates into business process efficiency, which is in turn translated into increased revenue. A CEM system which integrates NG-SLA management provides the tools for defining the Objective Level Agreements (OLA) in terms of user satisfaction and the necessary translation of their monitoring into the reporting to the CXO (Customer Experience Officer) level in terms of financial impact, as well as the tools for translating OLAs into internal and external provider's SLAs.

Extending the SLAs to the user experience domain is another promising research domain, with a number of positive business outcomes. User efficiency impact in the business process efficiency can be quite easily modeled in the context of the enterprise market, since the customer aspect of the user is not present. When considering the

[1] As already pointed out in Chap. 2, the notion of a utility function, which maps resource provisioning to the related customer value, is a fundamental concept in microeconomics where it is used e.g. for maximizing overall social value, see [8] for further details.

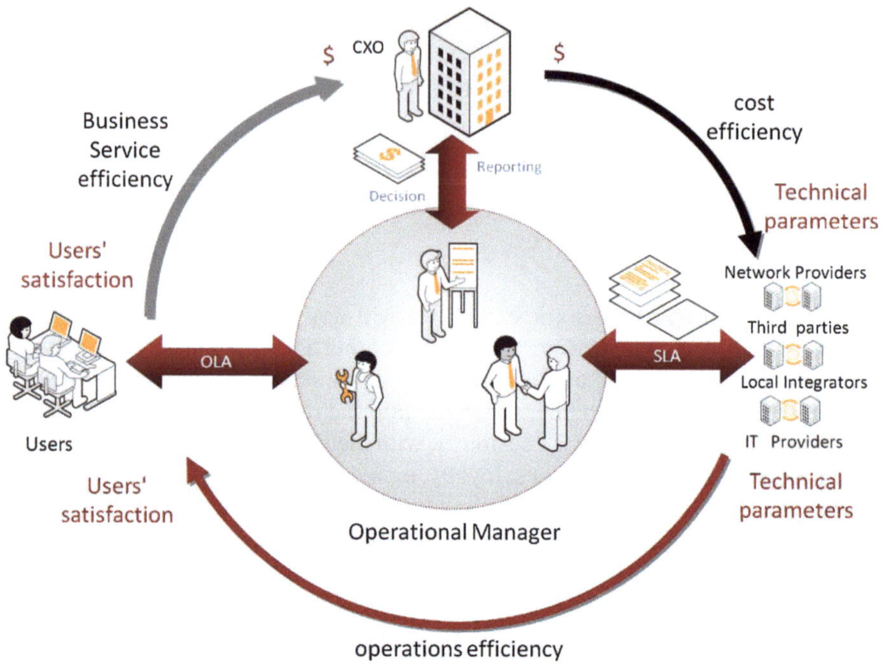

Fig. 7.4 Lifecycle of QoE-based service level agreements

individual user as a customer, the impact of satisfaction needs to take into account far more complex aspects as discussed above, making the outcome of a given experience into user's behavior more difficult to assess. This last element points out not only the difficulty of defining and managing user-experience SLAs, but also the importance of charging models as a way to influence the user behavior to obtain a given business outcome.

7.5 The Double Role of Charging and Corresponding Patterns of User Behavior

Having discussed the ecosystem, the system and the SLA level of QoE so far, in this section we will eventually focus on the question of how to charge for QoE. This issue is particularly important, as it is widely accepted that providing service quality is intrinsically tied to a corresponding differentiation of related pricing plans, and that lack of integration of the economic perspective into QoS architectures is among the key reasons for the notorious difficulties of introducing for instance the DiffServ concept into the current Internet [9, 10]. Thus, over the last couple of years the area of "Internet Economics" has succeeded in establishing itself as a vital research field

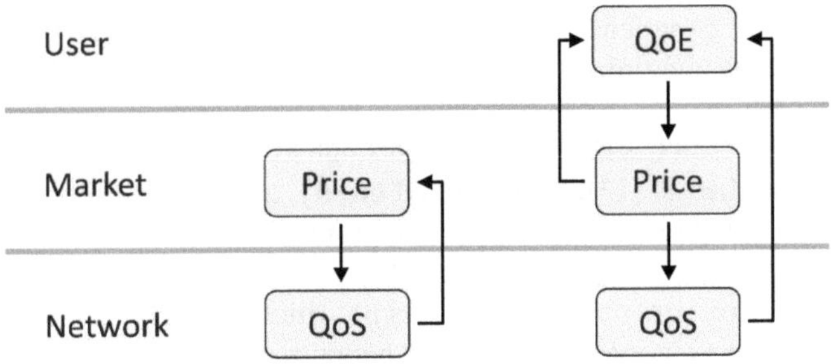

Fig. 7.5 QoS and QoE based charging

of its own, leading to a well-established set of proposals for pricing and charging differentiated services (for a comprehensive survey we refer to [11] and references therein).

However, while we consider the mentioned techno-economic ties to be invariant under the recent paradigm change from QoS to QoE, research on charging mechanisms for QoE-driven service differentiation, e.g. in the framework of a CEM system, is still in its infancies. One of the reasons for that is pointed out in [12]: the increasing complexity of the underlying techno-economic charging model, and specifically the double role of prices as sketched in Fig. 7.5. Simplifying the model proposed in [12], traditional QoS-based charging results in a feedback cycle between QoS provisioning on the networking layer and prices (Fig. 7.5 left), as on the one hand side higher prices reduce network load and at the same time increase provider revenues to be used for infrastructure investments, both effects leading to improved QoS, while on the other hand better QoS justifies to charge higher prices.

The situation is more complex for QoE-based charging (Fig. 7.5 right). Here, QoE is influenced by the QoS offered by the network, but also by a variety of other factors.[2] Hence, beyond its primary role as a parameter expressing the willingness-to-pay for the experienced quality, pricing becomes one of these additional context factors: the higher the tariff for a certain service quality, the higher the corresponding user expectations, which may automatically lead to lowering the Quality of Experience.

All in all, both charging models lead to fixed point problems whose solutions, however, are rather different from each other: as the mathematical analysis in [12] shows, the simpler case of QoS-based charging typically leads to two fixed points, a stable one for the case of high quality at high prices, and an unstable one where price and quality both are very low. In contrast, the above model for QoE-based charging typically leads to three fixed points: the first two are identical to the pair already identified above (low price–low quality, high price–high quality), however both of them being unstable, while the additional third fixed point is non-trivial and stable and can be interpreted as equilibrium where the price charged is in balance with the experienced

[2] Here, especially the manifold dimensions of the user context are worth being mentioned, including temporal factors and user characteristics, as pointed out already in [10]. Note, however, that for the purpose of clarity of the model, we restrict our present discussion to the role of pricing only.

quality (i.e. quality experience given the charge equals willingness-to-pay for this quality) and at the same time allows to provision sufficient QoS in order to enable the necessary QoE.

This analytical result has been validated through a comprehensive user trial, based on the setup described in [13]. Summarized briefly, a total of 40 representative test users have been confronted with a very fine granular selection of 20 video quality levels (in terms of transmission bit rates) linked to corresponding pricing levels. Users were given real money (10 Euros each) which they could freely spend for watching up to three videos (20 min each) in better quality or alternatively take home with them after the trial. In order to simplify the user interaction with the system, a jog wheel has been used which is well known e.g. from volume control and allows to rapidly switch between quality/tariff levels (which have been realized almost in real-time, i.e. with a delay of at most 1 s).

As a result of this experiment, during the initial free trial period of 5 min we have observed that all subjects used the possibility to adapt the video quality, usually between 10 and 50 times, with some individual cases going up to around 85 adaptations (i.e. one change every 3.5 s on average). More than 80 % of the users have been clearly exhibiting a behavior in line with the above fixed point model, and typically were following a convergence pattern closely reminding dampened oscillation. A more detailed investigation has revealed that users may be classified into three fundamental categories:

- User type "F" is characterized by a very fast convergence behavior: users sequentially climb up the quality/tariff levels until they reach their equilibrium level where they remain for the rest of the video.
- User type "S" exhibits slow convergence: users explore several times the entire space of quality/tariff options (hence a large amplitude of changes) before converging at a relatively late point in time.
- User type "R" follows a more regular pattern: user start with an exploration of the entire quality/tariff space, but immediately afterwards start reducing their amplitudes and step by step approach their equilibrium level.

Users were distributed almost equally over these three classes, except for around 15 % of them who either followed a free-riding behavior (best quality during free trial phase, worst quality without paying fees afterwards) or other inconsistent patterns (e.g. small initial oscillations plus a huge up and down later in the trial phase). For further details on this experiment, including a quantitative algorithm for automatic user classification based on root mean square deviations, we refer to [12].

7.6 Conclusions

The present chapter has been devoted to economic and business aspects of Quality of Experience, which we have addressed from different levels of abstraction. On the highest level, we have presented an ecosystem model which nicely clarifies the roles of different actors and stakeholders within the value chain of producing and consuming communication services. Based on that, we have discussed a couple of key aspects of related Customer Experience Management (CEM) systems and their ability to reflect user experience in a satisfying way. In a next step, moving even closer to the customer, we have analyzed the evolution of Service Level Agreements (SLA) from their traditional form of contractual description of QoS parameters towards the integration of user satisfaction and business efficiency. Finally, on the most fine-grained level, we have analyzed the double role of pricing in a QoE context where a price does not only reflect the value of a product or service, but at the same time has direct impact on user expectations and thus on the evaluation of the service quality as such.

Summarizing, it turns out that the paradigm change from QoS to QoE has a profound impact on all those layers, which clearly highlights the necessity of closely integrating the economic dimension into the general QoE research agenda. More specifically, future work will range from more detailed models of the overall ecosystem over using upcoming Big Data technologies for CEMs with a per-user per-service per-session granularity to further extending the range of NG-SLAs, while at the same time additional efforts will be necessary to better understand the role of user context factors (pricing being one amongst many others) on QoE valuation and evaluation.

References

1. Perkis A (2013) Quality of experience. SPIE Newsroom, 27 Feb 2013. http://spie.org/x92222.xml, doi:10.1117/2.1201302.004591
2. Le Callet P, Möller S, Perkis A (eds) (2012) Qualinet white paper on definitions of quality of experience–output version of the Dagstuhl seminar 12181. European network on quality of experience in multimedia systems and services (COST Action IC 1003). Lausanne, Version 1.1, June 2012
3. Perkis A (2013) QoE cross layer approach to model media experiences. IEEE COMSOC MMTC E-Letter 8(2):6–9. http://committees.comsoc.org/mmc/e-news/E-Letter-March13.pdf
4. Kilkki K (2008) Quality of experience in communication ecosystems. J Univ Comput Sci 14(5):615–624
5. De Moor K (2012) Are engineers from Mars and users from Venus? Bridging gaps in quality of experience research: reflections on and experiences from an interdisciplinary journey. Ghent University, Faculty of Political and Social Sciences, Ghent, Belgium
6. Ickin S, Wac K, Fiedler M, Janowski L, Hong J, Dey AK (2012) Factors influencing quality of experience of commonly used mobile applications. IEEE Commun Mag 50:48–56
7. Guerzoni R, Fontana C, Beker S, Soldani D (2013) A user centric troubleshooting framework for current and future networks. Wireless World Research Forum Meeting 30, Oulu, Apr 2013
8. Courcoubetis C, Weber R (2003) Pricing communication networks: economics, technology and modelling. Wiley, New Jersey
9. Jain R (2006) Internet 3.0: ten problems with current internet architecture and solutions for the next generation. In: Proceedings of IEEE/AFCEA MILCOM 2006, Washington, Oct 2006
10. Reichl P, Hammer F (2006) Charging for quality-of-experience: a new paradigm for pricing IP-based services. 2nd ISCA tutorial and research workshop on perceptual quality of systems, Berlin, Germany, pp 171–177

11. Reichl R (2010) From charging for quality-of-service to charging for quality-of-experience. Ann Telecommun (special issue on Quality of experience and socio-economic issues of network-based services) 65(3):189–199
12. Reichl P, Maillé P, Zwickl P, Sackl A (2013) A fixed-point model for QoE-based charging. In: Proceedings of ACM SIGCOMM 2013 workshop on future human-centric multimedia networking, Hong Kong, China, Aug 2013
13. Sackl A, Zwickl A, Reichl P (2013) QoE Alchemy 2.0: an improved test setup for the pecuniary bias of QoE. In: Proceedings of 5th International Workshop on Quality of Multimedia Experience (QoMEX' 13), Klagenfurt, Austria, July 2013

Chapter 8
Brain Activity Correlates of Quality of Experience

Jan-Niklas Antons, Sebastian Arndt, Robert Schleicher and Sebastian Möller

Abstract This chapter outlines common brain activity correlates that are known from neuroscience, gives an overview on established electrophysiological analysis methods and on the background of *electroencephalography* (*EEG*). After that an overview on study designs will be given and a practical guideline for the design of experiments using *EEG* in the research area of Quality of Experience (QoE) will be presented. At the end of this chapter we will close with a summary, give practical advice, and we will outline potential interesting future research topics.

8.1 Introduction

Experiences and therefore also *Quality of Experience* are subjective constructs (see Chaps. 2 and 3 for the QoE definition and its subjective nature) that are not directly observable by others. Most researchers nowadays agree that quality judgment processes happen inside and more specifically in the brain of the people who consume media. In the case of a qualitative experience this process which happens inside of the recipient is described in Chap. 2 as quality formation process (original description see Chaps. 2 and 3 of the Qualinet White Paper on QoE [1]). For these internal processes it is—as the brain is the central organ for information processing in humans—likely that changes in subjective experiences will also be reflected by

J.-N. Antons (✉) · S. Arndt · R. Schleicher · S. Möller
Quality and Usability Lab, Telekom Innovation Laboratories, TU Berlin, Berlin, Germany
e-mail: jan-niklas.antons@tu-berlin.de

S. Arndt
e-mail: sebastian.arndt@telekom.de

R. Schleicher
e-mail: robert.schleicher@alumni.tu-berlin.de

Sebastian Möller
e-mail: sebastian.moeller@telekom.de

S. Möller and A. Raake (eds.), *Quality of Experience*,
T-Labs Series in Telecommunication Services, DOI: 10.1007/978-3-319-02681-7_8,

certain brain activity patterns, be it neuronal or humoral (blood flow related). Please note that this chapter is entitled 'brain activity correlates', as we do not claim that changes on a physiological level are the foundations of subjective experience or even their 'true' representation. For the relation between psychological phenomena and physiological changes see Cacippio et al. [2].

This chapter deals with physiology following a distinction common in the field of psychophysiology between *central* nervous activity (i.e. the brain) and *peripheral* nervous activation, which summarizes all other possible recording sites [2]. The latter will be addressed in Chap. 9, while the present one focuses on brain-based measurements. There are various methods to monitor the neuronal activity of the brain, which can roughly be divided into *hemodynamic* measures that are based on changes in blood flow supposed to be indicative of changes in neuronal activity, and *electrophysiological* measures, which directly represent electromagnetical changes due to neuronal activity. A detailed overview of different neuronal measurement methods can be found in [3]. *Functional magnetic resonance imaging* (*fMRI*) and *positron emission tomography* (*PET*) are hemodynamic measures. While they allow for a three dimensional recording of the brain at work including its deeper structures with a high spatial resolution, their application requires substantial resources in terms of manpower and equipment. That is probably why they have, to our knowledge, not been used for quality-related research so far, next to the fact that lying in respective scanners and the loudness of the measurement itself are not a very well suitable setup to assess Quality of Experience. *Near-infrared spectography* (*NIRS*), another brain imaging technique, is less obtrusive and has a deficit that is persistent for all *hemodynamic* measures which is a low resolution in the temporal domain. Changes in response to e.g. fast changes in quality cannot be assessed. The trade-off between increased obtrusiveness and additional insights is apparently best met with electrophysiological measures, namely *electroencephalography* (*EEG*) in this context—and so the majority of studies regarding brain activity and *QoE* apply this method.

The remainder of this chapter is structured as follows; first an overview on the main established analysis methods and the background of EEG will be given in Sect. 8.2, followed by an overview of study designs and the presentation of a practical guideline for the design of experiments using EEG in the research area of Quality of Experience in Sects. 8.2.1.2 and 8.2.2.2, respectively. At the end of this chapter we will close with a summary of practical advice (Sect. 8.4) and will outline potential interesting future research topics (Sect. 8.5).

8.2 Electroencephalogram

The *electroencephalogram* (*EEG*) measures voltage variation due to neuronal activity in the brain by attaching electrodes to the scalp of a subject. Since its discovery by Berger in 1929, it has become a widely used method for investigating physiological correlates of perceptual and attentional processes [4, 5]. This measure has a rather limited spatial resolution—based on the fact that the brain is a wet conductor, the recorded signals by one electrode is a mixture of all existent sources—but an excellent temporal resolution with a precision of milliseconds. The corresponding data can mainly be analyzed in two different ways: on the one hand by having a closer look at the spectrogram of spontaneous activity, and on the other hand by analyzing so called *Event-Related-Potentials* (*ERP*) which are a time-locked reaction to an external stimulus in terms of a voltage change [6]. The latter approach can be used to analyze cortical potentials as well as voltage differences evoked in the brain stem. In this text we will focus on the cortical brain activity and just briefly mention brain stem measurements, because research on brain stem level is not yet fully usable in QoE-research in terms of degradation classes and length of stimuli. In addition, the signal-to-noise-ratio for this kind of measure is so high that the stimuli have to be presented numerous times, and due to the resulting experimental setup only few stimulus classes could be presented per experiment.

Beside the relevant information—brain activity—lots of unwanted information is recorded as well, e.g. voltage changes due to eye-movement, body movement and other unrelated signal sources. Due to the high noisiness of the signal, it is important to create highly controlled experimental setups. Clinical research guidelines for experimental designs already exist, and we want to outline important implications for research in the domain of Quality of Experience based on them [5].

In the following sections we will describe the principles of how to analyze *continuous and evoked EEG-data*; how the two ways of analyzing the EEG-signal are performed and how these techniques were already used for studies concerning Quality of Experience.

Lately new low-cost EEG devices have appeared on the market, such as the Emotiv-EPOC[1] and NeuroSky MindWave[2] headsets. Though these consumer products are comparably inexpensive, the data quality, i.e. precision and noisiness, of those products is worse compared to the devices used in clinical applications. However, these products have shown to capture useful information in the context of QoE-related research. Moldovan et al. [7] used the features provided by the Emotiv EPOC System to infer the level of frustration from the human observer caused by the quality of the played audiovisual excerpt. This level was determined by using a metric predefined by the headset manufacturer. In their study videos with different levels of quality were used, they manipulated the bitrate, frame rate as well as resolution of the presented video clips. Perez et al. [8] used the NeuroSky MindWave headset to measure brain activity and used the recorded data to classify the trials into high and low quality pictures.

8.2.1 Continuous EEG

In the *continuous EEG*, five main different frequency ranges are ascribed to specific states of the brain: *delta band* (1–4 Hz), *theta band* (4–8 Hz), *alpha band* (8–13 Hz),

[1] http://www.emotiv.com/

[2] http://www.neurosky.com/

beta band (13–30 Hz) and the *gamma band* (36–44 Hz) [9]. The *delta band* is present during deep sleep of subjects, the *theta band* during light sleep and is an indicator for decreased alertness. Activity in the *alpha band* is related to relaxed wakefulness with eyes closed and decrease in alertness. *Beta and gamma band* are ascribed to high arousal and focused attention [6].

Analyzing the power in the afore-mentioned frequency bands is widely done for assessing the *cognitive state* of car drivers. Lal et al. [10] for example showed that fatigued drivers had an elevated power in the delta and theta bands. Correlation between weighted combinations of the power of different frequency bands with subjective fatigue ratings was shown in [11].

Another reason to use frequency bands is to estimate the emotional state of subjects. Therefore, alpha values from frontal electrodes are being extracted. The asymmetry index, for example, is one way to obtain this information. It shows that higher values in the asymmetry index are the result of higher left frontal activity which is due to rather negative emotional processing [12].

8.2.1.1 Data Recording and Analysis

As the continuous signal is not related to one short single event, this method is suitable for stimuli of longer duration. Usually the analysis intervals are between 5 and 10 min long and set in relation to a baseline interval from 2 to 5 min, resulting in a baseline corrected power value.

For this analysis method a small set of electrodes is used, commonly up to 8 electrodes are sufficient and should be distributed at occipital/parietal scalp locations for attention and fatigue studies and frontal for asymmetry index studies following the *10/20 system*, which ascribes electrode position based on the ratio distance from the center point on the scalp [13]. Most interesting for a possible deployment in applied contexts is the use of a single electrode, minimizing preparation time and making the application more comfortable for the subjects. Less electrodes result in less information in terms of spatial distribution which also limits the possibility of dealing with noise (e.g. independent component analysis) and dipole source estimations based on the reduced spatial information.

To determine the asymmetry index, the relationship between left frontal and right frontal activity needs to be calculated; this is done by using the corresponding alpha proportions: $(ln(alpha_{right}) - ln(alpha_{left}))$ as proposed by Coan et al. [12].

8.2.1.2 Findings Related to QoE

Due to the possibility that more natural stimuli in terms of stimulus length can be used, it is possible to examine the effect of longer lasting media stimuli (>10 min) on the recipients (e.g. [14]). In this study, subjects were exposed to high quality and low quality sequences of auditory or audiovisual material. Their only task was to rate the content on a scale every few minutes, the rest of the time they should only focus on the presented content. In both setups, auditory and audiovisual, it was shown that higher values in the *alpha band* power resulted when being exposed to low quality stimuli as compared to higher quality (or reference) stimuli, which is ascribed to fatigue and impaired information processing [14, 15]. In an additional study, the impact of a high quality audio segment (5 min) inserted in a low quality audio stimulus (15 min) was assessed. Subjects got less fatigued due to the better audio quality as indicated by a lower *alpha band* power [16].

In another study alpha values extracted from frontal electrodes were also used in order to assess the emotional state of test participants. It could be shown that higher left frontal than right frontal activity, hence an increased asymmetry index, was present in case where subjects were exposed to low quality stimuli, which indicates a rather emotional negative processing of these stimuli, in contrast to higher quality stimuli. Respective correlations to subjective scores were shown as well [17]. The presented results indicate a correlation of the extracted parameter—brain pattern related to attention/fatigue (arousal) and pleasantness (valence)—and subjective QoE, such as subjective emotional self-assessment and quality ratings.

8.2.2 Evoked Potentials

In contrast to the continuous data analysis, a precise timing is essential when it comes to the analysis of *evoked potentials*. The commonly called *Event-Related-Potentials* (*ERP*) are the resulting brain activity following a certain stimulus or after a certain class of stimuli. The standard components of the *ERP* are named after their polarity ("N" for negative, "P" for positive) and the time of their occurrence after stimulus onset. For example the third positive component occurring after the stimulus is named *P3* or based on the passed time after stimulus presentation *P300* (approximately 300 ms after stimulus onset; see Fig. 8.1).

Components of *ERPs* are hard to distinguish and can only roughly be related to specific neural processing stages: Early differences are commonly based on sensory processing and later differences are due to cognitive processes, such as triggered during a detection task.

The *mismatch negativity* (*MMN*), which can be observed 100–250 ms after stimulus onset, is a measure of low-level visual and auditory memory (see [5] and [18]). Differences between the currently processed stimulus and previously ones are automatically detected by a momentary internal sensory reference [19].

The *P300* component and later ones are ascribed to higher cognitive processes and can be split into two parts: *P3a and P3b*. *P3a* is evoked whenever a mismatch between newly perceived information and internal memory copies on a cognitive level are noticed. The P3b component is related to processes of task-related attention. In general, the P300 is commonly elicited using the *oddball paradigm* where a deviant stimulus is presented among a series of more frequent "regular" stimuli, e.g., a high tone among a repeated series of low tones. Polichs [20] review gives

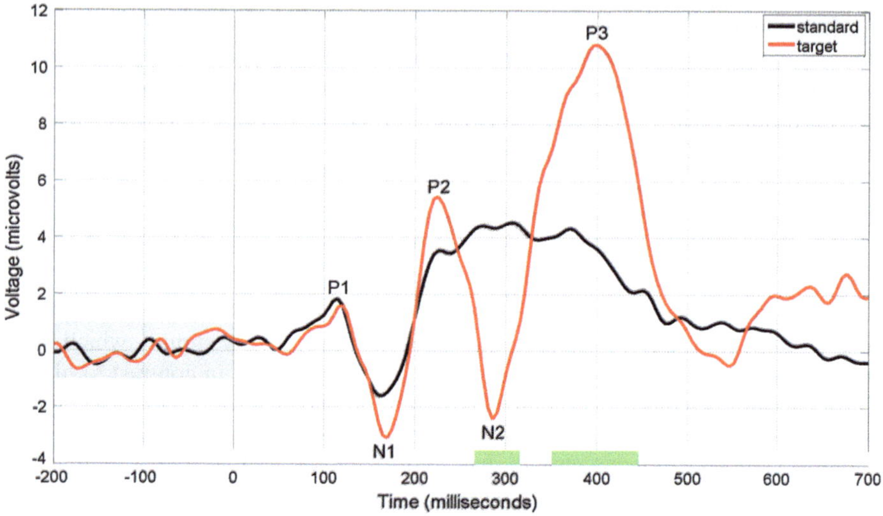

Fig. 8.1 Two *Event-Related Potentials (ERPs)* evoked by auditory stimuli (grand average, electrode Cz). *Oddball paradigm* with two tones as stimuli was used; standard (440 Hz beep tone for 150 ms, probability of 80 %, 480 repetitions) and target (1,000 Hz beep tone for 150 ms, probability of 20 %, 120 repititions), the subjects had to click a button as fast as possible deciding which tone they heard last. Inter-stimulus interval was set to 1,500 ms and sampling rate to 200 Hz. Data was band pass filtered (0.1–40 Hz), and only correctly identified trials were used for display. The *gray bar* indicates the time interval used for baseline correction (−200 to 0 ms: where 0 ms is the stimulus onset), and the *green* interval indicates the intervals where the two ERPs are significantly different, running t-test with adjusted $p < 0.05$. Components are marked as P1-3 for occurring positive peaks and N1-2 for negative peaks

background information on all relevant processes of *P300, P3a, and P3b* components [20]. Practical advice for measuring *MMN, P300, and N400* is given in [5].

8.2.2.1 Data Recording and Analysis

For the analysis of ERPs, a small set of electrodes can be sufficient, usually up to 8 electrodes; they should be distributed along the central line following the *10/20 system* [13], and for hemispheric differences equally distributed electrodes over the right and left hemisphere are advisable. More electrodes are needed for the analysis of more complex patterns e.g. spatial pattern distribution.

As evoked potentials depend on an exact timing, it is important that triggers are exactly synchronized to be able to average the signal while keeping the temporal information intact. *ERPs* cannot be observed in the raw *EEG*, as they are overshadowed by other, unrelated activity, which disappears when averaging several trials of single *ERP* recordings.

Usually 20–30 trials as minimum are needed for an average ERP per stimulus class; baseline averaged using the average value of the voltage in the interval usually up to 200 ms before the stimulus. This rather high number of trials compared to standard quality tests also explains the usually small number of subjects used for EEG-studies.

The aforementioned averaging methods are performed usually offline and as an average over a group of subjects. This average over all subjects is the grand average and is the result which is often plotted in these studies. Using classification techniques this can be transferred to an online analysis of incoming physiological signals, to decide whether the occurring brain activity of the proceeding stimulus was evoked by one special class of stimuli [21]. In the case of Quality of Experience an exemplary class of degradation can serve for that. With classification as a measure of separability, it can be distinguished between perceived stimulus classes. For a tutorial on single-trial *ERP* classifications see [22].

8.2.2.2 Findings Related to QoE

A first study using classes of degradations which are of interest for research in telecommunication industry, was conducted by Miettinen et al. [23] using *magnetoencephalography* (*MEG*), where they could show a significant increase in the measured amplitudes for distorted stimuli.

One of the first studies using EEG for *quality assessment* was conducted by Antons et al. [24] in the auditory domain, where signal-correlated noise was introduced in the stimuli, and the resulting signal-to-noise ratio was the independent and scalable variable [24]. Here, after each presentation subjects had to judge whether they perceived a distortion in the current stimulus or not. In this domain, the first paradigm using *EEG* in a *QoE* context was derived starting with meaningless syllables and developing it up to words. In each of the experiments it could be shown that the elicited *P300* gets significantly higher, the more distorted the stimulus is [25]. Additionally, the *P300* occurs earlier with stronger distortions. Furthermore, it could be shown that stimuli which were perceived as undistorted by the participants, but were distorted on the signal level, had a similar trend in the *ERP* as trials where the stimuli were rated as distorted by the participant. Thus, high classification rates for these trials could be obtained, and it was concluded that these degradations are presumably processed non-consciously as they do not penetrate up to the subjective behavior [24].

This measurement technique was further developed for (audio-)visual stimuli. In studies conducted by Arndt et al. [28], the *two-alternative forced choice* (*2AFC*) approach was used, which is a reduced implementation of the *double stimulus continuous quality scale* (*DSCQS*) method (see [26]). Here, a first video sequence with the reference stimulus was immediately followed by a possibly distorted one. As a distortion in these experiments artificial blockiness was introduced and varied in block length. After each trial, subjects had to tell whether they perceived a distortion in the second part or not. The findings from the previous auditory studies could be repeated, and the same relation for visual stimuli was shown: the *P300* is more distinct with more distorted stimuli [27]. In a next step bi-modal stimuli were introduced using the

2AFC paradigm [28]. Besides the already established relationship of P300 amplitude and distortion level, in this study a significantly high relation between *Mean Opinion Score* (*MOS*) [29] and obtained *P300* amplitudes was observed. In other, purely visually based studies Scholler et al. confirmed this finding and additionally could show that the *ERP* of stimuli not perceived as degraded on the subjective level could be identified similar to the ones perceived as degraded [30]. Another study using visual stimuli was conducted by Lindemann et al. [31]. Here, rather different kinds of distortions than the intensities were examined. They could also show high classification rates between distorted and undistorted stimuli with the obtained data.

New technologies such as e.g. 3D videos can also be examined regarding their quality and visual discomfort. This was done in a study by Li et al. [32], where the authors could show a higher visual discomfort (1) while watching 3D contents versus 2D contents and (2) while watching longer 3D sequences versus shorter ones [32]. The reproducibility of the results among independently working laboratories suggests that *EEG* is a reliable complementary measurement method to assess or underline *QoE* related judgments.

The correlation of the P300 component and MOS (as a QoE-related metric) was shown in several studies. Furthermore, there are brain patterns which are not correlated to QoE directly but indicate an effect on QoE influencing factors such as the cognitive state (see Chaps. 2 and 3).

8.3 Summary

The physiological basis of auditory and visual perception is well defined and guidelines for the neuronal measurement and analysis of such data have been established. In this chapter we showed two different paradigms: (1) frequency power analysis of the *continuous EEG*-data and (2) the analysis of components of event related EEG-data. For both approaches, initial successful applications to *QoE*-relevant stimuli were described and correlations of QoE and brain activation patterns were shown. To facilitate further deployment of this measure, we will now give practical advice based on our experience.

8.4 Practical Guidelines

1. Use short stimuli to get a clear picture of the *ERP* variations of interest when you start working with a new stimulus type. If possible, select the stimulus such that you have a good onset, meaning the onset of audio or video stimuli is in the beginning of a recording. Be aware that audiovisual speech stimuli have rarely a simultaneous beginning.
2. Start with a minimal stimulus set: (1) concerning stimulus length, perform tests with short stimuli before you aim for longer ones; (2) use only a reduced set of

speakers and sentences for auditory and a reduced set of scenes for (audio)visual experiments, respectively; (3) use one class of degradation rather than several classes.

3. Select only a few levels of degradation (e.g. three noise levels) instead of the full spectrum in order to reduce expenditure of time. Hence, determine individual levels of degradation intensity for each subject. It is best to aim for a similar percentage of detected versus non-detected levels of degradation for each subject.
4. Control the experimental environment closely. If available, use e.g. ITU Recommendations (such as [29]), and if appropriate reduce the suggested setup (see 1 and 3).
5. Use one of the established setups for presentation, e.g. oddball paradigm with short stimuli [33].
6. Adhere to established analysis paradigms in the beginning—data on brain activity tend to be overwhelming and polysemous, as they represent a variety of influences next to the ones you were interested in with your study. The established approaches developed over the years try to rule out as many of those variations as possible.

8.5 Future Research Topics

Studies so far have concentrated on signals after stimulus presentation. In addition it would be interesting if data—obtained by physiology measurement methods—can enable better quality prediction on a single-trial basis. This could be done by using the neuronal signal preceding the stimulus for an estimation of the impact on perceived quality. More specifically, the findings reported so far always focused on analyzing processes that occur within the subject *during* the perceptive and the descriptive event of quality assessment. While these methods can deliver useful additional information as shown, they do not take into account which impact the *QoE* influencing factors such as the *cognitive state* of the listener have on the judgment. Neuroscience studies show that it is possible to detect not only the emotion and neuronal response triggered by stimulus presentation but that the methods can also be used to assess the general cognitive state *prior to presentation*. In simple words it could be measured how the current state of a subject, be it emotional or cognitive, influences the process of forming a quality judgment. These results could lead to a better understanding on how the current state of subjects influences the subjective judgment.

One research field that was not extensively studied before in the context of Quality of Experience is the field of *brain stem* measurements which will be interesting for future research. Basically, this method is similar to the *ERP* procedure and measures voltage differences due to neuronal activity. In contrast to *ERP*, this activity is produced by the brain stem and is different to the recorded ERP signal in terms of (1) temporal behavior, as the signal emitted by the brain stem is measurable milliseconds after stimulus onset [34] and (2) in terms of strength, which is much smaller. Especially worth mentioning is the work of Nina Kraus' group who could show that for example musical experience has an influence on the processing of information already on the brain stem level [35]. It would be interesting to see whether quality expectations also come into play on this early level of perception. Another neurophysiological measure, *NIRS*, is showing promising preliminary results in the domain of *QoE* related research. Here, the differences of oxygenated and deoxygenated blood are recorded. In a first experiment using auditory stimuli significant correlations between recorded *NIRS* features and scored subjective ratings could be shown [36].

References

1. Le Callet P, Möller S, Perkis A (eds) (2013) Qualinet white paper on definitions of quality of experience. European network on quality of experience in multimedia systems and services (COST Action IC 1003), Lausanne, Version 1.2, Novi Sad, March 2013
2. Cacippio J, Tassinary L, Berntson G (eds) (2007) Handbook of psychophysiology. Cambridge University Press, Cambridge
3. Parasuraman R, Rizzo M (eds) (2008) Neuroergonomics: the brain at work. Oxford University Press, Oxford
4. Berger H (1929) Über das Elektrenkephalogramm des Menschen. Arch f Pschiatr 87:527–570
5. Duncan C, Barry R, Connolly J, Fischer C, Michie P, Näätänen R, Polich J, Reinvang I, Petten C (2009) Event-related potentials in clinical research: guidelines for eliciting, recording, and quantifying mismatch negativity, P300 and N400. Clin Neurophysiol 120:1883–1903
6. Coles MS, Rugg M (1995) Event-related brain potentials: an introduction. In: Coles MS, Rugg M (eds) Electrophysiology of mind: event-related brain potentials and cognition. Oxford University Press, Oxford
7. Moldovan AN, Ghergulescu I, Weibelzahl S, Muntean CH (2013) User-centered EEG-based multimedia quality assessment. In: Proceedings of international symposium on broadband multimedia systems broadcasting
8. Perez J, Deléchelle E (2013) On the measurement of image quality perception using frontal EEG analysis. International conference on smart communications in network technologies (SaCoNeT)
9. Pizzagalli DA (2007) Electroencephalography and high-density electrophysiological source localization. In: Cacippio J, Tassinary L, Berntson G (eds) Handbook of psychophysiology. Cambridge University Press, Cambridge
10. Lal S, Craig A (2005) Reproducibility of the spectral components of the electroencephalogram during driver fatigue. Int J Psychophysiol 55:137–143
11. Punsawad Y, Aempedchr S, Wongsawat Y, Panichkun M (2011) Weighted-frequency index for EEG based mental fatigue alarm system. Int J Appl Biomed Eng 4(1):36–41
12. Coan JA, Allen J (2004) Frontal EEG asymmetry as a moderator and mediator of emotion. Biol Psychol 67(1):7–50
13. American clinical neurophysiology society (2006) Guideline 5—guidelines for standard electrode position nomenclature. J Clin Neurophysiol 23(2):107–110
14. Antons JN, Schleicher R, Arndt S, Möller S, Curio G (2012) Too tired for calling? A physiological measure of fatigue caused by bandwidth limitations. In: Proceedings of the quality of multimedia experience (QoMEX)
15. Arndt S, Schleicher R, Antons JN (2013) Does low quality audiovisual content increase fatigue of viewers? In: Proceedings perceptual quality of systems (PQS)
16. Antons JN, Köster F, Arndt S, Möller S, Schleicher R (2013) Changes of vigilance caused by varying bit rate conditions. In: Proceedings of the quality of multimedia experience (QoMEX)
17. Arndt S, Antons JN, Gupta R, Laghari K, Schleicher R, Möller S, Falk TH (2013) The effects of text-to-speech system quality on emotional states and frontal alpha band power. In: Proceedings of the EMBS neural, engineering conference

18. Näätänen R (2008) Mismatch negativity (MMN) as an index of central auditory system plasticity. Int J Audiol 47:16–20
19. Garrido M, Kilner J, Stephan K, Friston K (2009) The mismatch negativity: a review of underlying mechanisms. Clin Neurophysiol 120:453–463
20. Polich J (2007) Updating P300: an integrative theory of P3a and P3b. Clin Neurophysiol 118(10):2128–2148
21. Mustafa M, Guthe S, Magnor M (2012) Single trial EEG classification of artifacts in videos. ACM Trans Appl Percept 9(3):1–15
22. Blankertz B, Lemm S, Treder MS, Haufe S, Müller KR (2011) Single-trial analysis and classification of ERP components—a tutorial. Neuroimage 56:814–825
23. Miettinen I, Tiitinen H, Alku P, May P (2010) Sensitivity of the human auditory cortex to acoustic degradation of speech and non-speech sound. BMC Neurosci 11(24):1471–2202
24. Antons JN, Schleicher R, Arndt S, Möller S, Porbadnigk AK, Curio G (2012) Analyzing speech quality perception using electroencephalography. J Select Topics Signal Proc 6(6):721–731
25. Antons JN, Porbadnigk A, Schleicher R, Blankertz B, Möller S, Curio G (2010) Subjective listening tests and neural correlates of speech degradation in case of signal-correlated noise. In: Proceedings of the audio engineering society (AES)
26. ITU-T Recommendation BT.500-13 (2012) Methodology for the subjective assessment of the quality of television pictures. International Telecommunication Union, Geneva
27. Arndt S, Antons JN, Schleicher R, Möller S, Scholler S, Curio G (2011) A physiological approach to determine video quality. In: Proceedings of the international symposium on multimedia (ISM)
28. Arndt S, Antons JN, Schleicher R, Möller S, Curio G (2012) Perception of low-quality videos analyzed by means of electroencephalography. In: Proceedings of the quality of multimedia experience (QoMEX)
29. ITU-T Recommendation P.800 (1996) Methods for subjective determination of transmission quality. International Telecommunication Union, Geneva
30. Scholler S, Bosse S, Treder MS, Blankertz B, Curio G, Müller KR, Wiegand T (2012) Towards a direct measure of video quality perception using EEG. IEEE Trans Image Process 21(5):2619–2629
31. Lindemann L, Wenger S, Magnor M (2011) Evaluation of video artifact perception using event-related potentials. In: Proceedings of the applied perception in computer graphics and visualization (APGV)
32. Li HC, Seo J, Kham K, Lee S (2008) Measurement of 3D visual fatigue using event-related potential (ERP): 3D oddball paradigm. In: 3DTV conference: the true vision-capture, transmission and display of 3D Video, pp 213–216
33. Luck S (2004) Ten simple rules for designing ERP experiments. In: Handy T (ed) Event-related potentials: a methods handbook. MIT Press, Cambridge
34. Roeser R, Valente M, Hosfort-Dunn H (2007) Audiology diagnosis. Thieme, Stuttgart
35. Wong P, Skoe E, Russo N, Dees T, Kraus N (2007) Musical experience shapes human brainstem encoding of linguistic pitch patterns. Nat Neurosci 10(4):420–422
36. Gupta R, Rehman K, Arndt S, Schleicher R, Möller S, O'Shaughnessy D, Falk T (2013) Using fNIRS to characterize human perception of TTS system quality, comprehension, and fluency: preliminary findings. In: Proceedings of the perceptual quality of systems (PQS)

Chapter 9
Evoking Emotions and Evaluating Emotional Impact

Robert Schleicher and Jan-Niklas Antons

Abstract This chapter gives an overview for Quality of Experience (QoE) practitioners on common setups in emotion research using audio (sounds), visual (pictures) and audiovisual (video clips) stimulus material to induce emotions. After presenting available databases for the different modalities, methods for subsequent as well as continuous self-assessment are discussed. Next to self-assessment, analysis of accompanying physiological changes is a common means to evaluate emotional responses. Here, typical measures of peripheral physiology are summarized. Finally, practical advices for including material with emotional content and recording physiological signals in experiments on audiovisual quality are given, and future research directions are outlined.

9.1 Introduction

Emotions have been mentioned several times in this book, for example in Chaps. 2 and 3 on QoE versus User Experience, or Chap. 4 on factors influencing QoE. There are at least two ways emotions can affect QoE: First of all, a stimulus may, as intended, evoke (among other things) an emotion in the recipient, and the QoE researcher intends to assess this emotional impact. Here, we can draw from the experience in related domains where standards to evaluate emotions and feelings have been established. Second, a stimulus may cause an emotion in the recipient due to its meaning, but this time it is not in the focus of the experiment, and in fact the emotion may influence other parameters relevant for the study, e.g. the assessment of image resolution (described in Chap. 6). In this case, it is important for researchers to know

R. Schleicher (✉) · J.-N. Antons
Quality and Usability Lab, Telekom Innovation Laboratories, TU Berlin, Berlin, Germany
e-mail: robert.schleicher@tu-berlin.de

J.-N. Antons
e-mail: jan-niklas.antons@tu-berlin.de

S. Möller and A. Raake (eds.), *Quality of Experience*,
T-Labs Series in Telecommunication Services, DOI: 10.1007/978-3-319-02681-7_9,

what stimuli can be considered as emotionally neutral, or even better, still measure its emotional connotation to account for it in later analyses. Thus, this chapter will focus on ways to systematically influence the subject's emotions via audiovisual stimuli, and methods to measure this effect. It is structured as follows: First, we will describe available stimulus material to evoke emotions. Next, common instruments for self-assessment of emotions are presented. After that, a brief summary on concomitant physiological changes is provided, as these are sometimes considered the more 'objective' way to measure emotions as they are difficult to manipulate voluntarily in contrast to self-assessment. Finally, we will give practical advises when studying emotions in the context of QoE and point out future directions of research.

9.1.1 Background

The study of emotions has a long tradition in psychology and related disciplines, going back to Wilhelm Wundt, one of the founding fathers of experimental psychology, who already proposed three dimensions to structure emotions [1], one of them being *quality*, which ranges from pleasure to displeasure—the resemblance to the Qualinet definition of QoE given in the Qualinet White Paper as "the degree of delight or annoyance of the user of an application or service" is perplexing (see Chap. 2).

However, as much as the controversy whether emotions are fundamentally organized along dimensions or in categories is still not settled among researchers, it is also not clear what has to be considered as the 'core' element of emotions: Is it the concomitant bodily changes, the upcoming intention to somehow react to the emotional stimulus, or its impact on our thinking, including the shift in attention? Animal research appears to focus on physiological changes and behavioral tendencies caused by an emotional stimulus, while for questions concerning Quality of Experience, conscious evaluation is of pivotal interest. Still, physiological changes could here be used to assess those. As the current book is targeting the QoE community, we will focus on this second aspect. We do so in a pragmatic way, which means that we do not intend to give final answers to the above mentioned questions on the fundamental nature of emotions, but describe which theoretical approaches and experimental paradigms have turned out to be useful when studying the emotional impact of multimedia material.

In Chap. 2, *perception* was defined as the conscious processing of sensory information. As a working definition, emotions can be understood as immediately evaluating this information as to what extent it is (expected to be) good or bad for the organism, and putting the organism in the state to react appropriately [2]. This includes devoting mental resources to further elaborate the situation up to creating a conscious representation of the own current emotional state, e.g. "I am scared/happy", which is called a *feeling*. Additionally, the own state is communicated to others via changes in facial expression, voice, etc.

At first glance, this 'evaluating' may sound similar to the quality assessment processes described in Chap. 2. However, it has to be pointed out that affective processes are predominately rooted in the evolutionary heritage of mammals,[1] and the reference point is irrevocably the organism's expected well-being in a biologically-inspired sense. Stimuli that are irrelevant in that regard do not cause an emotion, while also emotionally neutral stimuli may easily be judged with regard to their quality as understood in Chap. 2. A thorough discussion of emotion-related phenomena like moods, sentiments etc. can be found in Chap. 4.

What turned out to be good or bad for the organism and what might be proper affective reactions has been shaped by individual, but even more by evolutionary learning, which is the reason why emotional reactions to certain stimuli are quite consistent across people.

9.2 Media Stimuli

The need to make studies comparable has led to a couple of stimulus sets of which the emotional impact is known and that are more or less standardized. We will limit this description to stimuli that have been validated in a separate study prior to their usage to evoke emotions for a specific purpose (e.g. examine physiological changes), as otherwise the main research goal might have affected the emotional rating, and to databases that to our knowledge are available to other researchers.

9.2.1 Visual Stimuli

The most prominent set of stimuli that is presented to subjects to evoke an emotional reaction is probably the *International Affective Picture System* (*IAPS*) developed by Bradley et al. [3]. It consists of still images which depict pleasant (e.g. a smiling baby), unpleasant (e.g. mutilation scenes), and neutral scenes (e.g. a picture of a towel). For all stimuli, their values on the dimensions Bradley and Lang consider as basic for emotions, namely pleasantness (also called *valence*), arousal, and dominance are given based on the ratings of a sample of approximately $n = 100$ US students [3]. How these values were derived and further information on the meaning of the underlying dimensions is explained later in this text. The format of the images is jpeg with a maximum resolution of 1,024 $\times$ 768 pixels. These stimuli are used worldwide for various research purposes, including human-computer interaction (HCI) [4]. Meanwhile, a comparable database validated with a sample of German subjects exists, called *EmoPics* [5]. Their resolution is 800 $\times$ 600 Pixels in the jpeg-format.

---

[1] While we have heard colleagues jokingly speculating about a quality neuron which might be the foundation of quality judgements, we haven't heard anyone talking about a quality gene so far.

9.2.2 Audio Stimuli

Analogous to the IAPS, Bradley and Lang also issued the *International Affective Digitized Sounds (IADS)* library [6]. It consists of around 160 sounds of 6 s length with diverse content, including human laughter or thunderstorm noise. Their format is *.wav with varying bitrates. They also have been used successfully in the context of HCI [7]. The IADS includes brief excerpts of works of composers like Bach or Beethoven, but of course here longer pieces would be more appropriate to reflect the emotional impact. While emotion and music is an active field of research (for an overview, see [8]), no comparable standardized stimulus database exists to our knowledge.

Quite the opposite can be said for emotional speech stimuli. As emotion recognition is a major topic in speech analysis, there are numerous databases available in this domain to e.g. test classification algorithms. Schuller et al. [9] for example list eight databases with various languages for that purpose. They are usually not limited to valence and arousal values like the IADS, but also name a specific emotion the stimulus is representative for. There is one caveat with such databases if they are used for emotion *elicitation* though: the fact that a speech stimulus reveals a certain emotion of the speaker does not necessarily imply that it evokes the same emotion in the listener. Hearing an angry voice might for instance rather cause fear than likewise anger. One database that devoted special attention to this fact is the *Kiel Affective Speech Archive* (*KASPAR*),[2] where for a subset of the sentences the specific emotion they evoked in the listener were identified, including physiological changes.

9.2.3 Audiovisual Stimuli

Film clips meanwhile appear to be the most popular method to induce emotions in the laboratory [10] for a variety of reasons: They address both, auditive and visual modality, they are dynamic and thus have a high attentional capture, they allow to 'unfold' a complex emotional story or *narrative* as [11] calls it up to blending of several emotions within one stimulus. Still, they guarantee better comparability across subjects than mental imagery or personal recall of emotional scenes [12, 13].

Gross and Levenson [14] were among the first in 1995 to publish a list of scenes from commercially available movies that would evoke certain emotions in the viewer, and also offered detailed cutting instructions. An updated version can be found in [13], also linking to updated cutting instructions. This original list was furthermore extended by [15, 16] and the most recent set of film scenes from commercial films can be found in [12].[3] For all described films, a brief summary of the content plus the target emotion is given, followed by a 'hit rate', i.e. in how many subjects of

[2] http://www.stimmeundemotion.uni-kiel.de/Ressourcen.htm

[3] The link given in [12] is apparently outdated. The instructions etc. can now be found at [accessed 4.3.13]: http://www.ipsp.ucl.ac.be/recherche/FilmStim/.

the norm sample the desired emotion was triggered, and what other emotions where evoked, which might be interesting if someone intends to explicitly present such ambiguous stimuli. In addition, values referring to a dimensional model of emotions are available. These are all valuable information for researchers who intend to use the clips with a new sample of subjects. The audiovisual quality of the film scenes is however not specified. Schaefer et al. [12] report that the norm values were derived using copies on VHS videotapes, [13] refers in the cutting instructions to both, VHS and DVDs as source material. This is presumably done intentionally to leave it to the researcher, which type of media she can get hold of as they are all commercially available films. On the other hand it clearly shows that video quality appears of minor importance for the authors.

One question that arises when commercial films are used is whether it matters that subjects might have already seen the corresponding movie the excerpt is taken from. We did not find such an effect in an own study [17] based on film clips recommended by [16], and also Gross and Levenson report that if any, prior watching was associated with more intense feelings [14], which is stated in [13] again.

To summarize, while pictures and brief sounds are easy to classify according to their valence, but may be ambiguous with regard to what specific emotion they trigger (the static scene of a person attacking another one may cause empathy for the victim, fear of, or anger towards the attacker), film clips appear to be especially well-suited to evoke a specific emotion in a standardized way.

9.3 Assessment of Experienced Affect

Attempting to evoke an emotion usually goes hand in hand with assessing the effect of this intervention. In addition, it may in some cases also be desired to ensure that a certain stimulus did *not* cause an emotion, as pointed out in the introduction. Letting the users rate their emotion themselves to obtain a mean opinion score (MOS) analogous to the perceived quality may be the obvious way for QoE researchers. We will summarize the most common instruments for that purpose.

However, one fundamental difference of emotional stimuli as compared to other types of sensory processing, e.g. the perception of a color or a number, is that they trigger the intention to somehow react to them [18]. Bradley and Lang trace this back to phylogenetically old motivational circuits in the brain that ensure survival by letting the organism seek favorable conditions and avoid harmful ones. Therefore they guide the attention towards any stimuli of either kind, and also prepare the body for the corresponding reaction, appetitive or defensive [2]. While advocates of basic emotions doubt that there are just these two types of reactions, but argue for emotion-specific neural circuit as well as reaction patterns (e.g. [18]), they all agree that emotions are *embodied*, i.e. inevitably linked to physiological changes. This relationship lets physiological measures appear one of the major ways to measure emotional reactions. To account for this, one section will summarize findings regarding the *peripheral* as opposed to the *central* nervous system (i.e. the brain) which is covered in Chap. 8. As research is multitudinous in this area, we will focus on well-documented findings that are of practical relevance for QoE studies.

9.3.1 Self-Assessment

Bradley and Lang not only provide standardized stimulus material, they also offer the instrument they use to let subjects rate the experienced affect: the *Self-Assessment Mannikin* (*SAM*) [18] is a pictorial scale depicting a simple cartoon figure whose expression varies on the three dimensions *valence* from an unhappy to a happy face, *arousal* (a sleeping face to an exciting character whose whole body is trembling), and *dominance*. As the latter dimension ranges from *being controlled* to *being in control*, the cartoon is either very small in the picture or is covering more and more of it, almost bursting out of the frame. Due to its simplicity, the SAM can be applied efficiently to all kind of contexts, including a version on a mobile phone [7].

If a verbal assessment is preferred, the *Positive and Negative Affect Schedule* (*PANAS*) [19] is a common alternative: it consists of 2×10 emotional adjectives that people have to rate their feelings on using a five-point Likert scale, which ranges from 1 'not at all' to 5 'extremely'. The underlying dimensions are related to those of the SAM, but give credit to the fact that the stimulus distribution of IAPS and IADS follows a boomerang-shaped distribution: stimuli of high positive or negative valence tend to be associated with high arousal values. Thus the positive affect (PA) dimension of the PANAS represents increasing values of positive feelings concomitant with increasing arousal e.g. from *lethargic* to *enthusiastic*, while the negative affect (NA) dimension varies from low negative affect and low arousal (e.g. *calm*) up to highly arousing negative feelings like *anger* or *fear*. The underlying dimensions of SAM and PANAS can be mapped onto each other to some extent [20], but for the purpose of picture/film assessment, the PA dimension of the PANAS appears to be less well-suited [12].

To assess the emotional impact with regard to the evoked basic emotion, researchers mostly use self-developed surveys (e.g. [13, 15, 16]) or variants of the *Differential Emotion Scale* (*DES*) [21]. A newer instrument is PRemo which targets basic *emo*tions evoked by *PR*oducts using animated cartoon figures [22].[4] This type of emotional reaction is frequently addressed in the context of User Experience, which is discussed in more detail in Chap. 3.

9.3.1.1 Post-hoc Versus Continuous Rating

So far, the described instruments are all applied *subsequent* to stimulus presentation, which is completely adequate for still images or short sounds. If longer-lasting stimuli are presented, one final rating might not be sufficient to cover the whole emotional

[4] See also http://www.premotool.com/.

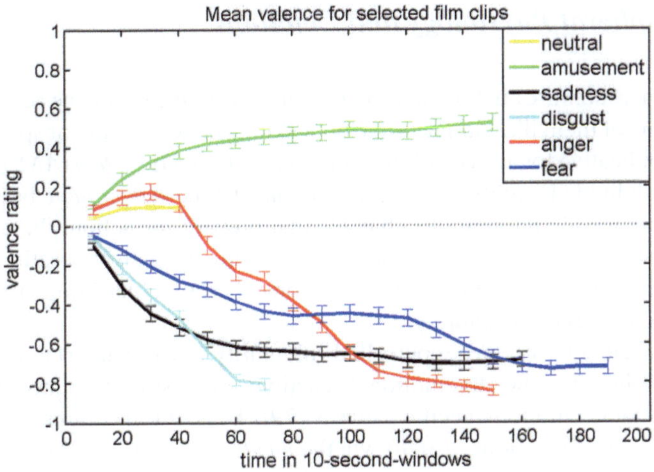

Fig. 9.1 Mean valence ratings (-1 = unpleasant to $+1$ = pleasant) averaged over 10 s during presentation of film clips of varying duration depicting different target emotions. Whiskers denote standard errors

episode though, and similar claims have been made for quality ratings, e.g. [23]. At the same time this continuous measurement should not distract the subject's attention from the stimulus under evaluation. For purely auditive stimuli this can easily be achieved by offering a device or graphical user interface (GUI) that is operated manually and monitored visually like done by [24] or [25] for music. For audiovisual material, the issue of divided visual attention is more pronounced. Nagel et al. [26] and Mauss et al. [27] used a hardware rating dial that, after a training phase, could be operated without having to look at it.

In our aforementioned study [17, 28], we adopted the interface developed by [25] such that it could present videos with thin xy coordinate axes overlayed. Subjects continuously rated their current emotion on the two dimensions of valence (x axis) and arousal (y axis) using a mouse. The mouse pointer was visible in the coordinate system as a small dot with a short tail to visualize the rating course of the last seconds. The average rating course for valence of $n = 60$ subjects (35 female, mean age 25+/$-$4.8 years) for the six clips of [17] that had turned out to be most effective in evoking the postulated target emotion is depicted in Fig. 9.1.

It can be seen that for one clip, namely anger (red line in Fig. 9.1) the valence changes from slightly positive in the beginning (when the later victims of violence are depicted as happy people) to negative throughout the later course of events, where they are being attacked. Such nuances in subjective evaluation would be lost when only relying on one aggregated value without temporal information. However, it also has to be noted that the values tend to be most pronounced at the end of the film clip, i.e. show the highest/lowest y values (see Chap. 10). We will get back to this fact later.

9.3.2 Peripheral Physiology and Emotions

Peripheral measures are all those that are not directly derived from the central nervous system. Some of them like cardiac activity or skin conductance are of special interest as they are indicative for activity of the *autonomous nervous system* (*ANS*), which can further be subdivided into its two major components: the *sympathetic* branch, which is responsible for all autonomous changes that prepare for action, therefore called *ergotrop*, and the *parasympathetic* branch, which influences all restorative processes. Both have in common that their activity is controlled by older brain structures and is difficult to influence voluntarily.

Facial muscle activity can be controlled voluntarily, but also has a large involuntary component when it comes to emotional reactions, at least in untrained persons, up to distinguishable emotion-specific patterns [29]. Thus, it is a common parameter to measure emotional reactions, usually via the *electromyogramm* (*EMG*).

For exposure to stimuli of short duration like the IAPS or IADS, certain relations are well established: emotional stimuli are attended faster and more extensively than neutral ones. They lead to a decrease in heart rate, an increase in skin conductance (indicative of sympathetic arousal), increased activity of the facial muscle *corrugator supercilii* (causing “frowning”) for unpleasant stimuli, and increased activity of the facial muscle *zygomaticus major* and *orbicularis oculi* (which both are involved in smiling) for pleasant stimuli [30]. In general, the reactions tend to be stronger to aversive than to pleasant stimuli, as the first may constitute an immediate threat to survival. However, all these changes occur within roughly five seconds after stimulus onset, and then parameters like heart rate tend to get back to their prior level [30].

Kreibig undertook the painstaking effort to review 134 publications that report autonomous changes due to specific emotions induced over longer time ranges, frequently with film clips [10]. For respiratory and cardiovascular parameters, certain emotion-specific patterns could be identified, however, the author also concludes that “Collecting valid data on autonomic responding in emotion has been and remains to be a challenge to emotion research.” (p. 411 in [10]). Compared to that, skin conductance appeared to be a quite stable indicator for arousal, as it was increased in most of the studies except for the emotions of sadness, contentment, and relief, which all share a tendency for passivity rather than the need for action [10].

A peripheral signal that reliably indicates the experienced valence also over longer time frames may be facial muscle activity—in a previous study, Kreibig et al. [30] used it together with self-assessment as a control variable to check successful induction of emotions, and we also found facial EMG to be able to differentiate between negative and positive emotions [28]. Some authors alternatively subsume facial expression under behavioral measures of emotion, an aspect we did not cover in this paragraph, and which also includes the changes in voice mentioned in the introduction.

Thus it may be a good moment to point out to the reader that this selection of emotion evaluation methods is rather the tip of the iceberg than an exhaustive listing. More detailed information on the topic can be found in alternative chapters of [13], or in [10, 31], which may serve as a starting point. However, there is always the danger of getting lost in the sheer amount of information. To avoid this, we will give some practical advices in the following that include an evaluative summary of the evaluation methods.

9.4 Practical Advices

The advices given in this section are based on our experience with presenting emotional stimuli and assessing them. The first and most straightforward advice is to use material of which the emotional content is known a priori, and not mix both research questions in a single study, i.e. present stimuli of unknown emotional content and unknown quality levels. As described in the section on future research, it is far from settled how emotion influences quality and vice versa.

The second advice is to use stimuli and validated questionnaires in the way described in the original source, and not pick single questions out of a set or discard the recommended instructions. One reason for the success of Bradley and Lang's material is surely that they are quite clear how to use the stimuli and how to rate them subsequently—for that purpose, they include instructions as well as SAM templates in their IAPS and IADS manuals, and ask researchers to report the numbers of the specific stimuli they used in their study. In psychological questionnaires, it is less the single item that matters, but the repeated and aggregated measurement or dimensional value it contributes to. So even if you consider single items redundant or needless for your use case (e.g. the *dominance* dimension of the SAM), it is better to present the survey in its original format, be it only for the purpose of later comparability with other studies. The dominance value might not be that crucial for passive viewing, but could be important for interactive settings, e.g. video conferencing. Next to pragmatic reasons, it is also important to remember that the scales you provide suggest the test subjects how to structure their experience internally—if dominance is not mentioned as a basic aspect of emotional experiences, they might try to map this aspect onto the items offered, which could in the worst case result in some kind of mis-attribution. In general, pictorial scales tend to be faster and easier to complete, at the expense that the information obtained may be less detailed with regard to the emotion experienced or specifics of the material presented.

When using standardized film clips that are targeting one single emotion each, it might be sufficient to collect a self-assessment subsequent to presentation: the subjective ratings in Fig. 9.1 show the most pronounced values at the end of the clip. This of course only holds for clips where the course of events is evolving in one direction—for excerpts of varying content, a continuous rating may reveal evaluative changes. We would recommend to ask for the emotional rating immediately after presentation and prior to any other queries, e.g. quality assessment. Emotions are a transient phenomenon, and the current state may be affected by subsequent appraisal processes [11]. While most people would agree that there are unpleasant movies of high audiovisual quality, subjects may be inclined to give a clip they just rated as 'low' with regard to quality also a less positive emotional rating to show internal consistency.

Physiological changes are best compared to a baseline level to account for interindividual differences. Here, the baseline should be similar to the later task, i.e. watching a neutral film clip instead of simply letting the subject rest in a chair. As physiological changes vary in their temporal course, it might be good to repeat short baseline phases between the presentation of emotional content to avoid carry-over effects of the current affective state to the next trial [13]. Compared to measures of self-assessment, bodily changes can reveal aspects of emotional processing that subjects are not conscious of, but similar to the difference between pictorial and verbal scales, the results might be more difficult to interpret unambiguously in the context of common QoE setups. Therefore it is always a good idea to ask the subjects at least once during the test if they observed anything unusual or have any other remarks that were not covered by the rating scales. These side notes can be very helpful to explain unexpected results and sometimes even lead to new research questions.

9.5 Future Research Directions

The most obvious reason for QoE researchers to pay attention to the emotional content of the material they are presenting is that it may affect the quality assessment. The question is, in what direction, and here the alleged relation might be that emotional content intensifies the quality rating, i.e. lower quality for unpleasant stimuli, and higher quality rating for pleasant ones. We are not sure whether this is the most conclusive relation: emotional stimuli tend to attract attention, including the need for appraisal how to deal with the situation, which at some point will withdraw attention from other aspects, e.g. the audiovisual quality of the presented material. Thus it may as well turn out that at least highly emotional arousing material smoothes quality ratings. To disentangle these dependencies is surely one future research direction.

Vice versa, it is also not clear how quality influences emotional assessment. Again, the simple relationship would be that the higher the quality (e.g. resolution, lighting), the more intense the emotion rating of a given clip. And again, we doubt that the influence is that simple: one common feature of the fear-inducing clips recommended by [16] is that they are set in the dark, and the immanent threat is present, but not identifiable. Here, the ambiguity increases the suspense as the actual appearance or extent of the threat is left to the imagination of the viewer. A well-lighted-scenery where all details are visible would probably decrease the emotion. In other cases, a lower quality might increase the credibility of the content: one of the reasons for the success of the movie *Blair Witch project* was that it was shot with a hand-held camera and thereby implied it depicted an amateur recording of a 'true' holiday that went terribly wrong. A similar phenomenon can be observed with youtube clips from crisis regions shown on news shows. Again, the obvious low quality rather adds to the perceived authenticity than decreasing it, and thereby might lead to higher affective reactions. As the definition of QoE given in Chap. 2 explicitly includes emotional impact, these aspects have to be taken into account, and one of the future research challenges will be to harmonize QoE models and research paradigms with their equivalents in emotion research.

References

1. Wundt W (1911) Grundzüge der physiologischen Psychologie, 3. Band, 6. umgearb. Leipzig, Engelmann.
2. Lang PJ, Bradley MM (2010) Emotion and the motivational brain. Biol Psychol 84(3):437–450
3. Lang PJ, Bradley MM, Cuthbert BN (2005) International affective picture system (IAPS): affective ratings of pictures and instruction manual. Technical report A-6, University of Florida, Gainesville.
4. Partala T, Surakka V, Vanhala T (2005), Person-independent estimation of emotional experiences from facial expressions. In: Proceedings of the 10th international conference on intelligent user, interfaces, pp 246–248.
5. Schönfelder S, Kanske P, Heissler J, Wessa M (2010) EmoPicS—Multimodale Evaluation neuen Bildmaterials zur neurophysiologischen Emotionsforschung, 36. Tagung Psychologie und Gehirn.
6. Bradley MM, Lang PJ (2007) The international affective digitized sounds (IADS-2): affective ratings of sounds and instruction manual, 2nd edn. University of Florida, Gainesville
7. Seebode J, Schleicher R, Möller S (2012) Affective quality of audio feedback in different contexts. In: Proceedings of 11th international conference on mobile and ubiquitous multimedia (MUM 2012), 0–3.
8. Juslin PN, Sloboda JA (2010) Handbook of music and emotion. Oxford University Press, New York
9. Schuller B, Zhang Z, Weninger F, Rigoll G (2011) Selecting training data for cross-corpus speech emotion recognition? Prototypicality vs generalization. In: Proceedings of Afeka-AVIOS speech processing conference.
10. Kreibig SD (2010) Autonomic nervous system activity in emotion: a review. Biol Psychol 84(3):394–421
11. Lazarus RS (2006) Stress and emotion: a new synthesis. Springer, New York
12. Schaefer A, Nils F, Sanchez X, Philippot P (2010) Assessing the effectiveness of a large database of emotion-eliciting films: a new tool for emotion researchers. Cogn Emot 24(7):1153–1172
13. Rottenberg J, Ray RR, Gross JJ (2007) Emotion elicitation using films. In: Coan JA, Allen JJB (eds) The handbook of emotion elicitation and assessment. Oxford University Press, New York, pp 9–28
14. Gross JJ, Levenson RW (1995) Emotion elicitation using films. Cogn Emot 9(1):87–108
15. Hagemann D, Naumann E, Maier S, Becker G, Lürken A, Bartussek D (1999) The assessment of affective reactivity using films: validity reliability and sex differences. Pers Individ Differ 26:627–639
16. Hewig J, Hagemann D, Seifert J, Gollwitzer M, Naumann E, Bartussek D (2005) A revised film set for the induction of basic emotions. Cogn Emot 19(7):1095–1109
17. Schleicher R, Galley L (2009) Continuous rating and psychophysiological monitoring of experienced affect while watching emotional film clips. Psychophysiology 46(1):51
18. Panksepp J (1998) Affective neuroscience. Oxford University Press, New York
19. Bradley MM, Lang PJ (1994) Measuring emotion: the self-assessment manikin and the semantic differential. J Behav Ther Exp Psychiatry 25(1):49–59
20. Watson D, Clark LA, Tellegen A (1988) Development and validation of brief measures of positive and negative affect: the PANAS scales. J Pers Soc Psychol 54(6):1063–1070

21. Yik MS, Russell JA, Barrett LF (1999) Structure of self-reported current affect–integration and beyond. J Pers Soc Psychol 77(3):600–619
22. Izard CE, Dougherty FF, Bloxom BM, Kotsch NE (1974) The differential emotion scale: a method of measuring the meaning of subjective experience of discrete emotions. Nashville.
23. Desmet P (2004) Measuring emotion: development and application of an instrument to measure emotional responses to products. In: Blythe MA, Overbeeke K, Monk AF, Wright PC (eds) Funology. Kluwer Academic Publishers, Dordrecht, From usability to enjoyment, pp 111–124
24. Gros L, Chateau N (2001) Instantaneous and overall judgements for time varying speech quality: assessments and relationships. Acustica 87:367–377
25. Schubert E (2004) Modeling perceived emotion with continuous musical features. Music Percept 21(4):561–585
26. Nagel F, Grewe O, Kopiez R, Altenmüller E (2007) EMuJoy–software for continuous measurement of perceived emotions in music: basic aspects of data recording and interface features. Behav Res Methods 39:283–290
27. Mauss IB, Levenson RW, McCarter L, Wilhelm FH, Gross JJ (2005) The tie that binds? Coherence among emotion experience, behavior, and physiology. Emotion 5(2):175–190
28. Hutcherson CA, Goldin PR, Ochsner KN, Gabrieli JD, Barrett LF, Gross JJ (2005) Attention and emotion: does rating emotion alter neural responses to amusing and sad films? Neuroimage 27(3):656–668
29. Bradley MM, Lang PJ (2000) Measuring emotion: behavior, feeling and physiology. In: Lane RD, Nadel L (eds) Cognitive neuroscience of emotion. Oxford University Press, Oxford, pp 242–276
30. Kreibig SD, Wilhelm FH, Roth WT, Gross JJ (2007) Psychophysiology Cardiovasc. Electrodermal and respiratory response patterns to fear-and sadness-inducing films 44(5):787–806
31. Mauss IB, Robinson MD (2009) Measures of emotion. A review. Cogn Emot 23(2):209–237
32. Schleicher R (2009) Emotionen und Peripherphysiologie. Pabst Science Publishers, Lengerich
33. Ekman P (2003) Emotions revealed. Henry Holt and Company, New York

Chapter 10
Temporal Development of Quality of Experience

Benjamin Weiss, Dennis Guse, Sebastian Möller, Alexander Raake, Adam Borowiak and Ulrich Reiter

Abstract Most research on Quality of Experience treats QoE as a static event. As a result, QoE is measured for stimuli of delimited length, and the QoE which is associated with the stimulus is considered to be stable along its duration. However, this rarely happens in reality where usage episodes extend over several seconds and minutes (e.g. a phone call), hours (e.g. a video film), or regularly over periods of weeks or months (when considering QoE of a subscribed service). In this chapter, we will discuss the cognitive processes involved when QoE is integrated over usage episodes, and describe corresponding assessment methods. We will also review models for estimating episodic and multi-episodic QoE from momentary QoE judgments or their predictions.

B. Weiss (✉) · D. Guse · S. Möller
Quality and Usability Lab, Telekom Innovation Laboratories, TU Berlin, Berlin, Germany
e-mail: bweiss@telekom.de

D. Guse
e-mail: dennis.guse@telekom.de

S. Möller
e-mail: sebastian.moeller@telekom.de

A. Raake
Assessment of IP-based Applications, Telekom Innovation Laboratories, TU Berlin, Berlin, Germany
e-mail: alexander.raake@telekom.de

A. Borowiak · U. Reiter
Department of Electronics and Telecommunications, Norwegian University of Science and Technology (NTNU), Trondheim, Norway
e-mail: adam.borowiak@iet.ntnu.no

U. Reiter
e-mail: ulrich.reiter@ntnu.no

S. Möller and A. Raake (eds.), *Quality of Experience*,
T-Labs Series in Telecommunication Services, DOI: 10.1007/978-3-319-02681-7_10,

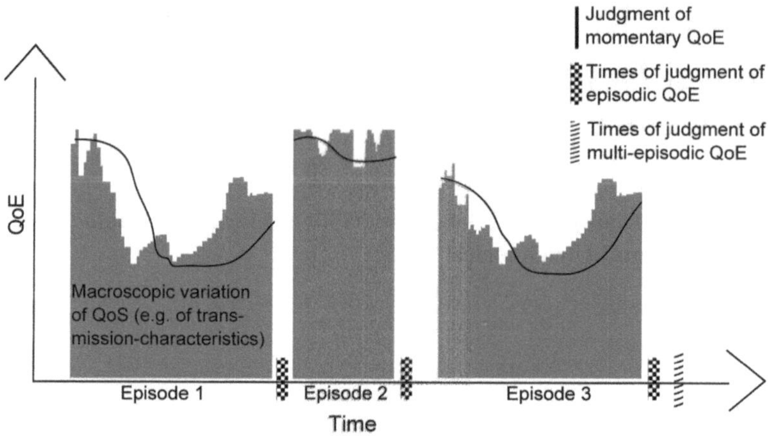

Fig. 10.1 Schematic illustration of QoE concepts

10.1 Introduction

With current communication services and networks, one major issue is the temporal variability of transmission characteristics as they are common in today's mobile and best-effort networks. From a Quality of Experience (QoE) perspective, a system with time-varying characteristics will have its impact on the user in terms of various aspects. Respective time spans have been defined for user experience [44], ranging from:

1. *momentary*/instantaneous experience of the system characteristics as a result of the current media quality level or of events like sudden changes in transmission characteristics,
2. over the retrospective appraisal of various events during an *episode* of usage like a call or video clip,
3. until the cumulative experience when evaluating a whole service after *multiple episodes*.

Of course, retrospective ratings can be asked for at any time during one episode. In this chapter, however, we consider only complete episodes. Confer Fig. 10.1 for an illustration of time spans and judgment events of momentary QoE, (remembered) episodic and multi-episodic QoE, and the macroscopic (see below) QoS variation.

These three time-spans have a close connection to Kahneman [31], who distinguishes between the *momentary*-based approach and the *memory*-based approach. This separation is made for the global field of human experience with concepts like joy or pain, and can be directly transferred to QoE (see also Chap. 2).

The momentary-based approach corresponds to momentary QoE and reflects direct instantaneous measures of experience. Variation over time of such measures corresponds to *macroscopic* changes of system characteristics, as these are actually perceived as being varying in quality, as opposed to *microscopic* system behavior (cf. [42]). Accumulating (e.g. averaging) ratings of momentary experience represents a useful abstraction, not actual experience, called *total utility* in [31]. Momentary QoE itself is an important measure for addressing the instantaneous user reaction to changes in Quality of Service (e.g. types of transmission degradations). Assessment methods of momentary QoE include those for truly instantaneous assessment, but also "sampled momentary" QoE, i.e. assessment for short-term samples which exhibit no macroscopic, but microscopic variation. These methods are presented in Sect. 10.3.

The memory-based approach is concerned with retrospective appraisal, i.e. remembered experience. This subsumes episodic and multi-episodic QoE, which can be differentiated regarding the time of assessment and the lengths of the experience: Episodic QoE is concerned solely with one episode and typically assessed directly after this one, whereas multi-episodic QoE might be assessed with or without temporal alignment to an episode, and the scope of the experienced quality is usually the whole service up to the event of assessment. This remembered QoE (*remembered utility* in terms of Kahneman [31]) reflects the users' integration processes and the establishment and development of attitudes towards a system or service (cf. Sects. 10.4 and 10.5).

Dependent on the research question, either momentary or (multi-)episodic QoE is in focus. But in order to study their relationship, instantaneous and retrospective ratings have to be analyzed together.

The focus in this chapter lies on the QoE as a result of time-varying transmission characteristics. We address this by presenting empirical results in a structured way, outlining assessment methods, and even presenting prediction models for auditory, visual and audio-visual quality. Waiting times are not a topic presented in this chapter, as they are covered in Sect. 22.3. Further information on cognitive aspects of QoE are presented in Chap. 2.

According to the three time spans presented, this chapter starts in Sect. 10.2 with cognitive processes related to the temporal development of QoE. Then, it provides an overview of methods to assess momentary, episodic and multi-episodic QoE (Sects. 10.3, 10.4, 10.5) and presents major principles of how retrospective appraisal is related to momentary QoE (mainly in Sect. 10.4). We conclude with an outlook on major issues in current research and applications (Sect. 10.6).

10.2 Cognitive Processes of Temporal QoE

One of the major topics in QoE research is the relationship between several momentary events and retrospective appraisal of the resulting QoE. Often, a weighted average of momentary ratings is correlated strongly with a single rating obtained directly after a session. Results for such weighted averages reveal a higher weight of the last ratings compared to the others. This observed effect is often called recency effect, implying a relation to cognitive processes of recall. The assumption seems to be, that human processes in recalling information from the working memory will ground such a retrospective appraisal, based on cognitive models [1].

Within the time range of the working memory of about tenths of seconds, typically 5–9 unrelated items (like object names or numbers) can be recalled right after the presentation or after a short break without distraction. The exact time span used for recall cannot be defined, as the working memory is not just an information storage, but a multi-component system used for complex cognitive tasks. For example, names or digits can be rehearsed within the phonological loop as long as there is no distraction. The actual performance and time span depend on the individual, content of information, motivation and attention to the task, or modality of distractor tasks (see also Chap. 2).

The main effect observed in most studies addressing free recall from the working memory is the likelihood to recall information better or worse depending on the position of presentation: The first and the most recent items are recalled with a higher probability in a free recall task. The first, so-called *primacy effect* lessens somewhat with more items (e.g., 10–20 and thus also with longer time to recall), whereas the *recency effect* can be reduced or eliminated by distraction or delay (e.g., 15–30 s) between presentation and recall of items. Recency can thus be viewed as the retaining of information, where the recall has not been deteriorated by subsequent information.

The likelihood to recall information better or worse depending on the temporal position within an episode can also be taken up by instrumental quality prediction models. The rationale is to include such a cognitive process to model the rating at the end of an episode on the basis of ratings for relatively short stimuli or calls with varying quality.

Interestingly, such positional effects can also be observed for longer time spans like several minutes to hours (e.g., remembering content of a talk), or even for a whole season or year, determined by the number of events, not the time elapsed. A different cognitive approach is not directly linked to actually recalling or retrieving information from memory, but judging individual, even long-term experience in retrospect based on memories, the memory-based approach presented in Sect. 10.1. The *peak-end rule* models the heuristic of appraisal of one's experiences in terms of valence and intensity only by two remembered moments [30]. That is, retrospectively judging long-term experiences (tens of minutes, but also time spans of, e.g., years) are dominated by the most extreme and the most recent experiences. It could be shown, that other information are not lost, but just not included into the retrospection. Here, a primacy effect is not apparent.

This heuristic and the underlying peak-end rule seem to be closer to the actual task of rating (multi-) episodic QoE than recall from the working memory. In fact, the task of an episode-final QoE judgment is quite different to an instantaneous judgment of momentary QoE, and thus does not precisely require to recall the experienced quality, compare, judge and describe it. Instead, remembered QoE is bettercharacterized by eliciting the *current* attitude towards the service based on those quality events in scope, may this be one or more episodes, and on which the peak-end rule is based on.

The relevance of intensity is already known for other judgment tasks based on heuristics, e.g. the topic of combining traits for interpersonal judgments. Here, a stronger impact of negative traits than positive ones is found [22], that can be modeled with a weighted average [32]. Accordingly, models for time-varying quality often take into account valence (i.e. special treatment of degradations) and variability itself (cf. Sect. 10.4).

10.3 Assessing Momentary QoE

Most of the existing mono- and multi-modal quality assessment techniques do not take into account fluctuations of the quality which happen to appear when a stimulus of extended duration is viewed (i.e. more than 10 s). The remembered quality rating provided after an episode (e.g. 10 min) should not be considered as an accurate measure of momentary QoE, which is continuously evolving. This is due to the fact that humans are more likely to make the overall quality judgment based on the most recent experiences which are assumed to be of a greater importance or significance (cf. Sect. 10.2). In order to overcome the mentioned inaccuracy two approaches can be applied: the long content can be divided (windowed) into a number of shorter episodes (e.g. 10 s each) and then evaluated separately (for an overview of parametric models for estimating QoE of such short samples, see Sects. 12.3, 14.3 and 19.2), or the quality can be judged on the fly, throughout the entire stimulus duration. In the first approach, standardized methods applicable for short sequences evaluation can be used. Examples of such methods, typically applied to 3–16 s long episodes, are: Double Stimulus Continuous Quality Scale (DSCQS) [26], Absolute Category Rating (ACR) [28] and Paired Comparison (PC) [28] (for more see [27, 29]).

However, the mentioned assessment techniques are lengthy in nature and hence impractical in real life applications, where e.g. content of 30 min duration needs to be evaluated. For this purpose, an appropriate, no-reference method (i.e. without a stimulus to compare) allowing for instantaneous quality evaluation should be employed. An example of such a method is the Single Stimulus Continuous Quality Evaluation (SSCQE) developed by the RACE MOSAIC project [19] and later incorporated into the ITU-R recommendation BT.500-7 [26], or the corresponding continuous assessment method for speech quality described in ITU-T Rec. P.880 [43]. The SSCQE allows participants to judge the perceived quality dynamically using a slider mechanism with associated interval scale (commonly from 0 to 100 with the range divided into five equal slots corresponding to the ordinal five-point quality scale). Although the method is capable of catching the quality variations instantaneously and over extended periods of time, it is not free from drawbacks and ambiguities. It has been reported that the continuous operation of the slider might divert the user's attention from the process of quality assessment [7] and that the differences in participant's reaction time to quality changes can reduce the accuracy of the method [41]. Recently, there has been increased interest in the development of alternative methodologies capable of tracking the quality changes in a continuous manner by using different types of rating devices, i.e. a glove [8], a steering wheel [37], etc. Nevertheless, except for the type of rating instrument, those methods do not bring any major methodological changes compared to the SSCQE. This is due to the fact that all of them use the same type of rating scale (or in some cases even simplified versions with reduced resolution) associated with each of the device. Moreover, the improvement in the rating instruments' performance over the slider mechanism has not been proven and an effect of stimulus duration on users' fatigue related to usage of these devices has not been verified.

Modification of another standardized method (ACR) has been suggested for a quasi-momentary assessment of a longer episode by providing judgments after time-intervals of fixed window size [18]. The quality judgments are made during the appearance of segments with no degradations in contrast to the gray segments used for this purpose in the ACR method. This way, according to the authors statement, the continuity of the sequence is preserved making the viewing conditions more realistic.

A different approach towards continuous quality evaluation has been proposed by Borowiak et al. [5]. Instead of providing a numerical representation of the perceptual experience, the user is allowed to actively adjust the quality to the most appreciated level in case degradation occurs. The improvement in the quality is achieved by means of an adjustment device (e.g. rotary knob) and based on perception of quality changes solely (no tactile feedback from the device). The scale assigned to the assessment instrument in fact is a direct representation of the quality levels used in the test, and a translation of the perceived quality into a numerical score or position of the rating device is not required. There are no physical limits in the rotation mechanism as witnessed in the previously mentioned methods, and the maximum quality can be overpassed if not recognized by the user, causing gradual decrease in quality. This is a reversible process, so the user can return to the reference quality by rotating the device in the opposite direction again.

With this new technique, a measure of momentary QoE is achieved in relation to the desired QoE. Eliciting the user's behavioral reaction to experienced quality by means of the quality adjustment approach allows for gathering the data with less cognitive resources required compared to typical assessment methods [5]. In consequence, subjects' attention is on the presented stimuli rather than on the usually challenging task of quality evaluation. Although new findings are possible with the method, it should not be treated as a replacement of the existing ITU-R rating scales based methodologies, but rather as an additional source of information with respect to the user's cognitive experiences.

In general, there has been relatively little attention devoted to the topic of continuous quality evaluation, resulting in a comparatively small number of related publications. However, some interesting research findings have been claimed with respect to the momentary quality assessment. In [33] it has been concluded that subjects react almost immediately when a change from good to poor quality occurs while in the reverse situation the adaptation process is much slower. This asymmetry in tracking the quality has been confirmed in [9] where changes in momentary speech quality were evaluated. Other studies [4, 6], in which longer duration content (30 min) was employed, prove the time dimension not being the main factor influencing quality perception. The authors discovered that quality expectations over extended periods of time are rather constant and that the same holds for the absolute quality level at which the change is usually discovered. Moreover, it has been found that sensitivity to quality variations highly depends on the awareness of the process of quality changes, and is higher when the subjects are in charge of the quality adjustment, than when the process is controlled externally. These findings hold, no matter which way the degradations appear in the presented material; whether stepwise in time or immediately, in one move [3].

10.4 Assessing Episodic QoE

In contrast to momentary QoE, quality attributed to an episode of usage is commonly judged directly after the end of the episode, may it be after watching a video clip or movie, or after finishing a telephone or video call. Such ratings of remembered QoE collected at the end-point of an episode may be different from the remembered QoE judged upon at a later point in time, where effects such as distraction or contextual transformation of QoE come into play. Still, such an episode-final quality rating may not be easily related to momentary QoE experienced during the episode: For the case of time-varying transmission characteristics, the averaged momentary QoE, i.e. the total utility (cf. Sect. 10.1), is in many cases not an appropriate estimate of the quality of the whole episode, as given by an episode-final judgment. Usually, such an average is too optimistic (cf., e.g., [13, 46] for speech quality, and [20] for video quality).

The recency effect between momentary and episode-final ratings was confirmed for speech quality [14] and video quality [20]. There is evidence that the impact of recency is smaller when momentary QoE is assessed continuously, i.e. with sliders [21], so it may be advisable to assess momentary and episodic QoE separately.

A method which aims at achieving both short-term and episode-final ratings is described in [12] for speech quality. The method consists of two separate tests which are carried out on a specific type of stimulus material. The material consists of several stretches of speech which have a duration of 4–8 s (thus the typical duration of short speech stimuli), and which are related to each other by their content. 5 to 6 of such stimuli form a storyline of a *simulated conversation*, i.e. a virtual exchange of information between two parties in which the stimuli represent the contributions of one party only. The stretches of speech are first presented in a standard test set-up according to [27], this way obtaining (position independent) short-term QoE judgments for each stretch. Secondly, the stretches are presented in their logical order, with pauses of approx. 8 s between the stimuli. During each pause, the test participant is asked to orally answer a content-related question, usually in a multiple-choice fashion, to incite her concentrating on the content and engaging in a conversation-like situation. At the end of the last stretch, the test participant is asked for an episode-final rating of the overall quality of the episode. This so-called simulated conversation test provides short-term and related episode-final QoE ratings, and thus allows to relate one to each other. The procedure has also been adapted to video calls (then putting the focus also on the visual modality [23]), and is currently considered for a future ITU-T Recommendation [25].

Although this method [12] is able to *simulate* conversations, it is methodologically difficult to assess momentary and episodic QoE in real interactive situations, as the duration of each turn cannot be controlled, and thus neither strength nor position of degradations can be systematically varied in order to obtain reliable averaged results. As a consequence, many results and models stem from data obtained in passive situations, and a valid transfer from judgments obtained this way to interactive quality cannot be guaranteed. For the example of speech quality, a recency effect found for passive listening was not replicated for the interactive session [16], and the method described above to simulate conversational structures obviously neglects the effects of echo and delay [12], which might be integrated differently from other sources of degradations like noise.

Another effect which was found to influence episodic QoE is the impact of extreme qualities [14, 20]. Consequently, models to describe episodic QoE with an integration of momentary ratings include weightings for each momentary rating, with stronger weighting for episode-final times of occurrence, and for stronger changes (or only stronger degradations, as mentioned in Sect. 10.2):

- In [10], from any individual rating of speech QoE, may it be continuous or a short-sample rating, a weighted mean is calculated, with a higher weight towards the end of the episode, and for extreme degradations.
- In [9] the asymmetric temporal delay to adjust to changes in momentary speech QoE observed by [14] (cf. Sect. 10.3) is integrated into a model of episodic QoE, that uses these estimated momentary ratings, integrates them by averaging, but also models a recency effect by taking into account the last significant degradation.
- In [12], there is also a two-step approach chosen for the speech domain. First, a weighted mean is calculated, taking into account a recency effect—in contrast to the two above with an absolute time window—and the impact of the strongest minimum is additionally subtracted as a difference to the episodic average, for estimating episode-final QoE.
- An alternative of the last model is presented in [46], using also the difference of momentary speech QoE to the average instead of absolute weighting values to obtain the weighted average.
- For picture QoE, [20] also propose a model incorporating a fixed recency effect and the strength of impairment to calculate a weighted mean.
- A simple regression model for picture QoE using continuous rating includes the contributions of the level of the extreme degradation only. Although the latest ratings (recency effect for the last 5 s) is also significant, its inclusion does not improve the model, and duration of the extreme degradation and residual mean quality do no contribute significantly at all [21].

- A prediction model for QoE of streamed video over mobile networks is referred to in Sect. 19.3.3 [40]. It uses parametric estimates of momentary QoE, which are adapted to cover context effects (e.g., [14]) and integrated to estimate overall, i.e. episodic QoE.

Applying the simulated conversation test method of [12], an evaluation of three of these models [10, 12, 46] could confirm their relevance also for scenarios with changing audio bandwidth and packet loss [36]. Applying these models unmodified even on audio-visual simulated conversation QoE, all three resulted in satisfying estimates of episodic ratings, with [10] and [46] performing better than [12] (cf. also Sect. 27.4). A comparable evaluation for audio-visual QoE including all models mentioned above showed similar results [2]. The models show slightly lower performance when estimates of momentary speech and/or video quality are used instead of subjective momentary quality ratings, depending on the type of model used for the momentary quality estimation (see e.g. [36] for a comparison, and Sects. 12.3,14.3 and 19.2 for an overview of models for estimating momentary QoE).

Incorporating also delay and echo as interaction degradations, [15] propose a simple model to estimate episodic QoE for speech telephony based on instrumental estimates of momentary QoE. In addition to echo and delay, noise and packet loss are taken into account as listening degradations. The method to elicit authentic conversational situations uses short conversation tests as described in [24, 38] to obtain episodic QoE for the interactive scenario. Although not dealing with time-varying QoE, this approach provides a method and subsequently a model to integrate listening degradations with echo and delay as a basis to study temporal aspects for realistic interactive scenarios.

As a summary, the averaged momentary ratings typically account for most of the variance explained. Based on [2, 36, 46], Pearsons' r are about 0.84–0.90 for the plain average. Adapting momentary ratings to context effects (e.g. [14] improves this a little bit (r increase of about 0.05). Accounting for recency and extreme qualities results in values typically about or even over 0.95. Of course, with instrumental estimates of momentary QoE, correlations are typically lower (r over 0.9). Still, for data which is covered already very well by the plain average, additional modeling does not improve the correlations that much further.

It seems that the recency effect itself is not as strong as the impact of the strength of a degradation [2, 20, 46], although such a conclusion is dependent on the media stimuli used, and therefore difficult to draw. Yet, both systematic effects seem to resemble the mechanism described by [30] for remembered utility, and it would be interesting to compare the models explicitly defined for episodic QoE with the peak-end rule incorporating only the most extreme experience in addition to that for the last portion.

Apart from the question of the cognitive processes involved in integrating momentary to episodic QoE, there is of course the issue of valid and reliable assessment methods. There is already much knowledge available for this area, resulting in a number of standards (e.g. [25]). Still, most methods define and recommend laboratory settings for quality assessment, although these do not represent typical (and thus ecologically valid) usage situations. For example, assessing video quality for an entire movie in the living room of actual test participants [45] revealed a systematic difference compared to the laboratory setting. The authors conclude, that their more authentic method and test environment provides more valid results, e.g. a lesser impact of picture degradations and a stronger impact of degradations interrupting the flow of the movie.

10.5 Assessing Multi-Episodic QoE

Quality of Experience should be considered over multiple episodes as continued usage influences the user's expectations and future behavior towards a system.

Interacting with a system for the very first time, the experience made by a user is determined by his prior expectations and experience, which are used to form his internal reference. The comparison between the experience and the internal reference leads to a quality judgment of the experience. The internal reference will then be updated according to the user's individual experiences, and will influence the perception of future interactions with the system. The update process of the internal reference happens during and after each interaction with the system (see Sect. 2.3).

The change of the user's internal reference will influence not only the QoE of future episodic interactions, but also the user's acceptance and thus behavior towards the system. This includes likeliness to use and attitude towards the system, but also task selection and task solving strategies (see Sect. 2.3).

Adaptation effects have also been found in research on User Experience. It could be shown that usage behavior of a user with a system changes over longer time periods [34]: in the beginning, new features are explored and the interaction is playful, whereas with time interactions become more task-driven and practical.

Methods to assess short-term QoE are neither designed, nor suited to study multi-episodic QoE. In fact, sequence effects are frequently balanced out by randomizing the order of stimuli over multiple participants. This is due to the targeted performance comparison of different systems like codecs or media network configurations. Furthermore, some experimenters try to replace an unknown, participant-specific internal reference, which is used in the quality judgment process, by priming the participants in the beginning of the experiment using a fixed set of anchor stimuli of pre-defined levels of quality. This is especially important if participants are habituated to "better" systems than the ones under study.

Assessing multi-episodic, and thus remembered, QoE is challenging for three reasons:

1. First, the order of episodic use and their perception is important due to the update process of the internal reference.
2. Second, the time-scales that must be taken into account are greater, and thus the experiment has to be longer, sometimes spanning over days, weeks or months. Using such long experimental periods, it is very difficult to control for other

(external) factors which might also influence the quality judgment. To validly study temporal effects on multi-episodic QoE also pauses between episodic usage periods must be considered, so that the update process of the internal reference can happen under realistic conditions.

3. Third, the usage behavior is dependent on the user himself, e.g. his personal preferences, socialization, context and attitude towards the system. This will influence the user's approach to use the system, including task selection and usage patterns, and ultimately his level of acceptance.

A practical issue occurs if a system is evaluated over a long period, where participants also use other comparable systems during the same period. The user's internal reference is influenced by all systems experienced during the usage period.

First work on multi-episodic QoE which is known to us has been conducted by Duncanson in 1969 for an oversea telephony system. He could show that the remembered QoE for multiple prior episodic uses is underestimated in comparison to the judgment to a just-finished episode with the same actual performance [11]. This suggests that low-quality episodes have a greater impact on remembered QoE over multiple episodes.

In [39] a method to study multi-episodic QoE with one system over multiple days is presented. A comparable system usage is achieved by providing task scenarios for each usage episode, so that not only each episodic use is similar but also the interaction lengths and timing are comparable. This reduces the impact of the participant's behavior in the quality evaluation process. Each participant has to perform several (in this case 24) usage episodes in fixed time intervals (here twice a day) within a certain usage period (here 12 days). Two types of questionnaires are used: one for assessing the QoE after each episodic use and one to assess the multi-episodic QoE after several days. In addition, an initial interview and a final interview are conducted.

Möller et al. [39] used this method in a field study over 12 days providing two tasks per day that should be fulfilled using a video telephony system. The system performance was controlled on a day-to-day basis. Overall, 5 system performance profiles were used with a total of 56 participants.

Two effects on QoE were found: first, a recovery effect after low performance episodes, showing that prior episodes influence the QoE rating of following episodes. The recovery interval was approx. 2 days long. Second, a general rise in QoE ratings over the usage period of 12 days was noticeable. In [39] also the integration of individual episodic QoE ratings into an overall QoE judgement for the system was studied: the multi-episodic QoE judgment could be estimated by the average of episodic QoE ratings of all prior episodes only for several days, whereas it was not a good predictor for the multi-episodic QoE judgment on day 12.

This method was used in [17] to study multi-episodic QoE in a multi-service scenario addressing audio-visual entertainment and telephony. In this study, the results of [39] have been confirmed. In addition, it was found that the impact of performance limitations depends on the type and use-case of the system.

Multi-episodic QoE is of special relevance for service providers, especially in telecommunication and entertainment, because those services are used very frequently and the switching costs for customers are low [35]. Thus, it is important to provide good QoE over longer usage periods in order to avoid customer churn.

10.6 Discussion and Conclusion

With time-varying Quality of Service, the primary issues for assessing QoE are different for the three time spans presented, *momentary, episodic* and *multi-episodic*.

For *momentary* QoE, the focus lies on the instantaneous assessment using, e.g., a slider, so that a relationship between instantaneous service performance and user-perceived QoE can be determined. As a drawback of continuous assessment, the validity for authentic service usage is not ensured by such a permanent and untypical secondary task, which directs attention from content to the quality judgment process. Here, reliability and validity of alternative methods have to be studied. An alternative would be to use, e.g., physiological or non-permanent methods, like assessing only extreme moments of QoE or even actively controlling quality like presented in Sect. 10.3.

Methods to assess *episodic* and *multi-episodic* QoE do not seem to be very different from each other. However, it is important to distinguish the "mere" integration process of an episodic experience when rating the (retrospective) quality from the attitude towards a service built within multiple episodes over a longer time span. This attitude towards the service is much more affected by individual needs and preferences and stronger linked to personal life. Therefore, authentic situations are much more important when assessing *multi-episodic* QoE compared to established and valid laboratory methods for *episodic* QoE. Multi-episodic QoE assessment thus requires field tests, although this results in less control over the test set-up, e.g. in creating specific quality profiles.

Relying on valid data might not only provide enough material, i.e. "profiles" of time-varying service performance, to build valid models, but will also provide insight into realistic distributions of time-varying performance to validate such models even better. Thus, the ongoing merging of service monitoring with QoE research is expected to solve several issues presented here.

Prediction models incorporating empirical results are at hand for short sample aggregations of momentary QoE and episodic QoE. However, there are still many relevant factors that are not considered to satisfy the demands of applications like monitoring or planning. The principal problem is the trade-off between incorporating factors like content, attention etc. in a reasonable, i.e. generic and scalable, way. And for multi-episodic QoE research related to modeling has just started. Here, especially context factors affecting attention (secondary tasks of users, interruptions, parallel usage of different devices) or the attitude towards the service (availability, mobility, and even more than for episodic usage: content) will also play an important role.

References

1. Baddeley A (2005) Human memory: theory and practice. Psychology Press, Hove (revised edn.)
2. Belmudez B, Lewcio B, Möller S (2012) Call quality prediction for audiovisual time-varying impairments using simulated conversational structures. Acta Acustica united Acustica 99:792–805
3. Borowiak A, Reiter U (2013) Long duration audiovisual content: impact of content type and impairment appearance on user quality expectations over time. In: Proceedings of 5th International Workshop on Quality of Multimedia Experience (QoMEX), Klagenfurt, pp 200–205
4. Borowiak A, Reiter U, Svensson UP (2012) Evaluation of audio quality requirements over extended periods of time using long duration audiovisual content. In: Advances in multimedia information processing. PCM, Lecture notes in computer science, vol 7674, pp 10–20
5. Borowiak A, Reiter U, Svensson UP (2012) Quality evaluation of long duration audiovisual content. In: Proceedings of the 9th annual IEEE consumer communications and networking conference. Special session on quality of experience (QoE) for multimedia communications, Las Vegas, pp 353–357
6. Borowiak A, Reiter U, Tomic O (2012) Measuring the quality of long duration AV content. Analysis of test subject/time interval dependencies. In: EuroITV—Adjunct Proceedings, Berlin, pp 266–269
7. Bouch A, Sasse MA (2000) The case for predictable media quality in networked multimedia applications. In: Procedings of ACM/SPIE multimedia computing and networking (MMCN), San Jose, pp 188–195
8. Buchinger S, Robitza W, Nezveda M, Sack M, Hummelbrunner P, Hlavacs H (2010) Slider or glove? Proposing an alternative quality rating methodology. In: Proceedings of the 5th international workshop on video processing and quality metrics for consumer electronics (VPQM), Scottsdale, Arizona
9. Clark A (2001) Modeling the effect of burst packet loss and recency on subjective voice quality. In: Proceedings of the internet telephony workshop (IPtel 2001), New York
10. Delayed Contribution ITU-T,D.064 (1998) Testing the quality of connections having time varying impairments. Source AT&T, USA (J. H. Rosenbluth) ITU-T Study Group 12 Meeting. International Telecommunication Union, Geneva
11. Duncanson JP (1969) The average telephone call is better than the average telephone call. Public Opin Q 33(1):112–116
12. ETSI TR 102 506: Speech Processing (2007) Transmission and quality aspects (STQ); Estimating speech quality per call. European Telecommunications Standards Institute, Sophia Antipolis
13. Gray P, Massara R, Hollier M (1997) An experimental investigation of the accumulation of perceived error in time-varying speech distortions. In: Proceedings of audio engineering society, 103rd convention, New York
14. Gros L, Chateau N (2001) Instantaneous and overall judgements for time-varying speech quality: assessments and relationships. Acta Acustica united Acustica 87:367–377
15. Guéguin M, Le Bouquin-Jeannès R, Gautier-Turbin V, Faucon G, Barriac V (2008) On the evaluation of the conversational speech quality in telecommunications. EURASIP J Adv Sig Proc 8:1–15
16. Guéguin M, Gautier-Turbin V, Gros L, Barriac V, Le Bouquin-Jeannès R, Faucon G (2005) Study of the relationship between subjective conversational quality, and talking, listening and interaction qualities: towards an objective model of the conversational quality. In: Proceedings of measurement of speech and audio quality in networks, Prague
17. Guse D, Möller S (2013) Macro-temporal development of QoE: impact of varying performance on QoE over multiple interactions. In: Proceedings of AIA-DAGA conference on Acoustics, Merano, Deutsche Gesellschaft für Akustik, Berlin

18. Gutierrez J, Perez P, Jaureguizar F, Cabrera J, Garcia N (2011) Subjective evaluation of transmission errors in IPTV and 3DTV. In: Proceedings of visual communications and image processing, Tainan
19. Hamberg R, de Ridder H (1995) Continuous assessment of perceptual image quality. J Opt Soc Am A 12:2573–2577
20. Hamberg R, de Ridder H (1999) Time-varying image quality: modeling the relation between instantaneous and overall quality. SMPTE Motion Image J 108:802–811
21. Hands D, Avons S (2001) Recency and duration neglect in subjective assessment of television picture quality. Appl Cognitive Psychol 15:639–657
22. Ito TA, Larsen JT, Smith NK, Cacioppo JT (1998) Negative information weighs more heavily on the brain: the negativity bias in evaluative categorizations. J Pers Soc Psychol 75:887–900
23. ITU-T Contr. COM 12-340 (2012) Methodology for the assessment of audiovisual quality for simulated video calls. ITU-T Study Group 12 Meeting. International Telecommunication Union, Geneva
24. ITU-T Contr. COM 12-35 (1997) Development of scenarios fort a short conversation test. ITU-T Study Group 12 Meeting. International Telecommunication Union, Geneva
25. ITU-T Contr. COM 12-38 (2013) Proposal for a subjective method for simulated conversation tests addressing speech and audio-visual call quality (P.ACQ). ITU-T Study Group 12 Meeting. International Telecommunication Union, Geneva
26. ITU-R Recommendation BT.500-7 (1996) Methodology for the subjective assessment of the quality of television pictures. International Telecommunication Union, Geneva
27. ITU-T Recommendation P.800 (1996) Methods for subjective determination of transmission quality. International Telecommunication Union, Geneva
28. ITU-T Recommendation P.910 (2008) Subjective video quality assessment methods for multimedia applications. International Telecommunication Union, Geneva
29. ITU-T Recommendation P.911 (1998) Subjective audiovisual quality assessment methods for multimedia applications. International Telecommunication Union, Geneva
30. Kahneman D (1999) Objective happiness. In: Kahneman D, Diener E, Schwarz N (eds) Well-being: the foundations of hedonic psychology. Russel Sage, New York, pp 3–25
31. Kahneman D (2000) Experienced utility and objective happiness: a moment-based approach. In: Kahneman D, Tversky A (eds) Choices, values and frames. Cambridge University Press, New York
32. Kenny DA (2004) PERSON: a general model of interpersonal perception. Pers Soc Psychol Rev 8:265–280
33. Kokotopoulos A (1997) Subjective assessment of a multimedia system for distance learning. In: Multimedia applications, services and techniques—ECMAST, Lecture notes in computer science, vol 1242, pp 395–408
34. Kujala S, Roto V, Väänänen-Vainio-Mattila K, Karapanos E, Sinnelä A (2011) UX curve: a method for evaluating long-term user experience. Interact Comput 23(5):473–483
35. Lee J, Lee J, Feick L (2001) The impact of switching costs on the customer satisfaction-loyalty link: mobile phone service in France. J Serv Mark 15:35–48
36. Lewcio B (2013) Management of speech and video telephony quality in heterogeneous wireless networks. Doctoral dissertation, Technische Universität zu Berlin, Berlin
37. Liu T, Cash G, Narvekar N, Bloom J (2012) Continuous mobile video subjective quality assessmentusing gaming steering wheel. In: Proceedings of the 6th international workshop on video processing and quality metrics for consumer electronics (VPQM), Scottsdale, Arizona
38. Möller S (2000) Assessment and prediction of speech quality in telecommunications. Kluwer Academic Publishers, Boston
39. Möller S, Bang C, Tamme T, Vaalgamaa M, Weiss B (2011) From single-call to multi-call quality: A study on long-term quality integration in audio-visual speech communication. In: Proceedings of interspeech, florence, International SpeechCommunication Association, pp 1485–1488
40. Next generation mobile networks alliance (2013). In: Wennesheimer M, Robinson D (eds) Service quality definition and measurement—a white paper, Frankfurt, Germany

41. Pinson M, Wolf S (2003) Comparing subjective video quality testing methodologies. Proc SPIE 5150:573–582
42. Raake A (2006) Short- and long-term packet loss behavior: towards speech quality prediction for arbitrary loss distributions. IEEE Trans Audio Speech Lang 14(6):1957–1968
43. Recommendation ITU-T,P.880 (2004) Continuous evaluation of time varying speech quality. International Telecommunication Union, Geneva
44. Roto V, Law E, Vermeeren A, Hoonhout J (Eds) (2011) User experience white paper: bringing clarity to the concept of user experience. Result from Dagstuhl seminar on demarcating user experience, 15–18 Sep 2010. www.allaboutux.org/uxwhitepaper
45. Staelens N, Moens S, Van den Broeck W, Mariën I, Vermeulen B, Lambert P, Van de Walle R, Demeester P (2010) Assessing quality of experience of IPTV and video on demand services in real-life environments. IEEE Trans Broadcast 56:458–466
46. Weiss B, Möller S, Raake A, Berger J, Ullmann R (2009) Modeling call quality for time-varying transmission characteristics using simulated conversational structures. Acta Acustica united Acustica 95(6):1140–1151

Chapter 11
Quality of Experience and Interactivity

Sebastian Egger, Peter Reichl and Katrin Schoenenberg

Abstract This chapter discusses the relation between interactivity and QoE. In this context, a definition of interactivity comprising human-to-human interaction as well as human-to-machine interaction is presented, and a description of a possible instrumentation is given. In terms of quality formation, a mediation layer between quality influence factors and perceived quality features is introduced that allows the inclusion of interactivity-related perception in the quality formation process. A discussion of commonalities and differences between interaction with a system and interaction with one or several other persons via a system identifies the open challenges for reliable and successful measurement of interactivity related aspects and the identification of relationships between these interaction measures and QoE.

11.1 Introduction

Quality of Experience is a multidimensional concept. While, in the course of this book so far, mainly the quality of experience of individual users has been addressed from a psychological, physiological and temporal as well as technological and eco-

S. Egger (✉)
Telecommunications Research Center Vienna (FTW), Vienna, Austria
e-mail: egger@ftw.at

P. Reichl
University of Vienna, Vienna, Austria
e-mail: peter.reichl@univie.ac.at

P. Reichl
Université Européenne de Bretagne/Télécom Bretagne Rennes, 2, rue de la Châtaigneraie
F-35576, Cesson-Sevigne, France

K. Schoenenberg
Assessment of IP-based Applications, Telekom Innovation Labs, TU Berlin, Berlin, Germany
e-mail: katrin.schoenenberg@telekom.de

S. Möller and A. Raake (eds.), *Quality of Experience*,
T-Labs Series in Telecommunication Services, DOI: 10.1007/978-3-319-02681-7_11,

nomical point of view, in this chapter we will focus on QoE issues that arise from the communication process between several entities and the way they interact with each other. Traditionally, this essential perspective has been addressed mainly in the context of human-to-human (H2H)—and, to a lesser extent, human-to-machine (H2M)—communication, while more recently also machine-to-machine (M2M) aspects have gained rapidly increasing relevance. Therefore, we will follow a rather broad approach and discuss the corresponding fundamental concepts and notions in an abstract way including all different basic scenarios.

In order to capture the intrinsic features of the mentioned communication processes, we resort to the notion of "interactivity" which we will formally define in the next section. It is however instructive to point out from the very beginning that this concept comprises a phenomenological (form) as well as a teleological (function) component. In terms of describing interactive phenomena in a formal way, this has led to the definition of "conversational temperature" (a term coined by M. Balinova almost a decade ago, cf. [1]) as a metric characterizing the intensity of the interaction process, while the underlying forces and motivations which trigger and influence the mutual behaviour of the communication partners have been characterized in [2] in terms of a game-theoretic equilibrium between respective user strategies optimizing the mutual exchange of information.

As far as underlying technology is concerned, it is mainly the intermediate communication channel—and more specifically its *two-way delay* characteristics[1]—which is responsible for the need to distinguish interactive from non-interactive QoE. Of course, this delay has a direct impact on the quality perception itself (as shown by [3]), but beyond that it may also massively influence the information sending/receiving behaviour of the individual communication partners involved in the communication.

Figure 11.1 depicts the fundamental structure of interactive communication. While the x-axis refers to time, we see how requests (REQ) and related responses (RES) are exchanged between a user A and a receiver B via the intermediate transmission channel with constant one-way delay. Messages are assumed to exhibit an underlying fine granular structure (for more details on the left dashed circle see Fig. 11.1). Requests can be initiated by both sides, and responses typically follow them in time, however, in certain cases (cf. right dashed circle) responses are started already before the end of the request transmission or are interrupted by additional arriving messages, see Fig. 11.3 for more details. Eventually, this can even lead to largely different perceptions with respect to the actual interaction pattern as pointed out in [4], leading for instance to the distinction between active and passive interruptions.

The remainder of this chapter is structured as follows: after providing a formal definition for interactivity in Sect. 11.2 and a brief overview of corresponding metrics in Sect. 11.3, we discuss the differences between static and interactive quality experiences in more detail in Sect. 11.4. Section 11.5 discusses specific aspects of

[1] Also other distortions in the communication channel such as e.g. echo or noise to impact the interaction behaviour of interactants. However, throughout this chapter we focus on transmission delay as most influential impairment for human mediated communication.

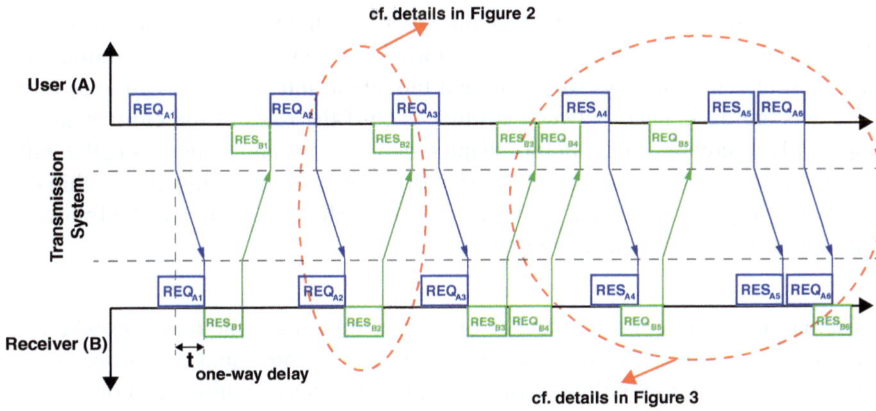

Fig. 11.1 An interactivity constituting request-response pattern

interacting with systems or persons. Section 11.6 concludes the chapter with a brief summary and outlook.

11.2 Interactivity Definitions

The topic of interactivity is a widely used concept which is rooted in several different research traditions. As the aim of this chapter is the assessment of interactivity and the identification of the effects interactivity exerts on QoE, as it is essential to differentiate between the different phenomena all identically labelled interactivity.

Common to all understandings of interactivity is the fact that interaction can only take place if certain interactive acts are performed by at least two actors communicating with each other. Nevertheless, the nature of the interactants (humans, machine, media) as well as the way how they interact with each other are a crucial point of differentiation between the existing concepts of interactivity. A classification along the most prominent categories of interactivity has been proposed by [5], distinguishing between:

Interactivity as Process: is interaction taking place between human subjects where subsequent messages consist of responses to prior messages or requests in a coherent fashion. Note that, in principle, the roles of the interactants are reciprocal and can be exchanged freely.
Interactivity as Product: occurs when a set of technological features allows users to interact with the system.

This classification already points towards the different scholar traditions of human interaction and human to system (computer) interaction. Human interaction researchers are rather strict in defining interactivity such as Rafaeli [6]. In their understanding, *true* interaction can only take place between human interactants when their roles (within the interaction) are 100 % reciprocal. In contrast, scholars in human-to-machine interaction are less stringent, and talk about interactivity as soon as interactive acts are exchanged between entities, even if the roles of the entities are not reciprocally interchangeable. In this chapter, however, we aim at analyzing the influence of interactivity on QoE for all types of different services as targeted within this book (including both H2H and H2M interaction).[2] Hence, we choose the following definition of interactivity as common ground:

> An interactive pattern is a sequence of actions, references and reactions where each reference or reaction has a certain, ex-ante intended and ex-post recognisable, interrelation with preceding event(s) in terms of timing and content.

Without loss of generality we restrict the further discussion throughout this chapter to request-response patterns which are considered to be the common ground for both H2H and H2M interactions.[3]

An exemplification of the above understanding of interactivity is depicted in Fig. 11.2. The most common feature constituting the exchange of interactive acts and thereby interactivity is the recurring characteristic of the request-response pattern. The user (A, top of Fig. 11.2) issues a request which is transmitted to the receiving side (B, bottom of Fig. 11.2). Now, the receiver (B) processes the request and starts responding by sending data back to the the user (A) again. In both directions, messages may exhibit a fine granular structure, which is shown in Fig. 11.2 as a sequence of arrows where different thicknesses are used to indicate the "semantic intensity" of the corresponding content. Following the model outlined in [2], one can for instance assume that the most important pieces of information (e.g. key answer facts in human conversation, HTML format instructions in Web traffic, etc.) are contained in the earlier parts of a response, while with ongoing message length, the corresponding information becomes less dense and/or less important. As a consequence, the receiver might be tempted to start her next action already before the entire message has arrived. While such a behaviour is typically observed in everyday communication, from an overall system behaviour it can also very naturally be interpreted as a Nash Equilibrium that maximizes the overall information exchange between both participants [2].

Hence, after a certain time, from the viewpoint of the user (A) the transmitted response leads to an intermediate rendering result which is considered already suffi-

---

[2] Due to space limitations we can only discuss interactivity for certain interactive services within this chapter, and hereby want to point out that our definitions of interactivity as well as the contribution of interactivity to the overall quality formation process are also valid for other interactive services such as sensory experiences and interactive gaming, as described in Chaps. 24 and 25, respectively.

[3] Human interaction scholars might argue that restricting interaction to request-response patterns is no longer an analysis of *true* interaction but rather *quasi* interaction (cf. [5–7]). However, as we target a broad range of services in addition to H2H interaction we are confident that this restriction is adequate for identifying the influence of interactivity on QoE for all of these services.

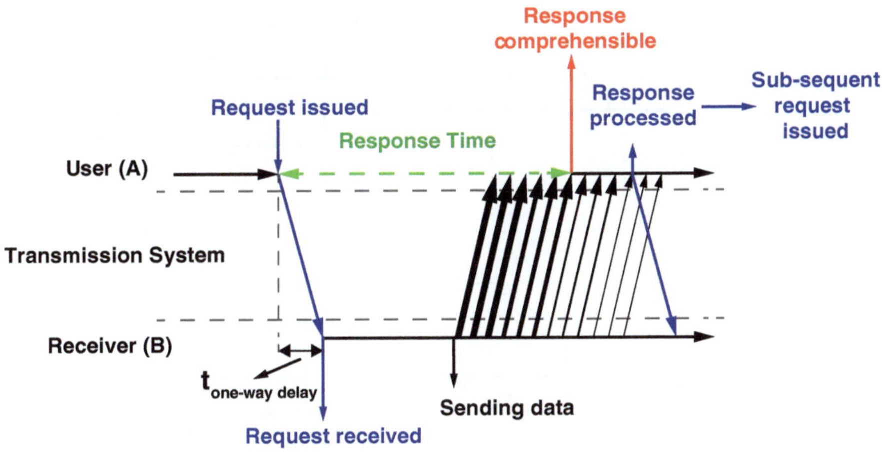

Fig. 11.2 An interactivity constituting request-response pattern taken from [13, 14]

cient by the user (we define this rendering state as "response comprehensible"). The user (A) starts now with processing the response. However, the receiver (B) might keep on with responding (e.g. in case of a long utterance or a heavy web page). After processing the response, the user (A) issues a sub-sequent request, thereby starting a new request-response cycle. Here it is important to understand that the issuing of the follow-up request by the user (A) does not necessarily take place after the complete response has been received at the user (A). He issues the request when he has acquired sufficient information from the (eventually technically incomplete) response and has processed it accordingly.[4] The essential characteristic here is a certain relation between the response and a preceding event (in the simple case only the relation to a single request). Together with the above definition of interactivity, it gets clear that *the request-response characteristic* is a *distinct feature of interactivity*. Considering the differentiation between H2H interaction and H2M interaction it can also be said that in terms of the receiving side (B), the nature (human, machine, etc.) of the entity answering the request is not essential for establishing an interactive request-response pattern, as it is, for example, the case in spoken dialogue systems, which makes it applicable to both interaction types and respective applications.

[4] This model is based on observations of H2H-communication interactions reported in [8–10] where users were interrupting the other person frequently, and observations of H2M interaction, where similarly users, while web-browsing [11, 12], were navigating further on a web page through clicking on a respective link before the web page was fully loaded. This lower bound of sufficient information (for issuing a subsequent request) might be defined in two ways: (1) with a relative or absolute amount of information (e.g. 70 % of rendered screen area, or fully rendered screen) (2) based on the considerations from [2] where the bound is reached after the entropy of user (A) gets smaller then the entropy of the response of user (B) in order to maximize the amount on information exchanged.

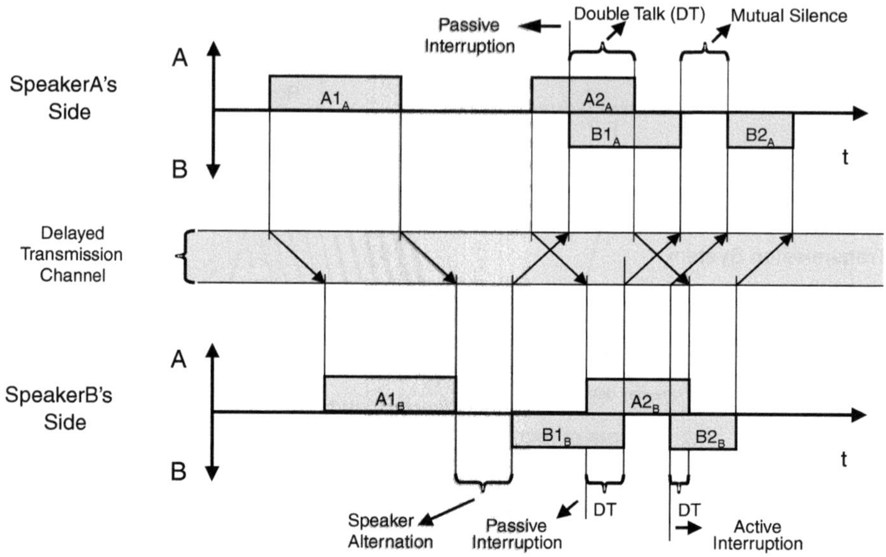

Fig. 11.3 Surface structure of an interactive conversation taken from [2, 9]

11.3 Measurement of Interactivity

Based on the definition of interactivity provided in the previous section, we are now discussing its instrumentation, i.e. how to make interactivity measurable. Deriving a formal interactivity metric has turned out to be a non-trivial task [4], although it does not seem too difficult to develop an intuitive understanding about different levels of interactivity, especially in the case of H2H conversations, which will therefore serve as our starting point. Hence, coming back to Fig. 11.1, we observe that the surface structure of an interactive conversation between two partners A and B may formally be described as a sequence of four states. These states are a result of each speaker alternating between active and passive periods which can be numbered according to their temporal order, see Fig. 11.3. Together with the delay added by the transmission channel, on each speaker's side this leads to a sequential chain made up of four states: "A" if speaker A is active (only), "B" if speaker B is active (only), "Double Talk" if both speakers are active, and "Mutual Silence" if none of them is active, see Fig. 11.4.

Note that, due to the mentioned transmission delay, the interaction pattern as observed by speaker A may substantially differ from the one observed by speaker B, while of course both of them are entangled in the same conversation. This leads to two immediate consequences: first of all, the notion of "interruption" becomes ambivalent, as we have to distinguish between "active interruptions" (where one of the speaker becomes active while she is still receiving an ongoing talk spurt) and "passive interruptions" (where an active speaker is interrupted by an arriving talk spurt which has not been intended as an interruption but has been subject to

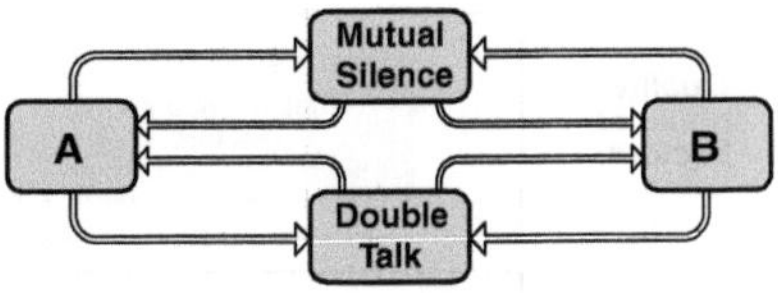

Fig. 11.4 Conversation states as a markov process according to [2, 15]

transmission delay), see Fig. 11.3. Moreover, while the mentioned sequence order of states may differ between the two speakers, any valid interactivity metric needs to be uniquely defined and hence must be independent of the underlying perspective, i.e. for the interactivity metric it must not make a difference whether it is calculated based on the interaction pattern as seen by A or B (symmetry criterion).

Figure 11.4 depicts the four states of an interactive conversation as well as the possible transitions between them as defined in [16]. In order to derive an interactivity metric, we may interpret the state sequence in terms of a Continuous Time Markov Chain with sojourn times equivalent to the durations of states "A", "B", "DT" and "MS", respectively. Together with some basic assumptions about limiting behaviour, normalization and monotonicity/first-order properties, this results in an exponential metric which shows a surprising strong analogy to the physical notion of temperature as defined by statistical thermodynamics, see [4] for further details. However, as demonstrated by Hammer [17], we can achieve similar accuracy also with less sophisticated means, for instance using the so-called *Speaker Alternation Rate* (*SAR*) which is defined as the average number of speaker alternations per minute (cf. Fig. 11.3). Note that both the "conversational temperature" as well as the SAR fulfill the stated symmetry criterion, at least to a sufficient degree [4].

Extending these approaches to the case of H2M interactivity, one could think of using click rates (for the case of web pages) or more generally the number of exchanged requests and responses over a certain time span as closely related interactivity metrics.

11.4 Quality Formation: Differences Between Static and Interactive Experiences

Existing approaches to explain the quality formation process within a person such as [18–20] mainly target media experiences on a single and static (in the sense of interactivity) input signal and do not consider actions by the experiencing person. As a result, these approaches do not account for recurring (inter) actions between two or more entities,[5] which result in the interactive request-response cycle described in Sect. 11.1 and its related signals. The quality formation process described in Chap. 2 integrates already certain (exploratory) actions by the person, however it still does not

[5] Thereby running several times through the respective perception and judgement processes.

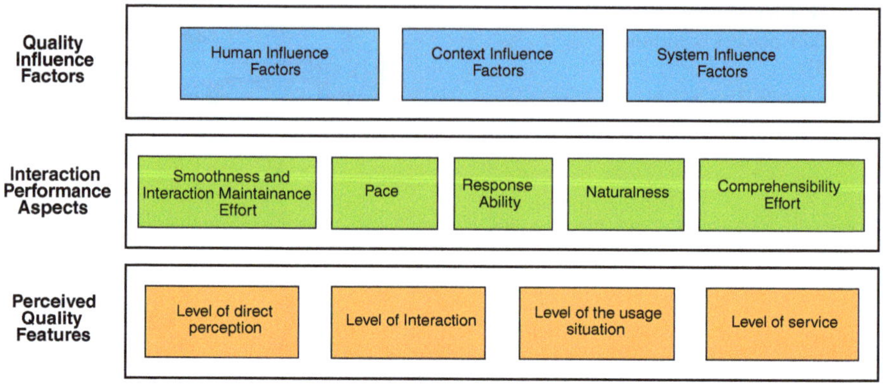

Fig. 11.5 Taxonomy of influence factors, interaction performance aspects and quality features, from [21] adapted with terminology from [19].

consider the (inter) actions between different interaction parties and their contribution to the formed quality.

An approach to overcome this shortcoming is outlined in the taxonomy proposed by Möller et al. [21, 22]. It incorporates the influence of the interaction process on the overall quality formation process by introducing an additional layer of *interaction performance aspects* which acts as mediation layer between *quality influence factors* and *perceived quality features*[6] as depicted in Fig. 11.5. This mediation layer spans over several stages of the quality formation process (cf. Chap. 2), therefore the relationships between these layers are not one to one and can vary in their strength depending on the system, user, or context (cf. [21]). Naturally, such a mediation layer does of course not fully integrate (inter) action between entities into the quality formation process, however it is a simple and efficient way to consider the influence interactivity exerts on QoE.

These interaction performance aspects result from the process of interaction between two or more entities and their perception of this process on several dimensions as depicted in Fig. 11.5 and described as follows:

Smoothness and Interaction Maintenance Effort: is how fluent and effortless the users experience the conversational flow. If normal interaction behaviour has to be adapted as a result of bad system performance in order to maintain the ongoing interaction as well as possible, the interaction will usually also be perceived as being less smooth. Typically, interaction has an inherent pace it establishes, thereby keeping the maintenance efforts of the interaction parties minimal. However, due to system impairments the interaction pace can be changed, thereby accordingly demanding additional user effort in order to adapt to the changed pace. For H2M interaction, this can severely impact the *flow experience* or the

[6] Note that the *quality influence factors* in the quality formation process described in Chap. 2 are considered in the lower left box that feeds in the sensory processing cirle in Fig. 2.3.

experienced smoothness, whereas for H2H interaction the conversational rhythm can be impaired (cf. [10]).

Pace: is the users' perceived promptness of the interactional acts and respective actions by the other entity.

Response Ability: denotes if it is possible to issue a response following a prior message or response from the system or other user. Response abilities through interruptions in H2H interactions can be severely obstructed by transmission delays, as interruptions may not arrive in time and are not able to interrupt in the way it was originally intended. In terms of browser-based applications, the modality type of the response can be a source of difficulty, but is rather caused by the web-site content in this application type.

Naturalness: is related to the inherent knowledge about how an interaction takes place in a non-mediated or ideal case.

Comprehension Effort: is required to understand either the other interlocutor (in case of H2H interaction) or needed to interpret the response from the machine. Comprehension can be distorted by e.g. double talk or non-rendered portions of the webpage which might be needed for navigation or information retrieval.

It has to be noted that the above aspects cannot be seen as disjunct factors, hence overlaps of the different concepts are possible.

In terms of quality formation, the output from this interactions performance-aspects layer is further translated into interaction-quality features, and then constitute an additional input to the comparison and judgement stage (cf. Fig. 2.3 in Chap. 2), where they are further processed in conjunction with the other (more media-related) quality features.

11.5 Interacting with Systems or Persons

An important question that needs to be answered when looking into interactivity within the scope of QoE is: Are we talking about an interaction *with* a system or an interaction with another person or group of persons *via* a system?

There have been profound works dealing with the quality formation for multi-modal systems, for instance, with smart-home or dialogue systems (see, e.g. [21, 23]). First insights have been described for the case of interaction with web-sites [12], too. Although there has been much work on QoE for two-party audio and video communication systems [20, 24–26], the aspect of interaction via such systems and its analyses has been limited to telephone communication [8, 10, 17, 27]. For multiparty communication, research is currently ongoing [28], but due to the high degree of possible diversity, QoE becomes even more difficult to predict (see Chap. 15). We can summarize that interaction behaviour plays a role in different research areas addressing both types of interaction—*with* and *via*—systems.

In the following, we are going to discuss major differences and similarities for the two cases, and the associated implications on QoE will be outlined.

In both cases, when interacting *with* or *via* a system, the initial situation comprises (a) a human (or humans) having perceptions, emotions, motivations and behaviours that are built on the human's personality, experiences and the current state and (b) a system providing a certain QoS and (c) a context (or contexts).

Furthermore, for both, the interaction quality level is fundamentally determined by the "speed or pace, conciseness, smoothness, and naturalness of the interaction" [21] and other interaction performance aspects as described in Sect. 11.4. Following from that, transmission delays and asynchronous transmission of different channels are critical.

To illustrate the role of interaction from the viewpoint of a user, the process of interaction can be understood as a tool for him or her to infer the quality provided by the system of interest. In the H2M context, the interaction quality directly provides information to the user on the quality of the system.[7] The gathered information, e.g. how comprehensible messages are, how easy it is to respond or how fluently an interaction can be maintained, can directly be transfered into a quality judgment. Interaction measures can therefore be considered to have a direct link to QoE.

In contrast, in the case of H2H communication, interaction problems experienced by the user can either be due to the person at the other end or due to the system. For some cases the identification is obvious; for instance, if background noises are falsely amplified, it may be difficult to maintain a fluent conversation, but it can easily be identified as a technical problem. The quality can be rated accordingly low. However, for other cases the reason, i.e. whether it is the other person or the system which is responsible for a low performance, is not clearly identifiable. When response times are rather long, for example, it can either be due to a long transmission delay or to a slowly responding person at the other end. Similarly, inappropriate timing of utterances can be due to either the system sending out messages at an incorrect time or to someone being inattentive or impolite and not getting the timing right. From face-to-face interaction, the most common and natural way of interaction, people are used to search for the reason of low interaction quality within the other person or the relationship to him. Therefore, it becomes difficult for the user to dissociate technical from human-related causes when interacting via a mediated communication system. As a result, QoE ratings can be distorted by wrongly attributed causes. How familiar interlocutors are with each other, the level of inherent structure of the conversation (e.g. storytelling vs. question-answer patterns), how well people are acquainted with possible malfunctions of the system, and if they are able to attribute them correctly, becomes very critical for the resulting QoE judgments.

In terms of predicting QoE, measuring interaction behaviour can serve as additional input for prediction algorithms [27]. The underlying idea is that certain interaction patterns help to understand why quality is perceived in a particular way. If, for example, a usually fast interactive conversation slows down dramatically and in addition quality judgments drop considerably as well, one can follow that people had to slow down their interaction pace due to some technical issues and additionally

[7] including all system parts such as e.g. the transmission path, the interface, amount of information stored etc.

reflected this in their judgments. Interpersonal factors can be excluded as an explanation for such effects, if the test design is chosen accordingly (e.g. randomized allocation of participants to groups).

When it comes to quality assessment, some more points need to be considered for mediated H2H interaction. First of all, the same interaction and usage situation is judged by at least two humans having different perceptions, emotions, motivations and behaviours based on different personalities, experiences and current states. Therefore, the very same interaction with identical QoS can lead to different quality perceptions and judgements. Hence, in these cases QoE evaluation results always require careful interpretation and a clear definition of the desired aggregation level of quality, for example, whether the focus is on the quality of each individual line or on an overall score (for more details on this issue, see Chap. 15). Second, more than one context influences the quality perception, since communication partners are usually distributed over remote locations. As a third point, the number of interlocutors may play an important role. The more interlocutors join a mediated call, the more attention might be needed to follow the discussions [29], and as a result, less attention is available for the evaluation of quality (for details, see Chap. 15).

Measuring communication behaviour becomes more complex with an increasing number of interactants, too. If multiple people are involved in one interaction, it needs to be specified if measures are based on an individual or a group level. Either individual performances are assessed and used to explain certain outcomes regarding quality, or group means (or medians) serve for the same purpose. As a third solution, for some measures it is possible to calculate an overall group value. For example, how often did someone (no matter who) misunderstand a prior utterance. The chosen level should be in line with the aggregation level of the measurement. If the aim is to relate interaction to individual quality ratings of different participants, individual interaction measures may be helpful to assess. When explaining outcomes on the overall quality (as e.g. the mean of all participants) of, e.g., a conference call, measures on the group level may be the better choice.

This discussion has shown that it is a long way to go towards reliably and successfully identifying relationships of interaction measures and QoE. However, discussing commonalities and differences of human-to-system (-machine) and human-to-human communication is a first step for acquiring in-depth knowledge on the role interaction metrics can play to predict QoE in the different cases of interaction *with* or *via* a system.

11.6 Conclusions

In this chapter, we highlight key QoE aspects in interactive mediated communications and interactive system use. We argue that the phenomenon of interactivity adds an essential new and complementary perspective to QoE as discussed in the previous chapters of this book, and thus needs both a careful definition and an efficient instrumentation, both provided in the first half of the chapter. Based on that, the specifics of quality formation in an interactive context is discussed and leads to the integration of a novel mediation layer in-between quality influence factors and perceived quality features, in order to handle interaction performance aspects. Finally, we address also the distinctive features of interaction with a system versus interaction via a system, and discuss related impact factors on QoE evaluation.

In general, our discussion shows that interactive phenomena may exert a non-negligible influence on QoE, while at the same time requiring a substantial extension of research and monitoring concepts. Especially the fundamental role of channel delay characteristics has become obvious and turns out to be a crucial issue for interactive services. Therefore, current and future work concentrates on investigating specific examples where interactivity-related impairments are measurable, for instance delay-induced interaction deficiencies in human-to-human interaction or distortion of flow in human-to-machine interaction contexts.

References

1. Reichl P, Balinova M, Hammer F (2005) Measuring non-spontaneous interactivity—an opera-related case study. In: Proceedings of 5th open workshop of musicnetwork—integration of music in multimedia applications, Vienna, Austria
2. Reichl P (2007) How to define conversational interactivity: a game-theoretic approach and its application in telecommunications. J Inf Technol Control (JITC) 3(No. 3–4/2006):18–24
3. Kitawaki N, Itoh K (1991) Pure delay effects on speech quality in telecommunications. IEEE J Sel Areas Commun 9(4):586–593
4. Hammer F, Reichl P, Raake A (2005) The well-tempered conversation: interactivity, delay and perceptual VoIP quality. In: 2005 IEEE international conference on communications (ICC 2005), vol 1, pp 244–249. doi:10.1109/ICC.2005.1494355
5. Stromer-Galley J (2004) Interactivity-as-product and interactivity-as-process. Inf Soc 20(5):391–394. doi:10.1080/01972240490508081. http://www.ingentaconnect.com/content/routledg/utis/2004/00000020/00000005/art00008
6. Rafaeli S (1998) Interactivity: from new media to communication. In: Hawkins RP, Wiemann JM, Pingree S (eds) Advancing communication science: merging mass and interpersonal processes. Sage Publications, Beverley Hills, pp 110–135
7. McMillan S (2005) Exploring models of interactivity from multiple research traditions: users, documents and systems. Handbook of new media 2:205–229
8. Egger S, Schatz R, Scherer S (2010) It takes two to tango—assessing the impact of delay on conversational interactivity on perceived speech quality. In: Interspeech, pp 1321–1324
9. Egger S, Schatz R, Schoenenberg K, Raake A, Kubin G (2012) Same but different?—Using speech signal features for comparing conversational VoIP quality studies. In: IEEE ICC 2012—communication QoS, reliability and modeling symposium (ICC'12 CQRM). Ottawa, Ontario, Canada
10. Schoenenberg K, Raake A, Egger S, Schatz R (2012) On interaction behaviour in telephone conversations under transmission delay. Speech Communication. Submitted June 2012
11. Egger S, Reichl P, Hoßfeld T, Schatz R (2012) 'Time is Bandwidth'? Narrowing the gap between subjective time perception and quality of experience. In: IEEE ICC 2012-communication QoS, reliability and modeling symposium (ICC'12 CQRM). Ottawa, Ontario, Canada
12. Egger S, Schatz R, Hoßfeld T, Müllner W (2013) ITU-T SG 12 contribution C-033: perceptual events in a page view cycle outcome from the interim meeting in Berlin 11/2012. Tech. rep, FTW, Geneva, Switzerland

13. Fiedler M (2004) Deliverable D.WP.JRA.6.1.1: state of the art with regards to user perceived quality of service and quality feedback. Tech. rep., EuroNGI (2004). http://eurongi.enst.fr
14. Egger S, Hoßfeld T, Schatz R, Fiedler M (2012) Tutorial: waiting times in quality of experience for web based services. In: IEEE QoMEX 2012, Yara Valley, Australia
15. ITU-T Rec. P.59: artificial conversational speech. International Telecommunication Union, CH-Geneva (1993)
16. Brady PT (1968) A statistical analysis of on-off patterns in 16 conversations. Bell Syst Tech J 47(1):73–91
17. Hammer F (2006) Quality aspects of packet-based interactive speech communication. Ph.D. thesis, Signal processing and speech communication laboratory, Faculty of Electrical and Information Engineering, University of Technology Graz, Graz, Austria
18. Jekosch U (2005) Voice and speech quality perception: assessment and evaluation. Signals and communication technology. Springer, Berlin. http://books.google.at/books?id=Ef3lHiSzq1QC
19. Möller S, Le Callet P, Perkis A (eds) (2012) Qualinet white paper on definitions of quality of experience—output version of the Dagstuhl seminar 12181: European network on quality of experience in multimedia systems and services (COST Action IC 1003), Lausanne, 1.1 edn
20. Raake A (2006) Speech quality of VoIP: assessment and prediction. Wiley, New York
21. Möller S, Engelbrecht KP, Kühnel C, Wechsung I, Weiss B (2009) Evaluation of multimodal interfaces for ambient intelligence. Human-Centric Interfaces Ambient Intell 347–370
22. Möller S, Engelbrecht KP, Kuhnel C, Wechsung I, Weiss B (2009) A taxonomy of quality of service and quality of experience of multimodal human-machine interaction. In: International workshop on quality of multimedia experience, 2009. QoMEX 2009, pp 7–12
23. Möller S (2010) Quality engineering—Qualität kommunikationstechnischer Systeme. Springer, Berlin
24. Möller S (2000) Assessment and prediction of speech quality in telecommunications, 1st edn. Springer, Berlin
25. Möller S, Berger J, Raake A, Wältermann M, Weiss B (2011) A new dimension-based framework model for the quality of speech communication services. In: 2011 Third international workshop on quality of multimedia experience (QoMEX), pp 107–112. doi:10.1109/QoMEX.2011.6065686
26. Yamagishi K, Hayashi T (2005) Analysis of psychological factors for quality assessment of interactive multimodal service, pp 130–138. doi:10.1117/12.586679
27. Raake A, Katrin H, Skowronek J, Egger S (2013) Predicting speech quality based on interactivity and delay. In: Interspeech 2013 (accepted)
28. Hoeldtke K, Raake A (2011) Conversation analysis of multi-party conferencing and its relation to perceived quality. In: 2011 IEEE international conference on communications (ICC), pp 1–5. doi:10.1109/icc.2011.5963021
29. Skowronek J, Raake A (2011) Investigating the effect of number of interlocutors on the quality of experience for multi-party audio conferencing. In: Proceedings of the 12th annual conference of the international speech communication association (Interspeech), pp 829–832

Part II
Applications and Methods

Chapter 12
Speech Communication

Nicolas Côté and Jens Berger

Abstract The goal of any speech service is the transmission and/or processing of speech signals. In this chapter we discuss the Quality of Experience (QoE) of speech communication systems, including networks, speech processing applications and terminals. We then give an overview of the methods employed to quantify and further estimate the QoE of speech communication services with a focus on diagnostic instrumental models. Such models provide indications on either the technical causes of degradations or the quality features impacted by a component in the speech communication system.

12.1 General Overview of Speech Communication

12.1.1 Quality of Experience in the Context of Speech Communication

As defined by Hardy [7], a voice [speech] service corresponds to a voice interaction through a telecommunication system.[1] Two types of speech services exist, namely (1) speech communication services, which imply a conversation between a talker and a listener (or several listeners in case of teleconferencing systems) in a near

[1] In the literature, the terms "voice service" and "speech service" are mostly used interchangeably. Here, we will refer to "voice" when the characteristics of the human voice are addressed, and to "speech" when both the signal carrier and the referred content are of interest.

N. Côté (✉)
Institute of Electronics, Microelectronics and Nanotechnology, ISEN Department, Lille, France
e-mail: nicolas.cote@isen.fr

J. Berger
SwissQual AG, a Rohde & Schwarz Company, Zuchwil, Switzerland
e-mail: jens.berger@swissqual.com

S. Möller and A. Raake (eds.), *Quality of Experience*,
T-Labs Series in Telecommunication Services, DOI: 10.1007/978-3-319-02681-7_12,

"real-time" manner, and (2) streaming services (e.g. recorded messages stored on a device). These services replace the air path between two interlocutors having a face-to-face conversation. Since the success of any service depends on its QoE, the quality assessment of the corresponding speech communication system or speech processing application is required for both the developers and the telecommunication providers.

12.1.2 Factors of Speech Communication QoE

Even if the quality of the transmitted speech is a factor determining the QoE of speech communication systems, user's satisfaction encloses many different aspects. According to the theoretical framework of QoE introduced in Chap. 4, the physical factors influencing the QoE are grouped into three categories: human influence factors, context influence factors and the system influence factors. The "human influence factors" here correspond to the talker's difficulties to produce an acoustic message (e.g. aphonia) and the listener's difficulties to understand this message (e.g. hearing impairments). Since humans can use speech services in very diverse situations, especially with the massive introduction of mobile terminals, the last category, "context influence factors", covers many heterogeneous environments (in terms of time and place). The "system influence factors" include all technical characteristics, physical equipment and computer programs, of the speech service. Section 12.2 describes mainly both subcategories, "network related system" and "device-related system" of the more general system influence factor category.

12.1.3 Features of Speech Communication QoE

The perceived quality of telephone systems has been studied for many decades [5, 9, 28, 40]. In these studies, auditory tests have been carried out where subjects had to judge the perceived quality of transmitted speech. It resulted from these studies that speech quality, like other perceptual magnitudes, is by nature a "multidimensional" object. Researchers introduced many quality features of speech signals: intelligibility, clearness, brightness, loudness, naturalness, nearness, spaciousness, etc. For instance, a good intelligibility of the transmitted and/or processed speech is a prerequisite for a maximum quality rating of the speech service. However, a perfect intelligibility of the talker's message at the listener's side is not sufficient to achieve high quality. For instance, the transmitted bandwidth can be restricted to the usual telephone bandwidth, while the intelligibility remains almost perfect.

According to Möller et al. [25], the QoE space of a speech communication service covers aspects of both speech perception and service usage. In Chap. 5, the QoE features were classified in terms of four levels from perception to service usage. In the field of speech communication, the first level of quality features called "level of direct perception" corresponds to the perception by the ear of the acoustic wave and the transmission of the resulting auditory information to the central nervous system. In a conversational situation, when two conversation partners interact, the QoE of the speech service is influenced by several other features classified in terms of the "level of interaction". For instance, this level includes the naturalness of the interaction between two interlocutors during a phone call. The third level of QoE features, the "level of the usage instance", includes all features related to the physical and social environment at the talker's and listener's side. For instance, the background noise or the room reverberation at the listener's side has an influence on the listening effort and thus on the QoE of the whole speech communication system [24]. Another example is the advantage of mobility with cordless terminals and mobile telephony. The last category of QoE features, called "level of service", covers aspects like stability over the entire duration of the communication, call set-up duration or interruptions of the connection. This organization of quality features in four layers shows that quality features are related to both instantaneous and multi-episodic experiences of the service. All of these features play a role in the long-term acceptability of the service and the averaging process is relatively complex (see Chap. 10).

Since many speech quality features exist in the literature, several authors developed perceptual spaces based on few orthogonal quality features referred to as "speech quality dimensions". The following section summarizes the speech quality spaces proposed in the literature. Wälterman et al. [41] combined two auditory methods to derive a speech quality space composed of the three following dimensions:

- **Discontinuity**: this dimension reacts to degradation in the time domain, i.e. an unpredictable variation over time of the signal.
- **Noisiness**: this dimension is affected by the amount of unwanted information added to the speech message (either noise or a second talker).
- **Coloration**: this dimension can be affected by the two following elements: (1) a deviation from a reference timbre (e.g. *dark* or *bright*) and (2) a bandwidth restriction.

These three dimensions are of the type vector model. In other words, the origin of the space defines the highest quality and the space is defined by positive values only.

However, all speech stimuli employed in Wältermann et al. [41] were adjusted to a fix listening level of 79 dB SPL. Consequently, Côté [2] proposed to include a fourth quality feature to the perceptual space; **loudness**. Indeed, loudness is considered as the main feature of speech services QoE [5]. A loudness impairment is introduced in the case of non-optimal listening level, that is, an attenuation or an amplification introduced by the entire communication system. Loudness thus is a feature of the "ideal-point model" type. The three perceptual dimensions described by Wältermann et al. [41] are considered as orthogonal. However, the perceptual dimension "loudness" can be correlated with the other dimensions. The loudness summation effect shows that the bandwidth of a sound has an impact on its perceived loudness [4]. In Côté et al. [3], the authors showed the converse effect; the coloration due to a speech communication system has an influence on the optimal listening level.

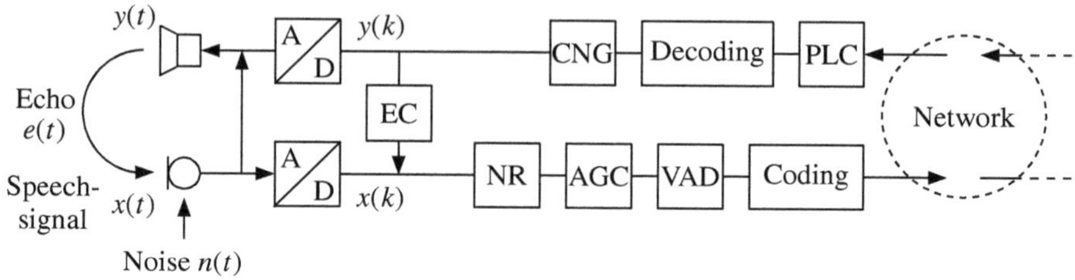

Fig. 12.1 Elements composing a speech communication system. A/D refers to analog to digital conversion, AGC to automatic gain control, EC to echo compensations, CNG to comfort noise generation, PLC to packet-loss concealment, VAD to voice activity detection and to NR noise reduction

These above described four features should reflect the whole perceptual quality space of transmitted speech. However, Sen [35] proposed a 5-dimensional space including noisiness and two sub-dimensions for each feature, coloration and discontinuity: slowly-varying and rapidly-varying discontinuities and low-frequency and high-frequency distortions. ITU–T Study Group 12 compares these two speech quality spaces within the work item P.AMD.

12.2 Speech Communication Systems

Nowadays, speech communication systems are composed of a multitude of components. The section below give an overview of the most important elements composing a speech communication system and their possible impact on its QoE. A typical example of such a system is depicted in Fig. 12.1. From the four types of speech processing systems described by Rabiner [30], only speech coding is introduced in the present chapter. Speech synthesis is covered by Chap. 13, speech recognition and speaker verification technologies are not covered by the present book.

Firstly, a telephone user talks and produces an acoustic signal, $x(t)$. This signal is received by the microphone of the talker's handset. However, this handset also receives sound from the environment, $n(t)$, produced by the sound sources surrounding the telephone user. The microphone converts the acoustic signal into an electrical signal, which is digitalized (i.e. sampled and quantized in $x[k]$, where k is the sample index) and *pre*-processed in order to remove the undesired signals (i.e. background noise, reverberation and echo). Then, this processed signal is encoded with a (low bit-rate) speech codec and sent to the transmission network. During the transmission to the handset of the conversation partner, the signal passes through several gateways and nodes. At the listener's side, a continuous electrical signal is decoded with the help of several digital "post-"processing algorithms. Then, the loudspeaker of the listener's handset converts the processed electrical signal into an acoustic signal,

$y(t)$. The listener's acceptability of the whole communication service based (mainly, but among other aspects) on the perception of the transmitted acoustic signal, $y(t)$, is the subject of the present chapter.

12.2.1 Telephony Networks

The traditional analog telephone network, referred to as Public Switched Telephone Network (PSTN), has been optimized for an almost perfect intelligibility of the speech message. For instance, the bandwidth of the transmitted speech corresponds to the transmission of the frequencies between 300 and 3,400 Hz that enables a comprehensibility of almost all phonemes. This bandwidth is, nowadays, referred to as Narrow-Band (NB). The PSTN is based on a "circuit-switched" network: the two interlocutors are connected by a physical circuit. In such a network, all physical parameters are well controlled to ensure a stable Quality of Service (QoS): the network accessibility is guaranteed and preserved over the whole call. During the last two decades, the deregulation of the telecommunications market led to heterogeneous transmission systems and speech processing algorithms. The first main transition was the introduction of digital transmission, the ISDN (Integrated Services Digital Network), which resulted in a decrease of circuit noise. Then, mobile phone networks, GSM (Global System for Mobile) and UMTS (Universal Mobile Telecommunication System) networks, have been broadly set up all over the world. The users of mobile telephony services are able to move from a quiet (house, office) to a noisy environment (street, train station) during a phone call. However, these networks are highly dependent on the characteristics of the radio channel between the mobile phone and the antenna. This air path leads to interferences, producing bit errors and frame losses, and handovers between two cells, two codecs and/or even two bandwidths, producing discontinuities in the transmitted signal. These quality variations in mobile networks result in a perceived instability of the communication system (see Chap. 27).

In addition to mobile telephone networks, speech communications over computer networks have been introduced. The Voice over IP (VoIP) protocol is based on a discontinuous transmission of packets of data, and the network is consequently referred to as a "packet-switched" network. Nowadays, the packet-switched network is the most widely used transmission path, because of its enhanced flexibility compared to the circuit-switched network. For instance, large audio bandwidths can be transmitted such as Wideband (WB, i.e. 50–7,000 Hz), Super-Wideband (S-WB, i.e. 50–14,000 Hz) and Full-Band (FB, i.e. 20–20,000 Hz) bandwidths. These wider bandwidths introduce less coloration of the speech compared to the the narrow telephone bandwidth and thus increase the QoE. For instance, a comparison of clean WB and NB transmissions shows an increased quality of 29 % in the WB case [27, 29]. However, VoIP transmissions may increase several quality impairments. For instance, these wider bandwidths may increase the influence of the environmental noise at the talker's side, and the packetization process lengthens the overall transmission delay. A long transmission delay may introduce an audibility of the talker's own voice (echo) and reduces talking quality and double talk capability.

A packet-switched network may introduces discontinuities in the transmitted speech message, too. This annoying degradation appears more frequently than in a circuit-switched network. These discontinuities have two origins: (1) the bit-rate allocation is not guaranteed over the whole call and (2) the packets can take different transmission paths that lead to a time-varying transmission delay. This variation in transmission delay is referred to as "jitter". To generate a continuous signal, a buffer is placed in front of the decoder. The size of this de-jitter buffer (e.g. 120 ms) defines the tolerated lengthening of transmission delay between two consecutive packets. However, the size of the jitter buffer increases the overall transmission delay and, thus, may affect the conversation effectiveness. In case the speech segment may be lost during the transmission or arrives too late to synthesize a continuous signal, an algorithm "reconstructs" the missing packets. This algorithm called Packet-Loss Concealment (PLC) reduces the discontinuities in the speech signal. Nowadays PLC algorithms uses time-scale modifications of the speech signals (also known as "time-warping") which enable a smooth reconstruction of the waveform and avoid any discontinuity in the speech signal.

12.2.2 User Interfaces

The physical interface between the customers and the transmission system can be a handset, a headset or a Hands-Free Terminal (HFT). Such acoustic terminals have an influence on the speech coloration. The timbre modification of the talker's voice is introduced by the electro-acoustic properties of the two transducers (microphone and loudspeaker). Therefore, QoE of user interfaces is determined by their sending and receiving frequency response characteristics. In addition, loudness is a main parameter for all acoustic interfaces. According to the "orthotelephonic reference position" [12], the output signal loudness of such acoustic terminals should be equivalent to the perceived loudness of two interlocutors having a face-to-face conversation at one-meter distance.

Nowadays, the handset manufacturers introduce new services to user terminals in order, for instance, to increase the mobility of the user. For instance, screens with haptic feedback are included in modern mobile phones. Place for transducers is consequently reduced and causes challenges for their acoustic design. Although they enable a greater mobility, these terminals include several digital processing systems such as noise reduction algorithm that may degrade the transmitted speech signal [26].

Table 12.1 Characteristics of NB speech coding algorithms

Codec	Codec type	Frame length (ms)	Bit-rate (kbits)	I_e
G.711	PCM	0.125	64	0
G.726	ADPCM	0.125	40	2
–	–	–	16	50
G.729	CS-ACELP	10	8	10
GSM-FR	RPE-LTP	20	13	20
GSM-EFR	ACELP	20	12.2	5

The value of I_e is expressed on the NB quality scale of the E-model [13], ranging from 0 to 100 [15]

Table 12.2 Characteristics of WB speech coding algorithms

Codec	Codec type	Frame length (ms)	Bit-rate (kbits)	$I_{e,WB}$
G.722	ADPCM	0.125	64	13
–	–	–	48	31
G.722.1	MLT	20	32	13
–	–	–	24	19
G.722.2	CELP	20	23.85	8
–	–	–	23.05	1
–	–	–	14.25	10
–	–	–	6.6	41

The value of I_e, WB is expressed on the WB quality scale of the E-model [14], ranging from 0 to 129 [15]

12.2.3 Speech Coding

A speech coding algorithm is a system that reduces the network rate used to transmit the speech signal. The speech coder produces a compressed signal from the input speech signal, referred to as the *bitstream*. After transmission over the network, the aim is to get a synthesized speech signal as similar as possible to the original speech. The impact of the speech codec on QoE depends on three physical characteristics: (1) the bit-rate expressed in *kbits*, (2) the frame length expressed in *milliseconds* (typical ranges of frame length are 5–30 ms), and (3) the paradigm employed by the coding algorithm. Tables 12.1 and 12.2 present the characteristics of several NB and WB speech coding algorithms. Almost all speech codecs have a flat band-pass within the allowed transmitted bandwidth (NB, WB or S-WB) and a low quantization noise resulting in a perfect intelligibility of the coded speech. However, they introduce audible non-linear degradations that decrease their perceived quality and affect automatic speech and/or speaker recognition algorithms. The parameter called "equipment impairment factor" (I_e), used in the E-model [13], quantifies the degradation introduced by the coding–decoding process. In addition, the coding–decoding process introduces a delay which impacts the conversation effectiveness. Nowadays, speech codecs use a simple model of human auditory perception [17], are scalable from NB to WB [18], and some modern codecs also allow for coding of both speech and audio signals [16].

12.2.4 Voice Quality Enhancement

Voice Quality Enhancement (VQE) algorithms are integrated into the network or even directly into the terminal to reduce the new impairments introduced by mobile or VoIP networks. These algorithms are, for examples, echo cancellation, noise reduction, de-reverberation and automatic gain control, see Fig. 12.1. Echoes of the talker's own voice is introduced either by an acoustic feedback at the listener's side or by an impedance mismatch at the interconnection between two networks. As already mentioned, the latter effect is exacerbated in packet-based networks due to longer transmission delay. Therefore, echo cancellation techniques are needed if the delay exceeds 15 ms. Noise reduction is another VQE algorithm that has been widely introduced in mobile terminals. It reduces the environmental noise at the talker's side transmitted by the network. This algorithm complemented by a de-reverberation algorithm and an echo canceller separates the desired signal components from the undesired ones. However, noise reduction algorithms based on spectral subtraction reduce the noise level but simultaneously introduce musical noises on the speech signal [33]. Therefore, Möller et al. [26] proposed to describe the speech degradations resulting from imperfect noise reduction and echo cancellation by two additional equipment impairment factors *Inr* and *Iec*.

12.3 Speech Communication QoE Measurement Methods

The following sections introduce the measurement methods employed to quantify and further estimate the QoE of speech communication services, i.e. speech transmitted through a network and/or processed by speech processing systems. However, voice and speech quality measurement methods are employed in very diverse scientific fields: medicine (e.g. the evaluation of voice-related problems), linguistics (e.g. cultural comparisons) or speech communication. Each field has its own assessment paradigm.

12.3.1 Auditory Methods

The most accurate auditory measurement method would be an assessment by customers in natural environments. In practice, such "in-field" tests are hardly implemented, and speech services QoE is assessed with artificial auditory quality

Table 12.3 5-point scales

Quality of the speech [22]	Score	Impairment [11]
Excellent	5	Imperceptible
Good	4	Perceptible, but not annoying
Fair	3	Slightly annoying
Poor	2	Annoying
Bad	1	Very annoying

tests carried out in laboratories where the perception process is "directed" by an experimenter. Many different auditory test methods are employed by the academic laboratories and the speech service industries. For instance, listening-only experiments are carried out to gather the most important QoE features. Their realism is lower than that of conversational tests, since only the transmission system influence factors are assessed. The P-Series of Recommendations published by the ITU–T describe a general framework of speech communication measurement methods. In a listening quality test (referred to as listening-only test by the ITU–T), the listeners rate on a measurement scale a set of short speech samples (4–8 s) transmitted by different speech communication systems. The most widely used measurement scale is the 5-point integral quality scale presented in Table 12.3 (left column [22]). Such methodologies are not suited to compare speech stimuli with small impairments. Consequently, high-quality speech processing systems are assessed by methodologies used in the audio world and published by the ITU–R organization [10, 11] (see Table 12.3, right column).

Most of the ITU–T and ITU–R auditory methods quantify the quality of a speech service with a single value. This value is often used as an estimation of the overall speech service QoE. In addition to these methods, more complex auditory test methods give diagnostic information about the assessed processing conditions. Such quality tests rely on either a multi-scale rating process or a multidimensional analysis of the auditory results. For instance, Voiers [39] developed a specific multidimensional scaling method called Diagnostic Acceptability Measure (DAM) which assesses quality features of speech samples. More recently, Wältermann [40] developed a similar method to assess the three speech quality dimensions discontinuity, noisiness and coloration. However, such multidimensional tests are expensive and time-consuming since the listeners are trained beforehand (experienced), and they employ several rating scales for each speech stimulus (see also Chap. 5).

12.3.2 Instrumental Methods

Auditory methodologies rely on judgments by test subjects who are asked to give their opinion about the quality of a speech stimulus. Since auditory tests are costly and time-consuming, instrumental methods have been developed. Instrumental methods have different applications such as the daily monitoring of transmission networks (e.g. VoIP) or the optimization of processing systems (e.g. speech codecs). They provide either a single estimated value that possibly represents the quality of the speech communication system (integral models), or a decomposition of the quality into several quality features (diagnostic models). In the following sections, we review the reliable models employed to predict the different aspects of speech communication QoE. Many building blocks have been developed such as the ITU–T Rec. P.863 [23] model which estimates the listening quality of transmitted speech. However, a tool that covers all aspects of the QoE and predicts the overall QoE of speech communication services in a single value is not available yet.

Richters and Dvorak [31] proposed a performance model based on seven quality criteria (speed, accuracy, availability, security, simplicity and flexibility) for each function of the service (sales, connection, billing, technical support, etc). This model is employed to assess the QoS of speech communication services and covers many aspects of the service usage. More recently, Möller et al. [25] organized all QoS parameters of speech communication services in a theoretical model which covers the four levels of QoE-features (perception, interaction, situation and service). For an example of an exhaustive evaluation of a speech communication service with such quality criteria, see Chen et al. [1].

Many models have been developed and standardized to estimate the quality of transmitted speech in a listening-only situation. Takahashi et al. [37] classified them in three different groups: parameter-based models that use parameters describing the elements of the system (e.g. ITU–T Rec. G.107 [13]), signal-based models that use the transmitted or processed speech signal (e.g. ITU–T Rec. P.863 [23]), and the packet-layer models that use information about the service operation (e.g. ITU–T Rec. P.564 [21]). For instance, the well-known Perceptual Evaluation of Speech Quality (PESQ) model includes a robust time-alignment algorithm useful for VoIP variable delay [32]. The PESQ is now superseded by a new listening-only signal-based model, called POLQA [23], that represents an intrusive speech quality model suitable for NB to S-WB connections, electro-acoustic interfaces and VQE algorithms. Most of these models provide an integral estimation of the quality. Recently, diagnostic models have been developed in order to either indicate (1) the technical causes of a single impairment or (2) describe the communication system QoE on few speech quality features. In the former case, diagnostic models provide useful information to system designers and operators that help them for maintenance purposes. For instance, the ITU–T Rec. P.502 [20] describes standard methods to assess each element of user terminals and network components. The corresponding test signals are described in a separate standard [19]. These methods assess characteristics such as (1) the frequency response, the sidetone, the harmonic distortion and the loudness ratings of the user terminals, and (2) the echo loss, the double talk capabilities and the background noise of the transmission networks. Even tough this first type of diagnostic models provides an exhaustive evaluation of the physical equipments, they do not help telecommunication providers to design a voice service optimized for their specific needs. The second type of diagnostic models describe a voice service in a simple quality space. They help the end-user to choose a voice service based on its cost and its QoE. The benefit of such a diagnostic model has initially been investigated by Quackenbush et al. [28]. More recently, two sets of quality-feature estimators have been developed from the perceptual quality space derived by Wältermann et al. [41]. Côté [2] improved estimators initially developed by Scholz et al. [34] and Huo et al. [8] into a signal-based model called Diagnostic Instrumental Assessment of Listening-quality (DIAL). This model provides values of the four dimensions "coloration", "discontinuity", "noisiness" and "loudness".Wältermann [40] proposed a diagnostic parametric model based on the E-model [13]. In parallel, Sen and Lu [36] derived four estimators for temporally localized (slow-jitter and fast-jitter) distortions and frequency localized (low-pass and high-pass) distortions.

So far, no instrumental model has been standardized for the estimation of the speech quality in a conversational situation. However, Guéguin et al. [6] proposed such a tool that combines estimations from three other models: PESQ and PESQM for listening- and talking-only situations, respectively, and the E-model for a delay impairment factor introduced by the transmission delay. Long-term quality estimation has been studied by more researchers. For instance, Weiss et al. [42] was able to estimate a long-term listening-only speech quality score (up to 2 min.) based on PESQ estimations for 4–8 s stimuli.

12.4 Conclusions and Future Trends in Speech Communication QoE

In this chapter we presented both the technical elements and quality features which are relevant for the Quality of Experience of speech communication systems. We reviewed the auditory and instrumental methods suitable for speech quality assessment with a focus on diagnostic instrumental models that provide one output per QoE dimension.

Over the last decades, instrumental models have been developed on either speech or music databases. The former ones estimate the QoE of speech services such as telephony, whereas the latter ones are dedicated to audio devices such as loudspeakers or headphones [38]. The new standard model POLQA [23] has been developed for the QoE estimations of speech communication systems only. However, both streaming and telephony services now employ similar packet-based networks. This new usages encourage the researchers to develop a common model that works with both types of input signals.

Current speech quality models do not cover the influence of the listener's acoustic environment. Indeed, listening through a handset in a noisy environment involves binaural hearing which is not covered by current models. Even though many studies have been published over the last two decades, the effects of binaural hearing are still unclear and difficult to include in such quality models.

Further work is thus expected in the development of reliable instrumental methods. However, such instrumental methods require, at first, auditory test results. Therefore, the community of researchers who works in the field QoE would appreciate collaborations with voice service providers and developers of speech processing systems to get access to databases including specific impairments and/or listening contexts.

References

1. Chen K, Huang C, Huang P, Lei C (2006) Quantifying skype user satisfaction. In: Proceedings of the conference on applications, technologies, architectures, and protocols for computer communications (SIGCOMM), pp 399–410. Pisa
2. Côté N (2011) Integral and diagnostic intrusive prediction of speech quality. Springer, Berlin
3. Côté N, Gautier-Turbin V, Möller S (2007) Influence of loudness level on the overall quality of transmitted speech. In: Proceedings of the 123rd AES convention, 7175, New York
4. Fastl H, Zwicker E (2007) Psychoacoustics: facts and models, 3rd edn. Springer, Berlin
5. Fletcher H, Galt RH (1950) The perception of speech and its relation to telephony. J Acoust Soc Am 22(2):89–151
6. Guéguin M, Le Bouquin-Jeannes R, Gautier-Turbin V, Faucon G, Barriac V (2008) On the evaluation of the conversational speech quality in telecommunications. EURASIP J Adv Signal Process. Article ID 185248
7. Hardy W (2003) VoIP service quality: measuring and evaluating packet-switched voice. McGraw-Hill, New York
8. Huo L, Wältermann M, Heute U, Möller S (2008) Estimation of the speech quality dimension "discontinuity". In: Proceedings of the 8th ITG-Fachbericht-Sprachkommunikation, Aachen
9. IEEE Standards Publication 297 (1969) Recommended practice for speech quality measurements. Institute of Electrical and Electronics Engineers, New York
10. ITU-R Recommendation BS.1116-1 (1997) Methods for the subjective assessment of small impairments in audio systems including multichannel sound systems. International Telecommunication Union, Geneva
11. ITU-R Recommendation BS.1284-1 (2003) General methods for the subjective assessment of sound quality. International Telecommunication Union, Geneva
12. ITU-T Handbook on Telephonometry (1992) International Telecommunication Union, Geneva
13. ITU-T Recommendation G.107 (2011) The e-model, a computational model for use in transmission planning. International Telecommunication Union, Geneva
14. ITU-T Recommendation G.107.1 (2011) Wideband e-model. International Telecommunication Union, Geneva
15. ITU-T Recommendation G.113 (2007) Transmission impairments due to speech processing. International Telecommunication Union, Geneva
16. ITU-T Recommendation G.718 (2008) Frame error robust narrow-band and wideband embedded variable bit-rate coding of speech and audio from 8–32 kbit/s. International Telecommunication Union, Geneva
17. ITU-T Recommendation G.722.1 (2005) Low-complexity coding at 24 and 32 kbit/s for hands-free operation in systems with low frame loss. International Telecommunication Union, Geneva
18. ITU-T Recommendation G.729.1 (2006) Based embedded variable bit-rate coder: an 8–32 kbit/s scalable wideband coder bitstream interoperable with G.729. International Telecommunication Union, Geneva
19. ITU-T Recommendation P.501 (2012) Test signals for use in telephonometry. International Telecommunication Union, Geneva
20. ITU-T Recommendation P.502 (2000) Objective test methods for speech communication systems using complex test signals. International Telecommunication Union, Geneva

21. ITU-T Recommendation P.564 (2007) Conformance testing for voice over IP transmission quality assessment models. International Telecommunication Union, Geneva
22. ITU-T Recommendation P.800 (1996) Methods for subjective determination of transmission quality. International Telecommunication Union, Geneva
23. ITU-T Recommendation P.863 (2011) Perceptual objective listening quality assessment. International Telecommunication Union, Geneva
24. Jung O (2012) Assessment of conversational speech quality inside vehicles, concerning influences of room acoustics and driving noises. Acta Acustica Acustica 98(3):461–474
25. Möller S, Berger J, Raake A, Wältermann M, Weiss B (2011) A new dimension-based framewrok model for the quality of speech communication services. In: Third international workshop on quality of multimedia experience (QoMEX), pp 107–112
26. Möller S, Kettler F, Gierlich HW, Poschen S, Côté N, Raake A, Wältermann M (2012) Extending the e-model for capturing noise reduction and echo canceller impairments. J Audio Eng Soc 60(3):165–175
27. Möller S, Raake A, Kitawaki N, Takahashi A, Wältermann M (2006) Impairment factor framework for wideband speech codecs. IEEE Trans Audio Speech Lang Process 14(6):1969–1976
28. Quackenbush S, Barnwell T, Clements M (1988) Objective measures of speech quality. Prentice Hall, Englewood Cliffs
29. Raake A (2006) Speech quality of VoIP—Assessment and prediction. Wiley, Chichester
30. Rabiner L (1995) The impact of voice processing on modern telecommunications. Speech Commun 17(3–4):217–226
31. Richters JS, Dvorak CA (1988) A framework for defining the quality of communications services. IEEE Commun Mag 26(10):17–23
32. Rix A, Hollier M, Hekstra A, Beerends J (2002) Perceptual evaluation of speech quality (PESQ), the new ITU standard for end-to-end speech quality assessment, part i-time alignment. J Audio Eng Soc 50(10):755
33. Scalart P, Filho J (1996) Speech enhancement based on a priori signal to noise estimation. In: Proceedings of the IEEE international conference on acoustics, speech, and signal processing (ICASSP-96), vol 2, pp 629–632
34. Scholz K, Wältermann M, Huo L, Raake A, Möller S, Heute U (2006) Estimation of the quality dimension "directness/frequency content" for the instrumental assessment of speech quality. In: Proceedings of the 9th international conference on spoken language processing (ICSLP), Pittsburgh, pp 1523–1526
35. Sen D (2004) Predicting foreground SH, SL and BNH DAM scores for multidimensional objective measure of speech quality. In: IEEE international conference on acoustics, speech, and signal processing (ICASSP'04), vol 1, pp 493–496
36. Sen D, Lu W (2012) Objective evaluation of speech signal quality by the prediction of multiple foreground diagnostic acceptability measure attributes. J Acoust Soc Am 131(5):4087–4103
37. Takahashi A, Yoshino H, Kitawaki N (2004) Perceptual QoS assessment technologies for VoIP. IEEE Commun Mag 42(7):28–34
38. Thiede T, Treurniet W, Bitto R, Schmidmer C, Sporer T, Beerends J, Colomes C (2000) PEAQ—The ITU standard for objective measurement of perceived audio quality. J Audio Eng Soc 48(1/2):3–29
39. Voiers WD (1977) Diagnostic acceptability measure for speech communication systems. In: IEEE international conference on acoustics, speech, and signal processing (ICASSP'77), Hartford, pp 204–207
40. Wältermann M (2013) Dimension-based quality modeling of transmitted speech. Springer, Berlin
41. Wältermann M, Raake A, Möller S (2010) Quality dimensions of narrowband and wideband speech transmission. Acta Acustica Acustica 96(6):1090–1103
42. Weiss B, Möller S, Raake A, Berger J, Ullmann R (2009) Modeling call quality for time-varying transmission characteristics using simulated conversational structures. Acta Acustica Acustica 95(12):1140–1151

Chapter 13
Text-To-Speech Synthesis

Florian Hinterleitner, Christoph Norrenbrock, Sebastian Möller and Ulrich Heute

Abstract In this chapter, we will address the quality experienced when listening to speech which is synthesized by state-of-the-art synthesis systems which generate artificial speech from text. Such systems are used, e.g., in information and navigation systems, but also for generating audiobooks. We describe both, auditory evaluation methods as well as instrumental models predicting perceived QoE. Besides overall perceived quality, we focus on perceptual quality features that can be used for diagnosis and system optimization.

From the bandpass-based Voder invented by Dudley in 1939 to the modern day Text-To-Speech (TTS) systems, synthetic speech has made tremendous progress. The most general type of system is able to generate artificial speech from written text. With the development of modern types of TTS systems,reminds listeners of robot-like voices from the 1980s but of real human speakers. This increase made it possible to use TTS in everyday services like email readers, information systems, and smart-home assistants. Especially, the boom in e-books and smartphones and the implied opportunity to synthesize the whole content of books and websites exposed an entire new user group to synthetic speech. These emerging new application areas demand a further constant improvement of TTS quality. The key to this progress consists of auditory quality evaluations with human participants.

F. Hinterleitner (✉) · S. Möller
Quality and Usability Lab, Telekom Innovation Laboratories, TU Berlin, Berlin, Germany
e-mail: florian.hinterleitner@tu-berlin.de

S. Möller
e-mail: sebastian.moeller@telekom.de

C. Norrenbrock · U. Heute
Digital Signal Processing and System Theory, CAU Kiel, Kiel, Germany
e-mail: cno@tf.uni-kiel.de

U. Heute
e-mail: uh@tf.uni-kiel.de

S. Möller and A. Raake (eds.), *Quality of Experience*,
T-Labs Series in Telecommunication Services, DOI: 10.1007/978-3-319-02681-7_13,

Although, the use of speech as an interaction modality still lacks behind touch at the time of writing, it can be expected that its spreading will increase in the mid-term, because speech has principle advantages over other communication modalities especially in specific situations, such as with parallel activities or when using small devices.

Parametric systems, like the formant synthesizers [23] that became popular during the late 1970s, were the first systems that could produce intelligible speech. Nonetheless, since this method uses a simplistic source-filter model of speech production and bandpass filters to reproduce the formants of speech sounds, the generated speech sounds artificial and robot-like. The first corpus-based synthesizers concatenated diphone units to build a speech signal. These voices mainly suffer from sonic glitches that occur at the conjunctions of two units. With the development of PSOLA (Pitch-Synchronous Overlap and Add) [30], transitions between units could be smoothed, leading to an increase in naturalness of diphone synthesizers; however, voice quality was still far from optimum. In the mid 1990s, unit-selection speech synthesis [3] was developed with the idea to select units from a large database of prerecorded speech with the goal to minimize two cost functions: target costs, which describe how well the units in the database fit to the text that is to be synthesized, and concatenation costs, which show how well two units join together. The latest development in synthesis was the HMM-synthesizer that was introduced in 2002 by Tokuda [36]. These systems are based on Hidden Markov Models (HMM) that are trained on excitation as well as spectral parameters of human-produced speech. During the synthesis phase, the synthetic speech signal is generated by a maximization algorithm that finds the optimal path through the HMM. Thus, HMM-synthesizers usually do not sound as natural as unit-selection synthesizers, however, they generate speech that does not suffer from prosodic glitches that are typical for concatenation-based systems.[1]

From the description of the diversity of synthesis techniques, it becomes obvious that the related quality suffers from different types of degradations. Thus, besides integral quality—which is important for selecting a synthesis technique for a given purpose—quality features, as they have been discussed in Chap. 5, are of paramount importance to characterize systems, to diagnose imperfectness, and to improve system quality. Thus, this chapter will concentrate on methods which fulfill both requirements. We will start with an overview of auditory evaluation methods, then report on experiments which identified relevant quality features, and conclude by presenting first approaches for the instrumental estimation of quality; approaches which are still in their infancy and which require further research, the directions of which are outlined at the end of the chapter.

[1] An extensive collection of speech produced by German speaking synthesizers can be found in [4].

13.1 Auditory Quality Evaluation

The primary goal of speech is to serve the communication of information. Thus, a key requirement to synthetic speech is that it is intelligible, i.e., that the linguistic information can be discerned by the listener. However, intelligibility is commonly not enough, and synthetic speech — even if it was 100 % intelligible — is not perceived as human-like, mostly due to a lack in naturalness, which largely determines the overall quality. In this section, we will describe functional tests to evaluate intelligibility and judgement tests to assess specific aspects of a TTS signal. Further quality features will be addressed in Sect. 13.2.

13.1.1 Functional Tests

Intelligibility is one of the big problems of parametric and diphone synthesizers, which were popular until the development of unit-selection synthesizers in the 1990s. Even though the development of corpus-based TTS synthesizers and the trend towards increasing corpora sizes made this problem less relevant, intelligibility assessment remains a common task for TTS, especially within the scope of HMM-synthesis.

Functional tests for intelligibility assessment can be classified into two categories: segmental tests on a word level, where single words are presented to the listener, and segmental tests on a sentence level, where complete sentences are evaluated. In both cases, the intelligibility can be expressed by a total error rate.

13.1.1.1 Intelligibility on Word Level

The test material of intelligibility tests on word level is mainly focused on consonants since they are more problematic to synthesize. In the following, the simple Diagnostic Rhyme Test (DRT), its successor, the Modified Rhyme Test (MRT), and the more complex Cluster Identification Test (CLID) are presented. The reader is referred to [10] for a detailed overview of these and further intelligibility tests.

Diagnostic Rhyme Test (DRT) The DRT [1] uses a fixed set of meaningful words to test for intelligibility of the initial consonant. The examined items are of the form CVC, i.e., an initial Consonant followed by a medial Vowel followed by a final Consonant. One auditory stimulus and one word pair are presented at a time. The word pair consists of two words which differ only in the initial consonant, e.g. *dune* and *tune*. The listener marks which of those two words he thinks was presented. For each of six categories (i.e., voicing, nasality,...), specific word pairs are chosen. The intelligibility is expressed by the total error rate or the percentage of correct initial consonants.

Modified Rhyme Test (MRT) The MRT is an extension of the DRT which is able to test for initial as well as final consonant intelligibility. The test items consist of sets of six one-syllable words. Half of the set differs in initial while the other half differs in final consonant, e.g., *bus*, *bug*, *but*, *buff*, *bun*, and *buck* (a set that differs in initial consonant). The listener has to identify which of the six items in the list was presented. The intelligibility is given as initial and final consonant error rate or as overall percentage of correct consonants.

Cluster Identification Test (CLID) The previous two approaches are fast, reliable, easy to administer, and no training of the participants is required. However, the intelligibility may be overestimated since the participants can chose words from the presented categories; thus, there is a probability to select the right word by chance. In addition, the words presented in a set are meaningful, but not equally frequent in a language; thus, there is an inherent distortion of the participants' responses, which is due to their knowledge of the language. A more balanced approach in intelligibility testing which overcomes the limitations of rhyme tests is the CLID test. On the basis of linguistic statistics gathered from speech databases containing monosyllables, an automatic word generator is used to create phonotactically correct monosyllables of the type C^iVC^j (where i and j represent the number of initial and final consonants, respectively). These, mostly non-sense words are evaluated in an open-response test where participants have to accomplish a task like:

> Please write down what you have heard in such a way that another person would read it aloud in the same way as you heard it originally. [24]

This guarantees that the participants are not biased by any given response categories. Subsequently, the recognition rates can be computed on word and on cluster level (initial, medial, and final consonant).

13.1.1.2 Intelligibility on Sentence Level

While intelligibility tests on word level lead to very diagnostic results, tests on sentence level are more similar to speech perception in normal communication situations.

Semantically Unpredictable Sentences (SUS) In the most common test methodology short semantically unpredictable sentences (SUS) [2] are used, i.e., they do not occur in real life. The advantage of SUS results from the fact that, even though the syntax of each sentence is correct, the whole sentence does not make sense, thus the listeners can not rely on a semantic context. This increases the importance of the acoustic characteristics of the TTS signal. A SUS test uses five different syntactic structures, e.g., *subject—verb—object* could yield the sentence "*The strong way drank the day*" [22]. Ten sentences for each of the five categories are produced and assessed in random order in a listening test.

13.1.2 Judgment Tests

While the previous section discussed evaluation methods focused on functional testing, i.e., intelligibility was measured by how well listeners correctly identify words and phrases, the current section addresses judgment tests. In these tests listeners are instructed to rate stimuli along a number of attribute scales determining specific aspects of a system and thus yielding very analytic results [10]. The most simple way would be to ask listeners to rate, e.g., the naturalness of TTS stimuli, on a 5-point absolute category rating (ACR) scale and to build a mean opinion score (MOS).

A more complex approach is specified in the ITU-T Rec. P.85 [18]. This method is recommended by the Telecommunication Standardization Sector of the International Telecommunication Union (ITU-T) to assess the quality of telecom services which provide synthetic speech output, be it via concatenating sentences, parts of sentences, or via TTS synthesis. A test according to this recommendation should include at least 5 different synthesis systems and at least one reference condition (e.g., natural speech corrupted with degradation or a known synthesis system). The test is designed in a way that requires the listener to concentrate on the content of each message, i.e., before rating each stimulus on 5-point opinion scales, the test participants have to answer several questions concerning the information contained in the stimulus. Thus, the messages should contain a fixed part that is specific to the addressed use case and a variable part that differs between stimuli. The duration of the stimuli should be between 10 and 30 s. A possible test phrase in a mail-order shopping scenario would be:

> Mr. Zimmerman, you have ordered running shoes, color: white, size: 41, price: 61€. They will be delivered to you in 10 days.

Here the name, the shoe color, it's size, the price and the delivery date are variable parts that can be requested from the listener.

Each stimulus is presented twice consecutively. After the first presentation the test participants answer questions on the information contained in the stimulus, and after the second presentation they judge the quality of the presented stimulus on different rating scales. Each listening test consists of 3 sessions: a training session, where the test participants should get used to the test procedure, the environment and get an impression of the quality range of the stimuli in the test, and two main sessions that either use scales concerning the intelligibility or the quality of the synthesizers. The intelligibility scales test the *listening effort*, *comprehension problems* and the *articulation* while the quality scales assess *pronunciation*, *speaking rate* and *voice pleasantness*. Furthermore, each session also includes the scales *overall impression* and *acceptance*. An extension of P.85 towards the assessment of synthesized audiobooks was proposed in [28]. Therefore, scales that are relevant when synthesizing the content of books were included, e.g., scales that assess *intonation* and *emotion*.

Given the complexity of this test the question arises why an evaluation via P.85 should be preferred over a simpler intelligibility test or a MOS test addressing overall quality. These three methods were compared in [35]. The results showed that high intelligibility ratings do not necessarily come together with high ratings on the naturalness and overall impression scales. Moreover, the best ranked synthesizer in the naturalness test did not get the best rating on the overall quality scale. Thus, while a simple MOS naturalness test can give a basic overview of the quality of synthesizers, the P.85 procedure yields far more fine-grained information about the performance of a system.

However, this evaluation protocol has also been heavily criticized. In [38] the authors suggested extensive modifications:

- natural speech-reference stimuli should not be included because they affect the mean ratings of TTS systems and thus tend to diminish the differences between them;
- items that assess *naturalness*, *audio flow*, and *ease of listening* should be included;
- the item *speaking rate* should be modified so that one end of the scale represents an optimal speed while the other end indicates extremely slow or extremely fast speech.

One of the main points that has already been addressed in [35] is the fact that many of the recommended scales are highly correlated. Thus, some of the scales mainly measure the same perceptual impression, when they should actually cover all perceptual quality features of synthetic speech. These features are not necessarily orthogonal but certainly exhibit smaller correlations than the scales from P.85.

13.2 Perceptual Quality Dimensions

Depending on the kind of TTS system, different degradations still diminish the overall quality impression: most PSOLA-based diphone synthesizers lead to artificial voices due to frequent concatenations of speech units, HMM-synthesizers can generate natural-sounding but also very noisy speech, and the quality of unit-selection systems mainly depends (1) on the size of the used speech corpus, (2) on how well the units fit together, and (3) on how well the units fit to the text that is to be synthesized. These impairments all sound differently, thus they degrade speech along different perceptual features. Hence, the quality of synthetic speech is of multidimensional nature.

In the following section we will present two studies that used different approaches to identify these perceptual quality features. The first one is based on the rating of stimuli on several attribute scales similar to ITU-T Rec. P.85 while the second approach uses a pairwise comparison test to create a stimulus space.

13.2.1 Semantic Differential

In [11] a Semantic Differential (SD) was used to evaluate perceptual quality dimensions of synthetic speech. The main idea was to use a set of attribute scales to measure the auditory impression of listeners. In order to find such a set, suitable for the assessment of synthetic speech, several pretests needed to be conducted. The goal of the first pretest was to collect a broad basis of attributes that describe auditory features of synthesized speech. Thus, expert listeners were instructed to write down nouns, adjectives, and antonym pairs that describe their auditory impression, and to rate the intensity of each item. The aim of the second pretest was to narrow down the set of attribute scales till it is small enough to be used in the main test while still precisely describing the quality stimulus space. The resulting scales were then used in the main test where the listeners rated stimuli produced by a variety of different TTS systems. A factor analysis with a subsequent oblique rotation revealed 3 factors. The first factor is related to the scales *accentuation*, *naturalness* and *rhythm*, therefore, it was labeled naturalness. The second factor can be associated with the scales *hiss*, *noise*, and *rasping sound*. This feature represents disturbances in the signal. The third feature is linked to the scales *polyphony* and *intelligibility* and thus indicates temporal distortions in the signal. Artifacts affecting the intelligibility especially occur in concatenation-based synthesizers at the transitions between two units. Connecting units with slightly different speaking rates sometimes even leads to the impression of two different voices speaking at the same time. Moreover, the scale *speed* indicated a fourth quality dimension which seemed to be of minor importance.

This study clearly produced results that improved the understanding of synthetic speech from the perceptual viewpoint. Nonetheless, since this approach uses global scales, the test participants rating is limited to them. Hence, perceptual impressions of the listeners that could not be expressed by the presented scales could not be captured.

In an attempt to come up with a similar set of attribute scales that is optimized for the evaluation of audiobooks synthesized by TTS systems, two further studies were conducted [12, 16]. Given the long stimuli duration and the special requirements for TTS audiobooks, e.g., the ability for emotional speech, several additional scales were developed. The listening tests contained TTS read books that were chosen with the attempt to cover a variety of different writing styles. A factor analysis with a subsequent oblique rotation resulted in 2 factors for both studies. While there were minor inconsistencies in the assignment of the attribute scales to the resulted factors, both studies yielded a *prosody and rhythm* dimension as well as a dimension associated with the *listening pleasure*.

13.2.2 Pairwise Comparison and Multidimensional Scaling

Given the drawbacks of the SD approach, this section presents a method to extract perceptual quality dimensions that is solely based on the unrestricted perceptual quality impression of the listener and not on given rating scales. The main idea is to scale dissimilarities between pairs of stimuli. These dissimilarities can then be transformed into a stimulus space in which the between-point distances correspond to the dissimilarities between stimuli. Via a Multidimensional Scaling (MDS) algorithm this stimulus space can be reduced in dimensionality until the solution is interpretable.

Thus, each stimulus in a set of n stimuli has to be compared to all $n-1$ stimuli resulting in $\frac{n(n-1)}{2}$ comparisons. In large object sets, this approach easily leads to way over hundred comparisons. Depending on the length of the stimuli, this would cause a test duration per subject of several hours. Therefore, listening tests with large object sets need to apply a method to derive dissimilarities without a full pairwise comparison test. For cases like these, Tsogo [37] proposed a sorting task where test participants are assigned to build groups of stimuli that are similar to each other while being different from the stimuli in other groups. This leads to an $n \times n$ incidence matrix per subject from which a dissimilarity matrix can be derived.

In [15], such a test has been conducted on a large set of different TTS systems. An MDS of the resulting dissimilarity matrix yielded 3 dimensions. Since MDS dimensions give no indication on their interpretation, the stimuli can only be analyzed along the identified dimensions via expert listening or an additional listening test. Thus, an interpretation is often a vague and highly subjective task. For this reason, the authors evaluated all stimuli on the scales that were developed during the SD experiment described in Sect. 13.2.1. Thereby, the correlations between the factor scores and the attribute-scale ratings gave indications on the interpretation of the dimensions. Stimuli with high ranks in dimension 1 sounded very human-like even if the speech was somehow distorted. These voices can be described as voices with personality and charisma. Thus, dimension 1 was labeled *naturalness of voice*. The second dimension is linked to the scales rhythm, fluency, and bumpiness. Therefore, this dimension represents the *prosody* as well as *temporal distortions* in the signal. Finally, the third dimension could be tied to the scales *speed and tension*. Stimuli with high values in this dimension were slowly speaking and relaxed, while voices with low values sounded stressed and restless.

13.2.3 Comparison of Perceptual Quality Dimensions

On first sight, the results from the Semantic Differential in Sect. 13.2.1 and the Multidimensional Scaling experiment in Sect. 13.2.2 seem to be contradictory. However, Table 13.1 reveals major similarities between the aforementioned studies. The feature *naturalness* from the SD experiment which seemed to be too broad to give useful information about a TTS system can be found in the dimension *naturalness of voice* and in the prosodic part of the dimension *temporal distortions* from the MDS test. Thus, it comprises the quality of the voice and the prosody of the generated signal. In contrast, the dimension *temporal distortions* from the MDS experiment combines prosodic features as well as characteristics that indicate the fluency and the intelligibility of a TTS signal. Remarkably, even though listeners could clearly distinguish between, e.g., noise and hiss in signals through the presented attribute scales, the feature *disturbances*, which was highly significant in the SD test, can not be found in the MDS experiment. We assume that this effect is most

<table>
<tr><th>Dimensions</th><th>Relevant Scales</th><th>SD</th><th>MDS</th><th>Audiobooks</th></tr>
<tr><td>Naturalness of Voice</td><td>naturalness
voice pleasantness</td><td rowspan="2">Naturalness</td><td>Naturalness of Voice</td><td>Listening Pleasure</td></tr>
<tr><td>Prosodic Quality</td><td>stress
rhythm
prosody
intonation</td><td rowspan="2">Temporal Distortions</td><td>Prosody & Rhythm</td></tr>
<tr><td>Fluency and Intelligibility</td><td>fluency
intelligibility
bumpiness
polyphony</td><td>Temporal Distortions</td><td></td></tr>
<tr><td>Absence of Disturbances</td><td>hissing
noise
rasping
disturbances</td><td>Disturbances</td><td></td><td></td></tr>
<tr><td>Calmness</td><td>speed
tension</td><td>Speed</td><td>Calmness</td><td></td></tr>
</table>

Fig. 13.1 Perceptual quality dimensions of synthetic speech

likely due to the nature of most TTS signals: even though TTS quality improved dramatically over the years, there are still major issues that catch the attention of listeners. These impairments mainly affect the features *naturalness* and *prosody*. They are so dominant that minor problems like disturbances, which most of the listeners are already used to via coding and transmission artifacts in cell phone or IP-based communication, might be masked. Thus, this dimension did not emerge from the MDS experiment. More similarities can be found in the dimensions *speed* and *calmness*. Even though they were labeled differently in each experiment, they both cover the same aspects of the signal.

Surprisingly, even the results from the audiobook-reading experiments can be assigned to the already discussed dimensions. The feature *listening pleasure* corresponds to the feature *naturalness of voice* while the feature *prosody and rhythm* is linked to the prosodic parts of the feature *naturalness* from the SD test and *temporal distortions* from the MDS experiment.

In summary, these four studies included stimuli for different use cases (e.g. short message reader, audiobook reader) produced by different kinds of state-of-the-art TTS systems (i.e., diphone synthesis, unit-selection synthesis, HMM-synthesis), that were tested with different types of listening tests (e.g. scale-based listening tests and tests that are solely based on the perceptual impression of the listener) but yielded a consistent and comprehensive image of the perceptual quality of synthetic speech. Thus, 5 more-or-less universal perceptual quality dimensions can be termed:

- naturalness of voice
- prosodic quality
- fluency and intelligibility
- absence of disturbances
- calmness.

13.3 Instrumental Quality Prediction

Evaluating synthetic speech as described in Sect. 13.1 is extremely cost-intensive as well as time-consuming. Thus, a continuous evaluation of a TTS system during its development process through auditory tests is almost not feasible. Therefore, developers of TTS systems would greatly benefit from instrumental methods that predict the perceived quality of TTS systems. Even though instrumental tools will not be able to supersede auditory tests, the continuous evaluation of a system could greatly support the development of high-quality voices. In this section we present an overview of different approaches towards instrumental quality prediction of synthetic speech. There are two categories of quality prediction models: reference-based measures, i.e., measures that use natural speech references to compute a distance value between the reference and the to-be-evaluated TTS signal, and reference-free measures that use, e.g., speech features or internal parameters of TTS systems to predict quality.

13.3.1 Reference-Based Measures

Several reference-based measures have been developed to predict distortions in natural speech introduced by transmission channels of telephone networks. These models use a clean speech reference, i.e., the signal before the transmission, and evaluate the perceptual distance to the signal after the transmission.

Synthetic speech as such can be considered as distorted speech which raises the question if instrumental measures developed for the quality assessment of coded speech can be used to predict the quality of TTS systems. In [5], Cernak et al. used the measure described in the ITU-T Rec. P.862 *Perceptual Evaluation of Speech Quality* (PESQ) [20] to evaluate single-word narrowband TTS signals. This study yielded correlations between the subjective MOS and the predicted MOS close to 1. These impressive results lead to further investigations [13] including the reference-based measures *Diagnostic Instrumental Assessment of Listening quality* (DIAL) [7] and the PESQ successor ITU-T Rec. P.863 *Perceptual Objective Listening Quality Assessment* (POLQA) [21]. In this study, synthesized sentences with a duration of 2–3 s were evaluated. The correlations achieved by all three measures were disappointing throughout all databases. Of course it has to be noted that all measures were used outside their original domain, therefore these results do not contradict their good performance on telephone-band coded speech. The main reason for the low correlations is most likely a poor time alignment between the reference and the synthesized signal which is due to non-linear distortions introduced by TTS algorithms. Thus, further research in this domain should include an improved Dynamic Time Warping to ensure an exact time alignment between reference and TTS signal.

13.3.2 Reference-Free Measures

In most practical cases, a natural speech reference of the "same speaker" as in the TTS system is not available, thus reference-free measures have to be applied.

In the approach by Chu et al. [6], an average concatenative cost function for unit-selection synthesizers is defined as a weighted sum of 7 sub-costs. These 7 components can be directly derived from the input text and the scripts of the speech database. The correlation between the cost function and the subjective ratings yielded −.87. Despite this impressing result, a cost-function of this type can only be computed for unit-selection synthesizers. To predict TTS quality independent of the synthesizer type, a more general approach has to be developed.

In 1993, Mariniak [25] introduced a framework for predicting the quality of synthesized speech. He proposed to construct a reference feature space by extracting signal-based features from a large number of different human speakers. Classifying synthetic speech patterns with respect to this feature set would yield a quality estimate of the TTS samples.

A similar approach, using a Hidden Markov Model (HMM)-based feature comparison, has been implemented by Falk et al. [8]. Given the fact that spontaneous spectral changes in natural speech can only occur about every 20 ms due to the inertia of the articulation organs, more frequent changes can thus be classified as distortions introduced by the speech synthesizer. In this approach, HMMs are trained on speech features, e.g., mel-frequency cepstral coefficients (MFCC), extracted from a given set of different natural speakers (not necessarily the same speakers as the to-be-evaluated TTS voices). In a second step, the same features are extracted from the TTS samples. Finally, via the forward-backward algorithm [34], a log-likelihood value can be derived which indicates the perceptual distance between the TTS signal and the HMMs. This approach led to fairly good correlations between subjective and predicted MOS on the evaluated databases. However, severe issues, especially with female voices, were reported when applying this algorithm to other databases [29].

A second approach towards quality prediction is based on features related to degradations introduced by the synthesis process. In [29], the internal parameters of the ITU-T Rec. P.563 [19] were investigated. These parameters capture characteristics that are typical for telephone-band coded speech signals, e.g., noise, temporal clipping, or robotization effects. A sequential feature selection was employed, followed by a factor analysis. The resulting factors were combined to a quality estimate via a simple linear regression model. The same approach was executed on a set of 1,495 general speech features [27] which provide a broad variety of information on vocal expression patterns that are useful when classifying human emotions. Both predictors led to correlations between subjective and auditory MOS of about 0.80 on 3 evaluated TTS databases and thus outperformed the HMM-approach in all cases. A combination of 3 different quality estimators (HMM-based predictor, P.563-model, predictor based on general speech features) via linear regression yielded more stable results. However, this first pilot study only evaluated 3 small TTS databases and did not perform any sort of cross-validation (CV) in the development process of the predictors.

With prosody being a major influence on the naturalness impression of listeners, features that model the intonation in a speech signal are of great interest. In [31], formal patterns of speech prosody were investigated. 18 purely acoustic markers were derived from the fundamental frequency (F_0) and vocalic/consonantal durations. These markers were analyzed individually and in conjunction via 3-fold CV regression models. Nonlinear parameters, based on the F_0 slope, proved most valuable. The regression models yielded correlations as high as .87.

With the promising results of prediction models based on prosodic features described in the previous paragraph, further research was conducted concerning the pitch contour in natural speech. In [14], features derived from the Fujisaki-model [9] which describes the F_0-contour of speech signals, were computed. A multiple linear regression was used to combine these features and tested for over-fitting by a leave-one-out CV. The correlations between the subjective MOS and the predicted MOS were around .60 for 3 of 4 databases.

In [32] further research on the performance of features that capture the intonational properties of human speech was conducted. A comparison between prosodic features of synthesized and human speech confirmed the assumption that prosodic variation systematically influences perceptual naturalness of TTS signals and imposes a major impact on their overall perceptual quality. Furthermore, F_0 dynamics were found to be substantially lower in TTS than in natural speech. The cross-validated prediction models yielded correlations between subjective and predicted dimension ratings of up to .83.

In [33], the prosodic features described in the previous paragraphs as well as MFCCs were investigated concerning their ability to model the perceived quality of TTS in audiobook reading tasks. As reported in Sect. 13.2.3, even though the use case differs from the TTS databases investigated in most of the presented studies (short messages deployed, e.g., in spoken dialogue systems), *naturalness of voice* and *prosodic quality* were also the main influences on perceived quality. Several approaches for perceptual modeling were investigated and compared. The cross-validated models yielded correlations as high as .87.

The latest study [17] again concentrated on the large-scale feature set from [27] for general speech features. 1,495 general speech features were extracted to build prediction models for 2 extensive TTS databases. Quality predictors for female and male voices were developed following two different approaches: a three-step feature selection followed by a stepwise multiple linear regression and a model based on support vector machines were implemented. The predictors were cross-validated via 3-fold CV and leave-one-test-out CV. A strict CV method, where the partitioning is realized prior to the feature scaling and feature selection steps, was applied. In the 3-fold CV case, correlations as high as .89 could be achieved. The more ambitious leave-one-test-out CV yielded correlations around .80 for the male speakers while the results for the female voices need further improvement.

13.4 Conclusions and Future Work

In this chapter, we addressed the quality experienced when listening to synthetic speech. Auditory tests can be divided up into two different categories: functional tests can be used to evaluate the intelligibility, and judgment tests are used to assess more subjective aspects of a TTS signal, e.g., its naturalness or the overall quality. Two different test methodologies to determine perceptual quality dimensions were introduced: either the ratings on the attribute scales presented in a Semantic Differential can be further processed via factor analysis to derive perceptual dimensions, or a Multidimensional Scaling algorithm can be used to transform the dissimilarities from, e.g., a pairwise comparison test, into interpretable perceptual quality dimensions. Tests of both kinds were carried out in different studies. A comparison of the results led to the set of 5 universal perceptual quality dimensions of synthetic speech: (1) *naturalness of voice*, (2) *prosodic quality*, (3) *fluency and intelligibility*, (4) *disturbances*, and (5) *calmness*. Moreover, different techniques to instrumentally predict the quality of TTS signals have been introduced. Even though reference-based measures have lead to a tremendous accuracy for quality predictions on word-level, the results for longer stimuli were disappointing. In addition to that, in most cases natural references of the same speaker are not available. Thus, reference-free measures have to be applied. These measures use features extracted from the synthetic voices to indicate distortions in the signal that are typically introduced by speech synthesizers. The results showed that cross-validated predictors can lead to correlations between subjective and predicted MOS of over .80. Nonetheless, all predictors still struggle with female voices.

In the future, further research will be done, especially with regard to female voices. Furthermore, predictors that compute quality estimates for the dimensions introduced in Sect. 13.2 are subject to further plans. Predicting individual quality dimensions will provide an insight into the characteristics of text-to-speech synthesizers and will help to further improve them.

Acknowledgments This work was supported by the Deutsche Forschungsgemeinschaft (DFG), grants MO-1138/11-1, MO-1138/11-2, HE-4465/4-1 and HE-4465/4-2.

References

1. ASA S3.2-2009 (2009) American national standard method for measuring the intelligibility of speech over communication systems. American National Standards of the Acoustical Society of America, Washington
2. Benoit C, Griceb M, Hazanc V (1996) The SUS test: a method for the assessment of text-to-speech synthesis intelligibility using semantically unpredictable sentences. Speech Communication 18(4):381–392
3. Black AW, Taylor PA (1994) CHATR: a generic speech synthesis system. In: COLING 1994, vol 2. pp 983–986
4. Burkhardt F (2013) Comparison of German TTS-systems. Cited 20 Apr 2013. http://syntheticspeech.de/index.html
5. Cernak M, Rusko M (2005) An evaluation of synthetic speech using the PESQ measure. In: Proceedings of forum acusticum, Budapest, Hungary, pp 2725–2728
6. Chu M, Peng H (2001) An objective measure for estimating MOS of synthesized speech. In: Proceedings of the 7th international conference on speech communication and technology (Eurospeech 2001), Aalborg, Denmark, pp 2087–2090
7. Côté N (2011) Integral and diagnostic intrusive prediction of speech quality. Springer, Heidelberg
8. Falk TH, Möller S (2008) Towards signal-based instrumental quality diagnosis for text-to-speech systems. IEEE Signal Processing Letter 15:781–784
9. Fujisaki H (1981) Dynamic characteristics of voice fundamental frequency in speech and singing. Acoustical analysis and physiological interpretations. In: STL-QPSR, vol 22. pp 1–20
10. Gibbon D, Moore R, Winski R (1997) Handbook of standards and resources for spoken language systems. De Gruyter Mouton, Berlin, Boston
11. Hinterleitner F, Möller S, Norrenbrock C, Heute U (2011) Perceptual quality dimensions of text-to-speech systems. In: Proceedings of the 12th annual conference of the international speech communication association (Interspeech 2011), Florence, Italy, pp 2177–2180
12. Hinterleitner F, Neitzel G, Möller S, Norrenbrock C (2011) An evaluation protocol for the subjective assessment of text-to-speech in audiobook reading tasks. In: Proceedings of the Blizzard challenge workshop, Florence, Italy
13. Hinterleitner F, Zabel S, Möller S, Leutelt L, Norrenbrock C (2011) Predicting the quality of synthesized speech using reference-based prediction measures. In: Proceedings of the 22nd Konferenz Elektronische Sprachsignalverarbeitung (ESSV 2011), Aachen, Germany, pp 99–106
14. Hinterleitner F, Norrenbrock C, Möller S (2012) On the use of fujisaki parameters for the quality prediction of synthetic speech. In: Proceedings of the 23rd Konferenz Elektronische Sprachsignalverarbeitung (ESSV 2012), Cottbus, Germany, pp 112–119
15. Hinterleitner F, Norrenbrock C, Möller S, Heute U (2012) What makes this voice sound so bad? A multidimensional analysis of state-of-the-art text-to-speech systems. In: Proceedings of the 2012 IEEE workshop on spoken language technology (SLT), Miami, USA, pp 240–245
16. Hinterleitner F, Norrenbrock C, Möller S (2013) Perceptual quality dimensions of text-to-speech in audiobook reading tasks. In: Proceedings of the 24th Konferenz Elektronische Sprachsignalverarbeitung (ESSV 2013), Bielefeld, Germany, pp 44–49
17. Hinterleitner F, Norrenbrock C, Möller S, Heute U (2013) Predicting the quality of text-to-speech systems from a large-scale feature set, Lyon, France, pp 383–387
18. ITU-T Recommendation P.85 (1994) A method for subjective performance assessment of the quality of speech voice output devices. International Telecommunication Union, Geneva
19. ITU-T Recommendation P.563 (2004) Single ended method for objective speech quality assessment in narrow-band telephony. International Telecommunication Union, Geneva
20. ITU-T Recommendation P.862 (2001) Perceptual evaluation of speech quality (PESQ), an objective method for end-to-end speech quality assessment of narrow-band telephone networks and speech codecs. International Telecommunication Union, Geneva

21. ITU-T Recommendation P.863 (2011) Perceptual objective listening quality assessment (POLQA). International Telecommunication Union, Geneva
22. Jekosch U (1993) Speech quality assessment and evaluation. In: Proceedings of Eurospeech, Berlin, Germany, pp 1387–1394
23. Klatt DH (1980) Software for a cascade/parallel formant synthesizer. Journal of the Acoustical Society of America 67(3):971–995
24. Kraft V, Portele T (1995) Quality evaluation of five German speech synthesis systems. Acta Acustica 3:351–365
25. Mariniak A (1993) A global framework for the assessment of synthetic speech without subjects. In: Proceedings of the 3rd European conference on speech processing and technology (Eurospeech), Berlin, Germany, pp 1683–1686
26. Mayo C, Clark RAJ, King S (2005) Listener's weighting of acoustic cues to synthetic speech naturalness: a multidimensional scaling analysis. In: Proceedings of the 6th annual conference of the international speech communication association (Interspeech), Lisbon, Portugal, pp 1725–1728
27. Minker W, Lee GG, Mariani J, Nakamura S (2010) Salient features for anger recognition in German and English IVR portals. Spoken dialogue systems technology and design. Springer
28. Möller S, Hinterleitner F (2013) ITU-T Contribution COM 12–37: proposal for an appendix to Rec. P.85 of the evaluation of speech output for audiobook reading tasks. Deutsche Telekom AG, ITU-T SG12 meeting 19–28 Mar 2013, Geneva
29. Möller S, Hinterleitner F, Falk TH, Polzehl T (2010) Comparison of approaches for instrumentally predicting the quality of text-to-speech systems. In: Proceedings of the 11th annual conference of the international speech communication association (Interspeech 2010), Makuhari, Japan, pp 1325–1328
30. Moulines E, Charpentier N (1990) Pitch-synchronous waveform processing techniques for text-to-speech synthesis using diphones. Speech Communication 9(5/6):453–467
31. Norrenbrock C, Hinterleitner F, Heute U, Möller S (2012) Instrumental assessment of prosodic quality for text-to-speech signals. IEEE Signal Processing Letters 19:255–258
32. Norrenbrock C, Hinterleitner F, Heute U, Möller S (2012) Quality analysis of macro-prosodic F_0 dynamics in text-to-speech signals. In: Proceedings of the 13th annual conference of the international speech communication association (Interspeech 2012), Portland, USA, pp 454–457
33. Norrenbrock C, Hinterleitner F, Heute U, Möller S (2012) Towards perceptual quality modeling of synthesized audiobooks. In: Proceedings of the blizzard challenge workshop, Portland, USA
34. Rabiner L (1989) A tutorial on hidden Markov models and selected applications in speech recognition. Proc IEEE 77(2):257–286
35. Sityaev D, Knill K, Burrows T (2006) Comparison of the ITU-T P.85 standard to other methods for the evaluation of text-to-speech systems. In: Proceedings of the 9th international conference on spoken language processing (Interspeech), Pittsburgh, USA, pp 1077–1080
36. Tokuda K, Zen H, Black AW (2002) An HMM-based speech synthesis system applied to English. In: Proceedings of 2002 IEEE speech synthesis workshop, Santa Monica, USA, pp 227–230
37. Tsogo L, Masson MH, Bardot A (2000) Multidimensional scaling methods for many-objects sets: a review. Multivariate Behavioral Research 35(3):307–319
38. Viswanathan M, Viswanathan M (2005) Measuring speech quality for text-to-speech systems: development and assessment of a modified mean opinion score (MOS) scale. Computer Speech and Language 19(1):55–83

Chapter 14
Audiovisual Communication

Markus Vaalgamaa and Benjamin Belmudez

Abstract Audiovisual communication has expanded rapidly over the last years on computers and mobile devices. This chapter discusses the key aspects of Quality of Experience of the audiovisual communication. We will give an introduction to audiovisual communication and explain technical elements and perceptual features which relate to the Quality of Experience. Main subjective and instrumental quality assessment methods will be presented. Finally, we specifically discuss a few key aspects impacting quality, namely, time-varying quality perception, audiovisual quality integration as well as the impact of overall delay and audiovisual synchrony, and give an outlook for future work.

14.1 Introduction

Early public video call experiments started in the 1930s at the German Post office which provided a service between Berlin and Leipzig [34]. In the 1960s, AT&T brought a first commercial product called the "Picturephone" to the market and continued the service until the 1970s.[1] During the beginning of the 21st century, the mobile communication industry made a significant investment to develop video telephony under 3GPP umbrella, resulting in the 3G-324M protocol enabling video communication over circuit switched networks. However, it took few years more before technology, infrastructure and business were ready and Voice over Internet Protocol

[1] http://www.corp.att.com/attlabs/reputation/timeline/70picture.html

M. Vaalgamaa (✉)
Skype Labs/Microsoft, Tallinn, Estonia
e-mail: markus.vaalgamaa@skype.net

B. Belmudez
Quality and Usability Lab, Telekom Innovation Labs, TU Berlin, Berlin, Germany
e-mail: ben.belmudez@gmail.com

S. Möller and A. Raake (eds.), *Quality of Experience*,
T-Labs Series in Telecommunication Services, DOI: 10.1007/978-3-319-02681-7_14,

(VoIP) solutions revolutionized video communications. Software-based VoIP clients enabled fast development and quick adaptation of new technologies such as real-time audio and video communication over public internet, efficient audio and video coding, and peer-to-peer technology. The broadband Internet with high bandwidth and speed, as well as the development of affordable personal computers and webcams, provided enhanced user experience in home environments. These factors combined with a new business model such as a free use of basic services created solutions that quickly gained a large audience. A few of the first notable solutions were proposed in the early 2000s such as Skype and Apple iChat.

Video communication was firstly seen as a natural evolution from voice calls towards face-to-face communication, however a few essential usage differences exist. Users of video solutions explain that video communication brings them closer and is socially more involving compared to an audio call [31]. Therefore video communication is highly popular among families and friends living far apart from each other. In addition, video chats are popular in work environments where people are regularly in close collaboration with colleagues at different locations. The addition of a video channel has also enabled new means to communicate for deaf, hard-of-hearing and speech-impaired people. On the other hand, video calls are not always the preferred choice as audio calls allow users to keep a social distance, e.g. when users do not wish themselves or their surroundings to be seen.

The audiovisual (AV) communication usage can be split into two segments: business and consumer usage. On the business side the solutions range from teleconferencing to telepresence solutions. The teleconferencing solutions are typically realized either with a software solution on a personal computer or with a dedicated device. Such solutions enable people to work effectively in multi-location business environments while minimizing travelling time and cost. These solutions often incorporate additional services such as meeting hosting and screen sharing in order to enhance the collaboration. The telepresence solutions expand teleconferencing to very high quality services with dedicated devices and even physical meeting rooms, such examples can be found from companies like HP, Tandberg, Cisco and Polycom. A common denominator for the business solutions is that users are willing to pay for good quality, reliability of the service and additional features, thus motivating solution providers to invest into specific hardware and if needed into dedicated network connections. Therefore, the companies developing these solutions have a high interest and motivation to measure and provide the best achievable user experience.

The usage of consumer video communication has grown rapidly over the past decade and nowadays millions of users share the same solutions. It has been estimated that VVoIP (Video and Voice over IP) communication creates more than one third of the world's international call duration in the beginning of 2013.[2] One of the reasons for this rapid growth has been free or very low-price subscriptions. These subscriptions have been possible as software-based solutions can run on the top of the users' existing personal computers, smartphones and access to broadband network

[2] http://www.telegeography.com/press/press-releases/2013/02/13/the-bell-tolls-for-telcos/index.html

connections. With a good audio and video setup, a device with enough processing power and a good broadband network connection, consumer products can achieve stunningly high audio and video quality. On the other hand, the variations of quality can be noticed in practice and in many cases software-based solutions provide the "best-effort" service by coping with constraints given by limited acoustic and video setups, computer resources and network limitations or temporal Quality of Service (QoS) variations. This means that in the consumer market there is a need for robust and high quality software and hardware solutions.

One of the growing areas over the past years has been video communication on mobile devices. The processing power of smartphones is now sufficient to simultaneously encode and decode video at a high spatial and temporal resolution during a real-time communication. In addition, the wide availability of broadband WLAN connections has enabled mobile video communication. The wireless mobile networks today such as 3G have issues with network speed, latency, quality and delays during the cell hand-over which can limit quality. However, the latest network technologies such as LTE and 4G are expected to significantly improve the quality and reliability of video communication on mobile devices.

In the upcoming years, we foresee that the quality of VVoIP solutions as well as the users' expectations will rapidly increase both on personal computers and mobile devices. This will keep the competition between the solution providers very active. As a consequence, mobile manufacturers, AV communication providers and network operators, who are able to provide reliable and high quality services with free or low price will dominate. The telepresence market still has room for even higher quality and innovative products. Compatibility and interoperability between business and consumer solutions will be some of the key issues to be addressed for an extended and successful user experience.

In summary, the success of future AV communication solutions will be defined by their ability to provide reliable and high-quality communication even in versatile and fluctuating mobile environments. This brings a specific need for effective methods and metrics to enable an accurate evaluation of AV communication quality. Considering the above-mentioned context of Quality of Experience (QoE) this chapter will concentrate on the following points. The next Sect. 14.2 will give an overview of technical aspects impacting the AV communication quality. Thereafter, we will discuss the key aspects of subjective and instrumental quality assessment in Sect. 14.3. In the following three Sects. 14.4–14.6 we focus on a few specific aspects: time-varying quality perception, audiovisual integration and interactivity factors. Finally, Sect. 14.7 provides conclusions and trends for the future.

14.2 Audiovisual Communication Systems

In this section we will give an overview of audiovisual communication systems and quality degradations that might occur while using them. The system chain and main algorithms will be presented. In addition, the typical audio, video and overall quality

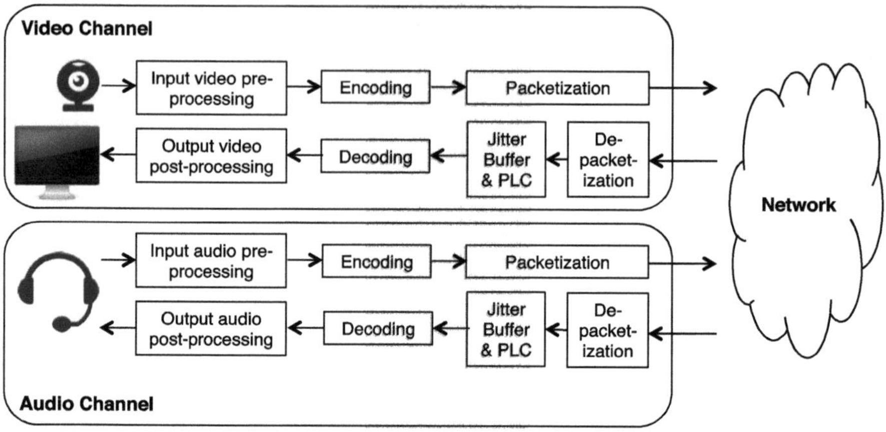

Fig. 14.1 Processing chain of a video communication solution. The video and audio processing chains are represented separately (i.e. no multiplexing of the video and audio signals before transmission over the network)

degradations are briefly described and a few practical aspects of the network usage during AV communication are given.

14.2.1 Real-Time Audiovisual Transmission

The quality of AV communication can be explored by dividing it into three parts: audio quality, video quality, and level of synchrony between these two modalities. In order to understand what affects the perception of these modalities, the processing chain of a video communication solution is shown in Fig. 14.1 and will be discussed below.

On the audio side, a microphone captures the speech of a user. The speech signal can be enhanced with pre-processing algorithms such as automated gain control to adjust the speech level, noise reduction and acoustic echo cancellation. The next processing step is speech encoding. A great variety of speech coders are used in AV communication solutions. The popular narrowband coders, with a 3.5 kHz audible bandwidth, are: ITU-T Rec. G.729A, ITU-T Rec. G.711, and 3GPP AMR-NB coders. Within the wideband (7.5 kHz audible bandwidth) and superwideband (11 kHz and beyond audible frequency bandwidth) codecs the popular ones are 3GPP AMR-WB (ITU-T Rec. G.722.2), the AAC-family codecs, ITU-T Rec. G.719, ITU-T Rec. G.722, and Silk.

On the video side, the webcam acquires the video signal that can be pre-processed with noise reduction and other adjustment algorithms. The video will be encoded to a certain frame rate, spatial resolution and bit rate while trying to optimize the perceived video quality. The most commonly used video codecs for AV communication are

ITU-T Rec. H.263 and H263++, MPEG-4 Part 2 (Xvid or DivX), ITU-T Rec. H.264, VP8 and VP9, and also ITU-T Rec. H.265 (HEVC) is foreseen to become popular on future systems.

Many of the above-mentioned audio and video codecs are also used in streaming, see Chaps. 16 and 19 for additional details. However, the key aspect that separates the usage of the codecs and algorithms is that real-time AV communication requires minimal latency. Therefore, all algorithms, codecs included, are tuned and optimized to achieve a low processing delay.

After the encoding stage, the audio and video streams are packetized and often encrypted for security reasons before being sent on the network. The majority of the video communication solutions today use public Internet for data transmission. On the network side, the solutions are based on peer-to-peer technologies or client-servers structures or even hybrid combinations of both. There is a variety of standards and proprietary protocols to enable communication. Calls are initialized through signaling, like SIP for instance, between clients or between clients and servers. Once the call has been initialized, real-time audio and video data streams can be transmitted through the network via protocols such as UDP, TCP and RTP. Special techniques and protocols are required to pass through firewalls, such as firewall traversing and network address translations (see STUN in RFC 5389, TURN in RFC 5766 and ICE in RFC 5245[3]). If packets are missing on the receiver side, techniques such as forward-error correction (FEC) and packet retransmission can be applied to minimize the corruption of the data stream.

Once received, audio and video packets will be buffered and reordered (if those arrived in a mixed order). Packets arriving too late might be discarded. The client could, for example, wait a while for the packets to arrive by increasing the buffer size and modifying the play-out speed of both audio and video, or replace the missing audio or video part with packet loss concealment algorithms. After decoding, the audio and video frames can be post-processed to further improve the overall quality before the playout via loudspeaker and display.

A major part of the algorithms mentioned above and their details are solution-specific and proprietary. However, there are a few algorithms which must be agreed upon in order to enable interoperability between different AV communication solutions. Such algorithms and protocols involve audio and video decoders and network protocols including packetization and de-packetization. One solution to enable the interoperability between different codecs is to transcode (in other words convert) one codec bit stream into another at the network. However, transcoding operations can cause a noticeable quality drop-off and it require a specific processing on the servers. For these reasons it is more beneficial to agree on common codecs and protocols beforehands instead of using transcoding between the different solutions.

[3] http://tools.ietf.org/html/rfc5389,http://tools.ietf.org/html/rfc5766, and http://tools.ietf.org/html/rfc5245.

14.2.2 Quality Degradation Factors and Features

Various quality elements (for a definition see Chap. 2) or technical factors may degrade audio and video qualities. A detailed list of the audio elements is presented in Chap. 16, and of the video elements in Chap. 19. The most common audio quality elements are network impairments, which can cause freezes in playback, packet loss artifacts, low bit rate distortions and acoustic impairments, particularly microphone noise and acoustic echo artifacts. In addition, on speakerphone and living room TV products, the reverberation and lack of clarity can be important technical factors. On the video side, the main elements impacting the quality are camera capturing quality, video encoding like encoder type, bit rate, frame rate, video resolution and network related factors like packet loss and jitter. In addition, the overall delay and synchrony between audio and video impact the overall quality.

These technical factors are closely linked to the perceptual artifacts of the AV signal. These are referred as quality features and describe the user's perception of the AV signal (see Chap. 2 for a definition). On the audio side, the common quality features are audible noise, echo, muting, lack of intelligibility, coding artifacts, limited audible bandwidth and temporal quality degradations. On the video side, common quality features are jerkiness, blurring, blocking, color distortions, image freeze, unnatural movement and noise. A long delay will impact the conversational quality by reducing the interactivity of the communication. If the relative delay between audio and video is large enough it can cause a perceptual asynchrony artifact, for example when lip movements do not match the perceived auditory signal.

14.2.3 Practicalities on AV Communication

Network quality and speed have drastically increased during the past years. The wide availability of broadband internet connection enables very high quality video communications. Towards low bit rates, improvements of coding and network technologies enable a sufficient video quality with a speed as low as 100 kbps. However, high-definition video with resolution of 720p or higher at 30 frames per second require a bit rate higher than 1 Mbps. In fact, it is not uncommon to see differences larger than a factor of 10 in the actual bit rates of AV communication. Also it is worth to note that high-quality audio coding clearly requires less bits than high-quality video coding. Thus, at rates of about 100 kbps it may be that audio uses around 30 % of the bit rate whereas video uses the remaining bit allocation. However, at higher rates the video coding may take 90 % or even more from the bit rate. We will return to the bit rate budget allocation in the Sect. 14.5.4 about the trade-off between audio and video for low bit rates.

14.3 Subjective and Instrumental Quality Assessment

AV communication quality can be described using perceptual dimensions. In practice, investigating all the dimensions and potential use cases through subjective user tests is very resource-intensive and time-consuming. Subjective experiments must be carefully designed so that the results can be trustfully exploited and also used for comparison between different experiments and laboratories. The presence of experimental biases can significantly affect the validity and reliability of the results; therefore, such experiments usually comply with international standards (e.g. ITU, EBU). Accuracy is also required in the usage of instrumental metrics; the acoustic and visual conditions as well as the usage area of the metrics can highly influence the validity of the prediction.

In the next Sect. 14.3.1, we will briefly introduce subjective assessment methods applicable for AV communication. In Sect. 14.3.2, we will focus specifically on conversational assessment. The modeling of AV quality as well as instrumental metrics are presented together in Sect. 14.3.3.

14.3.1 Subjective Assessment

Subjective experiments for assessing transmission quality, like the ones standardized by the ITU, are commonly used for evaluating speech quality (ITU-T Rec. P.800 [23]), video quality (ITU-T Rec. P.910 [26]), and audiovisual quality (ITU-T Rec. P.911 [27]). These standard methods describe the assessment of short stimuli (samples of 4–12 s for speech, and 10–15 s for video) in a passive viewing and listening situation. In turn, conversational quality can be evaluated using ITU-T Rec. P.805 [24] for speech-only, or ITU-T Rec. P.920 [28] for AV communication. These recommendations propose task-based conversational scenarios for pairs of conversing partners. The resulting conversations should last at least 2 min for audio-only, whereas a duration comprised between 3 and 5 min is required for the audiovisual counterpart. Further information on multi-party conversation scenarios and their associated assessment methods can be found in Chap. 15.

14.3.2 Conversational AV Communication Assessment

ITU-T Rec. P.920 proposes several types of task-based scenarios depending on the targeted application and on the factors to be evaluated. Five scenarios for AV conversations are described: name guessing, story comparison, picture comparison, block building and object description. Additional tasks are included to assess the impact of speech delay, audiovisual delay and audiovisual asynchrony on communication quality. This recommendation also provides general guidelines to modify existing scenarios or developing new ones:

1. The scenarios should allow test participants to primarily focus their attention on the audiovisual terminal.
2. The scenarios should be designed based on real-life audiovisual communication to ensure the validity of the results (to a sufficient degree).
3. If communication efficiency is measured, the task should allow "reproducible quantitative results".

The tasks have to be designed for a wide range of test participants (including elderly and hearing-impaired subjects). Additional recommendations can be set in order to obtain acceptable test results: the scenarios should be reproducible to ensure reliability (i.e. limited variations in the conduct of the conversational scenario), and a sufficient level of familiarity between conversing partners is advised in order to avoid additional uncontrolled hindrances to the communication process.

Aspects linked to the interactivity or to the usage[4] of the auditory and visual channels have to be taken into account when designing a scenario. For instance, if the impact of delay is under assessment, the resulting interactivity (e.g. number of turns, number of words, number of words per turn and presence of backchannels) will impact the subjective perception of delay. If the assessment goal is the quality evaluation of transmitted video, scenarios privileging the use of the visual channel will be better suited. Finally, a task-based conversation is likely to be cognitively heavy, thus leaving less resources to the user for the assessment task (unlike in a passive listening and viewing situation).

According to ITU-T Rec. P.920, subjective quality should be judged retrospectively, i.e. after each conversation is carried out. First, test participants are asked to rate the overall AV quality, then the video quality and at last the audio quality. This specific order is chosen to avoid a direct average of the audio and video qualities to form the AV score. Absolute Category Rating (ACR) scales are commonly used to assess conversational quality. Examples of such scales are the 5-point, the 9-point and the 11-point[5] ACR scales defined in ITU-T Recommendations P.910 and P.911.

14.3.3 Quality Modeling and Instrumental Measurement

Conversational quality of AV communication can be estimated using a parametric prediction model standardized as ITU-T Rec. G.1070 [18]. This model used for planning purposes is divided into three main modules: the speech and video quality estimation functions, and the multimedia quality function. The speech quality estimation function is a simplified version of the E-model [16, 17] available for narrowband and wideband services. The video quality estimation function has been primarily developed for typical videotelephony "head-and-shoulders" content and

[4] The channel usage refers to the extent to which subjects utilize a channel to transmit information relevant to carry out scenario-related tasks.

[5] The 11-point ACR scale has end-points which are verbally defined as anchoring points.

depends on application parameters (e.g. codec, display size, video resolution). The multimedia function takes into account the output from the audio and video quality modules, but also the absolute one-way delay and the asynchrony between the auditory and visual signals. The coefficients of the video and multimedia quality functions need to be trained through subjective testing. Their values depend on the conversational tasks used during the training phase. As a consequence, using the model for other types of tasks may lead to poorer performances. The Annex A of the recommendation describes a procedure to derive a set of coefficients for any particular set-up; some examples of coefficient values are provided in the appendix.

ITU-T Rec. G.1070 is a good framework for overall quality estimation of audiovisual communication based on the auditory and visual qualities. It allows to integrate those individual qualities into a multimedia quality taking into account the resulting audiovisual quality and the one-way transmission delay. Its prediction accuracy thus heavily depends on the individual performance of its speech and video quality estimation functions. This model was initially developed for planning purposes, however, in the case where the signal is available for diagnostic usage, one could replace the planning quality functions of the model by media-based models (quality metrics) which generally produce more accurate results.

The basis for standardized instrumental audio and speech quality metrics are defined primarily in the ITU-T[6] P and G series of Recommendations, and IEEE standard 269-2010.[7] Detailed and specific measurements can be found for example from 3GPP, ETSI, ITU, and TIA documentation, see more in Chap. 12. One of the potential tools is the state-of-the-art speech quality metric in ITU-T Rec. P.863 [25] also known as POLQA. It is a full-reference intrusive tool that mimics the human hearing system. This means that it compares the short (4–30 s duration) recorded speech file (called degraded file) to the original high-quality speech file (called reference file) using a complex quality estimation model based on the human hearing system, and as an output gives the estimated Mean Opinion Score (MOS). Therefore, POLQA can be very useful for a G.1070 type of framework in predicting audio quality MOS more accurately than the audio model of G.1070 does.

In the area of video quality tools, the above mentioned bodies added by the Video Quality Experts Group[8] have evaluated and recommended a number of metrics. As with the audio tools, many of the metrics do not provide direct estimation on an MOS scale. However, there are a few video metrics that do estimate quality on the MOS scale, such as ITU-T Rec. J.247 [21] and ITU-T Rec. J.341 [22]. Both of these tools are full-reference video quality metrics that compare the degraded video file to the original video file. In addition, there are various other full and reduced reference models to estimate video quality on the MOS scale, see more in Chap. 19. The predictions of these tools can be used in a G.1070 type of framework to estimate the quality of the video part within AV communication quality.

---

[6] http://www.itu.int/en/ITU-T/publications/Pages/recs.aspx

[7] http://standards.ieee.org/findstds/standard/269-2010.html

[8] http://www.its.bldrdoc.gov/vqeg/vqeg-home.aspx

14.4 Time-Varying Quality Perception

In error-prone networks, a constant level of quality cannot always be ensured by service providers, thus leading to a fluctuating quality. Time-varying distortions are the most characteristic type of degradations of Voice and Video over Internet Protocol. Transient degradations like packet loss are susceptible to affect in different ways the overall quality of a call, that can last from a few seconds to several hours. As a result, memory effects come into play in the process of describing and modeling the perception of an AV call.

14.4.1 Memory Effects

The evaluation of long AV stimuli implies a certain understanding of how humans construct affective experience over time. Studies in cognitive psychology revealed that "global evaluations of past affective experiences are not merely perceived or felt; they are constructed" [7]. Cognitive components like long-term memory, working memory and attentional processes have been reported to have an impact on judgments and evaluations [41]. In the course of an affective experience, some moments turn out to be found more meaningful than others. These moments receive greater weights in the global retrospective evaluation. As previously detailed in Chap. 10, peak affect intensity and the ending were defined as the moments that could serve as efficient proxies to evaluate an affective episode [7]. This is referred to as the "peak-end rule". Additional effects were brought to light like the "duration neglect", stating that the duration of negative episodes does not significantly impact the final judgement [6]. The trend of intensity change as demonstrated by Ariely [2] and Schreiber and Kahneman [38] can have an impact on the hindsight judgement especially when occurring towards the end of an affective episode. Finally, a recency effect was found to occur in several studies on speech quality evaluation [5, 8, 15, 40] and on video quality evaluation [1, 9, 10].

14.4.2 Conversational Time-Varying Quality Modeling

Methodologies to assess time-varying quality of long AV samples or in a conversational set-up have been described in Chap. 10. The methodology based on simulated conversational structures developed by Weiss et al. was extended to AV material by Belmudez et al. [4]. Quality scores (audio, video and audiovisual) for short samples were collected in a first session. In a second session, the short samples were concatenated into long samples and the overall AV scores for these were collected. Three types of temporal integrations were proposed: (1) using existing temporal integration models on the audiovisual MOS scores of short samples, (2) aggregating the audio and video MOS scores into an audiovisual MOS score before applying the temporal integration model, (3) applying the temporal models on the audio and video scores separately, and aggregate the final audio and video MOS scores into an AV score. The results showed that existing models exhibited a high correlation between individual sample ratings and end-call judgments (as an example, the model from Weiss et al. reaches a correlation of 0.97 against 0.93 for the plain average, with a decreased RMSE value on a validation dataset). Moreover, several perceptual effects were observed: the temporal location of strong degradations did not seem to play a major role, and a larger time constant for the recency effect was obtained. This method has the advantage of controlling the temporal location and the strength of the impairments, while simulating a conversational situation. It however differs from a real interactive conversation in the sense that the interpersonal aspects and the influence of the tasks remain missing factors. The assessment scope of this method is thus limited in terms of type of impairments (e.g. echo and delay cannot be included) and in terms of resemblance with a real conversation.

In summary, time-varying quality can be constituted in a simplified way from short-term (4–20 s) quality evaluations, like described in Sect. 14.3.3, which are aggregated to form a long-term evaluation. In the short-term interval, the quality will be highly dominated by the worst temporal artifact during that interval. In the long-term interval, according to the current models, the biggest contribution will be the average of the short-term qualities on MOS scale, followed by the worst quality (peak degradation) interval. The recency effect does have an impact, but less important than the previously mentioned parts.

14.5 Audiovisual Quality Integration

Audiovisual quality integration refers to the human process of combining or pooling different information from the auditory and visual modalities in order to get an estimation of the perceived AV quality. There are so far three possibilities for modeling AV quality: a quality-based approach like the one adopted in the G.1070 model which predicts the audiovisual quality on the basis of individual auditory and visual quality scores; an impairment-factor-based approach (ITU-T Rec. P.1201 [20] addressing HDTV video streaming) where each factor quantifies the quality-impact of different degradations; and a dimension-based approach like the one described in [43], where descriptive attributes are clustered into perceptual dimensions (e.g. aesthetic feeling and feeling of activity) and then pooled into an overall quality score.

14.5.1 Quality-Based Audiovisual Integration

Numerous studies support the theory that global AV quality can be predicted from the individual auditory and visual qualities separately. In that case, the auditory and visual signals are considered to be internally processed to produce separate auditory and visual qualities that are fused at a late stage to give a judgment of the overall perceived quality (late fusion theory). A commonly used function for integration is composed of a linear combination of the audio and video MOS, and of a multiplicative term, see Eq. (14.1). The ITU in Recommendation P.911 proposes to only use the multiplicative term between the audio and video qualities with an additive shift as an estimator of the AV quality, see Eq. (14.2):

$$MOS_{AV} = \alpha \cdot MOS_A + \beta \cdot MOS_V + \gamma \cdot MOS_A \cdot MOS_V + \theta \quad (14.1)$$

$$MOS_{AV} = \gamma \cdot MOS_A \cdot MOS_V + \theta \quad (14.2)$$

with MOS_{AV}, being the audiovisual quality, MOS_A, the audio quality, MOS_V, the video quality, and α to θ being constants. In Eq. (14.2), the recommended values of the formula coefficients are an average of the values obtained in different studies: γ is set to 0.11 and θ to 1.3 when MOS varies between 1 and 5. More generally, the values of the coefficients of Eq. (14.1) depend on various factors:

- testing methodology
- test conditions: type and range of impairments
- material, i.e. presentation devices
- audiovisual content
- experimental setup: passive listening/viewing or interactive; in a laboratory or in situ.

The extended range of variation and the combination of these experimental factors make it difficult to determine a unique integration function accounting for all possible scenarios [44]. A meta-analysis performed by Pinson et al. in [35] showed that the variation of the MOS range for both the auditory and visual modalities is of primary importance in order to get an unbiased integration function. The hypothesis is that the relative importance of both modalities should be comparable, as suggested by the results of their experiments.

14.5.2 Impact of the AV Content

The AV content has a major impact on the AV integration, as recent reviews [44] and [35] indicate. Studies performed with teleconferencing situations, including "head-and-shoulders" material or meeting room situations with several participants, suggest that both audio and video quality have a significant impact on AV quality integration. The situation seems to be different for TV and more generally streaming material, where experimental results show that the more attention the user dedicate to the visual channel, the higher is the relative weight for video quality compared to the one for audio quality in the overall integration.

Korhonen et al. [33] showed a practical study on how to find an optimal trade-off between audio and video packet-loss artifacts at bit rates varying between 360 kbps

Table 14.1 Cross-modal influences reported in the literature for different types of AV contents and evaluation methods

Experiment	Experimental context	Stimuli length	Method	A → V	V → A
[14]	Videotelephony passive	10 s	ACR	None	Weak
	Videotelephony interactive	5 min	ACR	None	Strong
[11]	Animation and narration	10 s	ACR	–	Medium
[37]	Videotelephony passive	6 s	ACR	Strong	Strong
[3]	Television	25 s	ACR	Weak	Strong
[29]	Television	30 min	SSCQE	Strong	None

The qualification "weak" refers to an impact smaller than 0.1 MOS, "medium" when it is comprised between 0.1 and 0.5, and "strong" above 0.5 MOS. The methods used in these studies are the absolute category rating (ACR) and the single stimulus continuous quality evaluation (SSCQE), see Chap. 10 for further details on these methods

and 1.4 Mbps. This study demonstrates the difference between news and opera type of material where the attention of the user is on audio, in contrast to football material as the other extreme having the attention primarily on video.

14.5.3 Cross-Modal Interaction

Cross-modal interaction refers to the impact of one modality to another, for example the impact of video quality on the perception of audio quality. In a passive experimental set-up, the general consensus is that cross-modal interactions affect perceived quality. However, experimental results substantially differ between studies depending on the context of application and on the AV content.

Main findings from the literature are summarized in Table 14.1. In the case of videotelephony, the influence of the audio quality level on the perceived video quality is relatively uncertain, as the two studies [14, 37] contradict each other. Additional information on the levels of degradation and intelligibility of the audio channel would help to understand this discrepancy. An influence of the video quality level on the perceived audio quality is found in three studies among the five studies examined. It can be argued that either the AV content of the experiment [11], or the assessment methodology used in it [29], might have made a difference in the measurement cross-modal perception.

14.5.4 Trade-off Between Audio and Video for Low Bit Rates

As it was discussed in Sect. 14.2.3, for a similar perceptual quality (i.e. the same MOS) audio coding requires less bits than the video coding. This also means that at low bit rates for a same increase of audio and video bit rates the audio quality raises quicker than the video quality. Therefore, at high bit rates (hundreds of kbps or more), the optimal AV quality can be achieved by giving audio coding close to the maximum bit rate it requires, and the rest (i.e. the majority of the bit rate budget) for the video, see Eqs. (14.1) and (14.2). Nevertheless, at high rates the audio coding uses still less than 1/10th compared to the video bit rate. At low bit rates the optimal trade-off between audio and video coding is slightly different. A study from Winkler et al. [42] showed that for a bit rate budget varying from 40 to 100 kbps, the optimal audio/video bit rate ratio was found to be around 30/70. The complexity of the content also plays a role: for complex scenes and higher bit rate budgets, more bits should go to audio compared to the optimal trade-off for simpler scenes and lower bit rates.

Investigating bit rates with total budgets comprised between 100 and 160 kbps, Jumisko-Pyykkö [30] came to similar conclusions: for a budget of 100 kbps, an audio-video ratio of 24/76 was preferred over 16/84, thus emphasizing the significance of the audio channel. For video contents with a high spatial complexity (sport and TV series), a ratio of 16/84 was judged equal to a ratio of 24/76, showing the relative weight of audio towards low bit rates.

These studies indicate that there seems to be a switch in user preference towards audio quality over video quality at low bit rates, and at more complex video scenes. As such, this is easy to understand as in low bit rate situations where the AV quality will be clearly compromised, users will focus on speech in order to understand the other party instead of weighting fidelity of the video. As Jumisko-Pyykkö indicated, the interesting question for future studies is to find the audiovisual quality threshold below which the basic audio quality, for example intelligibility, starts to dominate the overall quality.

14.6 Other Factors Impacting the QoE

14.6.1 Impact of Overall Delay

End-to-end delay occurring between two conversing partners may degrade the calling experience. A commonly used mapping between mouth-to-ear delay measured in ms and the quality scale (E-model rating scale) is given in ITU-T Rec. G.114 [19]. The mapping of ITU-T Rec. G.114 originates from the results by Kitawaki and Itoh [32] who used six different tasks ranging from the highly interactive "number exchange" to "free conversation". In summary, ITU-T Rec. G.114 states that on communication one-way delay of 150 ms or below does not degrade quality, however one-way delay above 400 ms is unacceptable. Several studies have indicated that whilst the G.114 mapping is applicable in the E-model for network planning purposes, the mapping is over-critical for assessing normal conversations. A review by Raake et al. [36] of recent studies proposes a modification to the E-model mappings, taking into account the interactivity of the conversation. Their proposal is based on the findings that everyday conversations are not as interactive as G.114 assumes, and that people adapt to the delay caused by the communication style by decreasing the interactivity. For these reasons, longer delays do not impact perceived quality as much as ITU-T Rec. G.114 indicates. Based on the proposed mapping in [36], the degradation caused by delay would be roughly half on the quality scale in a case of an everyday conversation compared to the current mappings of ITU-T Rec. G.114. The interplay between delay, interactivity and quality is further discussed in Chap. 11.

14.6.2 Audiovisual Synchrony

People are used to experience synchrony or slight audio delay/lag between audio and video in their everyday life. Therefore, a large asynchrony between audio and video is perceived as artificial, strange and annoying. Reasonable levels of asynchrony have been extensively studied in the broadcasting world, and ITU-R BT.1359-1 [12] sets the recommendations for the detection and acceptability. According to ITU-R Rec. BT.1359-1 the detectability thresholds for asynchrony are 125 ms when video leads the audio, and 45 ms when audio leads. The acceptability thresholds for broadcasting material are about 200 ms when video leads, and about 100 ms when audio leads. It is worth noting that compared to these values the often referred ITU-T Rec. J.100 [13] appears to propose over-engineered thresholds of 40 ms for video lead and 20 ms for audio lead.

In AV communication, the asynchrony is mainly perceived as a lip de-synchronization, that is a difference between lip movements and perceived speech. It could be assumed that if the lips are not properly visible, the asynchrony might not have an impact. Steinmetz's study [39] indicates that when lip synchronization is studied between head, shoulder and body views, there is a release on the detectability of asynchrony for the body view, however this difference is relatively small, in the order of a few tens of milliseconds.

14.7 Conclusion and Outlooks

In general, the following trends for AV communication can be drawn: First, AV communication will expand rapidly to mobile devices in the upcoming years. The expansion will be enabled by a wider availability of mobile networks that support high quality real-time communication, and smartphones that have enough processing power to run a high-quality AV communication solution. Second, we expect the QoE of mobile AV communication to rapidly increase, as better technical solutions and devices will be developed to overcome challenges given by limited bit rate, challenging use cases and available processing power. Third, the increased QoE will impact user expectations that will raise quickly following the best solutions on the market. This will create a tough competition between AV communication solutions to gain bigger user bases. Fourth, on the business side, we see an expansion of even higher-quality AV communication systems which support a wide range of mobile devices and consumer AV communication interoperability.

On the Quality of Experience side of AV communication, we expect developments in the following areas: first, there is a need to update the ITU-T E-models [16, 17] to capture the impact of delay on QoE in a more realistic way. In particular, this means that the E-model's one-way delay degradation part will be updated to incorporate the interactivity of communication into the model. Effectively, this will mean that the updated model with a normal setting will indicate a lower impact of the delay on overall quality compared to what ITU-T Rec. G.114 currently proposes.

Second, the increasing usage of mobile AV communication will raise a demand of fluent and duplex communication with mobiles using the speakerphone mode.[9] This would enable people to make AV chats outdoors and on the move while keeping a phone in front of them. This use case would specifically address challenges of acoustic echo cancellation due to the close proximity between microphone and loudspeaker and the high playback level that is required from the loudspeaker. Therefore, there will be a need for advanced methods to evaluate acoustic echo cancellation, loudspeaker reproduction and noise-suppression related quality factors.

Third, the quality evaluation of the mobile use case will also require specific tools and evaluation methods for "fair vs. poor" or hygiene quality. Hence, such tools should be tuned to measure and capture non-ideal speech and video quality to capture the essential parts of the quality degradations that are vital for the users. One of the needs is an intelligibility measure that is able to grasp the key parts of speech intelligibility in challenging situations. It should emphasize the intelligibility dimension of speech quality while down-weighting the dimensions related to the naturalness or fidelity.

Fourth, there will be a need for fast subjective and instrumental tools to capture the key parts of quality in AV communication. Such tools should be able to measure audio and video coding and packet loss concealment performance, as well as microphone and camera capturing and preprocessing qualities.

Finally, we will see that future mobile networks will also create time-varying network conditions. This means that there will be temporal quality variations. Therefore, there will be more attention paid to temporal quality integration and tools providing overall quality for temporally varying calls over both mobile and LAN networks.

There are numerous quality elements and factors that impact the quality of audiovisual communication. This increasing complexity will motivate a demand to identify the key elements and factors and concentrate on those, while at the same time acknowledging and re-evaluating the contribution of the remaining elements and factors to overall AV quality. Therefore, a suitable balance is needed between more time-consuming subjective tests and faster, but often less accurate, instrumental measures. If only one topic needs to be mentioned, we will foresee that the temporal quality which was discussed in Chap. 10 and the control of it will be the key for a successful audiovisual communication solution.

---

[9] This means that a loudspeaker of a device reproduces the sound from the other participant without a necessity to keep the device at the ear.

References

1. Aldridge R, Davidoff J, Ghanbari M, Hands D, Pearson D (1995) Recency effect in the subjective assessment of digitally-coded television pictures. In: Image processing and its applications, pp 336–339
2. Ariely D (1998) Combining experiences over time: the effects of duration, intensity changes and on-line measurements on retrospective pain evaluations. J Behav Decis Making 11:19–45
3. Beerends FE (1999) The influence of video quality on perceived audio quality and vice versa. J Audio Eng Soc 47(5):355–362
4. Belmudez B, Lewcio B, Möller S (2013) Call quality prediction for audiovisual time-varying impairments using simulated conversational structures. In: To appear in Acta Acustica united with Acustica
5. Clark AD (2001) Modeling the effects of burst packet loss and recency on subjective voice quality. In: Internet telephony workshop, pp 123–127
6. Fredrickson BL, Kahneman D (1993) Duration neglect in retrospective evaluations of affective episodes. J Pers Soc Psychol 65:45–55
7. Fredrickson BL (2000) Extracting meaning from past affective experiences: the importance of peaks, ends, and specific emotions. Cogn Emot 14(4):577–606
8. Gros L, Chateau N (2001) Instantaneous and overall judgements for time-varying speech quality: assessments and relationships. Acta Acustica Unit Acustica 87(3):367–377
9. Hamberg R, De Ridder H (1999) Time-varying image quality: modeling the relation between instantaneous and overall quality. SMTPE J 108:802–811
10. Hands DS, Avons SE (2001) Recency and duration neglect in subjective assessment of television picture quality. Appl Cogn Psychol 15(6):639–657
11. Hollier MP, Voelcker R (1997) Objective performance assessment: video quality as an influence on audio perception. In: Audio engineering society convention 103
12. ITU-R Recommendation BT.1359-1 (1998) Relative timing of sound and vision for broadcasting. International Telecommunication Union, Geneva
13. ITU-R Recommendation J.100 (1990) Tolerances for transmission time differences between the vision and sound components of a television signal. International Telecommunication Union, Geneva
14. ITU-T Contribution COM 12–61-E (1998) Study of the influence of experimental context on the relationship between audio, video and audiovisual subjective qualities. International Telecommunication Union, Geneva
15. ITU-T Contribution COM 12–D64-E (1998) Testing the quality of connections having time varying impairments. International Telecommunication Union, Geneva
16. ITU-T Recommendation G.107.1 (2011) Wideband E-model. International Telecommunication Union, Geneva
17. ITU-T Recommendation G.107 (2005) The E-model, a computational model for use in transmission planning. International Telecommunication Union, Geneva
18. ITU-T Recommendation G.1070 (2007) Opinion model for video-telephony applications. International Telecommunication Union, Geneva
19. ITU-T Recommendation G.114 (2003) One-way transmission delay. International Telecommunication Union, Geneva
20. ITU-T Recommendation P.1201 (2012) Parametric non-intrusive assessment of audiovisual media streaming quality. International Telecommunication Union, Geneva
21. ITU-T Recommendation J.247 (2008) Objective perceptual multimedia video quality measurement in the presence of a full reference. International Telecommunication Union, Geneva
22. ITU-T Recommendation J.341 (2011) Objective perceptual multimedia video quality measurement of HDTV for digital cable television in the presence of a full reference. International Telecommunication Union, Geneva
23. ITU-T Recommendation P.800 (1996) Methods for subjective determination of transmission quality. International Telecommunication Union, Geneva

24. ITU-T Recommendation P.805 (2007) Subjective evaluation of conversational quality. International Telecommunication Union, Geneva
25. ITU-T Recommendation P.863 (2011) Perceptual objective listening quality assessment. International Telecommunication Union, Geneva
26. ITU-T Recommendation P.910 (2008) Subjective video quality assessment methods for multimedia applications. International Telecommunication Union, Geneva
27. ITU-T Recommendation P.911 (1998) Subjective audiovisual quality assessment methods for multimedia applications. International Telecommunication Union, Geneva
28. ITU-T Recommendation P.920 (2000) Interactive test methods for audiovisual communications. International Telecommunication Union, Geneva
29. Joly A, Montard N, Buttin M (2001) Audio-visual quality and interactions between television audio and video. In: 6th International symposium on signal processing and its applications 2001, vol 2
30. Jumisko-Pyykkö S (2008) I would like to see the subtitles and the face or at least hear the voice: effects of picture ratio and audio-video bitrate ratio on perception of quality in mobile television. Multimedia Tools Appl 36:167–184
31. Kirk D, Sellen A, Cao X (2010) Home video communication: mediating 'closeness'. In: Proceedings of CSCW 2010, ACM. pp 135–144
32. Kitawaki N, Itok K (1991) Pure delay effects on speech quality in telecommunications. IEEE J Sel Areas Commun 9(9):586–593
33. Korhonen J, Reitel U, Myakotnukh E (2010) On the relative importance of audio and video in the presence of packet losses. In: 2nd International workshop on quality of multimedia experience (QoMEX)
34. Nature journal: public television in Germany, Nature, 127:391–391, 7th March 1936. http://www.nature.com/nature/journal/v137/n3462/abs/137391a0.html
35. Pinson MH, Ingram W, Webster A (2011) Audiovisual quality components. IEEE Sig Process Mag 28(6):60–67
36. Raake A, Schoenenberg K, Skowronek J, Egger S (2013) Predicting speech quality based on interactivity and delay. In: Proceedings of Interspeech, 25–29 August, Lyon, France
37. Rimell AN, Hollier MP, Voelcker RM (1998) The influence of cross-modal interaction on audio-visual speech quality perception. In: Audio engineering society convention 105
38. Schreiber CA, Kahneman D (2000) Determinants of the remembered utility of aversive sounds. J Exp Psychol Gen
39. Steinmetz R (1996) Human perception of jitter and media synchronization. IEEE J Sel Areas Commun 14(1):61–72
40. Weiss B, Möller S, Raake A, Berger J, Ullmann R (2009) Modeling call quality for time-varying transmission characteristics using simulated conversational structures. Acta Acustica Unit Acustica 95(6):1140–1151
41. Wickens CD (1992) Engineering psychology and human performance. HarperCollins
42. Winkler S, Faller C (2006) Perceived audiovisual quality of low-bitrate multimedia content. IEEE Trans Multimedia 8(5):973–980
43. Yamagishi K, Hayashi T (2005) Analysis of psychological factors for quality assessment of interactive multimodal service. Hum Vision Electron Imaging X (SPIE) 5666(1):130–138
44. You J, Reiter U, Hannuksela M, Gabbouj M, Perkis A (2010) Perceptual-based quality assessment for audio-visual services: a survey. Image Commun 25:482–501

Chapter 15
Multimedia Conferencing and Telemeetings

Janto Skowronek, Katrin Schoenenberg and Gunilla Berndtsson

Abstract With today's technical possibilities, in particular, packet-based data transmission and high processing power, telephone and videoconferencing systems celebrate increasing interest. However, the success of such systems is essentially determined by the quality provided and experienced when using them. This is why a high need of appropriate assessment methods can currently be observed. Given the broad range of possible solutions, assessing QoE of so-called telemeetings becomes very difficult and brings along the need for a high degree of variability regarding assessment methods. Since multiple participants usually communicate via such systems, it is required to also investigate aspects of the interaction process and their influence on QoE. Furthermore, the multiparty situation enables users to directly perceive asymmetries in the equipment and in qualities provided from different sites, which affects the perceptual situation as well. This chapter is intended to explain the described challenges in detail and to give first insights into how they might be handled.

15.1 Introduction

Already for several decades, telephone conferencing solutions have been used in business environments to connect more than two persons working at remote locations. It can be said that today most businesses could not work without conferencing

J. Skowronek (✉) · K. Schoenenberg
Assessment of IP-Based Applications, Telekom Innovation Laboratories, Technical University of Berlin, Berlin, Germany
e-mail: janto.skowronek@telekom.de

K. Schoenenberg
e-mail: katrin.schoenenberg@telekom.de

G. Berndtsson
Multimedia Technologies, Ericsson Research, Ericsson AB, Stockholm, Sweden
e-mail: gunilla.berndtsson@ericsson.com

S. Möller and A. Raake (eds.), *Quality of Experience*,
T-Labs Series in Telecommunication Services, DOI: 10.1007/978-3-319-02681-7_15,

applications. Besides telephone conferencing, videoconferencing plays an important role for many users too these days. The development started already in the seventies with the first picture telephone, to connect two interlocutors with an audiovisual connection. The big success unfortunately did not follow the invention right away. Newer technologies based on satellite transmission, and later ISDN, didn't seem to facilitate the breakthrough either. However, the unsurpassable argument of low costs compared to face-to-face meetings and the additional increasing globalization of businesses made the usage of conferencing systems indispensable. Thus, videoconferencing systems were used for business purposes, but audio-conferences constituted the bigger part. Presumably, with the quality available, the added value of the video was not high enough to exceed the cheaper and more flexible audio-conferencing systems. Only with packet-based transmission over the internet, sufficient bandwidths, the according easy way to use video, and the much lower costs, videoconferencing finally celebrated a breakthrough, also attracting the private sector this time. Thanks to the computer-based transmission, plenty of improvements could be achieved for audio-conferencing as well, in particular providing higher audio bandwidths.

Traditional telephone conferences are therefore augmented today by a large variety of systems, ranging from high-end audiovisual telepresence rooms to PC or mobile-phone based software solutions. Given the variety of telephone- and videoconference solutions and given the fact that such systems are used for both professional and private life, it is reasonable to use one term that covers all means of multiparty audio or audiovisual communication between distant locations. In this chapter, the term *telemeeting* is used to serve this purpose, as it broadens the scope of the conventional term teleconference by emphasizing that a telemeeting is considered to be more flexible and interactive than a traditional business teleconference.

Holding telemeetings—either for business or private purposes—bears a number of well-known advantages. Obviously, travel time and costs can be reduced if the number of journeys can be decreased by having telemeetings instead, which has a positive impact on the environment as well. It should be noted that in the past telemeetings have often been considered as alternatives to face-to-face meetings due to this travel-reduction argument. However, studies have shown that this is not really the case [2, 5–7, 22]. Instead, telemeetings should be seen as complementary to face-to-face meetings. Thus, a probably more important advantage is the possibility to hold telemeetings spontaneously, i.e. on short notice and for short meeting times.

The goal of future developments is to provide users with a telemeeting experience that is as close as possible to a face-to-face meeting, or, in case this is not possible (e.g. using mobile phones), to facilitate efficient meetings and pleasant experiences. In order to achieve this, proper assessment methods are needed that help system developers to improve or invent new systems. The primary goal of such assessments should then be to provide a system that is easy to use, is properly functioning and has a good quality.

Concerning QoE, the assessment of telemeeting systems is facing a number of challenges. First, QoE of multiparty telemeetings is currently not fully investigated. Although a first standardized recommendation on quality evaluation tests for multiparty telemeetings is available [16], there are still many detailed questions that require further study.

Second, the huge variety of telemeeting systems makes it difficult to use one common assessment method that is valid for all types of equipment. Nevertheless, this chapter will try to discuss the relevant questions to be dealt with when planning an assessment, following a similar approach as in [16]. The third challenge is the possibility that users may connect to the same telemeeting with individually different devices or via different networks, e.g. fixed phone, mobile phone, PC, videoconferencing or telepresence equipment. In contrast to a two-party call, having more than one other interlocutor makes a direct comparison of the different connections and devices possible. Due to the possibility of such asymmetric situations, it is necessary to assess the quality perception of all participants to obtain a good estimation of the overall quality of a telemeeting.

It can be concluded that the assessment of telemeetings is a complex matter that is strongly related to other aspects described in this book, e.g. speech, audiovisual, or audio transmission. The following will be in line with existing standards on assessing components [12–14, 16]. Even though most of the existing literature is not focused on the multiparty case and therefore does not cover a number of relevant aspects, the relevant publications serve as the base.

It can be summarized that the QoE of a telemeeting is more than one quality score. After clarifying which precise technical case is evaluated, there are further different targets that can be pursued. Due to this high degree of diversity, input from and co-operation with a number of different scientific fields would be beneficial to progress in the topic of quality evaluation of telemeetings.

15.2 Concepts and Definitions

An appropriate QoE assessment of any telecommunication system requires a clear specification of the actual goals, the target variables, and considered use cases and system configurations. Concerning multiparty telemeeting assessment, not only a very broad range of use cases and system configurations is possible, but also the actual goals and target variables can differ substantially between studies. Thus, a common understanding of the different perspectives and an appropriate technical language would facilitate a better exchange of ideas and knowledge as well as a proper comparison of study results. For that reason, this section provides concepts and definitions that assist investigators in the specification of their goals, target variables and use cases.

Concept No. 1: Definitions Around the Term Multiparty
Multiparty telemeetings can be held in principle between two or more than two interlocutors, who are located at two or more than two sites. To be more precise, ITU-T Rec. P.1301 [16] proposes to use a number of terms that disambiguate the meaning of multiparty telemeetings: First, multiparty is defined as "more than two

persons", regardless of the number of connected sites in the telemeeting. Second, to indicate whether the interlocutors are located at two or more than two sites, the terms "point-to-point" and "multi-point" are proposed. Third, to cover the special case of having exactly one person per site, the term "one-per-site" is proposed. With these terms, a precise differentiation is possible between, for example, a telemeeting connecting two locations with more than one person in at least one location (multiparty point-to-point), a telemeeting connecting more than two locations with exactly one person per location (multiparty one-per-site), and a telemeeting between more than two locations with more than one person in at least one location (multiparty multi-point).

Concept No. 2: Technical Asymmetry as Important Use Cases
The presence of technical asymmetry differentiates a multiparty telemeeting from a conventional two-party telephone or video-telephone call. Asymmetry means that the type of equipment used at each site or the connections between sites can be different from site to site; for example, one interlocutor uses a fixed-line telephone, one uses a PC-based software client, one uses a mobile phone. Obviously, also two-party telephone or video-telephone calls can be asymmetric, e.g. with a call between a mobile and fixed-line telephone or with different transmission capacities in each direction (upstream vs. downstream). The differentiating aspect in a multiparty setting is that interlocutors can directly perceive asymmetries because they can compare any differences between the other interlocutors in the same call. In a two-party setting, however, interlocutors do not have the possibility for such a comparison and cannot directly detect asymmetries, unless they are discussing any such impairments. At this point, it should be noted that also in a multiparty setting, it is possible that interlocutors are in a similar situation as in a two-party call: if one interlocutor connects via one type of device while all the other interlocutors use a second type of device, i.e. the first device is the asymmetric component in this setting, then this one interlocutor perceives all other interlocutors without any differences between them, i.e. he or she can not directly perceive asymmetry. As a result, a precise description of any asymmetric use cases and system configurations is necessary, because there are—from a perception point of view—two different implications of asymmetry: either the asymmetry can be directly perceived or not, depending on the particular setting and interlocutor.

Concept No. 3: The Communicative Situation as Part of Assessment Goals
The communicative situation is another main differentiator between a multiparty telemeeting and a conventional two-party call. The primary goal of a telemeeting is to facilitate an efficient communication between interlocutors, ideally irrespectively of the number of interlocutors. It is known in the field of computer-mediated group communication and remote collaborative working, e.g. [3, 4, 8, 19, 20, 25], that the ability of creating a common ground between interlocutors is a crucial aspect. That means, in a group-communication context, interlocutors do not only attempt to understand the shared information but they also strive for an understanding of who the others are, and they will adapt their responses accordingly. Furthermore, people use non-verbal signs (e.g. sounds, gestures, posture changes) [18] or use certain phrase

constructions and utterances [24] in a face-to-face meeting in order to negotiate, more or less subconsciously, who is speaking next. Consequently, the more a telemeeting system is supporting such group-communication aspects, the more users benefit from the system's performance. The implication is that telemeeting QoE can be investigated from two view-points: a technical view-point in terms of perceived quality of the system as such, and a more holistic view-point in terms of perceived quality of communicating via the system. That means investigators should be specific about which of these two view-points constitute their actual assessment goals.

Concept No. 4: The Aggregation Level of Quality as Target Variable
There are three different aggregation levels of quality possible in a multiparty context. The first and lowest aggregation level is the quality of the individual connections $Q_{i,j}$ between any pair of interlocutors i and j. At first glance, this would correspond to the quality of a conventional two-party connection. However, both from a technical and a perceptual point of view, this is not entirely true. Technically it is possible that a degradation occurring in one connection can also affect the other connections. For example, an acoustic echo at the device of one interlocutor does not only mean that the other interlocutors hear an echo of their own voices, but they also hear the echoed voices of all other interlocutors that are also fed back via that one echo causing device. Perceptually it is known that the context can influence quality judgments; hence it cannot be assumed that the quality perception of the considered individual connection in the multiparty call will be the same as the perception of the technically same connection in a two-party call. That means even for $Q_{i,j}$ an aggregation might take place in the form of some mutual influence of the other connections.

The second next higher aggregation level is the quality of the overall telemeeting Q_i as perceived by each individual interlocutor i. Obviously, Q_i is a function of the individual connection qualities $Q_{i,j}$, whereas this function has not been investigated in more detail yet. At first glance, some weighted linear addition would form an intuitive starting point. However, non-linear effects might need to be considered as well; for instance, one connection may be extremely degraded such that the overall judgment would be bad, independently on how good the other connections were.

The third and highest aggregation level is the quality of the overall telemeeting Q_{all} across all interlocutors. In analogy to the discussion above, Q_{all} is a yet unknown function of the individual qualities Q_i. However, there is one conceptual difference between determining Q_i as function of $Q_{i,j}$ and determining Q_{all} as function of Q_i : Q_i and $Q_{i,j}$ can be simultaneously assessed by the same interlocutors, e.g. they can form a judgment of individual connections and of the overall system during the same call. That means, the relation between Q_i and $Q_{i,j}$ can be determined from a perspective from "inside" the telemeeting. Determining Q_{all}, on the contrary, requires a perspective "outside" of the telemeeting, as now different perceptions of the same system need to be integrated.

These aggregation levels of quality constitute in fact different target variables for the assessment of telemeeting quality. Depending on the assessment goals, investigators should define the desired aggregation level(s) and they should design the assessment method accordingly.

Concept No. 5: The Quality Reference as Part of the Assessment Goals Quality judgments are a comparison between the observed characteristics of an entity with its desired characteristics. Thus, a proper interpretation of any assessment results requires that those desired characteristics, i.e the quality reference, are either explicitly known or at least can be reasonably assumed.[1] Regarding the assessment of a multiparty telemeeting system, two conceptually different quality references are of interest, depending on the actual assessment goals: Is the reference a multiparty but face-to-face communication, or is the reference a two-party communication via a telecommunication medium? The first reference would be appropriate to investigate for instance the advantage of holding a physical meeting compared to holding a telemeeting, which—as pointed out in Sect. 15.1—has been a discussion point in literature for a long time. The second reference would be appropriate to investigate for instance the performance of a new multiparty functionality of a telecommunication system compared to its performance in a two-party scenario. In other words, the quality judgment can be based on a comparison along two different dimensions: either comparing telecommunication with face-to-face communication or comparing two different types of telemeetings, for instance a multiparty conversation with a two-party conversation. Ideally, investigators should attempt to control for this aspect and they should be precise about this aspect in their reports. For instance, in subjective tests, a smart formulation of the test instructions could trigger the desired reference, which the corresponding report should explicitly mention.

15.3 Influence Factors

As already discussed, there is a broad range of telemeetings that can be very different in their character and the situation can be complex with several different types of equipment and different number of interlocutors at the sites. Furthermore, the perceived Quality of Experience of a telemeeting can be influenced by a number of factors. Chapter 4 provides a comprehensive overview of such aspects in terms of human, system and context influence factors. In this chapter we readdress some of them from the multiparty point of view. The influence of these factors depends on the type of meeting, thus not all factors mentioned are relevant for all types of meetings, and—as already discussed in Chap. 4—there is probably an interaction between several of these factors.

Human Influence Factors in Telemeetings. The personality of participants in telemeetings, or, specifically, the combination of different personalities of participants can have a major impact on the overall quality. For instance, the overall quality judgment might be dominated by the technical connection of "talkative" participants, because it is their connection that other participants are listening to most of the time.

[1] In utilitarian quality tests with naïve test subjects, one does not explicitly ask for the reference that the subjects are consciously or even subconsciously using. However, such tests assume that the subjects' quality references are similar, e.g. their experience with a normal landline telephone call.

Furthermore, this combination of personalities determines also the conversational structure, in terms of turn-taking behavior, single-, double- or multi-talk situations, etc., which in turn influences also the quality judgment. Other human influence factors, which are of high relevance in the telemeeting context, are the amount of experience with multiparty telemeetings and the voice character. Both aspects will be addressed in more detail in Sect. 15.4 when the profile of test participants is discussed.

System Influence Factors in Telemeetings. Chapter 4 subsumes under the term system influence factors essentially all technical aspects of the system that contribute to the quality judgment. Since it is possible to directly perceive asymmetries in a telemeeting, the particular individual combinations of end devices and network connections can have different impacts on the telemeeting QoE. Furthermore, telemeetings face a number of technical multiparty specific challenges that can influence the QoE. Here we refer to the fact that telemeeting systems can introduce additional artefacts that would not be present in comparable two-party calls. As an example, let us consider traditional telephone conference bridges. First, such bridges introduce an additional transcoding, as they decode all incoming streams, mix the audio signals together, and then encode them again for sending it out to the recipients. Furthermore, such bridges may also use voice activity detection (VAD) to mix only a limited number of signals with actual speech content together. While the goal of such processing is to reduce computational complexity and to prevent that small degradations (e.g. low but audible background noise) of individual connections can add up to large overall distortions, imperfect VAD performance can introduce artefacts such as speech clipping.

Context Influence Factors in Telemeetings. According to the categorization in Chap. 4, the acoustical and visual environment is part of the physical context. An influence of the acoustical and visual environment on a multiparty telemeeting that goes beyond the influence of a conventional two-party call can happen when multiple participants are in the same room, but located at different positions in it. First, the quality as assessed by those participants in that room depends on their actual position, i.e. the distance and angle from display(s) and loudspeaker(s). Second, the quality as assessed by the participants at the other remote sites depends also on the position of the participants sharing that same room, i.e. the distance and angle from camera(s) and microphone(s). As a consequence, the quality of the same call for participants in the same room can be different, even though the technical setup as such is exactly the same. Furthermore, the quality of the same call for the participants at the remote sites can also be different, depending on whom the remote participants are mainly listening to or looking at.

Another contextual influence factor that is particularly interesting for telemeetings is the different use case in terms of business or private meetings. If we assume that business telemeetings are mainly driven by a particular goal or to accomplish certain tasks ("We need to get this done.") and that private meetings are mainly driven by the desire to experience some feeling of presence or social connectivity ("It feels so good to see you."), the quality expectations of participants may be different: Participants might be less critical as long as the task at hand can be accomplished during the business telemeeting, while they would be more critical for the same technical condition when using the system in a private context. On the other hand, if the desired feeling of connectivity can be achieved in the private setting, e.g., through a good visual interaction, other aspects, such as a high intelligibility, that would be important for accomplishing a defined task may be less relevant.

15.4 Subjective Testing

The corresponding subjective test methods are as diverse as telemeetings can be. Therefore this section will not give exact guidance to one particular method, but it focuses on some aspects that need to be considered when choosing and setting up a multiparty telemeeting test. A more specific guidance to suitable test methods for telemeeting assessment can be found in the ITU-T Rec. P.1301 [16].

Communication and Test Mode. At first, it needs to be determined which communication mode is supported by the system under test. It can be audio-only, video-only (for hearing impaired persons), audiovisual, or—which is also quite common in telemeetings—a mix of these modes. Knowing this, it is necessary to decide which mode should be tested. This question is easy for audio- or video-only systems but not as simple for the audiovisual case. If, in this latter case, the audio quality is of main interest the visual quality should as far as possible be kept constant throughout the tests. A clear separation of the two modes will, however, not be possible because the interplay of the audio and visual channel will always affect the overall quality perception of a person. The situation in the case of mixed modes is even more complex, because in this case some participants have only the audio channel available while others have also visual information available. As an example, imagine a test design in which speech intelligibility is degraded. The visual information sent via a videoconferencing system will help to understand the content as well. As a result, the quality may not be affected as much as if it was tested within an audio-only system. For this case, it may help to define different assessment scales for the test subjects. For example, it would be possible to ask about the audio quality only and to explain the difference compared to an overall quality judgment when instructing the participants. Even though a conjunction of the audio and video information can never be fully avoided in audiovisual communication systems, the interacting effects regarding quality judgments can at least be minimized by selecting the right assessment scales.

Scales. When choosing the right assessment scale or scales, some other aspects have to be considered. First, it has to be decided if a conventional scalar quality value is of interest, e.g. the overall mean opinion score (MOS), or if a more diverse outcome using multiple scales is wished. For example, a single judgment of overall quality does not take into account aspects that are important in a telemeeting context, such as cognitive load, conversational structure, or the conversation goals. Second, one aspect to consider in this context is the expected differences that the test conditions will have on the quality scale. They may determine how fine-graded the quality-scale categories need to be set. If rather small differences can be expected, i.e. when comparing similar conditions in one test, the scale needs to be able to reflect those in contrast to a design where the differences are large, i.e. when comparing diverse conditions. In that case, categories could cover a wider range of qualities. Third, another aspect is the range that quality judgments of the selected conditions will span over the scale. When testing a high-end telepresence system, for example, quality scores can be expected to be rather high in general. Then participants should be able to tick different high values to reflect differences, in order to prevent any ceiling effects. Nonetheless, it is usually necessary to compare the outcomes to other configurations or systems. To be able to do so it is recommended to use one of the well-established quality scales (see e.g. [12–15]).

Test Design. Since people use telemeeting systems to interact with each other, the question of test task and the related type of quality, namely interactive or non-interactive is of high importance. Which type to choose mainly depends on the technical conditions that are tested and the conclusions in terms of validity that shall be drawn. Impairments, such as transmission delay, which highly affect the interaction, should always be tested with an interactive test task. This holds for both conventional two-party tests and multiparty telemeeting tests. On the other hand, the influence of listening or viewing related impairments can also be tested in a listening- or viewing-only test, having the advantage of lower effort and costs. However, also in this case, the highest validity can be achieved if the system is assessed most closely to the way it is to be applied, which implies an interactive test task. Due to the apparent importance of the communicative situation of multiparty telemeetings (see Sect. 15.2), the decision between interactive and non-interactive test is of particularly importance.

When selecting a test task, similar criterions should be used. The task should trigger a situation as close to the actual use-case as possible but be sensitive enough for the quality evaluation at the same time. To fulfill both of these needs can become difficult when a high number of sites are connected to one system. With increasing number of sites the number of interlocutors usually increases as well, and therefore the cognitive load required for following the conversational situation. As a result, less cognitive capacity may be available for judging the quality.

Related to this aspect of gathering reliable results, it is recommended to choose a suitable length for the entire test session and to include pauses for the test participants.

Test Participants. When the test design and scales are set, the desired profile of participants needs to be defined. Participants, in general, have different hearing and viewing abilities. As in conventional two-party tests, these should be tested prior to the test and be in a normal range (see e.g. [14]). Test participants will also have different personalities and different interaction styles. These will affect the conversation and in turn the qualities perceived. However, it is usually difficult to control for the personality. If participants are selected and allocated randomly to groups, hopefully the related effects are distributed equally. It is also important that all subjects perform all types of conversation tasks if possible to further diminish the influence of the personality on the test results. The communication, and therefore the perceived quality, can also be affected by the degree of familiarityof the interlocutors. Interlocutors that know each other tend to have more natural and fluent conversations and may detect abnormalities like longer response times or differences in voice characteristics faster and thus be more sensitive to impairments. However, in real-live usage, systems are not always used by participants who are familiar to each other. Similarly, it needs to be decided whether mixed-gender or only single-gender groups should run through the tests. On one hand, mixed groups are more likely to occur in real-life context, so results would show a higher degree of generalizability if mixed groups conduct the test. On the other hand, particular female or male voice characteristics can facilitate speaker separation in a multiparty call, which in turn can influence the cognitive load shared for conducting the test task and judging the quality.

As a last point, prior experience of the subjects with similar telemeeting systems is of major importance. Two strategies can be followed here. First, in case of a rather complex setting, for instance if many different devices are connected to one conference (high degree of asymmetry), it is recommended that participants have experienced a similar communication situation before in order to be able to judge the quality reliably and not to be overwhelmed by the technical possibilities. Second, if it is not possible to find such subjects, a longer training phase can be conducted prior to the actual tests to familiarize participants with the system and the communication situation. This training needs to be distinguished from a warm-up phase which is generally recommended to make participants comfortable with the lab situation and the test task. It is not recommended to include exactly the same configurations that are used in the training. If participants are confronted particularly often with certain degradation in a test context, an over-sensitization of the subjects can lead to over-critical quality ratings. On the other hand, if test participants are used to certain degradations in daily life, the opposite effect may happen: participants might be less critical because their quality expectations are low. As with most aspects in subjective test planning, it is a question of setting priorities to answer a particular question for a particular telemeeting system.

15.5 Instrumental Assessment of Telemeeting QoE

Concerning the instrumental assessment of telemeetings by means of quality-prediction models, at this moment in time, we are not aware of any published work that evaluates the performance of existing two-party methods in a multiparty case or that proposes a new method specifically developed for multiparty settings. For this reason, we can here discuss only some general aspects.

The first aspect is how such models try to predict quality in the conventional sense of existing models, i.e. modeling the perception of the system's technical performance. A first type of models could be based on existing models that follow the concept of different quality aggregation levels (see Sect. 15.2): Since the existing models are developed for two-party cases, they could be first applied to estimate the quality of the individual connections between participants $Q_{i,j}$. Then the aggregated levels Q_i and Q_{all} could be modeled by novel algorithms, which estimate the relations between $Q_{i,j}$, Q_i and Q_{all}. Another variant is to develop models from scratch, that is to directly estimate multiparty quality judgments (Q_i and Q_{all}) by means of appropriate data-fitting or machine-learning approaches.

An important issue for such models will be asymmetric conditions, because they constitute the main difference compared to conventional two-party cases [16]. Furthermore, asymmetry has some implications on the model development, because different interlocutors may experience different connections in the same call, which in turn requires smart approaches to properly organize the data accordingly to those different perspectives before the actual training of the model can be conducted. A discussion on the data complexity of asymmetric conditions and a first proposal for an adequate data processing is given in [27].

A second aspect is how models aim at a prediction of the quality in a broader sense including also the communicative aspects, i.e. modeling the perception of the system's performance to facilitate group communication. Those models could extend the more technical approaches above by incorporating communication-related measures, such as conversational parameters [10, 11], new measures that estimate task efficiency (e.g. based on conversation duration) or new measures that reflect the complexity of the communicative situation (e.g. based on the number of interlocutors). However, it may take some time until such models are available, because first more research is needed both on the influence of those communicative aspects on quality judgments and on the development of corresponding communication parameters that can be technically measured.

15.6 Specific Topics

In this section, there will be a focus on three very particular challenges that play a major role for multiparty telemeetings: asymmetry, delay, and speaker separation.

Asymmetry. As already mentioned, the problem of asymmetrical technical conditions arises unavoidably when evaluating different telemeeting use cases. The higher the number of participants connected to a telemeeting, the more likely a diversity of end devices and connection capabilities. Furthermore, not only may the equipment be different but also the number of participants at each end. As an example, this could mean that one person may be using a mobile phone, another one using a fixed-line telephone, while four participants in a third room are calling in via PC client running on a laptop. Such situations are very typical use cases and lead to completely asymmetric quality conditions for the different interlocutors. Berndtsson et al. [1] performed several different types of conversation tests using such rather complex settings. Some tests compared audio-only to audiovisual telemeetings and one-person to multi-person configurations. The results showed that participants preferred to use audiovisual equipment over audio-only connections and to be in a room with other interlocutors whilst being connected to the conference call than to be alone in a room. Another test described in that article revealed that it was preferred to synchronize the video delay to the audio delay than to keep the audio delay lower than the video delay (at least for delays shorter than 600 ms).

Delay. Especially in asymmetric settings, transmission delays are likely to occur due to different processing for the end-devices. However, until now little is known about the impact on the quality perception. Some results from the traditional telephony context give first insights. Back in the 1990s, Kitawaki and Itoh [17] found that the interaction speed of the task is a mediating factor regarding the impact of the delay on the quality judgment.

Delay is easily noticeable during highly interactive tasks, such as test subjects alternate to count from one to twenty back and forth as fast as possible, while it is less noticeable in free conversations. Even though delays in free conversations are often not noticed consciously until they are long, they can still affect the conversation quality as the conversation partner might be perceived to be unusually slow, hesitant or not interested [1]. Guéguin [9] showed that the impact of delay is much more severe when there is talker echo induced at the same time. Besides pure audio transmission delay, as mentioned, there might also be audio–video asynchrony in video-conferencing systems. This asynchrony has to be distinguished from the one that occurs in TV applications. In video-conferencing, users do not only watch a video but try to interact with each other via a more or less synchronous connection. It should be noted that interacting with a person whose gestures and mimic is asynchronous to his voice is a very artificial situation. It requires high concentration and effort to interact in this way which in turn may have severe impacts on the quality perception.

Speaker Separation. The third challenge is the issue of speaker separation in audio-conferences which naturally becomes more difficult with an increasing number of interlocutors. Luckily, new technical possibilities allow for a spatial rendering of the voices of participants during such calls, though a broad market introduction of such techniques has not been launched yet. A pilot study by Skowronek and Raake [26] investigated the relationship of number of interlocutors, cognitive effort and perceived quality. They found that the better the technical solutions, for example, using spatial representation with head-tracking compared to a non-spatial representation, the less cognitive effort was needed. The number of interlocutors, however, dominated the variation in cognitive-effort measures. Regarding transmission quality, the different system conditions showed a clear effect, while the number of interlocutors did not. These results give first insights into how the problem of multiparty interaction and the related higher cognitive effort could be diminished. A prior study by Raake et al. [23] showed a similar advantage of spatial representation in terms of perceived quality in three-party calls.

In sum, the status of the current work on the three challenges mentioned exemplarily shows that there is still considerable work to be done to improve the experience of users being connected via a multiparty telemeeting.

15.7 Future Applications

There are a number of strong trends in the development of new telemeeting technologies that will likely enhance the Quality of Experience, while, at the same time, bringing additional challenges for the proper assessment of such systems. Five such trends are: mobility, interoperability, ease of use, collaboration possibilities, and feeling of presence.

Mobility. More and more mobile solutions will be available, which means that users will have more possibilities to connect to a telemeeting from almost anywhere. This requires new assessment methods for two reasons. First, there is a lack of standardized recommendations on how to test the perceived quality of a telemeeting using handheld mobile devices, i.e. how to appropriately test quality aspects such as for instance image quality while the display is moving due to hand movements. Second, it is very difficult—if at all possible—to realize mobile test scenarios that are ecologically valid in laboratory test environments. Thus there is a strong need for methods to test telemeeting quality outside of a laboratory environment (field tests), which balance meaningful test cases and control over the test situation conditions.

Interoperability. All types of devices that can be used for telemeetings should also be technically capable to connect, which is often not the case. Obviously, standardization efforts are needed and ongoing to provide the necessary interfacing technology. However, the outcome of such standardization can have an impact on the QoE assessment: not only the quality of the device and its connection to the network, but also any limitations inherent in the standardized interfacing stages need to be considered. For example, in traditional telephony networks, the narrowband G.711 codec is often used as the interfacing component between operator networks, even though the ongoing proliferation of wideband telephony is changing this situation. That means, the possibility of a potential transcoding due to such interfacing standards should be checked when designing telemeeting tests with different devices.

Ease of Use. The mechanisms to establish telemeetings are often very cumbersome for users, even though more emphasis of manufactures to improve this situation can be observed compared to the past. Apparently, an easy call setup of a telemeeting is an important factor for users, but no research on the contribution of this aspect to the overall telemeeting QoE has been published and no standardized methods are recommended.

Collaboration. More and more telemeeting services allow the sharing of slides, pictures, and videos or writing in the same document, nowadays using web-based platforms such as webRTC (real time communication in the browser). For an overview of the different underlying conceptual models of collaboration, see for example [21]. Currently, no standardized assessment methods for such collaboration functionalities are available, and ITU-T Rec. P.1301 [16] can give only some general advices. Furthermore, the communicative aspects of telemeeting QoE, which have been already discussed earlier in this chapter, might be even more important in this context, given now the system's emphasis of supporting efficient collaboration.

Feeling of Presence. High-end telepresence systems use optimized equipment showing life-size videos of the participants on large screens in specially designed rooms. However, also other less expensive telemeeting solutions will likely strive for providing a certain feeling of presence in order to differentiate or compete in future markets. Two special technologies, which are available to enhance the feeling of presence, are spatial audio and 3D video (see Chaps. 17 and 20 for more details on the QoE of spatial audio and 3D video). With spatial audio users can be surrounded by the voices of the participants in a telemeeting. This does not only have a positive effect on speaker separation and cognitive load (see Sect. 15.6), but it also enhances the feeling of presence: Hearing individual voices at different positions in the auditory space is closer to the situation of a face-to-face meeting than hearing all voices coming from one direction. The use of 3D video in a telemeeting situation can provide a feeling of depth in the scene, which can make it easier for users to comprehend complex objects in a video-conference scene. This can lead to an enhanced feeling of presence in the meeting, which is closer to the situation in a face-to-face meeting than a conventional two-dimensional display technology can be. With the upcoming application of spatial audio and 3D video in telemeeting systems, appropriate test methods are needed that capture the added value of these technologies in terms of technical quality, ease of communication (spatial audio), visual comfort (3D video), and feeling of presence.

15.8 Conclusions

Although telecommunication between multiple parties has a history of several decades, the importance of telemeetings in both business and private life is still increasing, driven by a combination of economic and societal trends (e.g. global businesses, people living apart from friends and relatives) and the availability of new network and end-device technologies. One factor contributing to the success of telemeeting systems is the QoE that such systems provide, which brings us to the question of appropriate test methods. Obviously, the assessment of telemeeting QoE can and should build on the comprehensive amount of existing methods for conventional two-party telecommunication. However, at a number of points throughout this chapter, we discussed the lack of or need for specific methods that address some special characteristics of telemeeting QoE: First, telemeeting QoE depends not only on the audio or video signal quality (i.e. media quality), but also on a number of additional aspects that play an important role, such as ease of communication. This opens new questions on how to include such aspects into the quality measurement. Second, participants may directly perceive differences between interlocutors in case of asymmetric conditions, opening new questions on how to appropriately assess such situations. Third, the technical implementation of telemeeting systems can be very diverse, ranging from high-end telepresence rooms to mobile-device solutions, which makes it difficult to develop one method for all systems. Fourth, new technological trends, e.g. mobile solutions, spatial audio, or 3D video, require additional assessment methods in order to appropriately investigate the impact or added value of such technologies. To summarize, this chapter showed that some work on telemeeting QoE has been published and that some assessment approaches are already available, while there are still many open questions to be answered and corresponding assessment methods to be developed.

References

1. Berndtsson G, Folkesson M, Kulyk V (2012) Subjective quality assessment of video conferences and telemeetings. In: 19th international packet video workshop.
2. Blokland A, Anderson AH (1998) Effect of low frame-rate video on intelligibility of speech. Speech Commun 26:97–103
3. Clark HH, Brennan SE (1991) Grounding in communication. In: Resnick LB, Levine JM, Teasley SD (eds) Perspectives on socially shared cognition. American Psychological Association, USA
4. Daly-Jones O, Monk A, Watts L (1998) Some advantages of video conferencing over high-quality audio conferencing: fluency and awareness of attentional focus. Int J Hum Comput Stud 49:21–58
5. Doherty-Sneddon G, O'Malley C, Garrod S, Anderson A, Langton S, Bruce V (1997) Face-to-face and video-mediated communication: a comparison of dialogue structure and task performance. J Exp Psychol Appl 3:105–125
6. Dourish P, Adler A, Bellotti V, Henderson A (1996) Your place or mine? Learning from long-term use of audio–video communication. J Comput Support Coop Work 5(1):33–62
7. Fish RS, Kraut RE, Root RW, Rice RE (1993) Evaluating video as a technology for informal communication. In: Proceedings of the SIGCHI conference on human factors in computing systems, pp 37–48
8. Fussell SR, Benimoff NI (1995) Social and cognitive processes in interpersonal communication: implications for advanced telecommunications technologies. Human Factors 37:228–250
9. Guéguin M, Bouquin-Jeannès RL, Gautier-Turbin V, Faucon G, Barriac V (2008) On the evaluation of the conversational speech quality in telecommunications. EURASIP J Adv Signal Process 18524:1–15
10. Hammer F (2006) Quality aspects of packet-based interactive speech communication. PhD thesis, Signal processing and speech communication laboratory, faculty of electrical and information engineering, Graz University of Technology
11. Hoeldtke K, Raake A (2011) Conversation analysis of multi-party conferencing and its relation to perceived quality. In: IEEE international conference on communications (ICC).10.1109/icc.2011.5963021
12. ITU-T Recommendation P.800 (1996) Methods for subjective determination of transmission quality. International Telecommunication Union, Geneva
13. ITU-T Recommendation P.805 (2007) Subjective evaluation of conversational quality. International Telecommunication Union, Geneva
14. ITU-T Recommendation P.910 (2008) Subjective video quality assessment methods for multimedia applications. International Telecommunication Union, Geneva
15. ITU-T Recommendation P.911 (1998) Subjective audiovisual quality assessment methods for multimedia applications. International Telecommunication Union, Geneva
16. ITU-T Recommendation P.1301 (2012) Subjective quality evaluation of audio and audiovisual multiparty telemeetings. International Telecommunication Union, Geneva
17. Kitawaki N, Itoh K (1991) Pure delay effects on speech quality in telecommunications. IEEE J Sel Areas Commun 1991:586–593

18. Knapp ML, Hall JA (2010) Nonverbal communication in human interaction, 7th edn. Cengage Learning, Boston
19. Masoodian M, Apperley M, Frederickson L (1995) Video support for shared work-space interaction: an empirical study. Interact Comput 7(3):237–253
20. Olson GM, Olson JS (2000) Distance matters. Hum Comput Interact 15:139–178
21. Park KS (2003) Enhancing cooperative work in amplified collaboration environments. PhD thesis, Graduate college, University of Illinois, Chicago
22. Pye R, Williams EW (1977) Teleconferencing: is video valuable or is audio adequate? Telecommun Policy 1(3):230–241
23. Raake A, Schlegel C, Hoeldtke K, Geier M, Ahrens J (2010) Listening and conversational quality of spatial audio conferencing. In: AES 40th international conference
24. Sacks H, Schegloff EA, Jefferson G (1974) A simplest systematics for the organisation of turn-taking for conversation. Language 50(4):696–735
25. Sanford A, Anderson AH, Mullin J (2004) Audio channel constraints in video-mediated communication. Interact Comput 16:1069–1094
26. Skowronek J, Raake A (2011) Investigating the effect of number of interlocutors on the quality of experience for multi-party audio conferencing. In: Proceedings of the 12th annual conference of the International Speech Communication Association (Interspeech), pp 829–832
27. Skowronek J, Herlinghaus J, Raake A (2013) Quality assessment of asymmetric multiparty telephone conferences: a systematic method from technical degradations to perceived impairments. In: Proceedings of the 14th annual conference of the International Speech Communication Association (Interspeech)

Chapter 16
Audio Transmission

Bernhard Feiten, Marie-Neige Garcia, Peter Svensson and Alexander Raake

Abstract Audio with good quality is the essential fundament for all multi-media services. The transmission of audio signals relies on efficient encoding and decoding algorithms (codecs) that enable the reduction of the required channel capacity, but still provide an excellent audio quality, even when transmission errors occur. The most succesfull audio codecs are *mp2*, *mp3*, *aac* and *ac3*. The codecs employ sophisticated signal processing algorithms imitating properties of hearing. The processing may cause specific artifacts such as high frequency loss, narrow-band noise and pre-echoes. The final quality needs to be verified with statistically valid listening tests. Detailed procedures for conducting reliable speech and audio tests are defined in ITU Recommendations P.800, BS.1116, and BS.1534. Instrumental measurement methods such as BS.1387 replicate subjective tests allowing the estimation of the perceived quality. The ITU Recommendation P.1201 is a recently standardized method for estimating the audio quality of a transmitted signal without the need to have a reference signal available.

B. Feiten (✉)
Deutsche Telekom, Telekom Innovation Laboratories, Berlin, Germany
e-mail: bernhard.feiten@telekom.de

M.-N. Garcia · A. Raake
Assessment of IP-based Applications, Telekom Innovation Laboratories, TU Berlin, Germany
e-mail: marie-neige.garcia@telekom.de

A. Raake
e-mail: alexander.raake@telekom.de

P. Svensson
Department of Electronics and Telecommunications, Norwegian University of Science and Technology, Trondheim, Norway
e-mail: svensson@iet.ntnu.no

S. Möller and A. Raake (eds.), *Quality of Experience*,
T-Labs Series in Telecommunication Services, DOI: 10.1007/978-3-319-02681-7_16,

16.1 Introduction

Nowadays, media applications including audio are typically based on digital processing and transmission. After conversion from the analogue domain to the digital domain, the audio signals are further processed and compressed for achieving the required bit rate. For the encoding and decoding of the audio signals, a large number of different audio codecs have been developed and standardized. The codecs and their related coding artifacts are introduced in Sects. 16.2.1 and 16.2.2. Audio transmission is increasingly based on the Internet Protocol suite. For example, IP-based transmission is extensively used for IP-based Television (IPTV) as well as audio, or music streaming services. While the first digital broadcast services such as DAB and DVB-T used the MPEG-2 Transport Stream (MPEG-2 TS) directly, newer developments use IP protocols also on radio bearers. DVB-H, 3GPP MBMS and digital cable services employ IP, UDP and RTP protocols in combination with MPEG-2 TS.

At large, the quality of an audio transmission chain depends on five steps: (1) recording and processing, (2) encoding, (3) transmission, (4) decoding and error-handling, and (5) post-processing and reproduction. The quality related to the recording, (1), depends on aesthetic aspects and the choices made during production (cf. Chap. 2), as well as on the technical aspects of the recording equipment and acoustic environment. The signal is further processed, yielding a mix into a given number of audio channels possibly including meta-information. The corresponding impact on quality is significant; however, due to the focus on audio streaming, it will not be further discussed in this Chapter. Further information may be found in Chap. 17. The encoding, (2), may cause significant degradation, depending on the targeted compression and employed codec. Transmission across an IP-based network, (3), may lead to errors that have a severe influence on audio quality, for example audio distortion or crackle due to packet loss and corresponding concealment in case of unreliable transport via UDP, or stalling or interruptions of the audio stream in case of reliable transport using TCP. The respective effects depend on the subsequent handling of the loss- or delay-affected stream by the buffering and error-correction mechanisms, (4). Next, the decoded audio is reproduced via the device selected by the user, which may range from a high-end, and possibly multi-channel home or professional audio and loudspeaker system, to a cheap streaming device with low-quality mono loudspeakers. Considerations on the quality evaluation of reproduced sound can be found in [49, 57] and Chap. 17. The following text mainly deals with the quality impact due to audio coding and transmission errors. Sections 16.2.2 and 16.2.3 provide an overview of typical coding and transmission artifacts.

Audio quality is primarily assessed by listening tests with human listeners. One may rely on experts reporting their listening experiences, or on experiments conducted with a larger listening group. The ITU has released well elaborated recommendations on how to conduct such tests. They are introduced in Sect. 16.3.2.

To complement time-consuming and expensive auditory tests, instrumental methods for assessing audio quality have been developed. Besides basic measurement methods aiming for linear and non-linear distortion or the signal-to-noise ratio, more elaborate methods have been developed, which allow to estimate the quality ratings obtained from actual listeners for the signals under test. Most prominent are the full-reference models developed for audio and speech. The so-called Perceive Audio Quality (PEAQ) method is introduced in Sect. 16.4.1, as the most widely used model in this context.

For the planning of a service, and for monitoring the quality at arbitrary points in the transmission chain, an appropriate reference signal cannot easily be made available without substantial additional resources. For these cases, parametric, no-reference quality models can be applied. A parametric model for quality assessment in the context of network planning and monitoring has recently been standardized by ITU-T, and is described in Sect. 16.4.2.

The following sections will give insights into audio coding (Sect. 16.2.1), quality testing (Sect. 16.3.2) and quality perception modeling (Sect. 16.4). The text concentrates on unidirectional, non-interactive services. Conversational services and related delay aspects are handled in Chaps. 11, 12 and 14. Finally, in Sect. 16.5, the influence of the audio quality on the overall QoE is discussed.

16.2 Audio Coding and Transmission

16.2.1 Audio Coding Schemes

Audio signals can generally be transmitted as uncompressed or compressed (lossless or lossy) representations. Several lossless compression schemes are available. Because of the relatively small compression gain of 30–50 %, lossless coding is rarely used for transmission.

The majority of relevant audio coding schemes belongs to the lossy compression family. Lossless compression removes *statistical redundancy* from the audio signal (redundancy reduction), a process which can be reversed. On the other hand, perceptual, lossy audio compression removes *perceptually irrelevant* parts of the signal (irrelevancy reduction). A selection of common codecs is presented in Table 16.1.

Most codecs use a time-frequency analysis in terms of some transform to the frequency domain, often an MDCT (modified discrete cosine transform, see for example the review by Painter and Spanias [45]). Typically a *critically sampled* filter bank with perfect reconstruction is used. Earlier codecs, such as the MPEG-1, Layers I and II codecs [22], use a filter bank for sub-band processing, with 32 uniform bandwidth filters, and a block of audio samples is handled at a time. In a parallel path, a higher-resolution analysis is done using, for example, the Fast Fourier Transform (FFT). The spectral analysis is used to identify *tonal* and *noise-like* components. For these components, their respective masking properties are computed using a psychoacoustic model. A total *masking threshold curve* is derived. The effective level difference, within each sub-band, between the signal components and this masking threshold determines how much quantization noise may be allowed in that sub-band,

Table 16.1 Some common audio codecs. More details are given in the text

Codec	Special properties
MPEG-1, layer II (mp2)	Used in DAB and IPTV services
MPEG-1, layer III (mp3)	The audio codec that started the Internet audio revolution
MPEG-2/4 AAC	Advanced audio coding. Successor to mp3. Used in a range of streaming services. Supporting multi-channel
MPEG-4 HE-AAC	High-efficiency AAC. Uses spectral band replication and parametric stereo for efficient compression. Used in DAB+
MPEG-4 AAC-LD	Lower algorithmic delay, 20 ms, than AAC
MPEG surround	Exploits cross-channel correlations
MPEG USAC	Unified speech and audio coding. Incorporates audio and speech codecs for signal adaptive coding
APT-X	Very low compression gain, Algorithmic delay of 2 ms
Dolby digital (AC-3)	Multi-channel codec, used on DVD and IPTV services
DTS	Multi-channel codec used for cinema and DVD at higher bit rates
Ogg Vorbis	Open-source audio codec
Opus	Low-delay audio and speech codec standardized by IETF

without exceeding the masked threshold, that is, how many bits may be allocated to that sub-band. In the highly successful MPEG-1, Layer III codec (mp3) [6, 22, 23], a hybrid approach is used, where an MDCT is applied to the outputs of the sub-bands for increased frequency resolution in the signal path. Also, for enhancing the quality of transients in general audio signals, a switching between longer and shorter blocks of time samples is employed, based on the current properties of the signal. In the MPEG-2/4 AAC codec [5, 24], the sub-bands are abandoned, and the MDCT is used as the single transform step, similarly to how other codecs such as the Ogg Vorbis and AC-3 (Dolby Digital) codecs operate (see overview given by Herre and Dietz [19]). A low-delay version of the AAC codec, low-delay AAC (LD-AAC) uses a shorter MDCT transform, as well as some other design choices, to reach an algorithmic delay of 20 ms, as compared with a minimum of 55 ms for standard AAC [38].

The bit allocation is controlled by the desired bit rate selected for the encoding. Rather than always yielding inaudible quantization noise (by keeping the quantization error below the masked threshold curve), lower bit rates will aim at letting the noise be optimally distributed and ideally minimizing its audibility. From a service provider's perspective, a constant bit rate (CBR) is often attractive, but the signal properties typically favor a variable bit rate (VBR), since this leads to a constant audibility of the quantization noise. CBR is usually achieved by using a bit-reservoir, which involves an increased delay for a varying bit-allocation across blocks. This allows to spend more bits for audio frames containing transient signals than for the following frames.

The MPEG family codecs use a set of *tools*, that is, different functions, some of which can be manually chosen while others are used only in certain versions of the codecs [19]. For example, *Temporal Noise Shaping* is a function which hinders the quantization noise from spreading in time across a block, which otherwise could yield audible pre-echoes. Instead, quantization noise is temporally aligned with transient parts of the signal, which then mask the noise [18]. *Stereo Processing* handles the two channels jointly for increased coding efficiency. An important development in the MPEG family was the *Spectral Band Replication*, which uses the *waveform-replicating* approach described above only up to a certain frequency. The signal in the high-frequency range is then described in a parametric way. Harmonic components can be generated as an extension of the lower harmonics in the "baseband", with a simplified envelope shape description. Noise-like components can be described only by their envelope shape. This approach is used in the High Efficiency AAC codec (HE-AAC) [25]. The joint stereo processing is taken several steps further in the MPEG Surround codecs, for multi-channel audio signals [20]. In those surround codecs, the inter-channel relationships are described in a parametric way, exploiting correlation properties. Spatial audio coding is described further in the next subsection.

Coding audio signals at very low bit rates, such as 16 kbps, with reasonable speech quality, relies on modeling speech properties in the codec. Hence, the MPEG Unified Speech and Audio Coding (USAC) incorporates a Linear Predictive Coding (LPC) kernel that is adaptively activated for speech-like signals [44].

Alternatives to the transform-based codecs include a technique which uses a pre- and post-filter approach that employs a broad-band signal quantization. This approach can facilitate ultra-low algorithmic delay [55], as implemented in the Fraunhofer Ultra-Low Delay audio codec with an 8 ms algorithmic delay. Yet another ultra-low delay audio codec uses Adaptive Differential Pulse-Code Modulation (ADPCM) for very moderate compression ratios, reaching an algorithmic delay of less than 2 ms [38].

16.2.1.1 Spatial Audio Coding

Several multi-channel formats have been developed for applications such as home cinema, movie theater sound, video games, virtual reality, etc. Spatial audio formats that accompany a movie picture typically use a frontal set of loudspeakers for phantom sources, as in stereo reproduction, and a set of surround loudspeakers for environmental sound effects with less precise localization than the frontal phantom sources [21]. These formats include the 5.1, 7.1, 10.2 and 22.2 loudspeaker setups, where the number after the decimal point indicates the number of band-limited channels, for so-called Low-Frequency Enhancement. For each of these loudspeaker setups, several audio codecs are in use. The 5.1 format is the most common one, and popular codecs are Dolby Digital, DTS, and MPEG Surround. Several extensions of the first two codecs exist, such as Dolby Digital Plus and DTS-HD High Resolution Audio, with support for larger numbers of loudspeakers. Formats with many loudspeakers support more symmetrical rendering of spatial audio, that is, without a specific frontal sector. Computer games, virtual reality, and other applications might use the formats Vector Base Amplitude Panning (VBAP), Wave-field Synthesis (WFS) or Higher-Order Ambisonics (HOA), which can use arrays with any number of loudspeakers, as further described in Chap. 17. For these multi-channel formats,

Table 16.2 Audio artifacts obtained from user tests with descriptive methods and from expert classifications. Details are provided in the text

Processing	Artifacts
Coding	Quantization noise, binaural unmasking distortions, aliasing artifacts, timbre distortion (birdies), muffled audio (band-limitation), pre-echoes, rasping, metallic sound, tone trembling, sparkling, bubbling, change of stereo impression [12, 37]
Transmission	Interruptions, frame repetition [46], asynchrony

lossless compression such as Meridian Lossless Packing is often favored since the perceptual effects of applying lossy audio codecs are uncertain. The MPEG-H standards, which are under development, include a part for 3D-audio, where formats for multiple loudspeaker channels will be supported.

16.2.2 Coding Related Artifacts

Typical audio artifacts, that is, the *quality features* resulting from the encoding and decoding process are summarized in Table 16.2, row "Coding".

Non-linear distortion appears during quantization of the signal. Because of its random, fast fluctuating character, this distortion is perceived as noise. When a filter bank is used in the codec, and the filter bank output signals are quantized, the envelope of the amplitude spectrum of the quantization noise is shaped according to the signal so that the noise is optimally masked by the signal (see Sect. 16.2.1). The quantization noise is therefore not *white*, but *colored* noise. This colored noise is perceived as roughness when exceeding the masking threshold [58, 63].

The masking threshold can sometimes be lower when listening with two ears rather than one. In particular, the detection of a signal in noise improves when either the phase or level differences of the signal at the two ears are not the same as those of the masker. This phenomenon is called the *binaural masking level difference*, or *binaural unmasking* [37]. An implication of this is that the signal and masker appear to originate from different directions in space, which was found to make quantization noise more easily audible.

Aliasing errors [12, 37] are inherent to the use of critically sampled filter banks. Typically, perfect reconstruction filter banks are used, which provide aliasing cancellation, but the cancelling property may be reduced by sub-band quantization errors. The distorted harmonic structure is likely to be unmasked especially for signals containing harmonics [58].

High frequency loss and *band-limitation* may occur when the high-frequency bands of the filter-bank are cut due to bit rate constraints. Further, also sub-bands that are masked may be omitted for encoding. As a consequence, the sound may vary with respect to high frequencies and timbre. A cut of high frequencies results in *muffled audio* [15], and when single bands are switched on and off, sound artifacts may be heard that are referred to as *birdies* [37].

Pre-echo [12, 37] may occur for transient audio signals, especially when long frame sizes are used for transform coding. Long frame sizes are preferred for coding efficiency purposes. When an impulsive sound such as a castanet clap occurs in the middle or at the end of an audio frame, after decoding the quantization noise spreads over the entire frame. In such cases, a so called "pre-echo" before the clap may become audible. Such an artifact that occurs before a transient signal is more critical, as the temporal *pre-masking* of the human hearing system is much smaller than the *post-masking*. In speech, plosives and fricatives may be overlaid with reverberation and flanging artifacts. Flanging is a sound effect where the signal is processed with time-varying comb filters.

Codecs applying spectral band replication may not reconstruct the harmonic overtone structure correctly, which may be perceived as *rasping* and *metallic sound*. The simplified, parametrized spectral envelope over time may lead—in certain cases—to *tone trembling* or *sparkling* sound, which has also been referred to as *bubbling* [39, 40].

A *change of stereo impression* may occur when original level and delay differences between the transmission channels are not exactly reconstructed. For example, a two-channel stereo signal can be transmitted as a sum and difference signal of the two channels. The sum signal represents the mono information and the difference signal represents the stereo information. The preservation of the stereo image depends on how accurately the difference signal is maintained. In an extreme case, the stereo effect gets lost. More complex confusion of the stereo localization may occur, for instance with the Parametric Stereo (PS) module of the HE-AACv2 codec. This module reconstructs a stereo signal from the down-mixed mono signal according to the parameters extracted during the capture of the stereo input signal. This down-mixing procedure may result in the *loss of stereo image*, while the *tone leakage* artifacts, which refer to the leaking or vanishing of one channel to another channel, originate from the variability of mixing coefficients. These phenomena are perceived as *blurred spatial position* [37].

16.2.3 Transmission Related Artifacts

Transmission errors can have an effect at different levels of the protocol layers. For a transmission over IP using UDP for transport, an error typically leads to the loss of a full audio frame resulting in an audible gap between 20 and 30 ms, depending on the codec's frame size and sample rate.

The strength of the audible effect also depends on how the audio frames are connected to each other, that is, in how far coding information is inter-frame dependent. Usually, the frames overlap, and the overlapping parts are added together, weighted with a window function. For a lost frame this weighting has the positive side effect that the signal is faded-out and faded-in again. The MDCT provides a fading over half the window duration, while the polyphase filterbanks have a much smaller fade-in and -out. Hence a lost frame in the case of AAC results in a much smoother quality experience than a lost frame in the case of MPEG-1 Layer II. For MPEG-1 Layer III, the built-in bit reservoir may, in addition, cause a loss of significant frame information because, for complex frames, some of the allocated bits are transmitted with the following frame.

Transmission artifacts are summarized in Table 16.2, row "Transmission", and detailed in the following paragraph.

When a frame is lost, error concealment techniques can be applied to reduce the audibility of the loss. The most common method is to replace the lost frame(s) by the previous frame (*frame repetition*). When doing so in the compressed domain, the windowing of the codec provides a smooth blending between the frames. For longer gaps, silence (*interruption*) or a low noise is typically inserted. Alternatively, the missing frames may also be skipped. This results in a shorter audio stream, and may cause *asynchrony* in the case of an audiovisual signal. More complex concealment methods aim at higher order interpolation techniques between the previous and following frames [46]. For most of the mentioned codecs, the concealment methods are not standardized. Hence, for a given audio frame loss, the different concealment implementations may result in different levels of perceived audio quality.

In recent time, alternative transport mechanisms to unreliable UDP-based streaming with RTP have been introduced, which are instead based on the reliable TCP transport. Examples are web-radio services and music streaming services. In this case, loss of IP-packets are corrected by the underlying reliable transport mechanism of TCP. Here, degradations of quality may result when the play-out buffer of the client is filled below a given threshold, and the play-out is paused until a sufficient amount of media information is available in the buffer again. In this case, the listener perceives a stalling event, which may result in waiting until the play-out resumes. Work by Egger, Schatz and Reichl has indicated that the respective quality often follows a logarithmic behavior over the overall waiting time, in addition depending on the number of stalling events. This behaviour has been related with the Weber-Fechner law in [50]. In more recent work by Sackl et al. [53] stalling-related quality for audio and video streaming has been compared.

16.3 Subjective Quality Assessment

16.3.1 Room and Electro Acoustic Effects on Audio Quality

The previous subsection has described the impairments that are introduced due to audio coding and transmission. As mentioned in the introduction, the terminals, in this case loudspeakers or headphones for the receiving end, as well as the room acoustical environment can have a large impact on the perceived audio quality. This is especially true for loudspeakers that might range from miniature loudspeakers in

Table 16.3 ITU listening test methods and references to relevant tests employing these methods

Subjective test methods	Description
BS.1116 [26]	Methods for the subjective assessment of small impairments in audio systems employing a double-blind triple-stimulus with hidden reference. A continuous five-grade impairment scale is used for assessing the test item with respect to their *basic audio quality*, *stereophonic image quality* and/or *impression of surround quality*. Samples of conducted tests: [1, 61]
BS.1534 [29]	Method for the subjective assessment of intermediate quality level of coding systems employing a mUlti stimulus test with hidden reference and anchor (MUSHRA). This method is intended to give a reliable and repeatable measure of systems having audio quality with clearly noticeable artifacts. Samples: [2, 3]
P.800 [36]	Methods for subjective determination of transmission quality employing absolute category rating (ACR), degradation category rating (DCR) or comparison category rating (CCR). The most frequently used opinion category scale is *Excellent, Good, Fair, Poor, Bad*. Samples: [31, 36]

portable units with a very limited frequency range and dynamic range, to large high-quality loudspeakers in domestic or professional use. In many test situations, the terminals are treated as constants, or context factors. Yet, the subjective assessment of reproduced sound quality is a large field in itself, with standardized test methods that partly overlap with those addressed in the next subsection [4]. Quantities that affect the quality are the frequency range, the flatness of the frequency response, the non-linear distortion and additive noise. For loudspeakers, also the interaction with the room acoustical conditions has a substantial influence on the perceived quality. The same loudspeaker used in different listening environments can generate a substantially different perceived quality [59]. Another aspect is the interaction between coding artifacts and the acoustical conditions in the listening room: Coding-related artifacts have been shown to be partially masked by room acoustics [54], leading to less strong impairment when coded audio is listened to in reverberant spaces.

16.3.2 Assessment Methods

Subjective tests for assessing the perceived quality of audio transmission systems have to be designed very carefully as many factors may influence the outcome. Commonly used and established methods for subjective audio tests are the ITU-R Recommendations BS.1116, BS.1284, BS.1286 and BS.1534. For speech, ITU-T Recommendation P.800 is most often applied (see Table 16.3).

All these methods share common principles, but focus on different aspects. For instance, audio tests are typically conducted with expert listeners, while for speech tests non-expert ("naïve") listeners are preferred. While expert tests deliver better

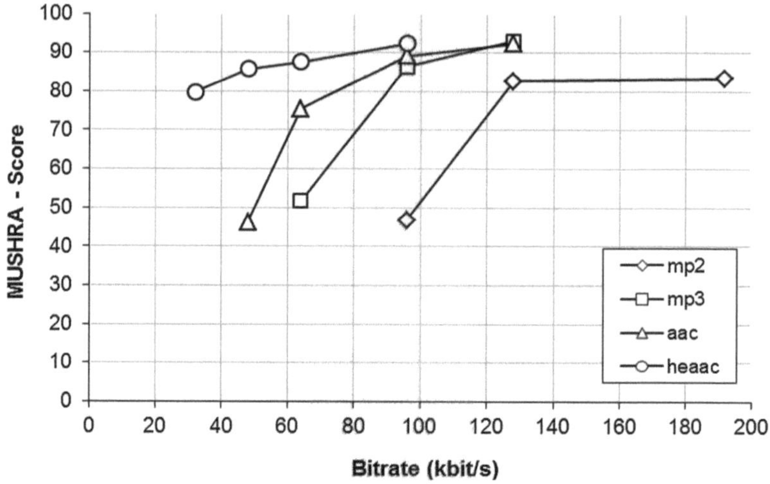

Fig. 16.1 Performance of frequently used audio codecs obtained with a MUSHRA test [14]

insights in technical deficiencies and show significant results already with a smaller number of listeners, the naïve listeners better represent the opinion of the targeted population. P.800 speech tests often apply single-stimulus methods with absolute category rating (ACR) or a degradation category rating (DCR). For audio tests paired or multi-stimulus tests with hidden reference are proposed. BS.1116 specifies a double-blind triple-stimulus with hidden reference test, while BS.1534 employs a MUlti Stimulus test with Hidden Reference and Anchor (MUSHRA). MUSHRA allows the direct comparison between a set of test stimuli. The stimuli can be compared to the clean reference, so that even small differences can be detected. With BS.1116 and BS.1534, test excerpts can be listened to multiple times. A typical result of a MUSHRA test for the most frequently used codecs is shown in Fig. 16.1. As expected from various listening tests [1–3, 61], the best performing codec at low bit rates is HE-AAC, followed by, in that order, AAC-LC, MP3, and MP2.

The selection of the test method also depends on the targeted application. For instance, in the context of the development of quality models for multimedia applications, audio, video, and audiovisual subjective tests have to be conducted. In that case, it may be required to align the listening test method to the video and audiovisual ones. This was the case for the development of the parametric audio quality model described in Sect. 16.4.2 for the IPTV scenario. In that case, an ACR test with single stimulus was selected and the 5-point categorical scale from ITU-T P.910 was adopted. This approach reflects quite well the normal user's situation: The user of a broadcasting service is listening a sequence only once and judges based on expectations. However, it was expected that for the upper quality range, where differences can only be perceived by direct and/or multiple comparisons, a different behavior of the test methods may show-up. The paired comparison and single stimulus methods may give different results when it is unclear if the perceptible difference between the reference and the test item is a quality degradation. Subjects may judge the degraded item in a single stimulus test as having superior quality. It was assumed that the ACR still would be appropriate to assess the full range of audio qualities and that the results of the ACR tests would be comparable to MUSHRA test results. With ACR, the assessment of codecs at high bit rates and under error-free conditions might fall in a very small range, and even the ranking of these codec conditions might not reliably be resolved anymore. Also, gradients, offsets, and saturation effects in the scale usage might be different. A comprehensive introduction into the biases in listening tests can be found in [62].

In a dedicated subjective test, MUSHRA and ACR were applied in parallel. It could be shown that the ACR method is sufficiently fine-grained to resolve the users' experience for high quality ranges as well. Only for a very few points and high bit rates ACR was not suited to identify the solution with superior quality [14]. The scales were compared in depth in [48].

A valuable parameter for assessing the quality of an audio transmission system is the so called *Coding Margin* [13], a way of describing inaudible artifacts. The coding margin can be determined in a subjective test by interactively amplifying the artifacts until they become audible for a test person. The coding margin then describes the headroom to the threshold of audibility of artifacts. A suitable method for amplification of the artifacts is the difference method. The difference signal of the time synchronous original and coded signal is amplified and added to the original signal. Changes introduced by the frequency response of the codec need to be compensated. Detection of the threshold of audibility is best performed with a forced-choice method. The definition and validation of the method to measure coding margin is described in [13]. An extension and application of the coding margin measurement in combination with other subjective testing has been implemented for a large crowd sourcing platform with the object to acquire quality comparisons of audio codecs on the Internet [56, 60].

16.4 Audio Quality Models

An overview of existing audio and speech quality models is provided in [51], covering both "full-reference" (also known as "double-ended" or "intrusive") and "no-reference" (also known as "single-ended" or "non-intrusive") methods, as well as signal- and parameter-based models. The current section concentrates on full-reference audio quality models (Sect. 16.4.1) and "no-reference" parametric audio quality models (Sect. 16.4.2).

16.4.1 Full-Reference Quality Models

Basic instrumental measurement methods for audio systems, such as Signal-to-Noise-Ratio (SNR) or Total Harmonic Distortion (THD), are well known to describe the technical quality of a system but are often not suited to express the quality the user perceives. The quality of codecs cannot be measured with basic methods because the codecs allocate their bits adaptively according to the characteristics of the content. Hence the perceived quality of a transmission chain is better measured using real test signals and ideally the measurement tools also reflect how the user perceives degradation by modeling the auditory system and the perception of sound.

A variety of approaches have been taken to consider hearing properties into the measurement. One of the most essential basic assumptions is to consider that the strength of a subjective sensation is proportional to the logarithm of the stimulus intensity (Fechner law). Another assumption is to weight the signals according to the spectral sensitivity. The essential psychoacoustic factors are categorized in loudness, sharpness, roughness and tonality and have been mathematically modeled in [63] based on comprehensive empirical experiments. The *loudness* models take the processing of the critical band levels of the audio signal into account and consider spectral and temporal masking. *Sharpness* describes how the spectrum of the signal is spread towards high frequencies, interpreted on the critical band scale. *Roughness* is a sensation that relies on the perception of fluctuation in the critical bands. *Tonality* models the sensation of composite tone mixture in relation to noise-like sounds. In addition modeling *binaural* effects is essential for considering spatial sounds, but also for explaining observed masking phenomena.

A full reference quality model aims at detecting noticeable differences between the reference signal and the transmitted audio. Typically, short-time *model output values* (MOVs) are calculated from the reference signal and the signal under test, employing *psychoacoustic modeling* as described above. The MOVs represent a time-varying multi-dimensional perception pattern. This pattern is processed further in a *cognitive detection model* aggregating and averaging the MOVs over time and deriving a single value estimation for the quality differences.

Several full-reference objective perceptual measurements for audio were developed. Six of these measurement methods, Disturbance Index (DIX), Noise-to-Mask Ratio (NMR), Perceptual Audio Quality Measure (PAQM), PERCEVAL, Perceptual Objective Measure (POM) and The Toolbox Approach were evaluated for ITU standardization. In a collaborative approach the best feature extractors were integrated into one single method, the so called PEAQ (Perceptual Evaluation of Audio Quality) standard ITU-R BS.1387 [30, 58]. PEAQ replicates subjective listening tests according to ITU-R BS.1116. It derives an *objective difference grade* (ODG) corresponding to the mean *subjective difference grades* (SDG) of the *basic audio quality* obtained in listening tests. The used test items were around 20 s long and contained different error types.

In a first step, PEAQ derives MOVs from comparisons between the reference signal and the signal under test. Examples for the selected MOVs are "modulation difference", "noise loudness of missing frequency components", "noise loudness with emphasis on introduced components", "linear distortions", "bandwidth of reference and signal under test", "total noise to mask ratio", "frequency bands containing significant noise components" and "harmonic structure". Most MOVs were studied for different averaging and threshold detection strategies.

The averaged MOVs are mapped to the ODG using an artificial neural network. The net is trained beforehand with optimization techniques that minimize the squared difference between the ODG distribution and the corresponding distribution of mean SDGs for a sufficiently large training data set. PEAQ performs best for high quality samples. For the validation set with unknown content it showed a correlation of $r = 0.851$ and only a few outliers. It has been proven that PEAQ generates both reliable and useful information for several applications [30, 58].

PEAQ cannot deal with stereo and does not perform well for coding tools such as spectral band replication. PEAQ also does not handle a degradation of quality caused by a transmission error such as a packet loss. Efforts have been made to enhance the psychoacoustic and the cognitive model and to support measurement of stereo quality for full reference audio quality model, summarized in [8]. A new version of the standard has not yet been released.

Alternative quality models for spatial audio exist as well, as reviewed and studied in the AABBA project [49].

16.4.2 No-Reference Audio Quality Models

In the case of single-ended audio quality models, the reference (non-degraded) audio signal is not available. The perceived audio quality is estimated for the degraded audio signal only, either from the audio signal itself, or from a parametric description of the transmission chain. With the latter approach, the quality of encrypted audio signal can also be estimated.

Most single-ended signal-based and parametric models described in the literature [7, 10, 35, 42] are dedicated to speech and especially to speech communication links such as Voice over IP (VoIP). However, the approaches related to the parametrization of the effect of packet-loss in VoIP [9, 47] may be used in the context of audio transmission as well. Single-ended signal-based audio models for estimating the perceived quality are not common.

The impairment-factor based approach of the E-Model (ITU-T G.107) [35] was adapted for single-ended quality estimation of IP-based audio transmission several times. In [17], the full-reference model PEAQ was used to derive the coefficients of the coding related impairment term. The transmission and packet loss impairment term was calibrated with the help of subjective testing. In [43], the model takes as inputs the audio bit rate and the percentage of lost audio packets. The model has been developed for two audio codecs (AAC and Lame MP3) and random loss error. Another single-ended parametric model has been proposed in [11] for AAC Low Complexity (AAC-LC). The model is suitable for both low and high bit rate applications, such as MobileTV and IPTV and takes as input the audio bit rate, the codec type, the sampling rate, the frame length, the packet-loss-frequency (the number of loss events), and the average IP packet burst-length (considering only IP packets containing audio). This model was validated against subjective test results and shows high performance results. However, as noted by the authors, the interaction between audio bit rate and packet-loss-rate is not considered.

The most up-to-date and thoroughly validated single-ended audio quality models are the audio models of the ITU-T P.1201.1 and P.1201.2 standards [32–34]. These models have been developed for low (e.g. MobileTV) and high (e.g. IPTV) bit rate applications respectively. They take as input parameters extracted from the Internet Protocol (IP) packet headers. Both models use the audio bit rate in order to capture the quality impact of audio compression degradation. The ITU-T P.1201.1 model captures the quality impact of audio packet loss with the aggregated length of lost audio frames normalized to the measurement duration. The ITU-T P.1201.2 uses instead the percentage of lost audio frames and a weighted average of the number of consecutively lost audio frames, also called burst length, which allows to take into account different loss distributions.

Both the ITU-T P.1201.1 and P.1201.2 models capture the interaction between audio compression and transmission error degradation. The burst-length related parameter of the ITU-T P.1201.2 reflects the observation that isolated losses yield better perceived quality than short bursty losses (e.g. two consecutively lost audio frames) and that long bursty losses (that is, from four consecutively lost audio frames onwards) are better perceived than isolated losses [16].

For both models, the model coefficients depend on the employed audio codec. Codecs covered by the P.1201.1 model are AAC-LC, AAC-HE, AMR-NB and—WB+, and the P.1201.2 audio model has been developed for the AAC-LC, HE-AAC, MPEG1-LII and AC-3 codecs. Formulas and coefficients of the models are reported in [32, 34]. Both models show high performance results on both known and unknown test databases, with a Pearson's correlation coefficient of $r = 0.94$ and a Root-Mean-Square-Error of $rmse = 0.35$ (on a 5-point scale) for the P.1201.1 model and $r = 0.949$ and $rmse = 0.34$ for the P.1201.2 model. Note that the test databases differ between P.1201.1 and P.1201.2 since the two models are not targeting the same application areas.

16.5 Discussion

Main efforts for developing methods for modeling the quality perception of audio transmission applications in the QoE domain are based on short-term sequences of 10–20 s so far. Short-term quality scores are not sufficient for estimating the QoE of audio transmission related services. It still needs to be found out how these measures can be aggregated to a validated long period QoE performance index. Long-term quality estimation is not trivial since new subjective test methods need to be developed and validated as well. The context in which the audio is listened to is essential for the QoE. The kind of applications and the chosen terminal, as addressed with the study of the room and electro-acoustic effects in Sect. 16.3.1, need to be taken into account. Finally, the user characteristics should also be considered. Steps to be undertaken towards QoE prediction are briefly reviewed in Chap. 19.

Furthermore, the tools for the estimation of the perceived quality still could be improved. The modeling of the auditory system, the perception and the reception of high quality audio is still an open field. The derived auditory features and the cognitive modeling are still insufficient for explaining the subjective quality judgments for a big part of audible coding artifacts and stereo presentation impairments.

The parametric models can be calibrated quite well to reproduce the subjective test results. But ideally, and since they provide different compression qualities and concealment features, each encoder and each decoder would need to be considered in the subjective testing.

If one succeeds to extract meaningful parameters representing compression efficiency, audio content complexity, and loss concealment characteristics, a hybrid audio quality model would be a promising approach.

References

1. EBU Document BPN 019 (1998) Report on the EBU subjective listening tests of multichannel audio codecs
2. EBU Document Tech3296 (2003) EBU subjective listening test on low-bitrate audio codecs
3. 3GPP TR 26.936 (2008) Performance characterization of 3GPP audio codecs. www.3GPP.org
4. Bech S, Zacharov N (2006) Perceptual audio evaluation. Theory, method and application. Wiley, New York
5. Bosi M, Brandenburg K, Quackenbush S, Fielder L, Akagiri K, Fuchs H, Dietz M, Herre J, Davidson G, Oikawa (1996) MPEG-2 advanced audio coding. In: Proceedings of the 101st Audio Engineering Society (AES) convention
6. Brandenburg K, Stoll G, Dehéry YF, Johnston JD, Kerkhof Lvd, Schroeder EF (1992) The ISO/MPEG-audio codec: a generic standard for coding of high quality digital audio. In: Proceedings of the 92th Audio Engineering Society (AES) convention
7. Broom S (2006) VoIP: quality assessment: taking account of the edge-device. IEEE Trans ASLP 14(6):1977–1983
8. Campbell D, Jones E, Glavin M (2009) Audio quality assessment—a review, and recent developments. Signal Process 89:1489–1500
9. Clark A (2001) Modeling the effects of burst packet loss and recency on subjective voice quality internet telephony workshop (IPtel)
10. Clark A (2003) ITU-T Delayed Contribution COM12-D105: Description of VQMON algorithm
11. Egi N, Hayashi T, Takahashi A (2010) Parametric packet-layer model for evaluation audio quality in multimedia streaming services. IEICE Trans Commun E93.B:1359–1366
12. Erne M (2001) Perceptual audio coders: what to listen for. In: Proceedings of the 111th Audio Engineering Society (AES) convention
13. Feiten B (1997) Measuring the coding margin of perceptual codecs with the difference signal. In: Proceedings of the 102nd Audio Engineering Society (AES) convention
14. Feiten B, Raake A, Garcia MN, Wüstenhagen U, Kroll J (2009) Subjective quality evaluation of audio streaming applications on absolute and paired rating scales. In: Proceedings of the 126th Audio Engineering Society (AES) convention

15. Gabrielsson A, Sjogren H (1979) Perceived sound quality of sound-reproduction systems. J Acoust Soc Am 65(4):1019–1033
16. Garcia MN, Raake A, Feiten B (2013) Parametric audio quality model for IPTV services—ITU-T P.1201.2 audio. In: Proceedings international workshop on Quality of Multimedia Experience (QoMEX)
17. Graubner M et al (2010) QoE assessment for audio contribution over IP (ACIP). In: Proceedings of the 38th AES international conference on sound quality evaluation
18. Herre J (2007) Temporal noise shaping, quantization and coding methods in perceptual audio coding: a tutorial introduction. In: Proceedings of the AES 17th international conference on high quality audio coding, pp 1–14
19. Herre J, Dietz M (2008) Standards in a nutshell: MPEG-4 high-efficiency AAC coding. IEEE Signal Process Mag 25:137–142
20. Herre J et al (2008) MPEG surround—the ISO/MPEG standard for efficient and compatible multichannel audio coding. J AES 56:932–955
21. Horbach U, Boone MM (1999) Future transmission and rendering formats for multichannel sound. In: Proceedings of the AES 16th international conference on spatial sound, reproduction, pp 409–418
22. ISO/IEC 11172–3 (1993) Coding of moving pictures and associated audio for digital storage media at up to about 1.5 Mbit/s—part 3: audio
23. ISO/IEC 13818–3 (1995) Generic coding of moving pictures and associated audio: audio
24. ISO/IEC 13818–7 (2006) Generic coding of moving pictures and associated audio: advanced audio coding
25. ISO/IEC 14496–3 (2006) Information technology—coding of audio-visual objects—part 3: audio
26. ITU-R Rec. BS.1116-1 (1994–1997) Methods for the subjective assessment of small impairments in audio systems including multichannel sound systems
27. ITU-R Rec. BS.1284 (1997–2003) General methods for the subjective assessment of sound quality
28. ITU-R Rec. BS.1286 (1997) Methods for the subjective assessment of audio systems with accompanying picture
29. ITU-R Rec. BS.1534-1 (2001–2003) Method for the subjective assessment of intermediate quality levels of coding systems
30. ITU-T BS.1387 (2001) Method for objective measurements of perceived audio quality
31. ITU-T GSTP-GVBR (2010) Performance of ITU-T G.718. Series G: transmission systems and media, digital systems and networks
32. ITU-T Recommendation P.1201 (2012) Parametric non-intrusive assessment of audiovisual media streaming quality
33. ITU-T Recommendation P.1201.1 (2012) Parametric non-intrusive assessment of audiovisual media streaming quality—lower resolution application area
34. ITU-T Recommendation P.1201.2 (2012) Parametric non-intrusive assessment of audiovisual media streaming quality—higher resolution application area
35. ITU-T Recommendation G.107 (2011) The E-model: a computational model for use in transmission planning
36. ITU-T Recommendation P.800 (1996) Methods for subjective determination of transmission quality
37. Liu C-M, Hsu H-W, Lee W-C (2008) Compression artifacts in perceptual audio coding. IEEE Trans Audio Speech Lang Process (ASLP) 16(4):681–695
38. Lutzky M et al (2004) A guideline to audio codec delay. In: Proceedings 116th Audio Engineering Society (AES) convention, Berlin
39. Mattila VV (2002) Descriptive analysis and ideal point modeling of speech quality in mobile communications. In: Proceedings of the 113th audio engineering society (AES) convention, USALos Angeles
40. Mattila VV (2002) Ideal point modeling of speech quality in mobile communications based on multidimensional scaling. In: Proceedings of the 112th audio engineering society (AES) convention, DMunich

41. Moller H (1992) Fundamentals of binaural technology. Appl Acoust 36:171–218
42. Möller S, Chan WY, Côté N, Falk TH, Raake A, Wältermann M (2011) Speech quality estimation, IEEE Signal Process Mag
43. Myakotnykh ES, Svensson UP (2010) Computational quality model for IP-based audio. In: Proceedings of the 38th AES international conference on sound quality, evaluation
44. Neuendorf M et al (2009) Unified speech and audio coding scheme for high quality at low bitrates.: In: Proceedings IEEE International Conference on Audio Speech and Signal Processing (ICASSP)
45. Painter T, Spanias A (2000) Perceptual coding of digital audio. Proc IEEE 88(4):451–515
46. Perkins C, Hodson O, Hardman V (1998) A survey of packet loss recovery techniques for streaming audio. IEEE Netw 12(5):40–48
47. Raake A (2006) Short- and long-term packet loss behaviour: towards speech quality prediction for arbitrary loss distributions, IEEE Trans ASLP 14(6):1957–1968
48. Raake A, Wältermann M, Wüstenhagen U, Feiten B (2012) How to talk about speech and audio quality with speech and audio people? J Audio Eng Soc 60(3):147–155
49. Raake A, Blauert J (2013) Comprehensive modeling of the formation process of sound-quality. In: Proceedings international workshop on Quality of Multimedia Experience (QoMEX), Klagenfurt, Austria
50. Reichl P, Egger S, Schatz R, D'Alconzo A (2010) The logarithmic nature of QoE and the role of the Weber-Fechner Law in QoE assessment. In: Proceedings IEEE International Conference on Communications (ICC)
51. Rix AW, Beerends JG, Kim D-S (2006) Objective assessment of speech and audio quality—technology and applications. IEEE Trans ASLP 14(6):1890–1901
52. Rumsey F (2002) Spatial quality evaluation for reproduced sound: terminology, meaning, and a scene-based paradigm. J Audio Eng Soc 50(9):651–666
53. Sackl A, Egger S, Schatz R (2013) Where's the music? Comparing the QoE impact of temporal impairments between music and video streaming. In: Proceedings international workshop on Quality of Multimedia Experience (QoMEX)
54. Schobben D, van de Par S (2004) The effect of room acoustics on MP3 audio quality evaluation. In: Proceedings of the 117th audio engineering society (AES) convention, USASan Francisco, 28–31 Oct 2004
55. Schuller G, Yu B (2002) Perceptual audio coding using adaptive pre and post-filters and lossless compression. IEEE Trans Speech Audio Process 10(6):379–390
56. Smirnoff S (2005) Difference level. An objective audio parameter. In: 118th AES-convention
57. Spors S, Wierstorf H, Raake A, Melchior F, Frank M, Zotter F (2013) Spatial sound with loudspeakers and its perception: a review of the current state. Proc IEEE 101(9):1920–1938
58. Thiede T, Treurniet WC, Bitto R, Schmidmer C, Sporer T, Beerends JG, Colomes C, Keyhl M, Stoll G, Brandenburg K, Feiten B (2000) PEAQ—the ITU standard for objective measurement of perceived audio quality. J Audio Eng Soc (AES) 48(1/2):3–29
59. Toole F (2008) Sound reproduction: the acoustics and psychoacoustics of loudspeakers and rooms. Focal Press
60. Website sound expert. http://soundexpert.org
61. Wüstenhagen U, Feiten B, Hoeg W (1998) Subjective listening test of multichannel audio codecs. AES Conv 105:P4813
62. Zielinski S, Rumsey F, Bech S (2008) On some biases encountered in modern audio quality listening tests—a review. J Audio Eng Soc 56(6):427–451
63. Zwicker E, Fastl H (1999) Psychoacoustics. Facts and models, 2nd edn. Springer, Berlin

Chapter 17
Spatial Audio Rendering

Matthias Frank, Franz Zotter, Hagen Wierstorf and Sascha Spors

Abstract Complementary to non-spatialized signals and their transmission, this chapter gives an overview of the quality of rendering methods that create spatial sound. Common methods and the underlying concept of a virtual sound scene are introduced and the herewith associated quality features. In particular, evaluation strategies and experimental results are presented in order to discuss spatial and timbral quality features of spatial audio rendering.

17.1 Introduction

Spatial audio rendering aims at mimicking our perception of spatial audio scenes by employing suitable processing techniques, loudspeakers, and headphones. As source material for rendering, an entire physical sound scene, or a single sound it contains, is captured by one or more microphones, see Fig. 17.1. The captured signals and additional spatialization parameters provide a *virtual sound scene* representation that is stored, played back, and modified during thinkable post production steps.

M. Frank (✉) · F. Zotter
Institute of Electronic Music and Acoustics, University of Music and Performing Arts Graz, Graz, Austria
e-mail: frank@iem.at

F. Zotter
e-mail: zotter@iem.at

H. Wierstorf
Assessment of IP-based Applications, Telekom Innovation Laboratories, TU Berlin, Berlin, Germany
e-mail: hagen.wierstorf@tu-berlin.de

S. Spors
Institute of Communications Engineering, University of Rostock, Rostock, Germany
e-mail: sascha.spors@uni-rostock.de

S. Möller and A. Raake (eds.), *Quality of Experience*,
T-Labs Series in Telecommunication Services, DOI: 10.1007/978-3-319-02681-7_17,

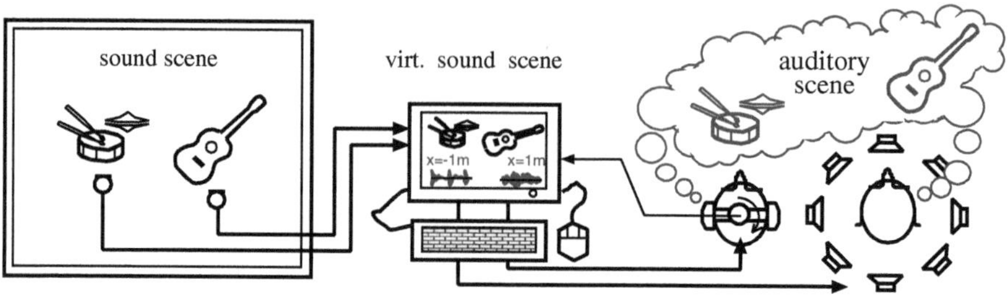

Fig. 17.1 Typical audio recording and rendering scenarios consist of a physical sound scene, a virtual one in which mixing and spatialization parameters of virtual sound objects might be specified, and finally a perceived auditory scene that is rendered on headphones or loudspeakers

Within the virtual sound scene, a *virtual sound object* is the representation that controls the rendering algorithms. A virtual sound scene is composed of several such virtual objects. Virtual sound objects can also be related to models of a physical source. Such an object can be, e.g., point-source-like, plane-wave-like, directional [2, 49], reverberant, diffuse, or wide [30, 34, 37, 54, 88]. Virtual sound scenes may also integrate virtual sound scene material produced for particular loudspeaker playback systems, e.g., two-channel-stereophonic recordings.

In order to translate the virtual scene into signals to be played back by loudspeakers or headphones, renderers use spatialization algorithms aiming at the listener perceiving a corresponding auditory scene.

How is the rendering quality assessed, and what is a good reference?

Not necessarily does spatial audio rendering reproduce an original, physical sound scene. Therefore a physical quality evaluation is often not rewarding. For instance, two-channel stereophony uses level and/or time differences between two horizontally arranged loudspeakers to create an auditory object from a virtual sound object that may only be located somewhere between the loudspeakers. How good the perceived quality features of this auditory object match those of the virtual sound object is best rated in relation to a perceptually resembling reference, preferably one that is no product of rendering. For stereophony, this reference is often a loudspeaker placed at the physical location of the virtual source. The sound field of this reference generally differs from the stereophonic sound field.

There are many auditory quality features that have no simple physical correspondence. For instance, quality features in a concert hall comprise apparent source width, listener envelopment, clarity, and mixing time, etc. [32, 40, 43], most of which being rather unrelated to physical extent or in a non-unique relation to the physical response.

While the perceptual reference for assessing virtual room acoustics or auralization [73] is still taken from the physical room or environment that is to be reproduced, spatial audio rendering might also evoke quality features that do not correspond to any physical reference outside, and which are described by properties of the virtual sound objects only, e.g. [54, 58, 64].

Some models exist for spatial audio quality features such as localization direction, [7], however, a comprehensive technical evaluation of all quality features, see e.g. [66], is not available yet. Therefore today, most spatial audio rendering methods still need to be evaluated perceptually.

This chapter gives a summarizing overview of current strategies and results assessing spatial audio rendering. After introducing fundamentals of spatial hearing, the typical spatialization techniques and their key features are outlined. Subsequently, ways are discussed of identifying and evaluating different quality features produced by such techniques. Next, we will see exemplary results about the spatial quality features such as perceived direction and width, and timbral fidelity for varying quality elements, e.g. number of loudspeakers. The chapter concludes with a discussion of the overall quality of spatial audio rendering.

17.2 Spatial Audio

In contrast to vision that mainly captures the look direction, our ears simultaneously receive a mixture of all sources surrounding us. Auditory sensation evokes auditory scene analysis so that we may focus on specific sounds and their provenance, mainly based on interaural differences and spectral cues. Current rendering methods attempt to provide immersion by surrounding the listener with sound, and they may directly utilize auditory cues related to spatial hearing.

17.2.1 Spatial Hearing

The direction of a single sound source in the horizontal plane is strongly related to interaural differences of level and time delay, so-called interaural level differences (ILDs) and time differences (ITDs). These differences are inherent in the head-related transfer functions (HRTFs) [6], the transfer functions between a sound source in a specified direction and the ear canals of the left and the right ear. Known as the *duplex theory*, ITDs are a dominating cue at lower frequencies, whereas ILDs are more important for perception at high frequencies [69]. Newer studies reveal that for broadband sources, low frequency ITDs are dominant [79] but ITDs are also important at high frequencies [46]. Nevertheless, differential cues only provide information about lateralization, i.e., how far a sound source is shifted to the left or right. For discriminating between front and back, up and down, spectral information in the HRTF is crucial [38], which is mainly due to the pinna. Vertical localization of sound sources is therefore less accurate and more individual than horizontal localization [6]. Up to some level, incoherence of the ear signals was shown to increase the perceived width [10] and spaciousness. This can be explained by the fact that the localization cues over time and frequency are less pronounced.

17.2.2 Rendering Methods

Neglecting the influence of other modalities, such as vision, vibration, or bone conduction, a perfect synthesis of the pressure at the eardrums would result in an indistinguishable (authentic) reproduction of the virtual scene. From the perspective of human perception, this technical goal does not need to be perfectly met in order to obtain a perfectly plausible impression of the virtual scene.

Binaural synthesis aims at reconstructing an auditory scene by generating ear signals. For this purpose, binaural synthesis reproduces the acoustic effect of the outer ears (including upper torso, head, and pinna) on the sounds in the audio scene. Playback primarily uses headphones, but may also use loudspeakers together with crosstalk canceling algorithms [36, 41]. There are two different techniques to capture the effect of the outer ear [50]: (1) direct recording of the signals at both ears or (2) measurement of the transfer functions from an acoustic source to both ears. The former technique (*dummy* or *original head stereophony*) involves the placement of microphones in the ears or the usage of a head and torso simulator. The advantages of such recordings are that they handle real-world mixes and do not require processing. Major drawbacks are that the head orientation in such recordings cannot be post-processed, and that the HRTFs might differ from those of the listener. The second above-mentioned technique filters the monophonic signals of the virtual source with the pre-captured HRTFs. These HRTFs are typically measured for a variety of head-orientations and/or source positions. In a fully virtual environment, head tracking and a database of HRTFs are used to involve an interactively changing head orientation [45], and provide the individually best matching HRTFs [51].

Various techniques employ loudspeakers for playback, which is considered as being more comfortable in many situations. Here, the target is not to involve a simulation of the outer ear but the synthesis of a suitable sound field that uses the natural cues of the auditory system. Loudspeaker playback can frequently handle a varying number of listeners, although the practically usable size of the listening area (sweet spot) may depend on the particular technique.

In *sound field synthesis*, the loudspeakers are used to produce a sound field that approximates the sound field of the virtual scene over an extended listening area. The best known representatives are *Higher-Order Ambisonics (HOA)* [1, 21, 23, 81] and *Wave Field Synthesis (WFS)* [4, 5, 68]. Regardless of their differences, these methods employ delays and gains, often also filters, to distribute the signal of a virtual sound object to loudspeakers.

A perfect reconstruction of a physical sound field for frequencies up to 20 kHz (the highest audible frequency) requires a loudspeaker spacing below 0.86 cm, in order to sample the corresponding wavelength at least twice. Even though practical implementations may employ a large number of loudspeakers, which can reach several hundred channels or even more [74], none of the current systems is capable of synthesizing a physically accurate sound field for the full frequency range of hearing. The implemented systems nevertheless prove that spatial sound reproduction following the principle of physical reconstruction is promising in producing the desired perception.

For instance, in WFS physical reconstruction suffers from spatial sampling artifacts above a critical frequency. Anyhow, as the perceived direction is dominated by low-frequency content, experiments prove that the perceived direction of the auditory object is still correct when using large loudspeaker spacings [78]. In addition, the artifacts are, to some extent, equally distributed across the entire listening area, which leads to the assumption that also perceptual localization is evenly consistent. Current research investigates the underlying hearing mechanisms and the overall quality as it may be affected by a larger loudspeaker spacing.

A much less complex loudspeaker-based spatial audio rendering is obtained by utilizing the ITD as the dominant binaural cue. For a listening position that is equally distant to all loudspeakers (sweet spot), a correct ITD is easily synthesized at low frequencies by amplitude panning, i.e. playback level differences of the loudspeakers. The introduced level difference controls the perceived direction of the auditory object for two-channel stereophony [39, 76].

For listening positions outside the sweet spot, the superposition is impaired, and the nearest loudspeaker dominates localization. Vector-base amplitude panning (VBAP [57]) allows to generalize stereophony from surrounding loudspeaker pairs on the horizon to triplets of loudspeakers including elevation. As the perceived width and coloration appears to modulate for moving sources [24], multiple-direction amplitude panning (MDAP [55]) or Ambisonic panning [86] are frequently used as an alternative. In Ambisonics, the order N determines the directional resolution of playback, which yields a technical sweet spot size of $\frac{N}{3}$ wavelengths.

17.3 Quality Evaluation

So far, there is no overview or recommendation about the comprehensive evaluation of spatial audio rendering available in the literature. A terminology and paradigm for the evaluation of the spatial quality is presented in [61] for multi-channel, stereophonic rendering (5.1 surround [16]). Evaluating these rendering techniques, the overall quality was found to be composed of timbral quality at 70 % and of spatial quality at 30 % in [62]. In turn, timbral quality and spatial quality are composed of multiple dimensions themselves.

Methods such as multidimensional scaling [47] can be employed to determine the dimensionality. In order to end up with an entire list of independent attributes, the collection of verbal descriptions for the different dimensions is required [18, 31, 82]. Another option is the repertory grid technique, where listeners respond to triads of stimuli and create their own attributes which describe a common feature of two stimuli that distinguishes them from the third one.

Known attributes can be rated entirely independently by a suited method. For example localization can be assessed by pointing methods [27, 65, 78]. Using such a method, the subjects point into the direction of the auditory object by using their head, a handheld device and/or a laser beam. For other attributes such as the width or the coloration of the auditory object one or more references are often included in the experiment to create reliable results. This can be done by a pairwise or a MUSHRA(-like) comparison of the stimuli. If no reference is or can be included, an extensive training phase can be required to give a reliable direct elicitation. For other attributes also a direct threshold measurement could be possible. In this case a standard alternative forced choice experiment can be employed.

17.4 Spatial Quality Features

Despite there are many more spatial quality features (spaciousness, listener envelopment, etc.), the most basic spatial quality features of an auditory object created by spatial audio rendering are its position and extent. The corresponding evaluation deals with perceived localization (direction and distance) and width, on which this section focuses. For an introduction to the assessment of more complex quality features, the example of presence in [71] is a good starting point.

17.4.1 Direction

For binaural synthesis via headphones, HRTFs, and a head tracker, the localization accuracy in the horizontal plane can be as high as it is the case for a physical source, i.e. roughly 1° for frontal directions [6]. In some cases, e.g. when there is no head tracking or for other than frontal directions, accurate localization depends on the usage of individual HRTFs and degrades when only using the HRTFs of a dummy head [65].

In order to facilitate the evaluation of WFS for multiple listening positions, binaural simulation of loudspeakers can be used. The study in [78] uses such a simulation to assess the localization of a frontal auditory object of a linear WFS loudspeaker array with a length of 2.85 m for three different loudspeaker spacings (0.19, 0.41, 1.43 m). To investigate the localization accuracy in the listening area, the listeners were seated at 16 different positions at two listening distances, 1.5 and 2 m. For the two smaller loudspeaker spacings the localization accuracy was equal in the whole listening area. In particular, the smallest spacing yields an accuracy that is indistinguishable from the accuracy when localizing a physical loudspeaker. The average absolute localization deviation is only 1° larger for a spacing of 41 cm. For a loudspeaker spacing of 1.43 m, corresponding to employing only three loudspeakers, localization accuracy is similar to what is achieved with amplitude panning. The deviation between desired and perceived direction is small for central positions and becomes very large off-center, leading to an average absolute deviation of 7°. At off-center positions, the listeners reported to always localize the auditory object from the nearest loudspeaker.

Similar results were found in [28] using 5th order 2D-Ambisonic amplitude panning on a ring of 12 loudspeakers and a radius of approximately 5 m. The median localization deviation for frontal sources was 3° at the central listening position. For a position that was one half of the radius off-center, the average deviation was 6°, which seems reasonable for plausible surround reproduction. For a lowered Ambisonic order, localization errors increase a little at the central listening position and become a lot worse at the off-center position.

The vertical direction of amplitude-panned virtual sources was evaluated in [75] for two loudspeakers in the median plane at elevation angles of $\pm 20°$ on a radius of 2.5^{m}. The different amplitude-panned virtual sources were created with 7 different inter-channel level differences (ICLD) of $\{\pm\infty, \pm 6, \pm 3, 0\}$ dB; in the $-\infty$ dB condition, only the lower loudspeaker is active, for ∞ dB only the upper one. The dependency of the perceived elevation on ICLD is monotonic and exhibits a $+6°$ bias. The perceived elevation is saturated towards the $\pm\infty$ dB ICLDs, and ± 3 dB yields a $\pm 10°$ shift of the perceived elevation, which is roughly 1.5 times more shift as for a frontal, horizontal loudspeaker pair. For these small ICLDs the perceived elevation subjectively varies twice as much as for the $\pm\infty$ dB conditions. This result agrees with the findings in [35, 56] that vertical amplitude panning is possible but with a larger subjective variation compared to horizontal panning.

Instead of pairwise panning, the study [11] evaluated the perceived direction using three-dimensional 1st and 4th order Ambisonic panning at the central listening position. The evaluation employed a hemispherical arrangement of 24 loudspeakers in three rings at $\{0, 30, 60\}°$ elevation and at a distance of 5 m. The vertical deviation from the desired directions is smaller for the 4th than for the 1st order reproduction. There is an overestimation of elevation for directions near the horizontal plane. This is because the Ambisonic representation spreads signals on the horizontal plane symmetrically to the upper and lower hemisphere. In playback, the energy of the lower hemisphere is largely preserved but lifted to the loudspeakers in the horizontal plane. On the other hand, there is an underestimation of elevation for 4th order virtual sources near the north pole. This effect is consistent with the typically underestimated elevation of physical sources there [6].

The $5°$ localization accuracy for the elevation of noise in the 4th order Ambisonics experiment is similar to the accuracy found using speech on single loudspeakers in [6]. It seems that 4th order Ambisonics is enough for an accurate reproduction of elevation, at least at the central listening position. One can estimate from the experimental setup that vertical loudspeaker spacings of $\leq 30°$ support the resolution of the perceived elevation angle.

There is no strict evidence that a deviation of less than $10°$ from the desired perceived direction is sufficiently plausible. Still the assumption is reasonable as the $1°$ resolution is restricted to localizing frontal azimuth directions. Typical standard deviations reach around $10°$ for localization from elsewhere, see [17]. Most of these values are optimistic in practice, as they were achieved for specific sounds and in laboratory environments.

17.4.2 Distance

Localization also comprises distance, whose perception depends on amplitude, direct to reverberant ratio, high frequency loss, as well as curvature of the wave fronts for nearby sources, see [83, 84]. In general, lateral distance perception is better than a frontal/dorsal one [15], and also better for broadband binaural sounds [14]. For spatial sound rendering, Völk [72] could show that the minimum audible free-field distance differences on a 96-channel circular wave field synthesis system are the same as for physical sources. Including room simulations, distance perception seems to be similarly good as in physical rooms, as shown for the LoRA loudspeaker system [22].

For binaural reproduction the phenomenon of inside-the-head localization can occur. This describes the case of an auditory object perceived with a distance equal or smaller than the radius of the head. Begault et al. [3] have investigated the influence of the usage of a head-tracker and adding reflections to the HRTFs on inside-the-head localization. Especially the last technique can enhance the perception out of the head. HRTFs for sources nearer than 50 cm change not only their amplitude but also their interaural level differences. This has to be considered for binaural reproduction, but can be simplified by obtaining near-field HRTFs from far field HRTF measurements [33].

In some recent works, audibility of the orientation of sources was shown [20, 60] and measured [53, 85], however, the particular attributes are not well-described yet, and directivity rendering is fairly new [19].

17.4.3 Width

The ratio between direct and reverberated sound not only influences the perception of distance but also the perception of spaciousness. An important aspect of spaciousness is the apparent source width (ASW) perceived in concert halls [32, 52]. In psychoacoustics, perceived width was shown to be inversely proportional to interaural coherence [9, 10] when using headphones. In spatial audio rendering, the ASW can be increased directly by adjusting the correlation of a pair of loudspeakers [34, 87] or across several loudspeakers [37] by a set of decorrelation filters. Whereas decorrelation filters can only increase the ASW, its lower limit is not an entirely free parameter. In amplitude panning, the rendering algorithm influences the minimal ASW [26, 48, 55]. If a single loudspeaker is active, the ASW is narrower than with simultaneously active ones for which the loudspeaker spacing determines the minimal ASW [25].

17.5 Timbral Quality Features

Spatial audio rendering frequently achieves spatial reproduction by using simultaneously active loudspeakers, and hereby sometimes introduce audible timbral changes of the sound, due to constructive and destructive interference. The audibility of such colorations depends on the spectra and correlation of the resulting ear signals. In many cases, binaural decoloration mechanisms [12] are able to suppress the coloration present in the individual ear signals. Differences in coloration are typically easiest to perceive with continuous signals of a uniformly excited spectrum and lengths of about a second. Pink noise signals are therefore often preferred in tests.

For WFS, the frequency response is generally impaired above the aliasing frequency. Whereas below the aliasing frequency, the wave fronts emitted by the loudspeakers merge to a single wave front, above additional, delayed wave fronts arrive at the listener position. Wittek [80] has investigated the coloration differences between ±30° stereophony and WFS on loudspeaker arrays with different spacings. He asked the subjects if they could perceive a timbral difference between a given reference with a flat frequency spectrum and the test stimuli. These differences were rated on a scale ranging from *no difference* towards *extremely different*. The subjects in the listening test were centrally seated at a distance of 1.5 m to the array, and pink noise bursts were presented. The test stimuli were generated via binaural synthesis [42], and different directions of the virtual sources were used. For a loudspeaker spacing of 3 cm, the coloration of WFS was rated as good as for stereophony. More coloration was perceived for a loudspeaker spacing of 12 cm. However, further increase of the spacing up to 48 cm did not increase coloration.

De Bruijn [13] investigated the variation of timbre with the position of the listener for speech shaped noise in WFS. He employed two linear loudspeaker arrays with loudspeaker spacings of 12.5 and 50 cm. Changes in coloration were clearly perceptible for the larger loudspeaker spacing and negligible for the smaller loudspeaker spacing.

For the reproduction of virtual sources with WFS that are located inside the enclosed playback volume, so-called focused sources, coloration often appears together with other audible artifacts such as chirps/clicks. In this case, the additional wave fronts arrive before the desired one. The geometry of the loudspeaker array and the position of the listener have a strong influence on the amount of coloration and artifacts that are perceptible. This is due to the fact that the time distance between the first additional wave front and the desired one has a strong impact on the perception. If the time difference is too large, artifacts and more than one object are perceptible. Thus smaller loudspeaker arrays seem to have lower coloration and artifacts. But they could impair the spatial quality due to the grouping of all wave fronts to one source, in which case the first wave front dominates the localization [77].

An example is shown in [29, 77], where focused sources reproduced using WFS are being assessed. Figure 17.2 shows a principal component analysis of the attribute ratings of one subject for castanets as focused source stimuli. The different stimuli conditions are marked by the black points. Points are labeled with the angle describing the shifted listener position and the length of the array. The listener was always looking into the direction of the focused source.

In [24], the perceived coloration changes of moving sources using amplitude panning were evaluated in a listening experiment. The experiment employed equidistant ring arrangements of 8 and 16 loudspeakers using the panning strategies VBAP, MDAP, and two variants of Ambisonic panning. Each condition was tested twice in a MUSHRA-like test procedure. The rotation speed of the source was 0.1° per ms using an interpolation time of 1ms, resulting in 3.6 s for a whole 360° movement

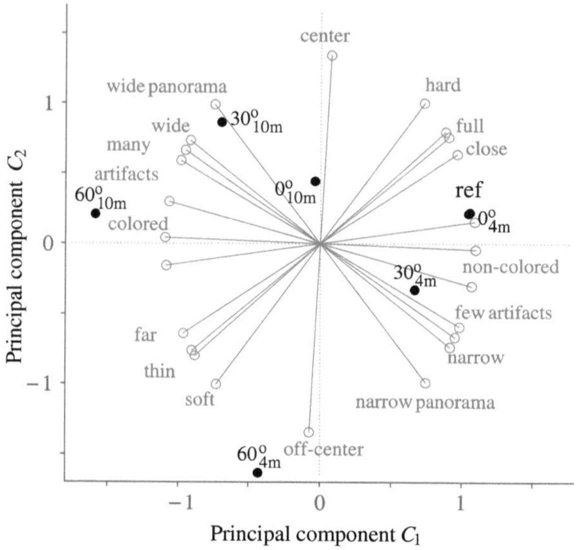

Fig. 17.2 Principal component analysis of the attribute ratings of a single listener for castanets synthesized as a focused source in Wave Field Synthesis. The black points indicate the position of the conditions given in the two-dimensional space determined by the two given components for each stimulus type. The angle indicates the listener position, were 0° was a center position and the index is the length of the used loudspeaker array. The gray lines show the arrangement of the attribute pairs in these two dimensions

around the listener. The listeners judged the coloration changes on a continuous scale from *imperceptible* to *very intense*. For both numbers of loudspeakers, VBAP yielded the strongest coloration changes. The smaller number of loudspeakers resulted in significantly less coloration changes. This result seems to contradict the findings in [70] that VBAP does not produce coloration in typical applications.

The above-mentioned experiment employed Ambisonic panning with the highest possible order for the respective number of loudspeakers. Spatial aliasing occurring outside the sweet spot covering $\frac{N}{3}$ wavelengths can be suppressed by increasing the number of loudspeakers. Despite this seems to bring a technical advantage, it yields a perceptual disadvantage: annoying coloration or auditory objects close to the listener's head were reported [67]. Thus, low order Ambisonic signals should be played back using fewer loudspeakers or in a reverberant-enough environment [63].

17.6 Conclusion

Obtaining overall quality ratings for spatial audio rendering techniques is a challenging task. Nevertheless, several results could be presented above evaluating specific quality features of auditory objects obtained by spatial audio rendering. At present, audio content produced for high-definition rendering systems such as WFS or binaural synthesis is not as common as for stereophony. One of the reasons is the greater complexity the production of a virtual sound scene requires, compared to a channel-based stereophonic production. Consequently, meaningful comparative quality ratings of entire sound scenes are still rare. And yet, we can learn a lot about the perception of spatial audio rendering with rather simple scenes of point sources with simple test signals.

A way to master the challenges of a quality evaluation of different spatial audio renderers is to investigate independent perceptual attributes such as localization and coloration, for which there are well-established methods, such as MUSHRA, AFC, or direct assessment. Establishing the definition of a standardized international list of attributes for evaluation is an ongoing challenge. For this reason, this chapter gave an overview of studies concerning attributes selected by the authors. Understandably, the question of how to estimate the overall quality from the single attributes remains to be answered.

One approach to overcome this uncertainty is to work not only with the concept of *authentic* reproduction, but let the listener rate if a perceived audio scene is *plausible*. *Plausible* means that the perceived features of the reproduced scenes show plausible correspondence with the listener's expectations in the given context, without necessarily being authentic [8, 44, 59].

References

1. Ahrens J, Spors S (2008) An analytical approach to sound field reproduction using circular and spherical loudspeaker distributions. Acta Acustica united with Acustica 94(6):988–999
2. Baalman M (2007) Reproduction of arbitrarily shaped sound sources with wave field synthesis-discretisation and diffraction effects. In: 122th convention of the audio engineering society, Vienna, Austria
3. Begault DR, Wenzel EM, Anderson MR (2001) Direct comparison of the impact of head tracking, reverberation, and individualized head-related transfer functions on the spatial perception of a virtual speech source. J Audio Eng Soc 49(10):904–916
4. Berkhout A (1988) A holographic approach to acoustic control. J Audio Eng Soc 36(12):977–995
5. Berkhout A, de Vries D, Vogel P (1993) Acoustic control by wave field synthesis. J Acoust Soc Am 93(5):2764–2778
6. Blauert J (1996) Spatial hearing: the psychophysics of human sound localization. MIT Press, Cambridge
7. Blauert J (ed) (2013) The technology of binaural listening. Springer, Berlin
8. Blauert J, Jekosch U (2003) Concepts behind sound quality: some basic considerations. In: Internoise, Jeju, Korea, pp 72–79
9. Blauert J, Lindemann W (1986) Auditory spaciousness: some further psychoacoustic analyses. J Acoust Soc Am 80(2):533–542
10. Blauert J, Lindemann W (1986) Spatial mapping of intracranial auditory events for various degrees of interaural coherence. J Acoust Soc Am 79(3):806–813
11. Braun S, Frank M (2011) Localization of 3D ambisonic recordings and ambisonic virtual sources. In: International conference on spatial audio
12. Brüggen M (2001) Sound coloration due to reflections and its auditory and instrumental compensation. PhD thesis, Ruhr-Universität Bochum
13. de Bruijn W (2004) Application of wave field synthesis in videoconferencing. PhD thesis, Delft University of Technology
14. Brungart DS (1999) Auditory localization of nearby sources. III. stimulus effects. J Acoust Soc Am 106(6):3589–3602

15. Brungart DS, Durlach NI, Rabinowitz WM (1999) Auditory localization of nearby sources. ii. localization of a broadband source. J Acoust Soc Am 106(4):1956–1968
16. BS.775, I.R.R., Multichannel stereophonic sound system with and without accompanying picture. International Telecommunications Union, Geneva
17. Carlile S, Leong P, Hyams S (1997) The nature and distribution of errors in sound localization by human listeners. Hear Res 114(1–2):179–196
18. Choisel S, Wickelmaier F (2007) Evaluation of multichannel reproduced sound: Scaling auditory attributes underlying listener preference. J Acoust Soc Am 121(1):388–400
19. Corteel E (2007) Synthesis of directional sources using wave field synthesis, possibilities, and limitations. EURASIP J Adv Signal Process 1:18
20. Dalenbäck BI, Kleiner M, Svensson P (1993) Audibility of changes in geometric shape, source directivity, and absorptive treatment-experiments in auralization. J Audio Eng Soc 41(11):905–913
21. Daniel J (2003) Spatial sound encoding including near field effect: introducing distance coding filters and a viable, new ambisonic format. In: 23rd international conference, Audio Engineering Society. Copenhagen, Denmark
22. Favrot S, Buchholz JM (2009) Distance perception in loudspeaker-based room auralization. In: Audio engineering society convention 127
23. Fazi F, Nelson P, Christensen J, Seo J (2008) Surround system based on three dimensional sound field reconstruction. In: 125th convention of the Audio Engineering Society. San Fransisco, USA
24. Frank M (2013) Phantom sources using multiple loudspeakers in the horizontal plane. PhD thesis, University of Music and Performing Arts Graz, Austria
25. Frank M (2013) Source width of frontal phantom sources: perception, measurement, and modeling. Archives Acoust 38(3):311–319
26. Frank M, Marentakis G, Sontacchi A (2011) A simple technical measure for the perceived source width. In: Fortschritte der Akustik, DAGA. Düsseldorf
27. Frank M, Mohr L, Sontacchi A, Zotter F (2010) Flexible and intuitive pointing method for 3-d auditory localization experiments. In: Audio Engineering Society conference: 38th international conference: sound quality evaluation
28. Frank M, Zotter F (2008) Localization experiments using different 2D ambisonics decoders. In: 25. Tonmeistertagung, Leipzig
29. Geier M, Wierstorf H, Ahrens J, Wechsung I, Raake A, Spors S (2010) Perceptual evaluation of focused sources in Wave Field Synthesis. In: 128th convention of the Audio Engineering Society
30. Gerzon MA (1992) Signal processing for simulating realistic stereo images. In: 93rd Convonvention Audio Engineering Society. San Francisco
31. Guastavino C, Katz BFG (2004) Perceptual evaluation of multi-dimensional spatial audio reproduction. J Acousti Soc Am 116(2):1105–1115
32. Hidaka T, Beranek LL, Okano T (1995) Interaural cross-correlation, lateral fraction, and low- and high-frequency sound levels as measures of acoustical quality in concert halls. J Acoust Soc Am 98(2):988–1007
33. Kan A, Jin C, van Schaik A (2009) A psychophysical evaluation of near-field head-related transfer functions synthesized using a distance variation function. J Acoust Soc Am 125(4):2233–2242
34. Kendall GS (1995) The decorrelation of audio signals and its impact on spatial imagery. Comput Music J 19(4):71–87
35. Kimura T, Ando H (2012) Listening test for three-dimensional audio system based on multiple vertical panning. In: Acoustics 2012, Hong Kong
36. Kirkeby O, Nelson PA, Hamada H (1998) The "stereo dipole": a virtual source imaging system using two closely spaced loudspeakers. J Audio Eng Soc 48(5):387–395
37. Laitinen MV, Philajamäki T, Erkut C, Pulkki V (2012) Parametric time-frequency representation of spatial sound in virtual worlds. ACM Trans Appl Percept 9(2): 8

38. Langendijk EHA (2002) Bronkhorst AW contribution of spectral cues to human sound localization. J Acoust Soc Am 112(4):1583–1596
39. Leakey DM (1959) Some measurements on the effects of interchannel intensity and time differences in two channel sound systems. J Acoust Soc Am 31(7):977–986
40. Lee D, Cabrera D (2010) Effect of listening level and background noise on the subjective decay rate of room impulse responses: Using time-varying loudness to model reverberance. Appl Acoust 71(9):801–811
41. Lentz T (2008) Binaural technology for virtual reality. Logos, Berlin
42. Lindau A, Hohn T, Weinzierl S (2007) Binaural resynthesis for comparative studies of acoustical environments. In: 122th Convention of the audio engineering society. Vienna, Austria
43. Lindau A, Kosanke L, Weinzierl S (2012) Perceptual evaluation of model- and signal-based predictors of the mixing time in binaural room impulse responses. J Audio Eng Soc 60(11):887–898
44. Lindau A, Weinzierl S (2012) Assessing the plausibility of virtual acoustic environments. Acta Acust U Acust 98(5):804–810
45. Mackensen P (2008) Auditive localization & head movements: about localization cues, head movements, and auralization methods. Verlag Dr. Müller
46. Macpherson EA, Middlebrooks JC (2002) Listener weighting of cues for lateral angle: the duplex theory of sound localization revisited. J Acousti Soc Am 111:2219–2236
47. Marins P, Rumsey F, Zielinski S (2008) Sound quality assessment: cardinal concepts. Proc Inst Acoust 30:297–305
48. Martin G, Woszczyk W, Corey J, Quesnel R (1999) Controlling phantom image focus in a multichannel reproduction system. In: Audio Engineering Society convention 107
49. Menzies D (2007) Nearfield synthesis of complex sources with high-order ambisonics, and binaural rendering. In: International conference on digital audio effects (DAFx). Montreal, Canada
50. Møller H (1992) Fundamentals of binaural technology. Appl Acoust 36(3–4):171–218
51. Møller H, Sørensen MF, Jensen CB, Hammershøi D (1996) Binaural technique: do we need individual recordings? J Audio Eng Soc 44(6):451–469
52. Okano T, Beranek LL, Hidaka T (1998) Relations among interaural cross-correlation coefficient ($IACC_E$), lateral fraction (LF_E), and apparent source width (ASW) in concert halls. J Acoust Soc Am 104(1):255–265
53. Otondo F, Rindel J (2004) The influence of the directivity of musical instruments in a room. Acta Acust U Acust 90(5):1178–1184
54. Potard G (2006) 3d-audio object oriented coding. PhD thesis, University of Wollongong
55. Pulkki V (1999) Uniform spreading of amplitude panned virtual sources. In: IEEE workshop on applications of signal processing to audio and acoustics, pp 187–190
56. Pulkki V (2001) Localization of amplitude-panned virtual sources II: two- and three-dimensional panning. J Audio Eng Soc 49(9):753–767. http://www.aes.org/e-lib/browse.cfm?elib=10179
57. Pulkki V (2001) Spatial sound generation and perception by amplitude panning techniques. PhD thesis, Helsinki University of Technology
58. Pulkki V (2007) Spatial sound reproduction with directional audio coding. J Audio Eng Soc 55(6):503–516
59. Raake A, Blauert J (2013) Comprehensive modeling of the formation process of sound-quality. In: Fifth International workshop on Quality of Multimedia Experience (QoMEX)
60. Ronsse LM, Wang LM (2012) Effects of room size and reverberation, receiver location, and source rotation on acoustical metrics related to source localization. Acta Acust U Acust 98(5):768–775
61. Rumsey F (2002) Spatial quality evaluation for reproduced sound: terminology, meaning, and a scene-based paradigm. J Audio Eng Soc 50(9):651–666
62. Rumsey F, Zieliński S, Kassier R, Bech S (2005) On the relative importance of spatial and timbral fidelities in judgements of degraded multichannel audio quality. J Acoust Soc Am 118(2):968–976

63. Santala O, Vertanen H, Pekonen J, Oksanen J, Pulkki V (2009) Effect of listening room on audio quality in ambisonics reproduction. In: 126th convention Audio Engineering Society
64. Scheirer E, Vaananen R, Huopaniemi J (1999) Audiobifs: describing audio scenes with the mpeg-4 multimedia standard. Multimed, IEEE Trans 1(3):237–250
65. Seeber B (2003) Untersuchung der auditiven Lokalisation mit einer Lichtzeigermethode. PhD thesis, Technischen Universität München
66. Silzle A (2008) Generation of quality taxonomies for auditory virtual environments by means of systematic expert survey. Shaker Verlag, Aachen
67. Solvang A (2008) Spectral impairment of two-dimensional higher order ambisonics. J Audio Eng Soc 56(4):267–279
68. Spors S, Rabenstein R, Ahrens J (2008) The theory of wave field synthesis revisited. In: 124th convention of the Audio Engineering Society
69. Strutt JW (1907) On our perception of sound direction. Lond, Edinb, Dublin Philos Mag J Sci 13(74):214–232
70. Theile G (1980) On the localisation in the superimposed soundfield. PhD thesis, Technische Universität Berlin
71. Västfjäll D (2003) The subjective sense of presence, emotion recognition. Virtual environments. CyberPsychol Behav 6(2):181–188
72. Völk F, Mühlbauer U, Fastl H (2012) Minimum audible distance (MAD) by the example of wave field synthesis. In: Fortschritte der Akustik (DAGA)
73. Vorländer M (2007) Auralisation: fundamentals of acoustics, modelling, simulation, algorithms and acoustic virtual reality. Springer, Berlin
74. de Vries D (2009) Wave field synthesis. J Audio Eng Soc
75. Wendt F, Frank M, Zotter F (2013) Application of localization models for vertical phantom sources. In: Fortschritte der Akustik, AIA-DAGA. Meran
76. Wendt K (1963) Das Richtungshören bei der Überlagerung zweier Schallfelder bei Intensitäts- und Laufzeitstereophonie. PhD thesis, RWTH Aachen
77. Wierstorf H, Geier M, Raake A, Spors S (2013) Perception of focused sources in wave field synthesis. J Audio Eng Soc 61(1):5–16
78. Wierstorf H, Raake A, Spors S (2013) Binaural assessment of multi-channel reproduction. In: Blauert J (ed) The technology of binaural listening, Chapter 10. Springer, Berlin
79. Wightman FL, Kistler DJ (1992) The dominant role of low-frequency interaural time differences in sound localization. J Acoust Soc Am 91(3):1648–1661
80. Wittek H (2007) Perceptual differences between wavefield synthesis and stereophony. PhD thesis, University of Surrey
81. Wu Y, Abhayapala T (2009) Theory and design of soundfield reproduction using continuous loudspeaker concept. IEEE Trans Audio, Speech Lang Process 17(1):107–116
82. Zacharov N, Koivuniemi K (2001) Audio descriptive analysis & mapping of spatial sound displays. In: Proceedings of the 2001 international conference on auditory display
83. Zahorik P (2002) Assessing auditory distance perception using virtual acoustics. J Acoust Soc Am 111(4):1832–1846
84. Zahorik P, Brungart DS, Bronkhorst AW (2005) Auditory distance perception in humans: a summary of past and present research. Acta Acust U Acust 91(3):409–420
85. Zotter F (2009) Analysis and synthesis of sound-radiation with spherical arrays. PhD thesis, University of Music and Performing Arts, Graz
86. Zotter F, Frank M (2012) All-round ambisonic panning and decoding. J Audio Eng Soc 60(10):807–820
87. Zotter F, Frank M (2013) Efficient phantom source widening. Archives Acoust 38(1):27–37
88. Zotter F, Frank M, Marentakis G, Sontacchi A (2011) Phantom source widening with deterministic frequency dependent time-delays. In: International conference on digital audio effects (DAFx). Paris, France

Chapter 18
Haptics

Rahul Chaudhari, Ercan Altinsoy and Eckehard Steinbach

Abstract Haptic communications refers to the ability to touch, feel and to physically manipulate objects in a remote (real or virtual) environment via technical means. The realization of convincing haptic interactions requires a solid understanding of both kinesthetic and tactile perceptual mechanisms and stimulation principles. This chapter starts with a concise overview of the current state of knowledge in these two areas. Then, we discuss the main performance parameters for haptic interaction systems, and point towards factors that may influence QoE in haptics. So far, the quality experienced by the human during haptic interaction has been mainly evaluated via time-consuming and costly subjective tests and only recently, first preliminary approaches for objective quality evaluation have surfaced. We briefly touch upon this topic and finish the chapter with a discussion of model-based prediction of haptic feedback quality.

18.1 Introduction

Remarkable technical innovation and tremendous growth in audiovisual communications have improved the quality of experience and productivity in networked interaction between distant people, e.g,. through video conferencing. State-of-the-art commercial teleconference solutions already simulate very convincing online environments where participants experience a high sense-of-togetherness with others

R. Chaudhari (✉) · E. Steinbach
Institute for Media Technology, TU Munich, Munich, Germany
e-mail: rahul.chaudhari@tum.de

E. Steinbach
e-mail: eckehard.steinbach@tum.de

E. Altinsoy
Chair of Communication Acoustics, TU Dresden, Dresden, Germany
e-mail: ercan.altinsoy@tu-dresden.de

S. Möller and A. Raake (eds.), *Quality of Experience*,
T-Labs Series in Telecommunication Services, DOI: 10.1007/978-3-319-02681-7_18,

(e.g., [15, 23]). Sophisticated audiovisual sensing/display devices, efficient audio/video coding standards and high-capacity communication networks have driven this progress.

The ability to *physically* interact with distant objects and humans, a feature that would contribute substantially towards ultimate immersion is, however, not yet fully realized. The rapidly rising field of haptics (the sense of *touch*) is widely thought to be a big step forward in delivering this promise [50, 52, 56]. Applications that involve physical interaction with environments that are remote, inaccessible, hazardous, or too big/small in scale for a human can benefit extensively from haptic technology. Some examples include on-orbit servicing for satellites [48], space exploration [53], tele-surgery [14], deep-sea exploration [49], safety-critical situations [19, 35], tele-manufacturing [18], tele-manipulation and tele-assembly [45]. Haptics also plays a key role in virtual simulations of real (e.g., collaborative assembly/design [28]) or fantasy environments (e.g., games [43]). Haptic communication is inherently *bidirectional*, i.e. humans not only feel haptic feedback—similar to audio/video—but also, physically act upon an environment. Accordingly, a bidirectional human-centric design and analysis of haptics technology is necessary [8].

18.1.1 Haptic Perception

From a functional perspective, haptic perception can be divided into two classes [37]—kinesthetic and tactile . The kinesthetic perception informs us about the current state (position and orientation) of body parts like the head, torso, limbs, etc., as well as their movements. Forces and torques imposed on the human body alter body states (e.g., by presenting resistance to motion and/or by changing its direction), and can thereby be sensed indirectly through kinesthetic feedback.

The tactile perception, on the other hand, informs us about cutaneous (tactile) stimuli acting on our body surface. These stimuli may be simple sinusoidal or multi-tone vibrations, or highly complex stimuli that encode various physical properties—like roughness, hardness, elasticity, viscosity, etc.—of objects or fluids.

Researchers study various aspects of haptic perception like the perception bandwidth, spatio-temporal resolution, detection, discrimination thresholds, etc. through psychophysical (relating the physical to the psychological or perceived) studies. In the following sections, we describe the haptic sensory system in some detail, and also present some psychophysical results that may bear upon the design and evaluation of technical systems involving haptic communication.

18.1.1.1 Kinesthetic Perception

The kinesthetic perception of the static and dynamic state of body limbs is supplied by peripheral mechanoreceptors in the muscles, tendons and joints. These mechanoreceptors include: (1) muscle spindles that are connected in parallel with muscle fibers,

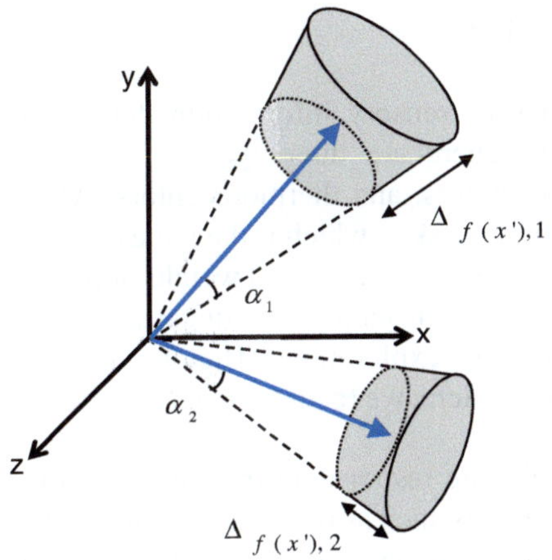

Fig. 18.1 Stimulus discrimination zones (*gray*) have been found to be non-uniform as a function of the direction of the force vector $\left(\Delta_{f(x'),1} \neq \Delta_{f(x'),2}, \alpha_1 = \alpha_2\right)$ [31]

and (2) Golgi Tendon Organs (GTOs) that are in series with muscle fibers at their connection with the skeleton. Muscle spindles encode changes in muscle length, whereas the GTOs encode changes in muscle tension. At higher levels, all these changes are integrated to form an estimate of the position/orientation of limbs. In addition to the afferent information flowing from the receptors towards the brain, the efferent information (commands flowing from the brain towards the muscles) is also considered in inferring the static and dynamic body posture [27].

A fundamental law of psychophysics—the Weber's law—states that the minimum *detectable* change in the magnitude of a stimulus is proportional to its original magnitude. This minimum change is also called the Just Noticeable Difference (JND) for that stimulus (JND—5–15 %, depending upon stimulus type and the limb/joint where it is applied [12]). Thus, a perceptual deadband exists around the stimulus magnitude, within which stimulus changes go unperceived. Human perceptual limitations captured by Weber's law have been exploited successfully in efficient transmission of kinesthetic signals over communication networks [24, 55, 56].

Today's picture of haptic perception is still not complete but with more and more investigations being performed, the individual results contribute towards a comprehensive understanding of the underlying mechanisms and limitations that can be exploited in technical systems. As an example, [31] has investigated human perceptual limitations for multidimensional kinesthetic (force) signals. Thresholds α and $\Delta_{f(x')}$ that capture human limitations in perceiving changes in force direction [9, 57] and amplitude [31, 46], respectively, were studied. It was found that the multidimensional extension of Weber's law for force signals is non-isotropic, which means that it is not uniform as a function of the direction of the force vector (see Fig. 18.1).

18.1.1.2 Tactile Perception

The tactile perception uses sensory information derived by cutaneous receptors embedded in the skin. There are three types of cutaneous sensory receptors: mechanoreceptors, nociceptors, and thermoreceptors. All three types of cells are located near the surface of the skin, which is the largest sensory organ in the body. In the average adult, it covers close to 2 m^2 and weighs about 3–5 kg [33, 47]. Hairless (glabrous) skin, which covers the palmar and fingertip regions of the body, plays the most important role in tactile explorations. Therefore, most of the research studies focus on the glabrous skin. Here, we focus on the properties of the mechanoreceptors alone.

Mechanoreceptor cells are responsible for the sensation of vibration, pressure, and object surface parameters such as roughness, shape and orientation of an object. They transduce mechanical energy into neural responses and can be grouped into two categories according to the rate of adaptation: rapidly adapting (RA) and slowly adapting (SA) mechanoreceptors. The four primary mechanoreceptors located in the glabrous skin are Pacinian corpuscles, Meissner corpuscles, Merkel cells, and Ruffini endings. RA mechanoreceptors, i.e. Pacinian corpuscles and Meissner corpuscles, only respond when the skin is moving [10]. Pacinian corpuscle (PC) fibers are found in the deep subcutaneous tissue. They sense vibrations also when the skin is compressed, but they are not sensitive to fine spatial discrimination and steady pressure. Meissner corpuscles are sensitive to low-frequency vibrations and they can detect and localize small bumps and ridges. SA mechanoreceptors consist of the Merkel cells and Ruffini endings. Merkel cells are responsible for the sensation of pressure and ruffini endings are sensitive to the stretching of skin. Being able to perceive stimuli at different frequencies is important and the frequency ranges of the receptors are [4]:

- Ruffini ending: 0.4–80 Hz,
- Merkel cell: 5–15 Hz,
- Meissner corpuscle: 10–60 Hz,
- Pacinian corpuscle: 50–1,000 Hz.

Similar to the human auditory system, the human tactile system is not equally sensitive to all frequencies. Our skin is sensitive to the frequency range from 0.4 to1000 Hz and the highest sensitivity is reached in the range of 200–300 Hz. The temporal resolution capability of the skin is high and the gap detection threshold is about 5 ms. The tactile system is capable of processing intensities up to 55 dB above the threshold. Face, tongue, and fingers are the most sensitive areas of the body. The point localization threshold of the fingertip is approximately 1 mm and lower than the two-point discrimination threshold (2–5 mm) [38]. The face has the smallest detection threshold for weight with 5 mg and highest tactile acuity for pressure.

18.1.2 Mechanical Signals that Stimulate Human Kinesthetic and Tactile Channels

In real-world interactions with objects in our immediate surroundings, we execute manual tasks and perceive object properties through direct touch. In this chapter, however, we are concerned more with how these interactions can be mediated by technical systems, so that natural or artificial haptic feedback can be delivered to humans. In the following, we describe which physical signals supply haptic information, and how they can be sensed and delivered to humans in a technology-mediated way.

18.1.2.1 Kinesthetic Signals

In technical systems, kinesthetic signals are represented by forces and torques (F/Ts) generated during physical interactions with objects. F/Ts affect the static/dynamic human body state, and can thereby be sensed indirectly through the kinesthetic sensory system.

A haptic device, such as the one in Fig. 18.2a or b, is typically used to sense human motion (position/orientation of the hand, and the corresponding velocities), and present kinesthetic (F/T) feedback in the reverse direction. The sensed human motion can be used to control either the motion of a virtual end-effector in a virtual environment (VE) (Fig. 18.2a), or that of a real end-effector at the end of a robot-arm (Fig. 18.2b). Whenever the controlled end-effector collides with an object, feedback F/Ts are either algorithmically computed (in case of a VE), or physically sensed (in case of real-world teleoperation). These F/Ts then drive the haptic device motors appropriately to display them to the human hand.

18.1.2.2 Tactile Signals

Tactile mechanical stimuli consist of vibration and pressure. Vibration is an oscillatory motion of a physical object or body that repeats itself over a given interval of time. Physical characteristics of vibration are described by amplitude (displacement), velocity, acceleration, and frequency. Each vibration can be regarded as an information carrier. Vibrations carry information on the texture of surfaces, mechanical system defects or material properties. Several researchers have concentrated on the design and construction of vibrotactile transducers for delivering vibrations to the human hand. Electromagnetic shakers, eccentric mass motors, piezoceramic actuators, bending wave actuators, surface acoustic wave (SAW) displays, and tactile pattern displays are used to generate tactile feedback. The feedback quality of the actuator is strongly influenced by the bandwidth of the device, frequency response, maximum feedback amplitude, resolution, and latency [5]. Therefore reproduction quality depends on the ability of a transducer to accurately reproduce a waveform.

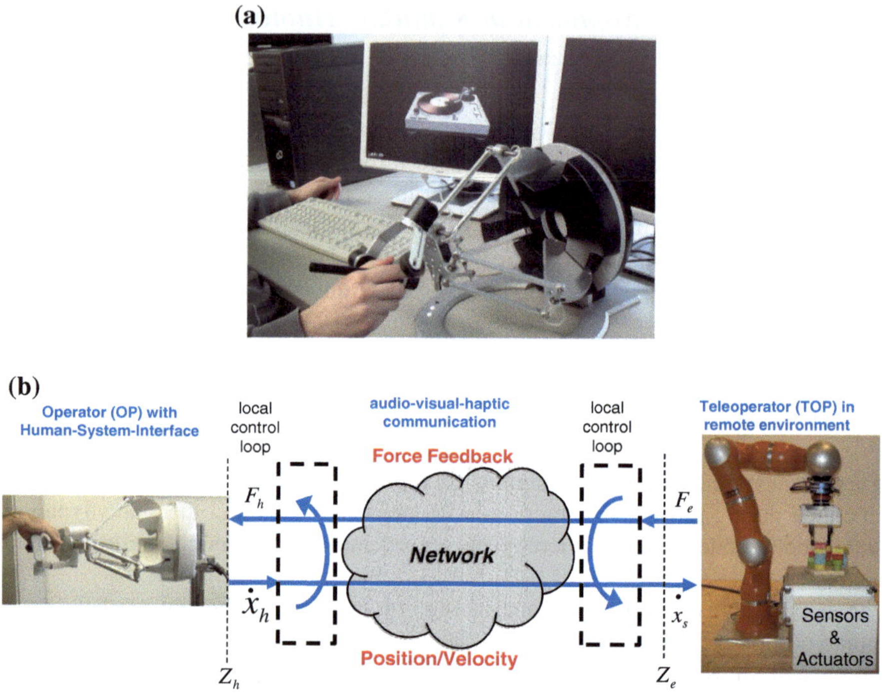

Fig. 18.2 **a** Haptic interaction with a virtual environment. A standard haptic (*force-feedback*) device, such as the one shown here, reads human motion commands through position sensors and in the other direction, displays force to the human hand via motors. **b** A haptic teleoperation system [17]. The human motion $\dot{x}_h$ is sensed at the human system interface (HSI) through the haptic device, and transmitted to the teleoperator robot over a communication network, where it serves as a reference for the robot's motion $\dot{x}_s$. When the robot comes in contact with its environment, forces/torques F_e are sensed and transmitted back to the human operator side, where they are displayed to the user through the haptic device as F_h

A detailed discussion of the reproduction characteristics of different actuator technologies can be found in [6]. Portability of the actuator and power consumption are two important design features. Users prefer generally smaller transducers because of the wear comfort.

The other physical property that is sensed by mechanoreceptors is pressure. The minimum activation force to stimulate the mechanoreceptors is 3.6 mN [60]. The excursion limit of the fingertip is approximately 3 mm. To simulate small-scale shapes, pin arrays are developed and used. Various actuator technologies are applied to develop pin arrays. Some of them are dc motors, servo motors, piezoelectric bimorphs, shape memory alloy actuators, solenoids, pneumatic actuators, and micro-electromechanical systems.

18.2 Performance Parameters and QoE in Haptic Systems

18.2.1 Performance Parameters

We classify the performance parameters of haptic systems into three categories—system-centric performance parameters (given technical limitations, what is the best a system can do), performance parameters of the system considering human psychophysical limitations (what is actually required of the system, given the limitations of human perception), and finally, human-centric performance parameters (what contributes towards the enhancement of Quality of Experience for human users when interacting with haptic systems).

18.2.1.1 System-Centric Performance

Kinesthetic haptics employs robotic hardware through which not only information, but also physical energy is exchanged between the human operator and the system. This necessitates a control theory-based treatment of design and analysis for haptic systems (which by itself ignores human factors). The control-theoretic performance parameter of *stability* is the most fundamental requirement for operability and human-safety of a haptic system [36]. With "stability", usually Bounded Input Bounded Output (BIBO) stability is implied, where a bounded input to a linear system is guaranteed to produce a bounded output from it.

The second most important parameter under this category is *system transparency* [40, 51]. Full system transparency requires that the corresponding signals (velocities and forces) on the human and the teleoperator sides be equal (see Fig. 18.2b) ($\dot{x_h} = \dot{x_s}$, $F_h = F_e$). Lawrence et al. [36] define transparency such that the mechanical impedance displayed to the human being be equal to the environment impedance ($Z_h = Z_e$). Here, the term "impedance"can be thought of as being an "opposition" to human motion. It is very frequently characterized as a mechanical mass-spring-damper model of a physical object.

Due to its fundamental trade-off with stability, transparency as defined above is impossible to achieve, especially when intermediate artifacts like communication delay are present. Based on the above requirements, typical performance metrics used are time- or frequency-domain integrals over position/force/impedance errors [34, 59, 62].

18.2.1.2 System Performance Considering Psychophysical Limitations

For kinesthetic feedback systems, Hirche et al. [25] were the first to account for limitations of human perception in evaluating system performance. They judged the telepresence system to be *perceived transparent*, if the difference between the impedance displayed to the human and the impedance of the remote environment is within the human Just Noticeable Difference $Z_h \in [Z_e - JND, Z_e + JND]$.

Also in tactile systems, an optimum interface should match (or possibly exceed) human sensory and control capabilities [5]. Psychophysical thresholds, such as tactile acuity, just noticeable level difference, and just noticeable frequency difference, define necessary information to obtain more realistic and compelling tactile interfaces. The methodologies, which are used in tactile acuity investigations, are based on classical psychophysical measurement methods, such as method of constant stimuli or method of adjustment (for a detailed overview of the methods [32]). Two key psychophysical measures are "Point of Subjective Equality" and thresholds.

18.2.1.3 Human-Centric Performance

The concept of *Presence* is the prime example of human-centric system performance characterization. It can be defined to be the feeling of "being there" or "feeling present or immersed" in a technology—(VEs or robotics in our case) mediated environment. The major factors that have been found to underlie Presence include: (1) richness of sensory information—how comprehensive, multi-modal, and consistent the sensory feedback given by the system is, (2) interactivity and responsiveness of the system, (3) the content of the experience, etc. (see [29] and the references therein). Overall, how effectively a technical system replicates (or even surpasses) our real-world experience decides how *present* a user feels in this system.

Presence being a subjective concept, it is best evaluated through psychometric tests. Traditional methods involve post-test rating scales and questionnaires (whereby the experience is had first, and then judged later), and online subjective evaluation (whereby the judgment happens during the experience). Subjective evaluation methods are often criticized for various reasons like response bias, the difficulty in constructing reliable questionnaires, attentional overload, etc. Hence, subjective results are usually supported by objective metrics like the amount of adjustment of body posture, physiological response measurement, reaction to distractions, social responses like facial expressions and gestures, etc.

18.2.1.4 Quality of Experience in Haptics

The previous three categories of performance parameters are characterized by an increasing degree of human-centricness. Thus, we predict that a good QoE metric for haptics would weight the parameters in Sect. 18.2.1.3 heavily as compared to the previous ones.

QoE Parameter Taxonomy for Haptic VEs

A comprehensive classification of QoE parameters for haptic-based VE applications has been presented in [20, 21], and the references therein. Both QoS influences and

User-Experience (UX) parameters have been considered in this taxonomy. While QoS parameters for haptics also typically involve delay, jitter, and packet loss, relevant UX parameters have been classified as: perception-related parameters, quality of haptic rendering, psychological, and physiological parameters. Perception measures reflect how the user broadly perceives the haptics-based application. The rendering quality jointly considers the rendering of graphics, audio, and haptics. The psychological and physiological parameters capture the subjective and objective user-states respectively. Examples of parameters that represent these classes are media synchronization (QoS parameter), fatigue and user intuitiveness (perception-related), haptic rendering (rendering quality parameter), and degree of immsersion (psychological).

Upto now, QoE in haptics has always been evaluated through subjective tests with the human-in-the-loop. Typically, subjects evaluate system artifacts on an Absolute Category Rating scale with gradations similar to "imperceptible", "perceptible, but not disturbing", "slightly disturbing", "disturbing", and "strongly disturbing". In the following, we discuss the limitations of subjective testing, challenges in objective quality evaluation (OQE), and a couple of OQE approaches.

18.3 Model-Based Prediction of Haptic Feedback Quality

Development of novel human-machine systems necessitates extensive subjective testing, which is difficult and very time-consuming. In haptics, it is especially so, since customized hardware makes it difficult to parallelize tests. Also, since subjects are typically not used to being delivered artifical haptic stimuli, extensive experimenter monitoring is required. This slows down the progress of technical developments. To circumvent this problem, algorithmic models of human perception are employed for performance evaluation. Here, we outline some challenges facing the development of human haptic models, and progress made in this area.

18.3.1 Methods and Models

18.3.1.1 Kinesthetic Feedback

Action-Perception Modeling

The uncertainty involved in a haptic system due to the bidirectional human-in-the-loop nature of haptics makes it nearly impossible to perfectly reproduce haptic signals involved in a task in two separate sessions. Therefore, it is not possible to have a fair comparison of haptic signals from two separate runs, making signal-based evaluation of effects of a system component on quality difficult. Hence, unlike audio and video, in addition to a perception model, another important challenge in haptic quality prediction is the development of a human action model that can replace the active human behavior from real experiments in a software simulation [13].

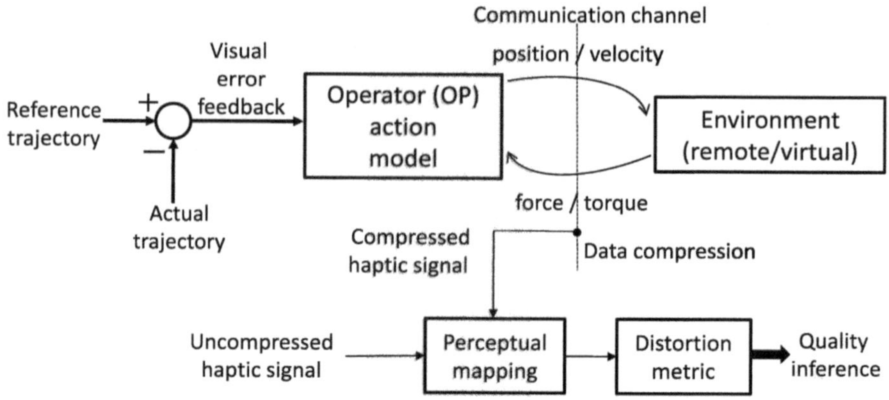

Fig. 18.3 Overview of the objective quality assessment framework introduced in [13]. Similar haptic experiences can be imposed on subjects by specifying the same reference trajectory to be followed in the compensatory tracking task for every subject. The trajectory error that they should try to minimize is displayed to them visually. With data from these tasks, values for the action model parameters are identified, the model is simulated, and the uncompressed and compressed versions of the haptic signals are recorded. These are then compared in the perceptual domain using the PMSE metric (Eq. 18.1)

A fully automatic objective[1] quality prediction framework for the compression of haptic signals is proposed in [13]. This framework is based on partial models for the central nervous system and the neuromuscular arm of the operator (action model), and a haptic perception model. These models are identified with data from manual control tasks (compensatory tracking) [22]. This (action) model allows us to simulate the entire telemanipulation experiment in software, and to predict haptic quality objectively using a perception model. Figure 18.3 illustrates this concept.

The work in [13] specificially attempts to do a model-based prediction of the quality of a compressed haptic signal relative to the uncompressed one. In this direction, haptic interaction is first simulated with compression of the haptic data turned OFF. The haptic sample sequence recorded here represents the undistorted reference signal. Then, the simulation is repeated with the haptic compression turned ON. The haptic samples recorded here represent the distorted signal. A perceptual comparison is then made between the two sequences using the Perceptual Mean-Square Error (PMSE) defined below, based on the Weber–Fechner law [61]:

$$PMSE = \frac{1}{N}\sum_{i=0}^{N-1}\left(S(i) - \hat{S}(i)\right)^2, \quad \text{where } S(i) = c \cdot ln\left(\frac{I(i)}{I_0}\right) \tag{18.1}$$

where N is the number of samples, S the sensation the human experiences as a function of the applied haptic stimulus, c a scaling constant that needs to be determined

[1] Here, "objective" implies algorithmic or mathematical prediction of subjective quality.

experimentally, I the magnitude of the applied stimulus, and I_0 the absolute detection threshold.

With the above models, the quality-prediction results show a (decreasing) quality trend similar to that from subjective tests, as the strength of the applied compression increases.

User-Experience Modeling

In [20, 21], a holistic system-level mathematical model for haptic QoE based on weighted linear combinations of QoS and User-Experience (UX) parameters (see Sect. 18.2.1.4) has been presented and validated:

$$QoE = \zeta \times QoS + (1 - \zeta) \times UX \tag{18.2}$$

where

$$QoS = \frac{\Sigma_l(\eta_l S_l)}{\Sigma_l(\eta_l)} \tag{18.3}$$

and

$$UX = A\frac{\Sigma_i(\alpha_i P_i)}{\Sigma_i(\alpha_i)} + B\frac{\Sigma_j(\beta_j R_j)}{\Sigma_j(\beta_j)} + C\frac{\Sigma_k(\gamma_k U_k)}{\Sigma_k(\gamma_k)} \tag{18.4}$$

where ζ can be used to prioritize QoS parameters versus user experience parameters. S_l, P_i, R_j, and U_k represent individual quality values for QoS measures, perception measures, rendering quality measures, and user state measures, respectively. $\eta_l, \alpha_i, \beta_j, \gamma_k$ are weighting factors which depend on the relative quality values of individual QoS and user experience parameters. Weightings A, B, C are determined empirically. Optimal weighting factors have been determined in [21], which have led to quality estimates with a high correlation with the subjective ratings (correlation coefficient 0.92, with $p < 0.005$).

18.3.1.2 Tactile Feedback

Quality judgments are based on physical (i.e., elements) and perceptual (i.e., features) layers. The quality elements and features of tactile interfaces are summarized and a quality model was presented in [5] (Fig. 18.4). However the weightings of the individual features on the quality judgment are context dependent and as until today there are very few empirical data on it.

User dependent factors, such as expectations, experiences, motivations, memories, emotions, attitudes, familiarity with the interface, and particularly fun, novelty, and ease of use play an important role on quality judgments.

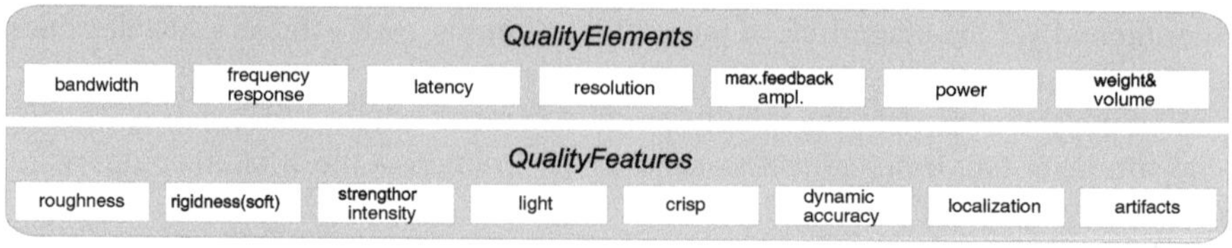

Fig. 18.4 The quality elements and features of tactile interfaces according to [5]

The use of touch sensitive displays and touch surfaces is just emerging and they are more and more replacing physical buttons [3]. A big disadvantage of such kind of displays is the missing tactile feedback which is required for the confirmation of the successful operation. The missing tactile feedback causes usage errors and quality problems. Recent studies revealed that tactile feedback enhances the usability and the quality of a handheld device with touch screen compared to a device without tactile feedback [11, 26, 41].

Improving the haptic exploration and identification of virtual shapes and objects provides great potential for numerous applications. However, many challenges exist because subjects experience various difficulties during the exploration and recognition process of virtual models when using a standard force-feedback device. The results of a recent study showed that additional tactile and auditory feedback can enhance haptic exploration and object identification [54]. The frequency alteration, temporal/spatial information and the amplitude modulation frequency are perceptually relevant cues for object recognition.

Apart from the interface quality, another important issue is the tactile "feel" or "appeal" of a product [4]. People are exposed to many forms of vibration from different products by using them, e.g., vibrations from vibrating tools (drill, electric-razor, hand mixer, vacuum cleaner, etc.), vibrations from vehicles, or vibrations from musical instruments (guitar, drum, etc.). Therefore, the research related to human response to product vibration is becoming increasingly important. Number of studies evaluated the product vibration quality [4, 7, 44, 58]. The identification of relevant descriptors, which characterize the tactile experiences, is pivotal in the tactile quality evaluation. Recently, an investigation was conducted to define the tactile verbal descriptors and establish a vocabulary [42]. The elicited descriptors for tactile experiences associated with human hand were tapping, prickly, tingling, strong (weak), pointed, rhythmic, constant, coming and going, pulsing, flowing, breeze, pulse of air, and dispersed. Frequency and temporal properties were observed to play an important role in the selection of attributes. A similar investigation was conducted for whole-body vibrations [2]. The results show that different perceptual properties were used depending on the frequency of the sinusoidal whole-body vibration signals. The attribute "bumpy" was found suitable for the low frequencies (8–30 Hz). The middle frequencies (up to 75 Hz) were characterized with the attribute "shaky" and high frequencies (75–300 Hz) were characterized with the attribute "humming". The attribute "rattling" was chosen as suitable for low modulation frequencies and the attribute "wavy" for high frequencies. For the impulsive signals, the attribute "beating" is suitable. Descriptor selection behavior of the participants shows that our experiences with vehicles play an important role. For example the property "bumpy" is used mostly to describe the whole-body vibrations which were generated by vehicles. Recent studies showed that acceleration level is insufficient to describe the product vibration quality [4, 7]. Therefore multidimensional characteristic of vibrations are taken into consideration to develop models based on abovementioned tactile features. However the investigation of the relationship between tactile features and signal properties is still necessary to develop effective prediction models.

Interest in human responses to whole-body vibration has grown, particularly due to the increasing usage of vehicles, e.g., cars, trucks, and helicopters etc. In recent years, a number of quality evaluation experiments were conducted on the vehicle seat and steering wheel vibrations [1, 7, 44, 58]. Another reason for the growing interest in recent years is the importance of the vibrations generated by the performance of music. The floor or the chair can vibrate because of the resonance or the structure-borne sound stimulated by instruments [16]. A recent study revealed that synchronous presentation of vertical whole body vibrations during concert DVD reproduction can improve the perceived quality of the concert experience [39].

18.4 Conclusion

In this chapter, besides offering the reader a concise background on haptics and haptic technology, we have elaborated on the objective and subjective performance of technical systems with haptics. We have also presented initial ideas on the characterization of a haptic QoE. We contend that the human factors as outlined in Sect. 18.2.1.3 weigh heavily in determining and enhancing QoE in haptics. Furthermore, we have described methods and models for predicting haptic quality objectively.

Some milestones in the characterization and evaluation of haptic QoE have been reached, e.g., the development of action and perception models for haptic interaction, identification of factors underlying QoE, etc. The state-of-the-art, however, needs to be extended further in several directions. For example, for kinesthetic feedback, the human action and perception models need to become more sophisticated to be able to close the quality-prediction gap to the results obtained from subjective tests. Research also needs to be concentrated on more realistic scenarios involving real-world telemanipulation systems. Finally, joint QoE evaluation methodologies for the audio, visual, and haptic modalities should be developed. On the tactile side, significant progress has been made in identifying and relating physical elements and perceptual features of tactile signals, and in integrating them to predict overall tactile quality. However, the need for further refinement of these relationships and models is still felt. Moreover, additional empirical data in a variety of contexts are needed to determine appropriate application-specific weighting of perceptual features in determining quality. In general, wide consensus on a QoE definition and the factors underlying it, specifically for haptics, does not yet exist. Future research should address both these issues.

Acknowledgments This work was supported, in part, by the German Research Foundation (project STE 1093/4-2) and, in part, by the European Research Council under the European Union's Seventh Framework Programme (FP7/2007-2013)/ERC under Grant 258941.

References

1. Ajovalasit M, Giacomin J (2007) Effect of automobile operating condition on the subjective equivalence of steering wheel vibration and sound. Int J Veh Noise Vib 3(2):197–215
2. Altinsoy ME (2011) What can we learn from psychoacoustics regarding the perception of whole-body vibrations? Tactile descriptors for whole-body vibrations. In: Proceedings of 2nd Polish-German structured conference on acoustics. The 58th open seminar on acoustics. Jurata, Poland
3. Altinsoy ME, Merchel S (2009) Audiotactile feedback design for touch screens. In: Haptic and audio interaction design. Springer, Berlin, pp 136–144
4. Altinsoy ME (2006) Auditory-tactile interaction in virtual environments. Shaker Verlag, Aachen
5. Altinsoy ME (2012) The quality of auditory-tactile virtual environments. J Audio Eng Soc 60(1/2):38–46
6. Altinsoy ME, Merchel S (2012) Electrotactile feedback for handheld devices with touch screen and simulation of roughness. IEEE Trans Haptics 5(1):6–13
7. Altinsoy ME (2013) Identification of quality attributes of automotive idle sounds and whole-body vibrations. Int J Veh Noise Vib 9(1/2):4–27
8. Aracil R, Buss M, Cobos S, Ferre M, Hirche S, Kuschel M, Peer A (2007) Advances in telerobotics: the human role in telerobotics. Springer, Berlin (Chap. 1)
9. Barbagli F, Salisbury K, Ho C, Spence C, Tan H (2006) Haptic discrimination of force direction and the influence of visual information. ACM Trans Appl Percept 3(2):135–143
10. Barlow HB, Mollon JD (1982) The senses. Cambridge University Press, Cambridge
11. Brewster S, Chohan F, Brown L (2007) Tactile feedback for mobile interactions. In: Proceedings of SIGCHI conference on human factors in computing systems, CHI '07. ACM, New York, pp 159–162
12. Burdea GC (1996) Force and touch feedback for virtual reality. Wiley, New York
13. Chaudhari R, Steinbach E, Hirche S (2011) Towards an objective quality evaluation framework for haptic data reduction. In: Proceedings of IEEE world haptics conference. Istanbul, Turkey, pp 539–544
14. Chauhan S, Coelho RF, Kalan S, Satava RM, Patel VR (2012) Robotic urologic surgery: evolution of robotic surgery: past, present, and future, Chap. 1. Springer, Berlin
15. Cisco Systems Inc (2011) TelePresence. http://www.cisco.com/en/US/products/ps7060/index.html
16. Daub M, Altinsoy E (2004) Audiotactile simultaneity perception of whole-body vibrations produced by musical presentations. In: Proceedings of CFA/DAGA
17. Ferrell W, Sheridan T (1967) Supervisory control of remote manipulation. IEEE Spectr 4(10):81–88
18. Fischer A, Barhak J (2001) Tele-design for manufacturing. CIRP Ann-Manuf Technol 50(1):77–80
19. Fujii Y, Usui H, Shinohara Y (1992) Development of multi-functional telerobotic systems for reactor dismantlement. J Nucl Sci Technol 29(9):930–936
20. Hamam A, Eid M, Saddik AE, Georganas ND (2008) A quality of experience model for haptic user interfaces. In: Proceedings of workshop haptic user interfaces (Ambient Media Systems), pp 1–6
21. Hamam A, El Saddik A (2013) Toward a mathematical model for quality of experience evaluation of haptic applications. IEEE Trans Instrum Meas 62(12):3315–3322

22. Hess RA (1980) Structural model of the adaptive human pilot. J Guidance Control Dyn 3(5):416–423
23. Hewlett-Packard Development Company, L.P (2011) Introducing Halo. http://www.hp.com/halo/introducing.html
24. Hinterseer P, Hirche S, Chaudhuri S, Steinbach E, Buss M (2008) Perception-based data reduction and transmission of haptic data in telepresence and teleaction systems. IEEE Trans Signal Process 56(2):588–597
25. Hirche S, Ferre M, Barrio J, Melchiorri C, Buss M (2007) Advances in telerobotics: Bilateral control architectures for telerobotics, Chap. 10. Springer, Berlin
26. Hoggan E, Brewster SA, Johnston J (2008) Investigating the effectiveness of tactile feedback for mobile touchscreens. In: Proceedings of SIGCHI conference on human factors in computing systems. ACM, New York, pp 1573–1582
27. von Holst E (1954) Relations between the central nervous system and the peripheral organs. British J Anim Behav 2:89–94
28. Iglesias R, Casado S, Gutirrez T, Garca-Alonso A, Yu W, Marshall A (2008) Simultaneous remote haptic collaboration for assembling tasks. Multimedia Syst 13(4):263–274
29. Ijsselsteijn WA, de Ridder H, Freeman J, Avons SE (2000) Presence: concept, determinants and measurement. Proc SPIE 3959:520–529
30. Jekosch U (2004) Basic concepts and terms of "quality", reconsidered in the context of product-sound quality. Acta Acustica united with Acustica 90(6):999–1006
31. Kammerl J, Chaudhari R, Steinbach E (2010) Exploting directional dependencies of force perception for lossy haptic data reduction. In: Proceedings of IEEE international symposium on haptic audio-visual environments and games (HAVE). Phoenix, AZ, USA, pp 1–6
32. Kingdom FAA, Prins N (2009) Psychophysics: a practical introduction. Academic Press, London
33. Klatzky R, Lederman S (2003) Handbook of psychology: touch, Chap. 2. Wiley, New York
34. Kron A (2004) Beiträge zur bimanuellen und mehrfingrigen haptischen Informationsvermittlung in Telepräsenzsystemen. PhD. thesis, Technische Universität München, Institute of Automatic Control Engineering
35. Kron A, Schmidt G, Petzold B, Zäh MI, Hinterseer P, Steinbach E (2004) Disposal of explosive ordnances by use of a bimanual haptic telepresence system. In: Proceedings of IEEE international conference on robotics and automation, vol 2, pp 1968–1973
36. Lawrence DA (1993) Stability and transparency in bilateral teleoperation. IEEE Trans Robot Autom 9(5):624–637
37. Loomis J, Lederman S (1986) Tactual perception: handbook of perception and human performance, Chap. 31. Wiley, New York
38. Louis D, Greene T, Jacobson K, Rasmussen C, Kolowich P, Goldstein S et al (1984) Evaluation of normal values for stationary and moving two-point discrimination in the hand. J Hand Surg 9(4):552–555
39. Merchel S, Altinsoy ME (2009) Vibratory and acoustical factors in multimodal reproduction of concert dvds. In: Haptic and audio interaction design. Springer, Berlin, pp 119–127
40. Minsky M (1980) Telepresence. Omni Magazine
41. Nashel A, Razzaque S (2013) Tactile virtual buttons for mobile devices. In: CHI'03 extended abstracts on human factors in computing systems. ACM, New York, pp 854–855
42. Obrist M, Seah SA, Subramanian S (2013) Talking about tactile experiences. In: Proceedings 2013 ACM annual conference on human factors in computing systems, CHI '13. ACM, New York, pp 1659–1668
43. Orozco M, Silva J, Saddik AE, Petriu E (2012) Haptics rendering and applications: the role of haptics in games, Chap. 11. InTech, pp 217–234
44. Parizet E, Amari M, Nosulenko V (2007) Vibro-acoustical comfort in cars at idle: human perception of simulated sounds and vibrations from 3-and 4-cylinder diesel engines. Int J Veh Noise Vib 3(2):143–156
45. Peer A, Unterhinninghofen U, Buss M (2006) Tele-assembly in wide remote environments. In: 2nd international workshop on human-centered robotic systems, pp 2–8

46. Pongrac H, Färber B, Hinterseer P, Kammerl J, Steinbach E (2006) Limitations of human 3d force discrimination. In: Human-centered robotics systems. Munich, Germany
47. Quilliam TA (1978) Active touch: the mechanism of recognition of objects by manipulation, vol 1. Pergamon Press, Oxford (Chap. The structure of finger print skin)
48. Reintsema D, Landzettel K, Hirzinger G (2007) Advances in telerobotics: DLR's advanced telerobotic concepts and experiments for on-orbit servicing, Chap. 19. Springer, Berlin
49. Ridao P, Carreras M, Hernandez E, Palomeras N (2007) Advances in telerobotics: underwater telerobotics for collaborative research, Chap. 20. Springer, Berlin
50. El Saddik A (2007) The potential of haptics technologies. IEEE Instrum Meas Mag 10(1):10–17
51. Sheridan TB (1992) Musings on telepresence and virtual presence. Presence: teleoperators and virtual environments 1(1):120–126
52. Sheridan TB (1992) Telerobotics, automation, and human supervisory control. MIT Press, Cambridge
53. Spudis PD, Taylor GJ (1992) The roles of humans and robots as field geologists on the moon. In Lunar bases and space activities of the 21st century, vol 1, pp 307–313
54. Stamm M, Altinsoy M (2013) The technology of binaural listening. Springer, Berlin, pp 449–475 (Chap. Assessment of binaural-proprioceptive interaction in human-machine interfaces)
55. Steinbach E, Hirche S, Kammerl J, Vittorias I, Chaudhari R (2011) Haptic data compression and communication for telepresence and teleaction. IEEE Signal Process Mag 28(1):87–96
56. Steinbach E, Hirche S, Ernst M, Brandi F, Chaudhari R, Kammerl J, Vittorias I (2012) Haptic communications. Proc IEEE 100(4):937–956
57. Tan H, Barbagli F, Salisbury K, Ho C, Spence C (2006) Force-direction discrimination is not influenced by reference force direction. Haptics-e 4(1):1–6
58. Vstfjll D (2013) Affect as a component of perceived sound and vibration quality in aircraft. PhD. thesis, Chalmers University of Technology
59. Wang X, Liu PX, Wang D, Chebbi B, Meng M (2005) Design of bilateral teleoperators for soft environments with adaptive environmental impedance estimation. In: Proceedings of IEEE international conference on robotics and automation. Barcelona, Spain, pp 1139–1144
60. Yang TH, Kim SY, Kim CH, Kwon DS, Book WJ (2009) Development of a miniature pin-array tactile module using elastic and electromagnetic force for mobile devices. In: EuroHaptics conference 2009 and 3rd joint symposium on haptic interfaces for virtual environment and teleoperator systems, IEEE, pp 13–17
61. Yilmaz H (1964) On the laws of psychophysics. Bull Math Biol 26:235–237
62. Yokokohji Y, Yoshikawa T (1994) Bilateral control of master-slave manipulators for ideal kinesthetic coupling-formulation and experiment. IEEE Trans Robot Autom 10(5):605–620

Chapter 19
Video Streaming

Marie-Neige Garcia, Savvas Argyropoulos, Nicolas Staelens, Matteo Naccari, Miguel Rios-Quintero and Alexander Raake

Abstract This chapter addresses QoE in the context of video streaming services. Both reliable and unreliable transport mechanisms are covered. An overview of video quality models is provided for each case, with a focus on standardized models. The degradations typically occurring in video streaming services, and which should be covered by the models, are also described. In addition, the chapter presents the results of various studies conducted to fill the gap between the existing video quality models and the estimation of QoE in the context of video streaming services. These studies include work on audiovisual quality modeling, field testing, and on the user impact. The chapter finishes with a discussion on the open issues related to QoE.

M.-N. Garcia (✉) · S. Argyropoulos · M. Rios-Quintero · A. Raake
Assessment of IP-based Applications, Telekom Innovation Laboratories, TU Berlin, Berlin, Germany
e-mail: marie-neige.garcia@telekom.de

S. Argyropoulos
e-mail: savvas@ieee.org

M. Rios-Quintero
e-mail: miguel.rios-quintero@telekom.de

A. Raake
e-mail: alexander.raake@telekom.de

N. Staelens
Department of Information Technology, Internet Based Communication Networks and Services, Ghent University—iMinds, Ghent, Belgium
e-mail: nicolas.staelens@intec.ugent.be

M. Naccari
BBC R&D, London, UK
e-mail: matteo.naccari@bbc.co.uk

S. Möller and A. Raake (eds.), *Quality of Experience*,
T-Labs Series in Telecommunication Services, DOI: 10.1007/978-3-319-02681-7_19,

19.1 Introduction

With the multitude of video transmitted across the internet infrastructure, video QoE is of large interest for users, internet service or content providers, and component manufacturers alike. As a consequence, video-related services have received a lot of attention by research and development activities over the past years. As with other media applications such as audio entertainment or speech communication, three principle types of quality assessment can be distinguished:

1. Explicit quality tests with users evaluating respective sequences in laboratory tests,
2. instrumental quality estimation algorithms, also called quality models, or
3. possible additional consideration of context and user behavior in conjunction with the assessment of technical performance parameters, for example capturing service and user data during large-scale service use [17].

As for the case of other media, test methods involving human viewers can be distinguished according to the presentation method (method of constants vs. method of adjustment, and use of single versus multiple stimuli), and the scale being used for judgment. For an overview of subjective test methods, the interested reader is referred to [98], Chap. 4, and to the ITU recommendations [19, 30, 31, 45–47]. In addition, discussions and comparisons of methods can be found in [10, 27, 72, 107].

Quality models have been developed to complement or replace time-consuming and expensive viewing tests. For an overview, cf. e.g. [9, 75, 97].

The categorization of algorithms based on the type and amount of information employed for the quality assessment is depicted in Fig. 19.1, modified from [75]. From this figure, it can be seen that the quality models can be categorized in terms of:

- the amount of reference information they employ: No-Reference (NR), where the models do not have access to the original non-degraded signal, Reduced-Reference (RR), where the models have access to features extracted from the original signal (see box "Feature extraction" in Fig. 19.1), and Full-Reference (FR), where the models have access to the original signal ("Original source" in Fig. 19.1),
- the type of information that is used for quality predictions: Signals ("signal-based model" in Fig. 19.1) and/or transmission-related parameters extracted from packet-header- ("Parametric model" in Fig. 19.1) or bitstream- information ("Bitstream model" in Fig. 19.1). Hybrid models take as input signal (pixel), bitstream, and/or packet-header information,
- the extent to which they include the explicit modeling of the human visual system.

This chapter mainly addresses quality models focusing on IP-based video streaming applications and the most widely used codec in this context, H.264 [38]. However, the scope of the presented models is broader: Signal-based models are usually not restricted to H.264, and they can be applied universally. Also, the structure of the other types of models make them generally adaptable to other codecs and network types.

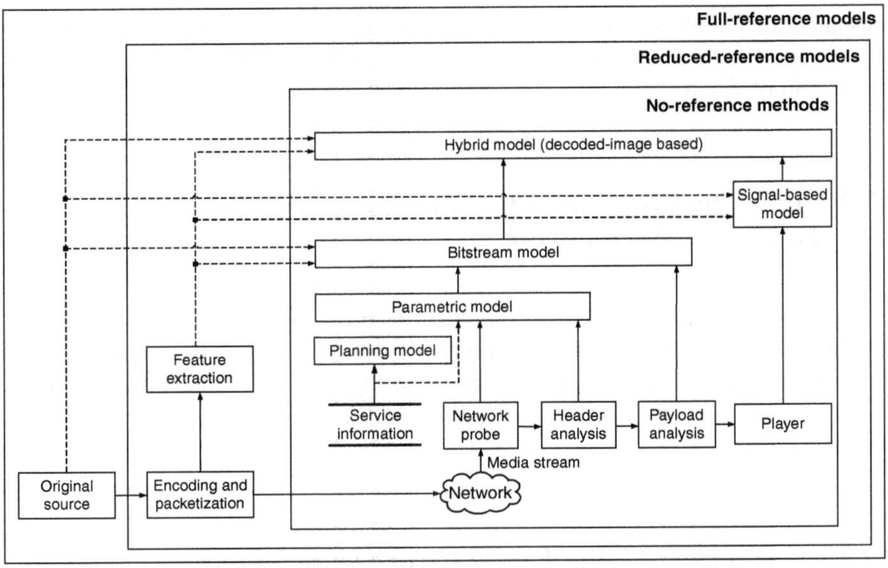

Fig. 19.1 Categorization of video quality assessment algorithms, adapted from [75]

Typically, video transmission over IP networks (e.g. for broadcast TV as in IP-based Television—IPTV) is performed using unreliable transport mechanisms, such as the Real Time Protocol (RTP) in conjunction with the User Datagram Protocol (UDP), to ensure limited delay and real-time operation. However, with the increase of available network bandwidth, multimedia content can now be delivered very efficiently using TCP and typically HTTP (e.g. in the case of Youtube or other so-called over-the-top services), which enables traffic reduction by the efficient usage of caches, for example in the vicinity of end-users. Different considerations for QoE in the context of UDP-based versus TCP-based transmission are discussed in Sects. 19.2 and 19.3, respectively.

The chapter is structured as follows: Sect. 19.2 summarizes the video degradations that are encountered in the case of streaming with unreliable transport, and respective approaches for instrumental assessment. These include the impact due to video coding as well as the packet-based transmission, possible packet loss and its concealment. Here, different types of models are outlined, including packet-header-based models for network planning and monitoring, bitstream-based models, pixel-based models and hybrid models. Sect. 19.3 discusses the differences between streaming over unreliable and over reliable channels, and how models initially developed for video streaming with unreliable transport can be used here, and which additional components are required to also handle adaptive streaming or re-buffering. In Sect. 19.4, the rather technical approach followed up to that point is re-considered, and current trends towards a more QoE-centric assessment of video streaming services are presented, including added modalities and audiovisual assessment, field rather than lab testing, and the impact due to the type of user. Finally, in Sect. 19.5, future work in the field of video streaming QoE assessment is discussed.

19.2 Video Quality Models for Streaming with Unreliable Transport Mechanisms

This section introduces the main degradations occurring due to compression or packet loss in the case of unreliable transport mechanisms. It also provides an overview of the different types of video quality models, with a focus on standardized models. Packet-, bitstream-, pixel-based, and hybrid models are addressed, as well as full-, reduced-, and no-reference models.

19.2.1 Video Coding and Packet Error Degradations

Blockiness, also referred to as *block distortion* or *tiling* [48], is a distortion of the image characterized by the appearance of an underlying block encoding structure. Block distortions are caused by coarse quantization. They comprise other identified degradations such as the *staircase effect*, *mosaic pattern effect* or the *DCT basis-image effect* [106]. In modern encoders, coarsely quantized block boundaries are usually filtered, reducing the visibility of the above artifacts but leading to *blurriness*, which is characterized by reduced sharpness of edges and spatial details [48].

In order to reduce the amount of video information that the system is required to transmit or process per unit of time, video frames may be skipped at the encoding stage. This may result in *jerkiness*, which is defined in [48] as a "motion which was originally smooth and continuous, but is now perceived as a series of distinct "snapshots" of the original scene". It is often observed in the case of high motion scene.

During the encoding process, video frames are assigned different types, which are called "I-frames", "P-frames" and "B-frames". The perceptual impact of packet loss depends on the type of the frame in which the loss occurs. Indeed, P- and B-frames are predicted from previous I- and P-frames, while I-frames are intra-coded and therefore do not depend on previous frames. As a consequence, if a loss occurs on an I- or a P-frame, the loss is typically propagated till the next I-frame. If a loss occurs on a reference B-frame, which is used in hierarchical coding, the loss propagates till the next P- or I-frame, i.e. it only affects the surrounding non-reference B-frames. There is no loss propagation if the loss occurs on a non-reference B-frame, since it is not referenced by other frames.

The perceptual impact of packet loss also depends on the packet loss concealment applied by the decoder. If *slicing* is applied as packet loss concealment, one packet

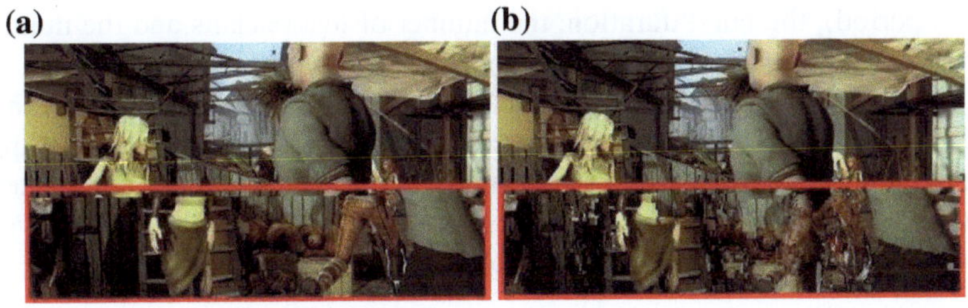

Fig. 19.2 Effect of packet loss when one slice per frame is used. **a** Loss occurred in the current frame. **b** Loss propagated from previous frames

loss results in the loss of the corresponding pixel-area as well as the pixel-area corresponding to the rest of the affected slice (see red rectangle in Fig. 19.2a). The decoder re-synchronizes its bitstream parsing process at the beginning of the next slice, using the slice header. Therefore, the spatial extent of the loss depends on the number of slices per frame. For instance, and as shown in Fig. 19.2a, the video was encoded with one slice per frame, the loss of a packet yielded the loss of the rest of the frame, and the lossy area was replaced by the same area in the last reference frame.

In the case of slicing, packet loss may also yield *blockiness* effect, as shown in the red rectangle area of Fig. 19.2b. Indeed, when loss occurs, content from the previous reference frame is usually copied to the lost portion of the hit frame. If there is motion in the sequence, this replaced content will not fit well the missing content. In the subsequent frames, the whole lost block of pixels will be displaced, resulting in a blocking artifact.

Another loss handling strategy is *freezing with skipping*. In this case, the frames affected by loss are completely discarded. According to the encoding process described above, the frames referencing the lossy frames are also discarded, and all discarded frames are replaced by the last unimpaired reference frame. The video is therefore perceived as frozen.

19.2.2 Packet-Based Models for Network Planning

In the case of network planning, parameters of the service to be deployed cannot be measured. Instead, planning assumptions are made. Typical model input parameters used for network planning are the average bitrate and the percentage of packet loss, as proposed by [91] for variable bitrate, random loss and motion-compensated transmission-error concealment. Since the impact of packet loss depends on the loss distribution [7, 21], burst-length-related parameters are commonly used, as presented in [89] for small video formats and in [21] for IPTV and High Definition (HD) video. In [104], the authors improved the ITU-T G.1070 model [36] by including a burstiness parameter computed from the burst density (fraction of lost or discarded packets in a burst period), the burst duration, the number of lost packets and the number of burst periods. Alternatively, the packet loss frequency may be used instead of the packet loss percentage and packet burst length [101]. Since the perceptual impact depends on the applied packet loss concealment (slicing or freezing with skipping) and on the number of slices per frame (in the case of slicing), both the packet loss concealment and the number of slices per frame should be considered by the model, as presented in [21] for the IPTV scenario.

19.2.3 Packet-Based Models for Service Monitoring

In the case of service monitoring, model parameters are extracted from the bitstream. The most applicable case of an encrypted bitstream is considered in this section. In this case, the model does not have access to the payload or to the pixel information.

The network planning models can be used in that case. However, additional information may be extracted from the encrypted bitstream such as the video frame boundaries and the video frame types (I-, P- or reference/non-reference B-frames). This allows a more accurate parametric description of the degradations and of the influence of the content.

Indeed, it is known that the impact of packet loss depends on the type of the frame in which the loss occurs [91]. In particular, a loss propagates till the next I-frame when it occurs in an I- or a P-frame (in contrast to a B-frame, see also Sect. 19.2.1). The number of impaired frames is therefore more appropriate than the percentage of packet loss for capturing the quality impact due to packet loss, as proposed in [100] and [102] for slicing, and in the ITU-T P.1201.2 standard [43], for both slicing and freezing with skipping.

As previously mentioned, the spatial extent of the loss depends on the slice size in the case of slicing. This spatial extent is computed in both the ITU-T P.1201.2 and in [20]. In the ITU-T P.1201.2 standard, the spatial extent of the loss is combined with the loss duration in a single parameter describing the whole degradation for directly predicting the perceived quality.

It has often been observed that the spatio-temporal complexity of content influences the quality impact due to coding artifacts and the visibility of the loss. For instance, slicing degradations are more visible in the case of panning or complex movements than in the case of almost static-content. In the case of an encrypted stream, this content complexity may be captured by the frame sizes and frame types, as proposed in [42, 43, 63, 100, 102, 103]. In particular, in the case of coding degradations, the I-frame sizes are used [42, 43], reflecting the observation that high I-frame sizes indicate low content complexity at low bitrates. The ratios between B-, P- and I-frame sizes may also be used [43, 103]. In [43], these ratios capture the observation that similarly small B- and P-frames sizes, compared to I-frame sizes, indicate low temporal complexity.

19.2.4 Bitstream-Based Models

Bitstream-based objective quality models predict video quality using the encoded video as it is transmitted through the network (see Fig. 19.1). These models parse the video bitstream without reconstructing the pixel information and are particularly interesting for monitoring video quality at any point in the distribution network (network-based monitoring). However, parsing the bitstream requires coder-specific implementations of each model.

By analyzing the encoded video bitstream beyond the packet headers as described in the previous section, additional information can be gathered at the frame level. For example, quantization and motion vector information can be obtained by parsing the video data [59, 82]. This provides a first indication of the video quality and video content characteristics [64].

In [65], quality is estimated in the compressed domain by considering the Quantization Parameter (QP), the motion and the bitrate allocation of inter macroblocks. Similar information has also been used in [59].

Besides predicting video quality, bitstream-based models have also been constructed to estimate the visibility of impairments due to, for example, packet loss in the network [2, 55, 56, 77, 87]. This approach can be used to verify if the delivery network is able to provide adequate QoE to the end-users [18, 37]. The bitstream features which are typically used to detect the visibility of loss are the duration and extent of the propagated error, the motion and the residual error of the degraded area in order to account for the spatio-temporal complexity of the video sequence and capture the perceptual effects caused by packet loss. Then, the prediction of visibility may be performed using classifications methods, such as generalized linear models [55, 56, 77], support vector regression [2], or binary trees [87].

Bitstream based quality assessment models are standardized in ITU-T Rec. P.1202 [32], where P.1202.1 [33] refers to lower resolutions (from QCIF to HVGA) and P.1202.2 [34] refers to higher resolutions (from Standard Definition (SD) to HD). In P.1202.1, compression artifacts are computed based on the QP, the key frame rate, the frame rate, and the motion vector magnitude. Similarly, in P.1202.2, the compression degradations are computed based on two parameters: the average QP of the sequence and a parameter which denotes the content complexity (computed based on the bits per pixel and the QP).

The perceived distortion of slicing degradations depends on the effectiveness of the employed error concealment technique. Thus, the *level of visible artifacts* is computed based on the motion information, the residual energy of the erroneous area, and the error propagation extent to indicate how annoying is the slicing artefact. This reflects the principle that the parts of the sequence which can be easily predicted (e.g. low texture, low motion) can be efficiently concealed.

The distortion caused by each freezing event is computed based on the freezing duration and a motion term to reflect the fact that a freezing in the fast moving part of the video results in larger jerkiness and therefore causes larger perceptual annoyance.

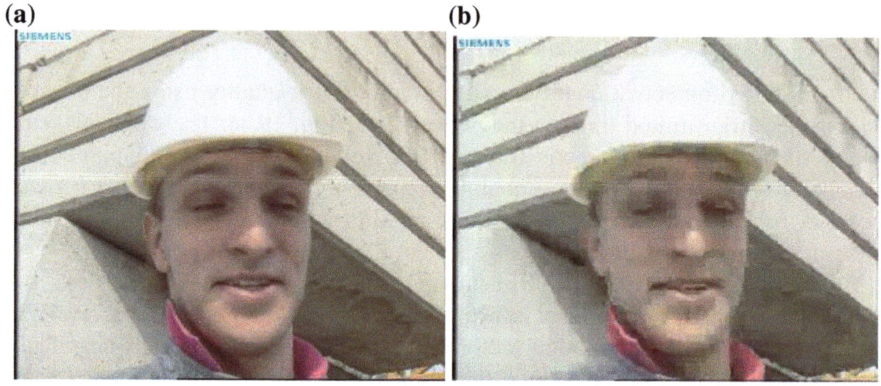

Fig. 19.3 Space varying sensitivity of the HVS to coding artifacts. The blockiness is more visible in flat areas than in high contrast areas as the building wall. **a** *Foreman* original frame. **b** Same frame coarsely quantized

Finally, the overall freezing degradation is computed as the square root of the sum of the individual degradations of each freezing event.

19.2.5 Pixel-Based Models

Pixel-based models estimate the video or image quality using features or information associated with pixel data only (see Fig. 19.1). These models are therefore taking as input the decoded video. They may in addition use the reference (non-degraded) video signal. The most popular metric in this category is the Peak Signal to Noise Ratio (PSNR), which is computed based on the Mean Squared Error (MSE) between the two (degraded and non-degraded) video signals. PSNR is quite popular due to its inherent properties: it is simple to compute, parameter-free and memory-less, and as a result, it can be calculated locally without consideration of other source samples. The MSE is, however, a signal fidelity measure and not a perceived quality metric. In fact, it shows poor correlation with perceived quality mainly because it does not take into account the properties of the human visual system [28].

The Human Visual System (HVS) is characterized among other things of contrast sensitivity and masking which lead to a space- and time-varying sensitivity of the artifacts associated with lossy coding and packet errors. As an example, Fig. 19.3 shows the first frame of the *Foreman* sequence in CIF resolution. The sequence has been coarsely quantized and therefore coding artifacts, namely blockiness, are visible. However, due to the space-varying sensitivity to distortion of the HVS, the blocking is more noticeable around the face, while it is less disturbing on the building wall. Therefore HVS-based models embed the HVS properties to weight more the artifacts present in image areas where the human eye is more sensitive and vice-versa. One of the first HVS-based model is the Visual Difference Predictor (VDP) [12], which uses the contrast sensitivity function of the HVS to weight differently the artifacts in image area. For a thoughtful review of HVS-based models, the interested reader is referred to [6].

Another well-known pixel-based model is the Structural SIMilarity (SSIM) index [93], which has been initially proposed for images and eventually extended for videos [94]. It is a full-reference model and for each pixel computes the distortion as the contribution of three terms: luminance, contrast and structure. The terms are then multiplied together and the final SSIM score is obtained as the average over all pixels. The work in [90] estimates the Mean Square Error (MSE) induced by channel errors by performing a maximum-a-posteriori estimation to detect the location of corrupted pixels. After these pixels are detected, the NORM algorithm ([66], see also Sect. 19.2.6) is run to estimate the final MSE. Considering also the temporal component of videos, the work in [73] proposes a general reduced-reference Video Quality Model (VQM). The VQM uses features computed over the pixels of the original and processed videos and combines all of them according to a linear weighting. The features used by VQM are related to quality degradation introduced by lossy coding or channel errors. The VQM has been standardised as quality model in the set of models specified in the ITU-T J.144 [39]. Furthermore, VQuad-HD is a full-reference video quality model for high definition video signals, which was selected by ITU-T as Recommendation J.341 [40]. It is based on the computation of the following quality features: blockiness, slicing, blurring, and jerkiness, as defined in Sect. 19.2.1.

19.2.6 Hybrid Models

Hybrid video quality assessment models employ a combination of packet information, bitstream information and the decoded reconstructed video sequence (see Fig. 19.1). In general, in a hybrid video quality assessment algorithm the features extracted or calculated from the bitstream (e.g. motion vectors, macroblock types, transform coefficients, quantization parameters, loss duration, etc.), and the information extracted from packet headers (e.g. bitrate, packet loss, delay, etc.) are combined with the features extracted from the decoded and reconstructed images in the pixel domain. Since the reconstructed image can be obtained from the decoding device, this type of model ensures that the error concealment method of the decoder is taken into consideration. Within the Video Quality Experts Group (VQEG), efforts are ongoing towards the joint construction and validation of novel objective hybrid video quality models [86].

The V-Factor [97] model inspects different parts of the encoded bitstream and extracts information from the packet headers and the encoded and decoded videos to model the impact of packet loss during video streaming. In [62], coding parameters are extracted during the decoding process of the video to construct a hybrid bitstream-based quality model. The model is based on a linear combination of quantization parameters, bitrate and boundary strength parameters. The latter parameter influences the intensity of the deblocking filter in order to minimize blockiness artifacts. In [57], the authors further extended an existing bitstream-based objective video quality model [58] by including pixel-based features. These features are based on detecting bluriness, blockiness, motion continuity and other aspects describing video quality. Their results show that the hybrid model outperforms the bitstream-based model. Similarly, a hybrid no-reference model for H.264/AVC encoded sequences is proposed in [13]. It is based on the quantization parameter and a pixel difference contrast measure.

The work in [78] estimates the MSE induced by channel errors between the reconstructed signal at the encoder and decoder. The model targets MPEG-x and H.26x codecs and is designed in three different versions denoted as Full-Parse (FP), Quick-Parse (QP) and No-Parse (NP). The FP version estimates the square error on the luma component for each pixel and then provides the MSE at the required level of granularity (e.g. macroblock, slice, frame, etc.). The input data to the FP algorithm require entropy decoding and inverse quantisation only. The QP estimates the MSE at slice level using bitstream parameters such as packet headers, thus without requiring any decoding operation. Finally, the NP estimates the MSE at sequence level using a linear relationship between the packet loss rate and the MSE. As may be noted, the three different versions lead to a different trade-off between model complexity and accuracy, with the FP being the most complex and accurate version. Finally, the work in [66] describes a NO-Reference video quality Monitoring (NORM) model which estimates the MSE induced by channel losses for H.264/AVC coded videos at the macroblock level. The NORM algorithm models the channel distortion as the result of three contributions: lack of motion vectors, lack of prediction residuals and error propagation from previous frames due to motion compensation. The contribution to the MSE due to specific coding tools of the H.264/AVC standard (i.e. intra prediction and deblocking filter) is also addressed. The NORM estimate has also been used to devise a reduced-reference quality model based on SSIM [93].

19.3 Video Quality Models for Streaming with Reliable Transport Mechanisms

This section begins with a short description of the progressive download and adaptive streaming mechanisms and of the corresponding degradations. Subsequently, an overview of models developed for the quality assessment of progressive download and adaptive streaming services is presented.

19.3.1 *Progressive Download and Adaptive Streaming Mechanisms and Degradations*

There are two main types of video streaming over HTTP: (a) progressive download and (b) HTTP adaptive streaming (HAS). In progressive download, the client may begin the playback of the media before the whole media is downloaded. However, if the available throughput is lower than the bitrate of the media, the player will stall until enough data have been downloaded. This is perceived by the end users as *freezing without skipping*, which is typically called *rebuffering* or *stalling*. To avoid stalling during playback and enable smoother flow of video, HAS methods adapt to the available network conditions. In HAS applications, the video is encoded in multiple quality versions, called "representations", which are segmented in short intervals, typically between 2 and 10 s long. The adaptive client periodically requests segments of the video content from an HTTP server, and then decodes and displays this segment. The client may switch between different representations at each request depending (mainly) on the available bandwidth. The aim is to provide the best quality of experience for the user by avoiding stalling events. However, the perceived artefact in this case is the fluctuating quality of the video sequence and quality models should consider the impact of temporal quality adaptation. The interested reader is referred to [83] for more information on standardized HAS methods such as the Dynamic Adaptive Streaming over HTTP (DASH).

19.3.2 *Progressive Download Models*

An audiovisual quality model has been proposed in [26] for rebuffering degradations.[1] The model takes as inputs the number of stalling events and the average length of a single stalling event. It was developed based on the results of laboratory and crowdsourcing tests. Video sequences had different video resolutions, but no extremely small or high definition resolutions. The video durations were typical of Youtube videos (on average 5.54 min and up to 15 min).

Another model has been proposed in the ITU-T P.1202.1 Recommendation [33] for capturing the quality impact due to rebuffering degradations, for low video resolutions (up to HVGA). The model is based on the ratio between the rebuffering duration of the sequence normalized to the total duration of the sequence (including the rebuffering duration). The ITU-T P.1202.1 model has been developed based on the results of subjective laboratory tests. The model was validated on video sequences of 16 s and up to 30 s. However, the rebuffering ratio parameter is normalized to the sequence duration and is therefore in principle applicable to longer sequences.

[1] Note that the focus was so far on visual stimuli and, therefore, video quality models. Due to its impact on the scientific work dedicated to rebuffering models, and although it is audiovisual, the model of [26] is presented in this section.

The rebuffering quality model of the ITU-T P.1201.1 Recommendation [42] has been developed using the same test databases as the P.1201.2 model. Similarly to [26], it uses as input parameters the number of stalling events and the average length of a single stalling event. In addition, the average distance between two rebuffering events is used as an input parameter to capture the distribution of the rebuffering events in the sequence.

Both the ITU-T P.1201.1 and P.1202.1 are addressing the lower video resolution application area. No rebuffering quality model has been so far standardized for higher resolution applications such as IPTV. However, the ITU-T is currently planning to develop a parametric model for progressive donwload ("P.NAMS-PD") valid for both lower and higher video resolution application areas. This model will also capture the quality impact due to initial rebuffering. Note that the quality impact due to initial rebuffering has also been studied in [25], with initial delays up to 30 s for 60 s video duration. This impact was found negligible compared to the quality impact due to stalling events occurring during the playback of the video.

19.3.3 Adaptive Streaming Models

The main feature of video sequences transmitted using HAS is the change in quality over time. Most of the existing quality assessment algorithms of Sect. 19.2 assume constant base (i.e. without packet loss) quality over the whole sequence and have been designed for short durations, typically between 10 and 20 s. With HAS services, the base quality is varying over time, and the quality adaptation typically lasts more than 20 s. In addition, these services facilitate the switching between devices and video resolutions (TV screen, tablet PC, smartphones) during playback. Quality models have therefore to be adapted to estimate quality for longer duration sequences (from a few seconds to several minutes), for fluctuating quality within the sequence, and even for switching between devices. Since quality models are preferably developed based on the results of subjective tests, a revision of the existing standardized subjective test methods [29, 45] is needed.

The impact of time-varying quality on human perception was investigated in [23], and it was shown that subjects react with different time constants to large sudden quality degradations or improvements. Moreover, it was shown that the location of a quality degradation or improvement influences subjects' overall judgements, revealing a recency effect. On the video domain, user perception of adapting quality was studied in [11, 68], revealing that the perception of spatial distortions over time can be largely modified by their temporal changes. Moreover, in [80], an hysteresis effect was observed in the subjective judgment of time-varying video quality. The video adaptation scheme of [99] for modeling the impact of bitrate and frame rate adaptation on perceived quality suggests that for video sequences with high temporal complexity, adaptation on frame rate will result in better quality than adapting QP; on the other hand, adaptation of QP is beneficial for video with low motion or fine texture details.

Quality assessment models for HAS can also employ existing models to assess the quality of short intervals (e.g., approx. 10 s) and then combine these scores into a single quality estimate using temporal pooling techniques. There are several temporal pooling algorithms ranging from simple approaches which consider the maximum, minimum, or mean video quality to those which integrate the temporal properties of human perception, memory effects, and transient properties. For example, in [61], temporal pooling is based on motion, while in [69], a content adaptive spatial and temporal pooling strategy is proposed which takes into consideration the severity of the quality degradations. In [81], an evaluation of the most popular pooling techniques using PSNR and SSIM as quality predictors concluded that the plain average of individual quality scores can achieve comparable results with the most sophisticated pooling methods. Further extending this study, a temporal pooling scheme based on an auto-regressive moving-average (ARMA) model to simulate the adaptation of perceived quality over time was presented in [67]. It is based on the computation of standardized video quality models, such as J.144, J.341, and P.1201.2, on video chunks of short duration, typically from 5 to 15 s. Moreover, a penalty parameter was introduced into the model to take into consideration the abrupt quality degradation within the sequence and the frequency in representation switches. In conclusion, the aforementioned approaches indicate that objective estimates of short sub-sections of a video sequence can be efficiently pooled into a single score as long as the memory and recency effects are taken into account.

19.4 From Video Quality Towards QoE

The models presented in the previous sections estimate the video quality of short sequences, typically 10 s, in the context of laboratory testing, where the viewing environment as well as the task given to the user deviate from typical viewing conditions. These models are therefore not estimating the QoE, and several aspects need to be taken into account to achieve a more QoE-centric assessment approach. These aspects include audio-visual quality, field testing, and user impact characterization. An overview of theses aspects is provided in this section.

19.4.1 Audiovisual Quality Models

Several studies on audiovisual perception, summarized in [60], have been conducted in the 1980s. However, the first audiovisual quality models to be found in the literature appeared as late as in the 1990s. At this time, these models addressed either analog degradations, such as audio and video noise [3, 35, 49], or compression artifacts, such as blockiness [8, 24, 50, 52]. For an overview of audiovisual quality models covering analog and compression degradations, see [105]. The interest in modeling audiovisual quality has risen again in the past ten years, reflected for instance by standardization activities such as the ITU-T Recommendations P.1201.1 and P.1201.2 [42, 43] or the Audiovisual High Definition Quality (AVHD) project of VQEG, which intends to evaluate audiovisual quality models for multimedia applications and HD resolution. Audiovisual quality models for mobile applications have been developed in 2005 and 2006 [79, 96], but the reported model versions do not cover the effect of transmission errors. This latter point is problematic since in the case of the time-varying degradation due to transmission errors, the impact of audio and video quality on the overall audiovisual quality as well as their interaction might differ from the case of compression artifacts. This influence of the degradation type has been studied in [22] for higher resolution application areas such as IPTV. In particular, the authors show that the impact of the audio quality on the overall perceived audiovisual quality is higher in the case of audio packet loss than in the case of audio coding degradations. In parallel to this finding, an audiovisual quality model has been proposed which captures the impact of the audio and video degradation types (coding vs. transmission-error degradations). Both compression and transmission-error degradations are covered as well in [4] for interactive scenarios and small video resolutions. Based on an extensive review of the literature, both [22] and [105] highlight that video quality generally dominates the perceived audiovisual quality, but that this dominance depends on the semantic audiovisual content. Finally, an overview of existing audiovisual quality models is provided in [70]. The authors show that a simple model based on the product of audio and video quality terms is valid for a wide range of scenarios and applications. When a small amount of data is available for a wide range of applications, it is indeed a safe choice to use a model as simple and with as few coefficients as possible to avoid overtraining.

19.4.2 Ecologically Valid Testing

As detailed in Chap. 10, subjective methodologies for assessing momentary QoE provide detailed guidelines describing how to conduct such experiments. These recommendations define different methods for presenting and rating video sequences. Typically, short duration (10–15 s) video sequences are presented to the test subjects and rated immediately after watching. In the case of longer video sequences (up to 30 min), continuous quality evaluation is recommended where subjects rate quality while watching the video [29]. Specific instructions are provided to the test subjects on how to evaluate the video sequences at the beginning of the experiment. The assessment methodologies also specify requirements for the environment in which the subjective experiment is conducted. These requirements are formulated in terms of room illumination, subject seating position, screen calibration, etc. As such, subjective quality assessment experiments are usually conducted in controlled lab environments. These assessment methodologies are still actively used for measuring pure video or audiovisual quality.

The broad availability of high speed Internet access and growing number of multimedia-capable devices (such as smartphones and tables) enable watching video content at anyplace, anytime. Thus, the environment in which video is actually consumed does not necessarily comply with the recommended controlled lab setting. Furthermore, according to its definition ([41, 74] and Chap. 2), Quality of Experience (QoE) is influenced by user expectations, context, and personal preferences. Therefore, capturing and understanding end-users' QoE go beyond purely measuring video quality [76]. This calls for new subjective studies and methodologies [5, 15, 16, 71, 92] enabling episodic and multi-episodic subjective quality evaluation in more realistic and ecologically valid environments (cf. Sect. 10.4).

A first effort towards assessing QoE of IPTV services in real-life environments has been made in [84, 85]. This study involved conducting subjective experiments in subjects' own home environment under realistic viewing conditions. Comparing the results obtained with subjective tests conducted in a controlled lab highlighted the importance of the assessment environment, primary focus, and immersion on impairment visibility and quality perception. It is found that immersion has a major impact on impairment tolerance and overall QoE.

Field testing can provide new insights and findings which cannot necessarily be discovered in a controlled lab setting. In this respect, field testing should complement lab testing rather than replace it.

19.4.3 User Impact

The viewing and listening environmental set-up is thoroughly controlled in standardized video quality test methods [30, 44]. These systematic methodologies reduce significantly the amount of noise in the results and enable the comparison of test results between labs. However, the outcome of these tests cannot be easily extrapolated to more realistic scenarios due to variable factors such as the context of use.

As presented in Chap. 4, the main limitation in the development of more realistic tests is the complex interaction between all determinant influential factors of the ecosystem. Hence, researchers from the field of social sciences and economy tried to analyse how factors such as user demography, overall service quality and context of use interact and influence the perception. New approaches were developed, such as the Theory of Acceptance Model (TAM) [14], and more recently the Unified Theory of Acceptance and Use of Technology (UTAUT) [1]. These theories intend to understand the development of intention of use of a service, based on strong behavioural elements such as external influential variables, perceived usefulness and perceived ease of use.

First uses of these theories can be seen in [51, 95], where modified versions of TAM are utilized to predict the adoption of IPTV services. The results are surprisingly aligned to studies in the context of mobile television [53], demonstrating that the content offered, the technological knowledge of the user, his/her attitude towards technology, and socio-economic factors are crucial for a positive experience of the service. The above-mentioned findings suggest that users with similar characteristics show similar behaviour, an observation that, in spite of being studied in other services for decades, has been little explored in the study of multimedia services. Among the most relevant factors, the degree of expertise of the users with the service greatly influences the way in which they perceive and evaluate quality. However, this type of classification only gives a general idea concerning the main factors influencing the users' evaluation process. It has therefore been necessary to combine cognitive and psycho-perceptual methods that provide the user with the freedom to assess the service attributes in their own words, as seen in [54, 88].

19.5 Discussion

The literature is rich in video quality models, with different model types according to the application needs. The first models were developed in order to address compression artifacts. Then, new models appeared for covering packet loss degradations yielding slicing or freezing with skipping degradations in the case of unreliable network such as RTP over UDP. Other models have recently been proposed for addressing the progressive download scenario.

These models all target short-term video quality predictions and ignore which user and in which context the user utilizes the video streaming service. Some studies have already been conducted for addressing longer term quality predictions. Other subjective tests have been conducted to address the context- and user-impact. However, this topic needs much more investigation, especially with the new emerging types of video streaming applications such as adaptive streaming and complex scenarios such as portable TV, where the user can watch TV in different locations and on different screens

It is still open which subjective tests should be conducted to address these complex scenarios and to identify which factors, in addition to the perceived video quality, influence the overall QoE. Also open is the type of measurement tools to be targeted. Indeed, a "measurement-window" approach, with which quality scores are output for example every 10 s, is traditionally used in service monitoring. With the diversity of degradations, and in order to better capture the long-term QoE prediction, a "remembered-event" approach may become more appropriate, where an event represents any kind of degradations of any duration. In that case, efforts should be spent on the identification and characterisation of these "events", as well on the weighting of their contribution to the overall QoE.

References

1. Al-Qeisi KI (2009) Analyzing the use of UTAUT model in explaining an online behaviour: Internet banking adoption. Ph.D. thesis, Brunel University
2. Argyropoulos S, Raake A, Garcia MN, List P (2011) No-reference video quality assessment for SD and HD H.264/AVC sequences based on continuous estimates of packet loss visibility. In: International workshop on Quality of Multimedia Experience (QoMEX), pp 31–36

3. Beerends JG, Caluwe F (1999) The influence of video quality on perceived audio quality and vice versa. J Audio Eng Soc 47:355
4. Belmudez B, Möller S, Lewcio B, Raake A, Mehmood A (2009) Audio and video channel impact on perceived audio-visual quality in different interactive contexts. In: IEEE international workshop on Multimedia Signal Processing (MMSP)
5. Borowiak A, Reiter U, Svensson UP (2012) Quality evaluation of long duration audiovisual content. In: IEEE Consumer Communications and Networking Conference, pp 337–341
6. Bosc E, Le Callet P, Morin L, Pressigout M (2013) Visual quality assessment of synthesized views in the context of 3D-TV. In: Zhu C, Zhao Y, Yu L, Tanimoto M (eds) 3D-TV systems with depth-based-image-rendering architectures, techniques and challenges, Springer, New York
7. Boulos F, Parrein B, Callet PL, Hands D (2009) Perceptual effects of packet loss on H.264/AVC encoded videos. In: International workshop on video processing and quality metrics for consumer electronics
8. Chateau N (1998) ITU-T Contribution COM 12–61 Relations between audio, video and audio-visual quality. International Telecommunication Union (ITU-T), Geneva
9. Chikkerur S, Sundaram V, Reisslein M, Karam LJ (2011) Objective video quality assessment methods: A classification, review, and performance comparison. IEEE Trans Broadcasting 57(2):165–182
10. Corriveau P, Gojmerac C, Hughes B, Stelmach L (1999) All subjective scales are not created equal: The effects of context on different scales. Signal Process 77(1):1–9
11. Cranley N, Perry P, Murphy L (2006) User perception of adapting video quality. Int J Hum Comput Stud 64(8):637–647
12. Daly S (1992) The visible difference predictor: an algorithm for the assessment of image fidelity. In: SPIE human vision, visual processing and digital display, vol 1666
13. Davis AG, Bayart D, Hands DS (2009) Hybrid no-reference video quality prediction. In: Proceedings of IEEE international symposium on broadband multimedia systems and broadcasting, pp 1–6
14. Davis F, Bagozzi R, Warshaw P (1989) User acceptance of computer technology: a comparison of two theoretical models. Manage Sci 35:982–1003
15. De Moor K, Ketyko I, Joseph W, Deryckere T, De Marez L, Martens L, Verleye G (2010) Proposed framework for evaluating quality of experience in a mobile, testbed-oriented living lab setting. Mob Netw Appl 15(3):378–391
16. De Pessemier T, De Moor K, Joseph W, De Marez L, Martens L (2013) Quantifying the influence of rebuffering interruptions on the user's quality of experience during mobile video watching. IEEE Trans Broadcast 59(1):47–61
17. Dobrian F, Awan A, Joseph D, Ganjam A, Zhan J, Sekar V, Stoica I, Zhang H (2013) Understanding the impact of video quality on user engagement. Comm ACM 56(3):91–99
18. DSL Forum Technical Report, TR-126 (2006) Triple-play services quality of experience (QoE) requirements, DSL Forum
19. EBU-SAMVIQ (2003) SAMVIQ subjective assessment methodology for video quality. Report by the EBU Project Group B/VIM, European Broadcasting Union, Geneva
20. Frossard P, Verscheure O (2001) Joint source/FEC rate selection for quality-optimal MPEG-2 video delivery. IEEE Trans Image Process 10(12):1815–1825
21. Garcia MN, Raake A (2010) Parametric packet-layer video quality model for IPTV. In: International conference on information science, signal processing and their applications
22. Garcia MN, Schleicher R, Raake A (2011) Impairment-factor-based audiovisual quality model for IPTV: influence of video resolution, degradation type, and content type. EURASIP J Image and Video Process 2011:1–14
23. Gros L, Chateau N (2001) Instantaneous and overall judgements for time-varying speech quality: assessments and relationships. Acta Acustica 87(3):367–377
24. Hands D (2004) A basic multimedia quality model. IEEE Trans Multimedia 6(6):806–816
25. Hoßfeld T, Egger S, Schatzand R, Fiedler M, Masuch K, Lorentzen C (2012) Initial delay vs. interruptions: between the devil and the deep blue sea. In: International workshop on Quality of Multimedia Experience

26. Hoßfeld T, Schatz R, Biersack E, Plissonneau L (2013) Internet video delivery in Youtube: from traffic measurements to quality of experience. In: Biersack E, Callegari C, Matijasevic M (eds) Data traffic monitoring and analysis: from measurement, classification and anomaly detection to quality of experience. Springer, Berlin
27. Huynh-Thu Q, Garcia MN, Corriveau FS, Raake PJ (2011) Study of rating scales for subjective quality assessment of high-definition video. IEEE Trans Broadcast 57(1):1–14
28. Huynh-Thu Q, Ghanbari M (2012) The accuracy of PSNR in predicting video quality for different video scenes and frame rates. Telecommun Syst 49(1):35–48
29. ITU-R Recommendation BT.500 (2012) Methodology for the subjective assessment of the quality of television pictures. International Telecommunication Union, Geneva
30. ITU-R Recommendation BT.500-12 (2009) Methodology for the subjective assessment of the quality of television pictures. International Telecommunication Union, Geneva
31. ITU-R Recommendation BT.710-4 (1998) Subjective assessment methods for image quality in high-definition television. International Telecommunication Union, Geneva
32. ITU-R Recommendation P.1202 (2012) Parametric non-intrusive bitstream assessment of video media streaming quality. International Telecommunication Union, Geneva
33. ITU-R Recommendation P.1202.1 (2012) Parametric non-intrusive bitstream assessment of video media streaming quality–lower resolution application area. International Telecommunication Union, Geneva
34. ITU-R Recommendation P.1202.2 (2012) Parametric non-intrusive bitstream assessment of video media streaming quality–higher resolution application area. International Telecommunication Union, Geneva
35. ITU-T Contribution COM 12–27 (1994) Extension of combined audio/video quality model. Bellcore, USA
36. ITU-T Recommendation G. 1070 (2007) Opinion model for video-telephony applications. International Telecommunication Union, Geneva
37. ITU-T Recommendation G.1080 (2008) Quality of experience requirements for IPTV services. International Telecommunication Union, Geneva
38. ITU-T Recommendation H.264 (2013) Advanced video coding for generic audiovisual services. International Telecommunication Union, Geneva
39. ITU-T Recommendation J.144 (2001) Objective perceptual video quality measurement techniques for digital cable television in the presence of a full reference. International Telecommunication Union, Geneva
40. ITU-T Recommendation J.341 (2011) Objective perceptual multimedia video quality measurement of HDTV for digital cable television in the presence of a full reference. International Telecommunication Union, Geneva
41. ITU-T Recommendation P.10/G.100 Amd 2 (2008) Vocabulary for performance and quality of service. International Telecommunication Union, Geneva
42. ITU-T Recommendation P.1201.1 (2012) Parametric non-intrusive assessment of audiovisual media streaming quality–lower resolution application area. International Telecommunication Union, Geneva
43. ITU-T Recommendation P.1201.2 (2012) Parametric non-intrusive assessment of audiovisual media streaming quality–higher resolution application area. International Telecommunication Union, Geneva
44. ITU-T Recommendation P.910 (1999) Subjective video quality assessment methods for multimedia applications. International Telecommunication Union, Geneva
45. ITU-T Recommendation P.910 (2008) Subjective video quality assessment methods for multimedia applications. International Telecommunication Union, Geneva
46. ITU-T Recommendation P.911 (1998) Subjective audiovisual quality assessment methods for multimedia applications. International Telecommunication Union, Geneva
47. ITU-T Recommendation P.920 (2000) Interactive test methods for audiovisual communications. International Telecommunication Union, Geneva
48. ITU-T Recommendation P.930 (1996) Principles of a reference impairment system for video. International Telecommunication Union, Geneva

49. ITU-T Contr. COM 12–20 (1993) Experimental combined audio/video subjective test method. Bellcore, USA
50. ITU-T Delayed Contribution COM 12 D.038 (1998) Results of an audio-visual desktop teleconferencing subjective experiment. NTIA/ITS, USA
51. Jang YH, Noh JM (2011) Customer acceptance of IPTV service quality. Int J Inf Manage 31(6):582–592
52. Jones C, Atkinson D (1998) Development of opinion-based audio-visual quality models for desktop video-teleconferencing. In: IEEE International workshop on quality of service
53. Jumisko-Pyykkö S, Häkkinen J (2008) Profiles of the evaluators: impact of psychographic variables on the consumer-oriented quality assessment of mobile television. In: Proceedings of SPIE, vo 6821, article id. 68210L, p 14
54. Jumisko-Pyykkö S, Strohmeier D (2013) Cognitive styles and visual quality. IS&T/SPIE Electron Imaging 4(4):86670K–86670K
55. Kanumuri S, Cosman P, Reibman A, Vaishampayan V (2006) Modeling packet-loss visibility in MPEG-2 video. IEEE Trans Multimedia 8(2):341–355
56. Kanumuri S, Subramanian S, Cosman P, Reibman A (2006) Predicting H.264 packet loss visibility using a generalized linear model. In: IEEE International Conference on Image Processing (ICIP), pp 2245–2248
57. Keimel C, Habigt J, Diepold K (2012) Hybrid no-reference video quality metric based on multiway plsr. In: Proceedings of the 20th European signal processing conference, pp 1244–1248
58. Keimel C, Habigt J, Klimpke M, Diepold K (2011) Design of no-reference video quality metrics with multiway partial least squares regression. In: International workshop on Quality of Multimedia Experience (QoMEX), pp 49–54
59. Keimel C, Klimpke M, Habigt J, Diepold K (2011) No-reference video quality metric for HDTV based on H.264/AVC bitstream features. In: IEEE International Conference on Image Processing (ICIP), pp 3325–3328
60. Kohlrausch A, van de Par S (2005) Audio-visual interaction in the context of multi-media applications. In: Blauert J (ed) Communication acoustics. Springer, Berlin, pp 109–138
61. Lee K, Park J, Lee S, Bovik A (2010) Temporal pooling of video quality estimates using perceptual motion models. In: IEEE International Conference on Image Processing, pp 2493–2496
62. Lee SO, Jung KS, Sim DG (2010) Real-time objective quality assessment based on coding parameters extracted from H.264/AVC bitstream. IEEE Trans Consum Electron 56(2):1071–1078
63. Liao N, Chen Z (2011) A packet-layer video quality assessment model with spatiotemporal complexity estimation. EURASIP J Image Video Process 5:1
64. Lin TL, Kanumuri S, Zhi Y, Poole D, Cosman P, Reibman A (2010) A versatile model for packet loss visibility and its application to packet prioritization. IEEE Trans Image Process 19(3):722–735
65. Lin X, Ma H, Luo L, Chen Y (2012) No-reference video quality assessment in the compressed domain. IEEE Trans Consum Electron 58(2):505–512
66. Naccari M, Tagliasacchi M, Tubaro S (2009) No-reference video quality monitoring for H.264/AVC coded video. IEEE Trans Multimedia 11(5):932–946
67. Next Generation Mobile Networks (2013) Service quality definition and measurement. White Paper. http://www.ngmn.org/uploads/media/NGMN-P-SERQU_Service_Quality_Definition_and_Measurement_-_A_Technical_Report_by_NGMN_Alliance__v1_0_4_.pdf
68. Ninassi A, Le Meur O, Le Callet P, Barba D (2009) Considering temporal variations of spatial visual distortions in video quality assessment. IEEE J Sel Top Signal Process 3(2):253–265
69. Park J, Seshadrinathan K, Lee S, Bovik A (2013) Video quality pooling adaptive to perceptual distortion severity. IEEE Trans Image Process 22(2):610–620
70. Pinson M, Ingram W, Webster A (2011) Audiovisual quality components. IEEE. Signal Process 28:60–67

71. Pinson M, Janowski L, Pepion R, Huynh-Thu Q, Schmidmer C, Corriveau P, Younkin A, Le Callet P, Barkowsky M, Ingram W (2012) The influence of subjects and environment on audiovisual subjective tests: an international study. IEEE J Sel Top Signal Process 6(6):640–651
72. Pinson M, Wolf S (2003) Comparing subjective video quality testing methodologies. In: Proceedings of SPIE
73. Pinson MH, Wolf S (2004) A new standardized method for objectively measuring video quality. IEEE Trans Broadcast 50(3):312–322
74. Qualinet White Paper on Definitions of Quality of Experience-Output Version of the Dagstuhl Seminar 12181 (2012). In: Möller S, Le Callet P, Perkis A (eds) European network on quality of experience in multimedia systems and services (COST Action IC 1003), 1.1 edn, Lausanne
75. Raake A, Gustafsson J, Argyropoulos S, Garcia MN, Lindegren D, Heikkilä G, Pettersson M, Feiten B, List P (2011) Monitoring the quality of IP-based mobile and fixed network audiovisual media services. IEEE Signal Proces Mag 28(6):68–79
76. Redi J (2013) Visual quality beyond artifact visibility. In: SPIE human vision and electronic imaging XVIII
77. Reibman A, Kanumuri S, Vaishampayan V, Cosman P (2004) Visibility of individual packet losses in MPEG-2 video. IEEE International Conference on Image Processing (ICIP). 1:171–174
78. Reibman AR, Vaishmpayan VA, Sermadevi Y (2004) Quality monitoring of video over a packet network. IEEE Trans Multimedia 6(2):327–334
79. Ries M, Puglia R, Tebaldi T, Nemethova O, Rupp M (20058) Audiovisual quality estimation for mobile video services. In: 2nd IEEE symposium on wireless communication systems
80. Seshadrinathan K, Bovik A (2011) Temporal hysteresis model of time varying subjective video quality. In: IEEE International Conference on Acoustics, Speech and Signal Processing (ICASSP), pp 1153–1156
81. Seufert M, Slanina M, Egger S, Kottkamp M (2013) To pool or not to pool: a comparison of temporal pooling methods for http adaptive video streaming. In: International workshop on video processing and quality metrics for consumer electronics
82. Shi Z, Chen P, Feng C, Huang L, Xu W (2012) Research on quality assessment metric based on H.264/AVC bitstream. In: International conference on anti-counterfeiting, security and identification, pp 1–5
83. Sodagar I (2011) The MPEG-DASH standard for multimedia streaming over the internet. IEEE Multimedia 18(4):62–67
84. Staelens N, Van den Broeck W, Pitrey Y, Vermeulen B, Demeester P (2012) Lessons learned during real-life QoE assessment. In: EuroITV–10th European conference on interactive TV
85. Staelens N, Moens S, Van den Broeck W, Mariën I, Vermeulen B, Lambert P, Van de Walle R, Demeester P (2010) Assessing quality of experience of IPTV and Video on Demand services in real-life environments. IEEE Trans Broadcast 56(4):458–466
86. Staelens N, Sedano I, Barkowsky M, Janowski L, Brunnström K, Le Callet P (2011) Standardized toolchain and model development for video quality assessment–the mission of the joint effort group in VQEG. In: International workshop on Quality of Multimedia Experience
87. Staelens N, Van Wallendael G, Crombecq K, Vercammen N, De Cock J, Vermeulen B, Van de Walle R, Dhaene T, Demeester P (2012) No-reference bitstream-based visual quality impairment detection for high definition H.264/AVC encoded video sequences. IEEE Trans Broadcast 58(2):187–199
88. Strohmeier D (2012) Open profiling of quality: a mixed methods research approach for audiovisual quality evaluations. SIG Multimedia Rec 4(4):5–6
89. Tao S, Apostolopoulos J, Guerin R (2008) Real-time monitoring of video quality in IP networks. IEEE/ACM Trans Networking 16(5):1052–1065
90. Valenzise G, Magni S, Tagliasacchi M, Tubaro S (2012) No-reference pixel video quality monitoring of channel-induced distortion. IEEE Trans Circ Syst Video Technol 22(4):605–618

91. Verscheure O, Frossard P, Hamdi M (1999) User-oriented qos analysis in MPEG-2 video delivery. Real-Time Imaging 5(5):305–314
92. Waltl M, Timmerer C, Hellwagner H (2009) A test-bed for quality of multimedia experience evaluation of sensory effects. In: International workshop on Quality of Multimedia Experience, pp 145–150
93. Wang Z, Bovik AC, Sheikh HR, Simoncelli EP (2004) Image quality assessment: from error visibility to structural similarity. IEEE Trans Image Process 13(4):600–612
94. Wang Z, Lu L, Bovik AC (2004) Video quality assessment based on structural distortion measure. Signal Process Image Commun 19(2):121–132
95. Weniger S (2010) User adoption of IPTV: a research model. In: International bled e-conference
96. Winkler S, Faller C (2006) Perceived audiovisual quality of low-bitrate multimedia content. IEEE Trans Multimedia 8(5):973–980
97. Winkler S, Mohandas P (2008) The evolution of video quality measurement: from PSNR to hybrid metrics. IEEE Trans Broadcast 54(3):660–668
98. Wu HR, Rao KR (2006) Digital video image quality and perceptual coding. Taylor & Francis, Boca Raton
99. Xue Y, Song Y, Ou YF, Wang Y (2013) Video adaptation considering the impact of temporal variation on quantization stepsize and frame rate on perceptual quality. In: International workshop on video processing and quality metrics for consumer electronics, pp 70–74
100. Yamada T, Yachida S, Senda Y, Serizawa M (2010) Accurate video-quality estimation without video decoding. In: International Conference on Acoustics, Speech, and Signal Processing (ICASSP)
101. Yamagishi K, Hayashi T (2008) Parametric packet-layer model for monitoring video quality of IPTV services. In: IEEE international conference on communications
102. Yamagishi K, Okamoto J, Hayashi T, Takahashi T (2012) No reference video-quality-assessment model for monitoring video quality of IPTV services. IEICE Trans Commun 95:435–448
103. Yang F, Song J, Wan S, Wu H (2012) Content-adaptive packet-layer model for quality assessment of networked video services. IEEE J Sel Top Signal Process 6(6):672–683
104. You F, Zhang W, Xiao J (2009) Packet loss pattern and parametric video quality model for IPTV. In: International conference on computer and information science
105. You J, Reiter U, Hannuksela M, Gabbouj M, Perkis A (2010) Perceptual-based quality assessment for audio-visual services:a survey. J Signal Process Image Commun 25:482
106. Yuen M, Wu H (1998) A survey of hybrid MC/DPCM/DCT video coding distortions. EURASIP J Signal Process 70:247
107. Zieliński S, Rumsey F, Bech S (2008) On some biases encountered in modern audio quality listening testsa review. J Audio Eng Soc 56(6):427–451

Chapter 20
3D Video

Pierre Lebreton, Marcus Barkowsky, Alexander Raake and Patrick Le Callet

Abstract 3D video has been considered as the next step in television for some time. The transition from 2D to 3D is frequently seen as comparable to the transition from monochrome to color. The introduction of this new dimension adds new challenges regarding the question of its relates with Quality of Experience (QoE). This chapter mainly focuses on presenting the particular challenges related with stereoscopic 3D video quality. This includes the difficulty to evaluate QoE of 3D video, taking into account all relevant factors. In particular, traditional approaches fail to capture aspects such as the added value in terms of QoE due to 3D depth or quality-issues brought by 3D-specific artifacts and their effect on visual comfort, so that alternative solutions for evaluation are required. As a consequence, the aim of this chapter is to address 3D-specific aspects of visual perception. The employed technology is another aspect of high influence on 3D video QoE. The chapter addresses the different issues related with content creation, transmission and representation, to help the reader understand the differences to a 2D transmission chain, and how technology affects the perception and the construction of the general judgment of 3D video QoE.

P. Lebreton (✉) · A. Raake
Assessment of IP-based Applications, Telekom Innovation Laboratories, TU Berlin, Berlin, Germany
e-mail: pierre.lebreton@telekom.de

A. Raake
e-mail: alexander.raake@telekom.de

M. Barkowsky · P. Le Callet
University of Nantes and IRCCyN, Nantes, France
e-mail: marcus.barkowsky@univ-nantes.fr

P. Le Callet
e-mail: patrick.lecallet@univ-nantes.fr

S. Möller and A. Raake (eds.), *Quality of Experience*,
T-Labs Series in Telecommunication Services, DOI: 10.1007/978-3-319-02681-7_20,

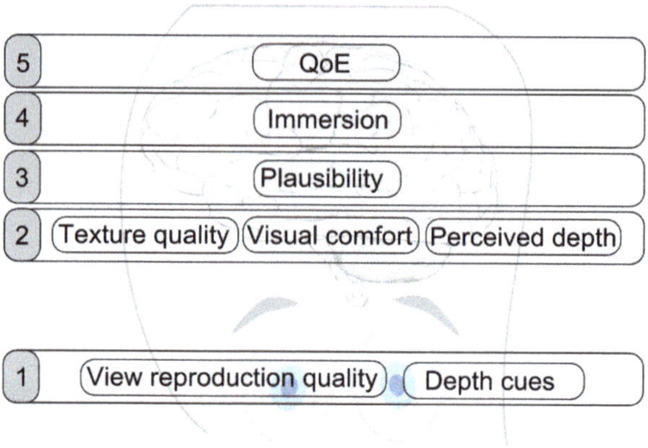

Fig. 20.1 Analysis of QoE in different layers

20.1 Introduction

Quality of Experience in 3D video is of multi-dimensional nature, and many factors contribute to establishing the overall experience. The purpose of this chapter is to provide information about the differences between 3D QoE and 2D QoE. A simplified layered scheme of factors leading to or influencing QoE is depicted in Fig. 20.1. The general motivation of the construction of this model is explained in Sect. 20.2. Section 20.3 describes the perceptual aspects related with the 3D QoE model. Section 20.4 emphasizes the differences between a 2D and 3D transmission chain, and how these affect the different levels of the QoE model. Section 20.5 presents methods for evaluating the QoE, both in perception tests and using prediction algorithms. The final section discusses key issues to be considered in future research.

20.2 Model of 3D Video QoE

It is commonly assumed that the visual experience of 3D, as presented by Seuntiëns [46], may be measured and modeled as a complex combination of three main factors: texture quality, depth perception, and visual comfort. These three factors represent the second level of the layered QoE model depicted in Fig. 20.1. They stem from sensory input coming from the outside world via the eyes of the observers, which are included in the first layer.

The link between overall QoE and the three stated main factors is difficult to establish. Traditional approaches to evaluate QoE by means of single stimulus methodologies fail to capture the multi-dimensionality of 3D video. They seem to be mainly evaluating one component, namely pictorial quality, as can be seen in [33, 55]. In these studies, it can be observed that 3D QoE is not rated higher than 2D QoE. Moreover, studies have shown that when evaluating quality on an absolute scale, 2D and 3D are not rated differently [26, 46], whereas when using other methodologies such as Pair Comparison [35] or another evaluation concept [31, 46], a difference of QoE is clearly visible. An explanation may be related with the expectation of the observers: in case of a subjective experiment with coding conditions, it can be assumed that their expectation is driven by pictorial quality. This is also enhanced by the fact that the proposed test conditions do not investigate the depth and quality dimensions separately. This results in an inability to account for depth, also due to the absence of a clear reference. Similar restrictions apply to all three main factors, rendering their isolated measurement difficult.

To tackle this issue, intermediate steps of the construction of 3D video QoE can be considered to link the main factors in level 2 to the global QoE in level 5, as shown in Fig. 20.1 [46]. These higher-level concepts include "naturalness" or "plausibility" and "immersion"(level 3 and 4 in Fig. 20.1).

Linear models between QoE and the factors of level 2 and those in level 3 and 4 are proposed in [30, 60]. The subjects were asked for judgments of "visual experience", defined in [46] by a model taking into account "the diverse set of image attributes which contributes to the overall perceived quality of 3D-TV images", and was evaluated in the presence of white Gaussian noise and Gaussian blur [30]. As the study shows, these judgments depended mostly on "pictorial quality" (level 2), and to a lesser degree on "immersion" (level 4). In the absence of image noise and blur, depth and visual comfort were found to be of similar importance in the study reported by Chen et al. in [60] for the "visual experience", defined as " the overall quality of experience of the images in terms of immersion and the overall perceived quality". As opposed to the study in [30], the effect of image quality was found to be negligible in this study. These results provide some information concerning the contribution of the different factors to the "visual experience", which is better approaching the notion of "3D video QoE".

20.3 Perception

As further described in the 3D QoE video model of Sect. 20.2, 3D video QoE depends on pictorial quality, visual comfort and perceived depth (level 2 in Fig. 20.1). While the perception of pictorial quality can be assumed to not differ significantly from the 2D case [30, 46], visual comfort and perceived depth do. A detailed analysis of the origins of these two factors with regard to the level 1 input is required. This will prepare the explanation of distortions which will be presented below in the context of the technical considerations regarding the 3D video transmission chain.

20.3.1 Depth Perception (Level 2)

The introduction of 3D video enabled adding stereoscopic depth perception. The general perception of depth comes from different cues, which can be classified into

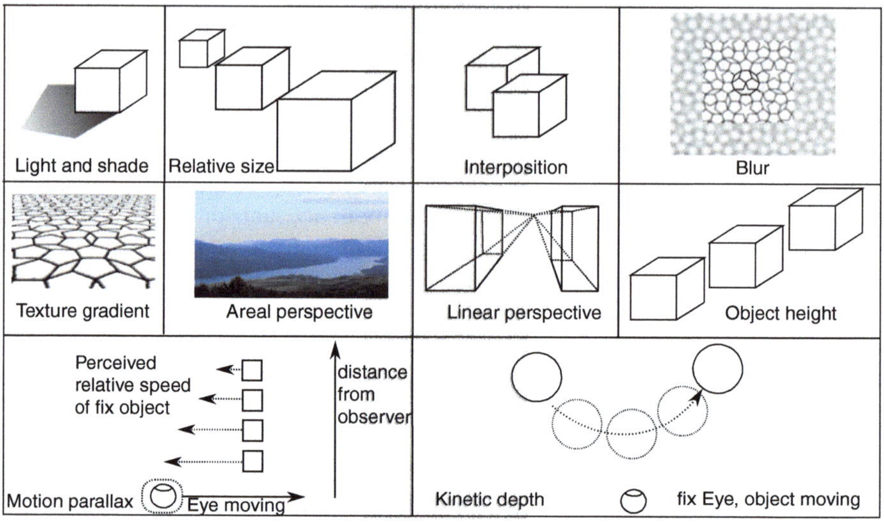

Fig. 20.2 Monocular depth cues (figure adapted from [32])

two main categories: binocular and monocular depth cues [14]. These cues originate from the level 1 of Fig. 20.1. The monocular depth cues are illustrated in Fig. 20.2, that is, the cues that provide information about the depth using only information from a single view. The binocular depth cues (also of level 1 in Fig. 20.1), depicted in Fig. 20.3, are based on the fact that healthy humans have two eyes, and each of these eyes provides a slightly different view. The binocular vision is based on two aspects, the convergence of the eyes to the object of interest, called "vergence", and the focus on the considered object using the variable lens-system of the eyes, referred to as "accommodation". The retinal images are processed by the brain to understand the position in depth of the elements in the scene.

Besides the source of depth information from monocular and binocular cues, it should be considered that the depth perception itself has two different aspects: the *depth quantity* and the *depth quality* [60].

The *depth quantity* describes how much depth is perceived due to the 3D effect, the *depth quality* provides information on the extent to which the depth rendering appears plausible. In both cases, there are strong interactions between the monocular depth cues and the binocular depth cues with regard to the construction of the perception of depth. Indeed different combinations of monocular and binocular depth cues can result in different amounts of the two dimensions of depth perception: for example, the effect of blur was studied in [56, 58], the effect of light in [39, 43], of relative size in [53], and of texture gradient in [20]. No general model has been unanimously defined so far to model depth perception from monocular and binocular depth cues. However, since studies have found that depth-cue pooling refers to a statistical inference problem, maximum likelihood estimation (MLE) has been used to

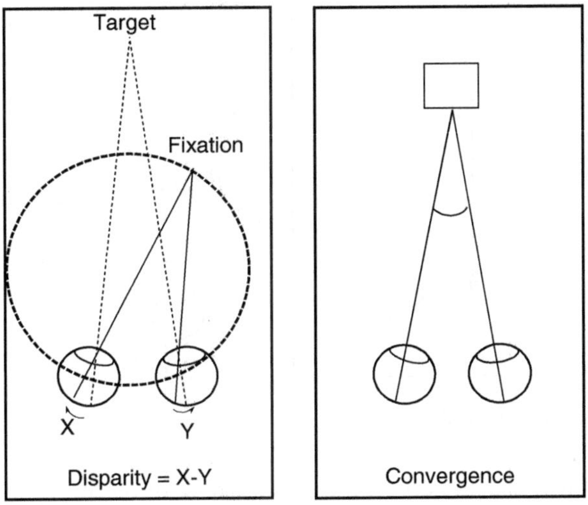

Fig. 20.3 Binocular depth cues

consider the reliability of the depth cues when pooling information from individual cues [32, 34, 39].

The depth quality is based on the monocular depth cues, the binocular depth cues, how they are perceived and how they agree, taking also prior knowledge about object shapes into consideration. Indeed, to achieve a realistic depth rendering, the camera capturing has to consider how the sequences will be reproduced [11]. For example, the "cardboard effect", which is characterized by objects to appear unnaturally flat, is caused by a contradiction between the monocular depth cues or general knowledge of the object's shape and the binocular depth cues that indicate that the object is flat, when it actually is expected to have depth. This may happen, when the camera distance was chosen for a reproduction on a cinema screen, but the video is displayed on a smaller television screen. See [11, 62, 63] for the complete description of the phenomenon and a measure of distortion based on shooting and rendering condition.

20.3.2 Visual Comfort (Level 2)

A major issue regarding 3D video reproduction is visual (dis)comfort (level 2 in Fig. 20.1). It is widely assumed that beside artifacts due to poorly generated 3D material (strong vertical misalignment, temporal misalignment, view inversion, window violation effect, lighting and color misalignment etc. [4, 13]), the source of visual discomfort is due to the range of binocular disparity used. First, to ensure binocular comfort, it is usually assumed that disparity values should remain smaller than 60 min of arc. The second major problem is due to the vergence/accommodation conflict, which is depicted in Fig. 20.4: in the case of a real object, the convergence

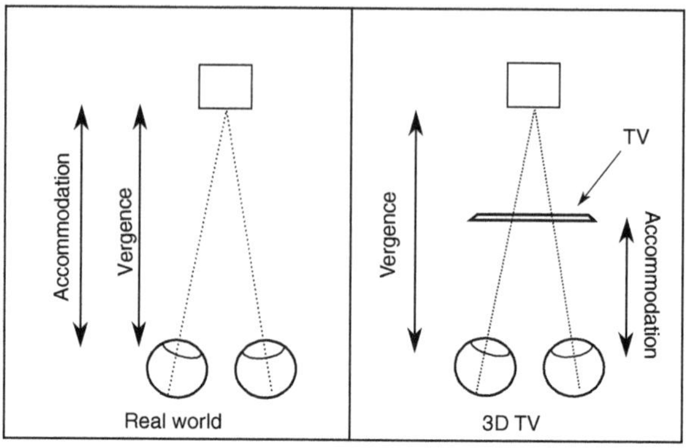

Fig. 20.4 Vergence/accommodation conflict

of the two eyes to a specific object (vergence) is synchronized with the adaptation of the lens to provide the focus on the light reflection of the examined object (accommodation). However, in case of a 3D screen, the position of the display, the light emitting source, remains at the same planar location. Thus, the accommodation remains fixed, whereas the vergence changes with the visual exploration of the depth of objects in the scene by the observers. The visual discomfort is produced by the stress due to the resulting rupture of synchronization between the vergence and accommodation. The position of the objects in depth, their motion, and the display size influence the visual discomfort significantly [22, 38, 49]. The limit of acceptable vergence/accommodation difference has been related to the depth of field (DOF) of 0.2 diopters [64]. This defines a depth range around the sharp depth plane where the image can remain sharp. This is motivated by the fact that the human visual system does not always adapt the vergence and accommodation, and there is an area which enables adaptation of the vergence without updating the accommodation.

Several other possible sources of visual discomfort need to be considered, notably the conflict between depth cues such as the window violation, where objects "popping out" of the screen appear cut by the frame of the TV. These is an example of interposition depth cue conflicting with the binocular depth cues.

20.4 Technical Aspects

Technically, 3D video has often been implemented as an extension of existing 2D video chains. However, there are several differences [21]. Figure 20.5 depicts a typical transmission chain. In every step of the transmission chain, new types of artifacts can be found due to the specificities of 3D. A classification of these new artifacts based on their origin can be found in [4]. In the following, links between the artifacts and the respective cues and layers as of Fig. 20.1 will be discussed.

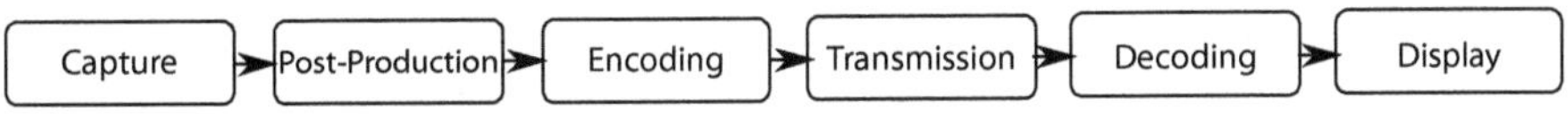

Fig. 20.5 Typical transmission chain

20.4.1 Capture

The production of high quality 3D material is one of the most challenging tasks for 3D video. The use of multiple cameras has induced new kinds of artefacts that may significantly affect quality, such as keystone, or depth quality as related with the cardboard effect, the puppet theater effect, etc. [4]. In addition to these quality aspects, inappropriate combinations of shooting and display conditions can result in highly uncomfortable contents [11], for example due to too high values of disparities.

In addition, other factors due to the use of multiple cameras can impact the comfort, such as vertical misalignment, brightness and color misalignment, temporal offset, different focus point, window violations etc. These factors must be monitored and minimized during the acquisition process [16].

20.4.2 Post Production

In the post-production process, a lot of effort is spent to conceal the artifacts from capture such as brightness, color, vertical alignment between stereoscopic views, correction of geometrical distortion due to lenses. However, these steps cannot solve the issues around the quality of the depth reproduction. Moreover, any processing to make the content available for a different screen size than the original target one may result in a distortion of the depth rendering, due to conflict between monocular and binocular depth cues. The format conversion from the separate views to one of the 3D formats such as "frame compatible" or the "Multi-view plus depth" (MVD) format [52] can also induce new kinds of distortion: loss in spatial resolution for the first one, and synthesis distortion for the latter. The synthesis errors are, at this particular step of post-production, due to the high complexity of the depth estimation from multiple views, if it is required to estimate depth. Indeed the problem of stereo correspondence has not been solved yet, in spite of its long history. The interested reader in the development of such algorithms can refer to [45] where respective work from the early eighties is referred to. Traditional approaches employ assumptions about the physical world, the piecewise-smoothness of the surfaces and assumptions on the camera calibration and epipolar geometry [45], to enable finding solutions to this highly underconstrained problem.

20.4.3 Encoding

The effect of coding on 3D video QoE is multiple. Depending on the representation format and coding algorithm used, transmitted as simulcast, MVC, frame compatible or depth image based rendering (DIBR) [52], loss of spatial resolution may occur. Coding itself has also been found to affect the perceived amount of depth, in addition to the impact on perceived quality [61].

Moreover, one issue specific to MVC and Simulcast approaches is the possibility to perform asymmetric coding, which will result in different perceived quality, depending on the kind of artefact: for example, in case of coding, image quality is perceived as the average of the quality of the individual stereoscopic views [46]. In contrast, in case of blurring, image quality is perceived as the maximum quality between the quality of the two stereoscopic views [46, 50]. As a consequence, downscaling before encoding and transmission has been found as an efficient approach to increase overall quality [55]. However, these studies only consider short sequences, and the long-term effect of such approaches is still unknown, and may be related with the effect of discomfort and visual fatigue. Results in this direction were shown by Seuntiëns, where highly asymmetric coding conditions have shown a significant increase of Eye-strain [46].

MPEG is in charge of the development of the 3D extension of the HEVC standard for 3D video encoding based on depth image based rendering (DIBR) [52]. One of the issues at the coding level is the allocation of the bitrate between texture and depth [7]. Different allocation scenarios will result in different kinds of distortions, either in the texture or in the depth rendering of objects. These distortions are mainly visible at the texture quality level (layer 2 of Fig. 20.1). This was studied in [46], where it was shown that both image quality and depth are affected by coding, but the effect on image quality is much stronger than on perceived depth.

20.4.4 Transmission and Decoding

The effects due to the transmission of 3D video do not principally differ from the 2D case. However, packet loss during transmission has a significantly different effect on the overall quality than for 2D, since in addition to pictorial degradations, it adds binocular rivalries, which make contents painful to watch [33]. These binocular rivalries are due to asymmetric distortion of the pictures. In this particular case, switching back to a 2D presentation might even be preferred by the users, when transmission error only affect one view [2]. Other alternatives for the concealment of 3D video transmission errors may also be considered, for example looking into redundancies between the stereoscopic views [40].

In the case of DIBR, one particular issue is the one of decoding, since this process requires the synthesis of the views from the transmitted texture and depth information. Different algorithms are available for this purpose, as described and evaluated in [6].

This process is particularly error-prone, since it requires inpainting for filling occluded areas which were not available during acquisition, for example due to the specific viewing angle of each camera during shooting. This results in a new kind of distortion of the pictures.

20.4.5 Display

3D displays do not directly show the decoded signal to the user, but carry out some processing before reproduction. Displays have different sizes and different resolutions. Depending on the technology employed (passive, shutter, autostereoscopic [41]) and the display size, the displays have different 3D rendering abilities [1, 12, 28]. In addition, displays can produce new kinds of artifacts such as crosstalk, with specific implications: ghosting, the picket fence effect, flicker, shear distortions, etc. [4], which may strongly affect the perception. To evaluate the quality of the displays, standardization efforts on the characterization of the displays are underway by the International Committee for Display Metrology (ICDM) [23]. The produced standard provides an in-depth description of the methodologies for achieving the characterization of the display properties such as crosstalk, contrast, luminance, effect of head-tilt, etc. [47].

20.5 Evaluation

In Sect. 20.2, a general model of 3D video QoE has been discussed, and the need to introduce new evaluation concepts regarding "naturalness" and "immersion", to capture the multi-dimensionality of 3D QoE. The purpose of this section is to provide additional information on evaluation methods using perception tests and instrumental algorithms.

20.5.1 Subjective Evaluation

Assessing the added-value of 3D as compared to 2D video in quality tests is challenging. The evaluation of multidimensional quality perception as experienced in 3D using single-scale methods such as absolute category rating (ACR) tests was found to fail. An evaluation using a single perceptual scale cannot well capture such high-level concepts as shown on the top layer in Fig. 20.1, "Quality of Experience" [30, 33, 46]. Lower-level concepts, up to the level of "naturalness", "immersion" and viewing experience may however be measurable in perception tests using single scales, if one experiment per scale is targeted [30, 46, 60].

For tackling the overall QoE, however, other evaluation concepts or methodologies may succeed. Explorative studies using small groups of test participants may be used for the identification of the dimensions involved with 3D QoE. Sensory profiling techniques such as the Open Profiling of Quality can be used, too, to let the test participants define the evaluation concepts they find relevant [51]. Other evaluation paradigms using multiple stimuli allowing comparisons between different presentations such as the SAMVIQ methodology combined with an appropriate evaluation concept can also be considered [60]. This enables the test participants to sequentially compare different test conditions, and help them to perform the evaluation by having direct comparisons between the sequences. In addition, the ability to adapt scores after several inspections of the different test conditions can also contribute to achieving higher accuracy in the subjective ratings [29]. Still, the problem persists that participants understand and use adjective scale terms such as "more natural" in different ways. Considering that some of the tasks asked for 3D video QoE evaluation may be difficult for the test participants, the use of paired comparison methodologies can be particularly meaningful. It transforms a difficult question, which takes into account many factors, into a simple binary decision of preference. This is particularly useful for measuring complex concepts such as 3D video QoE in terms of a global measure taking into account all the dimensions involved, such as display size [24], pictorial quality, visual comfort [37], or depth and comfort [35, 36]. This approach also is closer to the concept of acceptance of one system compared to another [24, 35]. Performing a paired comparison experiment is usually time-consuming, but methods have been adopted that allow for significantly reducing the number of pairs [15, 37]. The pairwise comparison data can then be analyzed to estimate perceptual scores using the Bradley-Terry model or the Thurstone-Mosteller model [17], providing the distance between the test conditions on the resulting perceptual quality scale.

However, these evaluation methods are time consuming and costly, and cannot be applied to monitor real-time services. As a consequence, there is the need for instrumental measurement tools, which will be presented in the next subsection.

20.5.2 3D Video QoE Prediction

Key issues that must be considered for 3D QoE prediction are the consideration of which dimensions are to be predicted, and in which application scenario. In many cases, 2D perceptual evaluation algorithms called "C4" [3] and defined in [10] already show high performance as compared to subjective test data obtained for evaluating traditional image coding algorithms, such as JPEG or JPEG2000, yielding a Pearson correlation of 0.92 and RMSE of 0.36 on the 5-point ACR-scale [3]. Asymmetric coding in case of JPEG or JPEG2000 compression can also be estimated well using C4 by averaging the quality estimates for the two stereoscopic views, with resulting Pearson correlation of 0.94 [9]. However, pictorial quality measured by C4 is only one aspect, and visual discomfort can affect the QoE ratings. Obviously, these algorithms only focus on pictorial quality, and only results from quality tests that target this quality dimension can be matched. Full QoE including the preference of 3D over 2D cannot be addressed with these algorithms, as discussed in more detail below. Benoit et al. showed that 3D image quality prediction performance with the Structural SIMilarity image quality metric [54] (SSIM) can be improved by considering the distortion on the disparity maps in addition to the pictorial distortion, however, such improvement was found to be minimal for C4 [3]. Improved approaches for the consideration of pictorial and depth degradation in the context of H.264-based coded video have been considered by Jin et al., by representing and evaluating the distortion in a 3D space [25]. Their method decomposes the signal using a 3D Discrete Cosine Transformation (3D-DCT) to consider different aspects in the distortion evaluation process, such as frequency-based masking, the contrast sensitivity function (CSF) of the human visual system (HVS), and luminance masking. This model shows a high performance in comparison to subjective test data with a Pearson correlation of 0.92.

3D video QoE prediction is however not solved, since with a new application scenario, the DIBR video coding scheme, traditional 2D objective metrics completely fail to capture 3D QoE [5, 57]. From the twelve quality prediction algorithms compared in [5], none achieves a higher Pearson correlation than 0.4. This is due to the nature of the new kind of distortions induced by image synthesis, which are not captured by traditional prediction approaches, and first alternative algorithms have been proposed for this particular scope [5, 18].

In addition, the prediction algorithms mentioned in this subsection were compared with quality scores obtained from an ACR-based test which, as explained in Sect. 20.2, fails to capture the differences between 2D and 3D. The assessment of evaluation concepts such as "naturalness" and "immersion", which show the QoE improvement of 3D over 2D, requires the prediction of the other factors at the level 2 of Fig. 20.1 in addition to pictorial quality: visual discomfort and depth. Prediction algorithms for these two factors have been developed, namely for comfort [19, 42, 48], and for depth [32, 44, 65]. However, no instrumental prediction for the overall 3D video immersion or naturalness are available. This topic will require further study, and should be a direction for future research on QoE prediction algorithms.

20.6 Discussion

Splitting the QoE concept for 3D video into multiple dimensions, and organizing the different perceptual constructs into different layers, allows each of the aspects that are involved to adequately be tackled. Each aspect may have different technical causes, and its impact on the global QoE may depend on the severity of other aspects, leading to complex dependencies. As a consequence, quality test methodologies, the understanding of the human visual system (HVS), and the technical algorithms, both on the service exploitation as well as on the measurement side, co-exist, interact, and advance based on their results. Further understanding of the HVS will refine the rules and guidelines that may lead to improved technology, while technical advances allow for further isolating the influence factors on the HVS and the resulting QoE. An example is the current stereoscopic display technology, which has largely advanced from high crosstalk, anaglyphic reproduction which were used in some cases for TV such as in the United Kingdom in November 2009 [59], to better view separation, Full-HD active shutter technology. Another step forward may be expected with the upcoming Ultra-High Definition (UHD) technology, which not only enables line-interleaved polarized displays at Full-HD resolution, but also autostereoscopic displays with reasonable resolution per view. Hence, the "4K" (3,840 × 2,160 pixels, i.e. 4 times HD resolution) will not only be of interest for higher definition 2D, but especially is expected to significantly improve 3D display technology. OLED TV will also provide the ability to enhance the overall experience by enabling higher picture quality, which will also benefit 3D.

Currently, 3D display manufacturers and costumers are cautious with promoting and accepting 3D content, since health issues have not been completely negated. Visual discomfort, as the immediate sensation of an unusual and often unpleasant state, may be captured from individual observer's opinion or through questionnaires such as the "Simulator of sickness" [8, 27]. There may, however, also be long-term effects, such as the dry-eye syndrome known from 2D screens, that have been subsumed under the term "visual fatigue". Visual fatigue may be diagnosed in a medical sense from various factors that may be reported by observers, or be measured objectively. So far, only limited knowledge exists about the long term effects of watching video on 3D screens, and adaptation effects may overlap with fatigue. Both visual discomfort and visual fatigue may be predicted using objective evaluations of the content shown on the screen. In the future, the measurement of QoE in 3D video needs to take into consideration all the aspects shown in Fig. 20.1. When the technological advances reach a similar level as in 2D, it is expected that the 3D video QoE will be appreciated by the viewer, clearly outperforming the 2D video experience.

References

1. Andrén B, Wang K, Brunnström K (2012) Characterizations of 3D TV: active vs passive. In: SID symposium digest of technical papers
2. Barkowsky M, Wang K, Cousseau R, Brunnström K, Olsson R, Callet PL (2010) Subjective quality assessment of error concealment strategies for 3DTV in presence of asymmetric transmission errors. In: 18th international of Packet video workshop (PV)
3. Benoit A, Callet PL, Campisi P, Cousseau R (2008) Quality assessment of stereoscopic images. In: IEEE international conference on image processing. ICIPSan Diego, California, pp 1231–1234
4. Boev A, Hollosi D, Gotchev A, Egiazarian K (2009) Classification and simulation of stereoscopic artifacts in mobile 3DTV content. In: Proceedings of the SPIE 7237, stereoscopic displays and applications XX, vol 7237
5. Bosc E, Pepion R, Callet PL, Köppel M, Ndjiki-Nya P, Pressigout M, Morin L (2011) Towards a new quality metric for 3-D synthesized view assessment. IEEE J Sel Top Sign Process 5(7):1332–1343

6. Bosc E, Pepion R, Callet PL, Pressigout M, Morin L (2012) Reliability of 2D quality assessment methods for synthesized views evaluation in stereoscopic viewing conditions. In: 3DTV-conference: the true vision capture, transmission and display of 3D video (3DTV-CON)
7. Bosc E, Riou P, Pressigout M, Morin L, (2012) Bit-rate allocation between texture and depth: influence of data sequence characteristics. In: 3DTV-conference, (2012) the true vision capture, transmission and display of 3D video. Zurich, Switzerland
8. Brunnström K, Wang K, Andrén B (2013) Simulator sickness analysis of 3D video viewing on passive 3DTV. In: Stereoscopic displays and applications XXIV
9. Campisi P, Callet PL, Marini E (2007) Stereoscopic image quality assessment. In: European signal processing conference
10. Carnec M, Callet PL, Barba D (2003) An image quality assessment method based on perception of structural information. In: Proceedings of the IEEE international conference on image processing (ICIP 03), vol 2. Barcelona, Spain, pp 185–188
11. Chen W, Fournier J, Barkowsky M, Callet PL (2011) New stereoscopic video shooting rule based on stereoscopic distortion parameters and comfortable viewing zone. Stereoscopic displays and applications XXII. Proceedings of the SPIE, vol. 7863, pp 786310–786313
12. Chen W, Jérôme F, Marcus B, Patrick LC (2010) New requirements of subjective video quality assessment methodologies for 3DTV. In: Video processing and quality metrics 2010 (VPQM). Scottsdale
13. Consortium, 3D@home.& the MPEG Industry Forum 3DTV Working Group: Glossary for video & perceptual quality of stereoscopic video. Available online at: http://www.3dathome.org (2010)
14. Cutting JE, Vishton PM (1995) Perceiving layout and knowing distance: the integration, relative potency and contextual use of different information about depth. In: Epstein W, Rogers S (Eds) Perception of space and motion. Handbook of perception and cognition (2nd ed.), pp. 69–117. San Diego, CA, US: Academic Press
15. Dykstra O (1960) Rank analysis of incomplete block designs: a method of paired comparisons employing unequal repetitions on Pairs. Biometrics 16(2):176–188
16. Grau O, Müller M, Kluger J (2011) Tools for 3D-TV programme production. British Broadcasting Corporation, technical report
17. Handley JC (2001) Comparative analysis of Bradley-Terry and Thurstone-Mosteller model of paired comparisons for image quality assessment. In: Proceedings of the IS&T's image processing, image quality, image capture, systems conference
18. Hanhart P, Simone FD, Ebrahimi T (2012) Quality assessment of asymmetric stereo pair formed from decoded and synthesized views. In: Fourth international workshop on Quality of Multimedia Experience (QoMEX), pp 236–241
19. He S, Zhang T, Doyen D (2011) Visual discomfort prediction for stereo contents. In: Proceedings of the SPIE 7863, stereoscopic displays and applications XXII
20. Hillis JM, Watt SJ, Landy MS, Banks MS (2004) Slant from texture and disparity cues: optimal cue combination. J Vision 4:967–992
21. Huynh-Thu Q, Callet PL, Barkowsky M (2010) Video quality assessment: from 2D to 3D–challenges and future trends. In: 17th IEEE Quality of Multimedia Experience (QoMEX), Hong Kong, pp 4025–4028
22. Ijsselsteijn W, Ridder HD, Freeman J, Avons SE, Bouwhuis D (2001) Effects of stereoscopic presentation, image motion, and screen size on subjective and objective corroborative measures of presence. Presence Teleoperators Virtual Environ 10:298–311
23. Information Display Society: Display measurements standard. In:http://icdm-sid.org/Public/DMS/ICDM-DMS.html
24. ITU-T Contribution COM 12–C192-E (2011) Comparison of the ACR and PC evaluation methods concerning the effects of video resolution and size on visual subjective ratings. In: ITU. SG12 Meeting, Geneva
25. Jin L, Boev A, Gotchev A, Egiazarian K (2011) 3D-DCT based perceptual quality assessment of stereo video image. In: 18th IEEE International Conference on Image Processing (ICIP), pp 2521–2524

26. Kaptein RG, Kuijsters A, Lambooij MTM, IJsselsteijn WA, Heynderickx I (2008) Performance evaluation of 3D-TV systems. Image quality and system performance V. Proceedings of SPIE, vol 6808. pp 1–11
27. Kennedy RS, Lane NE, Berbaum KS, Lilienthal MG (1993) Simulator sickness questionnaire: an enhanced method of quantifying simulator sickness. Int J Aviat Psychol 3:203–220
28. Kim JS, Banks S (2012) Effective spatial resolution of temporally and spatially interlaced stereo 3D televisions. In: SID symposium digest of technical, pp 879–882
29. Kozamernik F, Steinmann V, Sunna P, Wyckens E (2005) SAMVIQ—a new EBU methodology for video quality evaluations in multimedia. SMPTE Mot Imag 114:152–160
30. Lambooij M (2011) IJsselsteijn W, Bouwhuis DG, Heynderickx I (2011) Evaluation of stereoscopic images: beyond 2D quality. IEEE Trans Broadcast 57(2):432–444
31. Lambooij M, IJsselsteijn W, Heynderickx I (2011) Visual discomfort of 3D-TV: assessment methods and modeling. Displays 32:209–218
32. Lebreton P, Raake A, Barkowsky M, Callet PL (2012) Evaluating depth perception of 3D stereoscopic videos. IEEE J Sel Top Sig Process 6:710–720
33. Lebreton P, Raake A, Barkowsky M, Callet PL (2011) A subjective evaluation of 3D IPTV broadcasting implementations considering coding and transmission degradation. In: IEEE international workshop on Multimedia Quality of Experience, MQoE11. Dana Point
34. Lebreton P, Raake A, Barkowsky M, Callet PL (2012) Perceptual depth indicator for S-3D content based on binocular and monocular cues. In: Asilomar. Pacific Grove
35. Lebreton P, Raake A, Barkowsky M, Callet PL (2013) Perceptual preference of S3D over 2D for HDTV in dependence of video quality and depth. In: IVMSP workshop: 3D image/video technologies and applications. Seoul, Korea
36. Lee JS, Goldmann L, Ebrahimi T (2012) Paired comparison-based subjective quality assessment of stereoscopic images. Multimedia Tools Appl 1–18.
37. Li J, Barkowsky M, Callet PL (2012) Analysis and improvement of a paired comparison method in the application of 3DTV subjective experiment. In: IEEE International Conference on Image Processing (ICIP), Orlando
38. Li J, Barkowsky M, Wang J, Callet PL (2011) Study on visual discomfort induced by stimulus movement at fixed depth on stereoscopic displays using shutter glasses. In: 17th international conference on digital signal processing. Corfu, Grece
39. Lovell PG, Bloj M, Harris JM (2012) Optimal integration of shading and binocular disparity for depth perception. J Vision 12:1–18
40. Matthias CC, Kunter M, Knorr S, SikoraT (2004) A hybrid approach for error concealment in stereoscopic images. In: 5th international workshop on image analysis for multimedia interactive services
41. Pastoor S, Wopking M (1997) 3-D displays: a review of current technologies. Displays Technol Appl 2(17):100–110
42. Richardt C, Swirski L, Davies IP, Dodgson NA (2011) Predicting stereoscopic viewing comfort using a coherence-based computational model. In: Computational aesthetics in graphics, visualization, and imaging
43. Robinson TR, Toronto BA (1896) Light intensity and depth perception. Am J Psychol 7(4):518–532
44. Ross MG, Oliva A (2010) Estimating perception of scene layout properties from global image features. J Vision 10(1):2, 1–25
45. Scharstein D, Szeliski R (2002) A taxonomy and evaluation of dense two-frame stereo correspondence algorithms. Int J Comput Vision 47:7–42
46. Seuntiëns PJ (2006) Visual experience of 3D-TV. Ph.D. Thesis, Eindhoven University
47. Society for information display (2012) Information display measurement standard. International Committee for display metrology
48. Sohn H, Jung YJ, Lee S, Man Y (2013) Predicting visual discomfort using object size and disparity information in stereoscopic images. IEEE Trans Broadcast 59(1):28–37
49. Speranza F, Tam WJ, Renaud R, Hur N (2006) Effect of disparity and motion on visual comfort of stereoscopic images. In: Stereoscopic displays and virtual reality systems XIII, vol 6055

50. Stelmach L, Tam WJ, Meegan D, Vincent A (2000) Stereo image quality: effects of mixed spatio-temporal resolution. IEEE Trans Circuits Syst Video Technol 10(2):188–193
51. Strohmeier D, Jumisko-Pyykkö S, Kunze K (2010) Open profiling of quality: a mixed method approach to understanding multimodal quality perception. Adv Multimedia 2010:1–28
52. Vetro A, Tourapis AM, Müller K, Chen T (2011) 3D-TV content storage and transmission. IEEE Trans Broadcast Spec Issue 3D-TV Horizon: Contents Syst Visual Percept 57(2):384–394
53. Wagemans J, van Doorn AJ, Koenderink JJ (2011) Pictorial depth probed through relative sizes. i-Perception 2:992–1013
54. Wang Z, Bovik AC, Sheikh HR, Simoncelli EP (2004) Image quality assessment: from error visibility to structural similarity. IEEE Trans Image Process 13(4):600–612
55. Wang K, Barkowsky M, Brunnström K, Sjöström M, Cousseau R, Callet PL (2012) Perceived 3D TV transmission quality assessment: multi-laboratory results using absolute category rating on quality of experience scale. IEEE Trans Broadcast 58:544–557
56. Wang J, BarkowskyM, Ricordel V, Callet PL (2011) Quantifying how the combination of blur and disparity affects the perceived depth. In: Proceedings of the SPIE. Human vision and electronic imaging XVI, vol 7865. pp 78650K–78650K-10
57. Wang K, Brunnström K, Barkowsky M, Le Callet P, Sjöström M, Tourancheau S (2013) Stereoscopic 3D video coding artifacts quality evaluation with 2D objective metrics. In: Stereoscopic displays and applications XXIV, vol 8648
58. Watt SJ, Akeley K, Ernst MO, Banks MS (2005) Focus cues affect perceived depth. J Vision 5(10):834–862
59. 3D Week 2009.10. 11-18. Retrieved 2009–11-18. Glasses that will work for Channel 4's 3D week are the Amber and Blue Colour Code 3D glasses
60. Wei C, Fournier J, Barkowsky M, Callet PL (2012) Exploration of quality of experience of stereoscopic images: binocular depth. In: International workshop on video processing and quality metrics for consumer electronics (VPQM). Scottsdale
61. Yamagishi K, Karam L, Okamoto J, Hayashi T (2011) Subjective characteristics for stereoscopic high definition video. In: Third international workshop on Quality of Multimedia Experience (QoMEX). Mechelen, Belgium
62. Yamanoue H, Okui M, Yuyama I (2000) A study on the relationship between shooting conditions and cardboard effect of stereoscopic images. IEEE Trans Circuits Syst Video Technol 10(3):411–416. doi:10.1109/76.836285
63. Yamanoue H, Okui M, Okano F (2006) Geometrical analysis of puppet-theater and cardboard effects in stereoscopic HDTV images. IEEE Trans Circuits Syst Video Technol 16:744–752
64. Yano S, Emoto M, Mitsuhashi T (2004) Two factors in visual fatigue caused by stereoscopic HDTV images. Displays 25:141–150
65. Yasakethu S, Silva DD, Fernando W, Kondoz A (2010) Predicting sensation of depth in 3D video. Electron Lett 46(12):837–839

Chapter 21
Crowdsourcing in QoE Evaluation

Tobias Hoßfeld and Christian Keimel

Abstract Crowdsourcing enables new possibilities for QoE evaluation by moving the evaluation task from the traditional laboratory environment into the Internet, allowing researchers to easily access a global pool of subjects for the evaluation task. This makes it not only possible to include a more diverse population and real-life environments into the evaluation, but also reduces the turn-around time and increases the number of subjects participating in an evaluation campaign significantly by circumventing bottle-necks in traditional laboratory setup. In order to utilise these advantages, the differences between laboratory-based and crowd-based QoE evaluation must be considered and we therefore discuss both these differences and their impact on the QoE evaluation in this chapter.

21.1 Introduction

Quality of Experience (QoE) evaluations are usually performed in specially equipped laboratories according to established protocols and methods. The QoE evaluation in this standardised environment is well understood, allowing for reproducible and reliable results. Unfortunately, there are also several disadvantages: the number of simultaneous participants is limited, the demography of the subjects is often not representative of the diversity encountered in the general population, the evaluation environment does not reflect the majority of real-life environments in which the stimuli are consumed and lastly, depending on the location, a possible reimbursement of the subjects can introduce an additional financial burden.

T. Hoßfeld (✉)
Institute of Computer Science, Chair of Communication Networks,
University of Würzburg, Würzburg, Germany
e-mail: hossfeld@informatik.uni-wuerzburg.de

C. Keimel
Institute for Data Processing, TU Munich, Munich, Germany
e-mail: christian.keimel@tum.de

S. Möller and A. Raake (eds.), *Quality of Experience*,
T-Labs Series in Telecommunication Services, DOI: 10.1007/978-3-319-02681-7_21,

Crowdsourcing provides an alternative to this traditional approach, elevating many of these issues by using the Internet to assign evaluation tasks to a group of online workers. QoE evaluations are thus no longer performed in a laboratory, but conducted via the Internet with participants from all over the world, in a multitude of environments, representing real-life conditions. This not only allows us to recruit the subjects from a larger, more diverse group, but also to either reduce the financial expenditures significantly or alternatively hire more subjects, leading to more representative results. Moreover, the global worker pool allows for studies targeting different demographics and cultures, usually not possible in evaluations limited to the location of an evaluation laboratory. Another advantage of the distributed evaluation with crowdsourcing is the faster turn-around time and instead of days, test campaigns can be completed within hours.

Moving from the laboratory to the crowd, however, is not as straight-forward as simply generating a web-interface for an existing test. There are significant differences between laboratory-based and crowd-based evaluation with respect to conceptual, technical and motivational aspects that need to be considered when performing the crowd-based QoE evaluation. But if the unique properties of the crowdsourcing environment and their impact on the QoE evaluation are considered appropriately, crowd-based QoE evaluation provides an efficient, simple, cheap and more representative alternative to traditional laboratory-based QoE evaluation

In this chapter, we provide a short introduction on crowdsourcing for QoE evaluation (Sect. 21.2), followed by highlighting the differences between the crowd-based evaluation and the evaluation in the laboratory (Sect. 21.3). We then discuss in detail the impact of the crowdsourcing environment on the QoE assessment task (Sect. 21.4). Finally, we provide the conclusions and lessons learned from this chapter (Sect. 21.5).

21.2 Background on Crowdsourcing for QoE Evaluation

Before we compare advantages and disadvantages of laboratory-based and crowd-based QoE evaluation, this section provides a brief introduction into the crowdsourcing-principle, common crowdsourcing platforms and reviews existing crowd-based frameworks for QoE evaluation.

21.2.1 Crowdsourcing Concept

Crowdsourcing is a further development of the outsourcing principle, where the granularity of work is reduced as well as the administrative overhead. In outsourcing, tasks are performed by designated workers or subcontractors, whereas in crowdsourcing the task is submitted to a huge crowd of anonymous *workers* in the form of an open call. Crowdsourcing tasks can be accomplished within a few minutes to a few hours and do not require a long-term employment. Tasks are often highly repetitive e.g. image annotation or speech recognition, and are usually grouped in larger units, referred to as *campaigns*. Most *employers* submitting tasks to an anonymous crowd use a mediator who maintains the crowd and manages the employers campaigns. These mediators are called crowdsourcing platforms offering web-based access and services. Some platforms allow restrictions of the anonymous crowd for certain tasks e.g. based on their country of residence.

Amazon's Mechanical Turk (MTurk) [2], Microworkers [22], and Facebook [10] are typically used Crowdsourcing platforms. MTurk and Microworkers are commercial platforms with their own worker crowds also denoted as *human cloud*. For the successful execution of a task, the worker gets paid by the platform on behalf of the employer defining successful task completion. MTurk is one of the largest crowdsourcing platforms and is often used in research studies and in commercial third-party applications. It provides an API, several filters and qualification test mechanisms are available and the main workforce of this platform is located in the USA and in India [24], but only US residents or companies can legally submit tasks to the MTurk platform. In contrast to MTurk, Microworkers also allows international employers, its worker are more diverse [12], and the redirection of workers to own servers or test applications is permitted. Additionally, the employer can define specialized groups of workers, based on certain skills, for example, workers from Germany with fluency in French or based on the experience with the workers from previously conducted campaigns. Microworkers does not offer some features of MTurk yet, especially no API and neither elaborated test nor qualification campaigns, leading to a more difficult identification of trustworthy workers.

Besides these commercial providers, Facebook and other social networks can be used to recruit test users for free. However, redesigning a user test to be suitable for a social network environment imposes a significant amount of additional work and is not always possible. Also participants recruited from a social network might be biased in terms of expectations or test behaviour and need to be provided with the right incentives for participating such as gamification [25].

21.2.2 Existing Crowdsourcing Frameworks for QoE Evaluation

Crowdsourcing tests for QoE evaluation require the presentation and assessment of the different stimuli in a suitable web-interface. Instead of implementing an appropriate interface separately for each QoE test, existing frameworks can be utilised and we briefly discuss two examples of QoE evaluation frameworks for Crowdsourcing: the *Qudrant of Euphoria* by Chen et al. [5] and *QualityCrowd* by Keimel et al. [18].

Chen's *Quadrant of Euphoria* provides an online service for the QoE evaluation of audio, visual, and audio-visual stimuli. It allows for a pairwise comparison of two different stimuli in an interactive web-interface, where the worker can judge which of the two stimuli has a higher QoE. Additionally, the platform provides some rudimentary reliability assessment based on the actual user ratings under the assumption that the preferences of users are transitive relations: if a user prefers the test condition A to B and B to C, the user will also prefer A to C. If this condition is not met for a certain number of triplets, the user is rejected.

The second example, the *QualityCrowd* framework, is not an online service, but a complete platform designed especially for QoE evaluation with crowdsourcing. QualityCrowd is an open-source project that can be installed and modified with relatively low effort on any suitable web server [28]. Using this framework a test can consist of any number of questions, be compromised of videos, sounds or images, or a combination thereof and it facilitates the use of different testing methodologies, e.g. single stimulus or double stimulus, and different scales, e.g. discrete or continuous quality or impairment scales. In its latest iteration QualityCrowd2, QualityCrowd provides a simple scripting language allowing the creation of test campaigns with a high flexibility, e.g. to combine video and still image evaluations, choose from different testing methodologies, specify a training session and introduce control questions to identify reliable user ratings and ensure high data quality.

21.3 Comparison of Crowdsourcing and Laboratory QoE Studies

Crowdsourcing provides a compelling alternative to the traditional QoE evaluation in the laboratory. The main advantage of crowdsouring is the vastly larger pool of subjects with a significantly more diverse background compared to usual laboratory-based evaluation, where the demographic of the test subjects is often rather limited. Diversity in this context focuses on the cultural background of the subjects and the resulting differences in the experienced quality between different countries. Although the last disadvantage of the laboratory-based evaluation can easily be avoided by recruiting the subjects more selectively, this issue is usually neglected and especially in an academic setting often students with a similar background e.g. only engineering students are selected for convenience. There are, however, also fundamental differences between a crowd-based and laboratory-based QoE evaluation in *conceptual*, *technical* and *motivational* areas [17] as listed in Table 21.1.

Conceptual differences arise mainly from the fact that on the one hand crowdsourcing tasks are usually much shorter than comparable laboratory tests, and on the other hand the test supervisors have significantly less control over the participating test subjects. In general, crowd-based QoE evaluation should be in the order of minutes, whereas laboratory QoE evaluations are in the order of tens of minutes. Hence, it is not always possible to map the structure of an existing laboratory evaluation directly to a crowd-based evaluation, but the structure needs to be split into multiple smaller task. This also implies that unlike in the laboratory, not all test conditions will be assessed by all subjects, making common statistical outlier detection not applicable e.g. ITU-R BT.500 [16]. Moreover, the test supervisors have much less control over the subjects and therefore it is more difficult in crowd-based

Table 21.1 Differences in QoE studies in the laboratory and with crowdsourcing

Differences	Crowdsourcing	Laboratory
Conceptual		
Test duration	5–15 min	30–60 min
Outlier detection	Common statistical methods often not applicable	Common statistical methods
Training	Can not be ensured	Training with feedback
Technical		
Environment and equipment	Real-life environment	Standardised and artificial
Stimuli	Mostly limited to web-supported audio and video	Multi-sensory
Design	Limited by internet access, web-browsers and devices	No limitations
Motivational and subjects		
Demography	Global and diverse	Local and limited
Incentives	Mostly financial	Financial and altruistic
Cost	Cheap	Expensive

evaluations to ensure a proper training of the subjects, in particular as no direct feedback between supervisors and subjects is possible.

Technical differences are related to the web-based nature of crowdsourcing. In contrast to the standardised environment and equipment in a laboratory-based evaluation, the crowd-based evaluation is performed in a more real-life environment using consumer devices, reflecting the everyday experience of people. This, however, also implies that evaluations requiring explicitly a controlled environment e.g. for determining the thresholds of just noticeable differences of stimuli are not suitable for crowd-based evaluation. Web browsers and consumer internet devices, however, limit both the complexity and the stimuli e.g. no eye tracking or haptic stimuli, respectively, are possible. Further the Internet transmission of the test contents has to be taken into account in the implementation of the test.

Motivational differences are caused by the usually pure financial incentive for the crowd-workers, unlike test subjects in laboratory evaluations that often have a higher intrinsic motivation due to an interest in the evaluation's goals. Hence it is not uncommon that workers on the one hand try to cheat by providing bogus results, or on the other hand do not evaluate the test conditions as diligently as subjects in the laboratory, as the more evaluation tasks a worker performs, the higher his financial gain is and thus the less time spent on a task, the better.

Differences in the reliability of the results between crowd-based and laboratory-based evaluations are mainly due to the motivational issues described above, but also due to conceptual and technical differences to the laboratory-based evaluation, as we can neither ensure that the workers are properly trained, nor are we able to ensure that the stimuli are presented as intended.

Considering these differences, the limitations and reliability issues caused by the conceptual and technical differences can be overcome by appropriately designing the QoE evaluation with crowdsourcing in mind. The last issue related to the motivational difference, however, is not as easily overcome as it is related to the workers and not the infrastructure provided by the evaluations' designers. Nevertheless with sufficient checks, reliable results can be gained as discussed and demonstrated in the next section.

21.4 Impact of Crowdsourcing Environment on QoE

In order to understand the impact of the Crowdsourcing environment on the QoE assessment task, it is illustrative to repeat QoE assessment tests in the Crowdsourcing environment and compare the achieved results. We discuss the unique influence of the Crowdsourcing environment and the corresponding task design on the example of two different subjective QoE assessment tests: one video and one image quality evaluation. For both examples, we use data from existing, well-documented laboratory experiments and re-implemented the tests as close as possible in the Crowdsourcing environment. Both tests were implemented with an interactive web-interface, providing a slider for continuous quality assessment similar to the slider used in the laboratory set-ups.

21.4.1 Task Design and Incentives

The task design in crowdsourcing has to take into account the distributed and remote test environment in order to ensure high data quality. The actual user ratings are affected because of the QoE influence factors which are additionally emerging from the remote setting and which are not directly controlled [11]. Thus, it is necessary to monitor the users' environment in order to identify additional influence factors on the QoE assessment. For instance the effect of the viewing environment on quality of subjective rating for QoE evaluation of video having coding distortions is well known like the impact of devices [4, 23] or viewing conditions [3]. Reliability mechanisms are necessary to identify and filter out ratings from unreliable users or wrong test conditions, whereas incentive mechanisms aim at increasing the data quality. There are several reasons why some user ratings are not reliable and need to be filtered out in order to avoid false QoE results: wrong test conditions may occur due to errors in the web-based test application or due to incompatibilities of the test application with the subject's hard- and software. Unreliable user ratings may also be caused by unclear or too complex test instructions. Similarly, language problems may occur with international users. Furthermore, there may also be *cheating* users who try to submit invalid or low quality work in order to maximize their received payment while reducing their own effort. This is even the case if the expected gain is very small [27].

Reliability Mechanisms. Numerous efforts have been made in order to improve the quality of the results submitted by the workers and to detect cheating workers. Hoßfeld et al. [14] add different elements to check the reliability of the user in the task design, consisting of consistency tests [1, 7], content questions [21], gold data [15], and application-layer monitoring. Examples of consistency tests include, but are not limited to the repetition of the same test condition twice, simple questions that are unrelated to the test content like human computation of simple text equations ('two plus 3=?') or knowledge questions like 'In which continent lies Italy? Europe, Asia, New Zealand'. Content questions take the content in the actual test contents into account by asking, for example, 'Which type of sport was shown in the video? Tennis, football or horse riding'. Lastly, application-layer monitoring analyses measurable properties like the response time of the user or the viewing time of a video. However, all mechanisms mentioned above should not influence the true QoE assessment test.

General Task Design. Beyond the integration of reliability mechanisms, a task should be designed in such a way that there is no incentive for the user to cheat. Kittur et al. [20] concludes that a task should be designed in such a way that cheating takes approximately the same time as faithfully completing it. Eickhoff and de Vries [9] discourages cheaters instead of detecting them by appropriate task design. Tasks that require creativity or abstract thinking decrease the ratio of cheaters, as money-driven workers prefer simple tasks over creative ones. However, too difficult questions or too complex tasks may also discourage test participants. Therefore, a good trade-off has to be found in practice. Also long tasks should be split into smaller tasks, as the task duration has a severe impact on the cheater ratio. Workers' share of previously accepted submissions as provided by Crowdsourcing platforms, however, is not a robust measure of worker reliability [9].

Incentives. Incentives play a key role in the successful use of crowdsourcing. They ensure high data quality and are complementary to the reliability mechanisms [13]. While monetary interests may be the key driver in commercial crowdsourcing, other incentives address social aspects, entertainment and altruism [26]. Altruistic crowdsourcing is carried out by volunteers with a desire to help in, for example, scientific research or community work. Gamification is the concept to develop incentives aiming at entertainment and fun for the subjects. Therefore games with a purpose enable human contributors to carry out tasks as a side effect of playing online games, for example, computation tasks [25] or data/image labelling [8, 30]. Gamification is strongly task related and there are no general guidelines on how to design a game especially for QoE assessment. Nevertheless, the results using the gamification approach are very promising. For example, [8] shows that gamification reduces fake ratings significantly by a factor of five and that innovative, creative tasks are less likely to invite cheating, increase data quality and efficiency. In summary, gamification has the potential to make crowdsourcing an even more powerful tool for quality testing.

Example: **Video Quality Test Results**. We illustrate the influence of different incentives and task designs on the user ratings on the example of a video quality evaluation. The video quality test is based on a laboratory test performed by De Simone et al. [6] and the corresponding *EPFL-PoliMI* data set consists of six video sequences, compressed with H.264/AVC at different bitrates and transmission errors, resulting

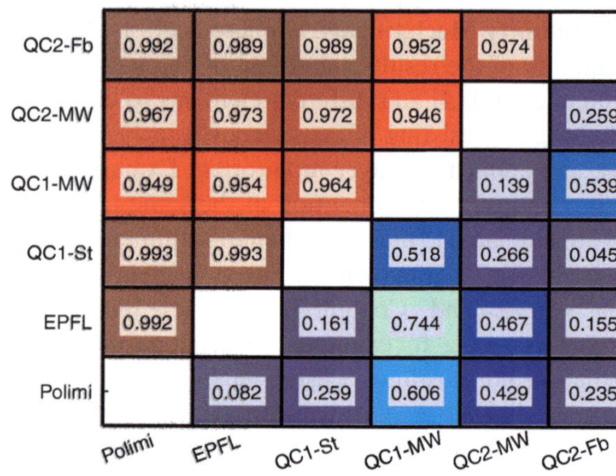

Fig. 21.1 Differences of the MOS values between two video quality studies in terms of the **a** pearson linear correlation coefficient in the upper part of the matrix and **b** mean squared error (MSE) in the lower part. Additionally, a linear regression curve is provided

in 78 different processed videos with corresponding MOSs. One unique feature of this data set is that the test was performed in two different laboratories, at *EPFL* and *PoliMi*, allowing us to also compare the results from Crowdsourcing with the inter-lab agreement in this test.

The experiments were re-implemented with two different task designs with the QualityCrowd framework [18, 19] and different demographies. The crowd used in the test is indicated by the two ending characters ('-ST': students; '-MW': Microworkers; '-Fb': Facebook) in Fig. 21.1. The first adaptation of the video test to the Crowdsourcing environment is a straight-forward task design, where each task consists of exactly one video to be assessed, denoted as *QualityCrowd1* ('QC1'). In the second adaptation, each task consists of five videos including content questions and additionally each worker was required to participate in a training task, denoted as *QualityCrowd2* ('QC2').

Figure 21.1 shows the Pearson correlation of the MOS values between any two video quality studies in the upper part of the matrix, and the lower part quantifies the difference of the MOS values in terms of the mean squared error. Figure 21.2 compares the MOS values for certain test conditions with the results from the 'EPFL' laboratory study and we can observe the following:

1. Crowdsourcing leads to similar results as in the lab (e.g. 'QC2-Fb' and 'EPFL').
2. Crowdsourcing frameworks with reliability checks (QC2) lead to better results.
3. Users with the same incentives, altruistic users in 'QC1-St' and 'QC2-Fb', paid users in 'QC1-MW' and 'QC2-MW', and lab users in 'Polimi' and 'EPFL' provide similar MOS ratings. However, subjects with different incentives can not fully replicate the behavior of all users as also concluded in [23].
4. Paid Crowdsourcing users providing unreliable ratings have a severe impact on the MOS as seen in 'QC1-MW'. Consequently, reliability and screening mechanisms must be included in both test and the analysis.

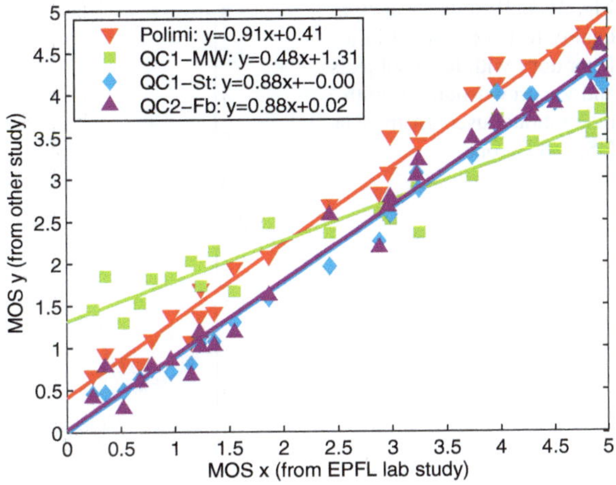

Fig. 21.2 Impact of test environment (laboratory or crowdsourcing) and incentives (paid vs. altru-istic crowdsourcing) on the MOS values in the video quality tests

5. Results from laboratory- and crowd-based evaluations may differ absolutely, possibly caused by different incentives or context and users, but the shift of the curves can easily be taken into account by a normalisation procedure.

In summary, the results achieved with workers acquired in a social network show similar results to the inter-lab correlation between different evaluation laboratories, regardless of the chosen task design and the integrated reliability mechanisms. Yet, workers hired with purely financial incentives on common Crowdsourcing platforms provide results that clearly depend on the task design: for our simple design in *QualityCrowd1*, without any worker training and control question, the results are significantly worse than the results achieved with the two-stage design [13] of training and content questions in *QualityCrowd2*.

21.4.2 Broadening QoE Research

Crowdsourcing offers unique possibilities for QoE research, allowing the investigation of research questions so far not considered or not feasible to consider. Using the example of image quality evaluation, we will sketch how Crowdsourcing gives researchers a powerful tool for QoE assessment. Perceived quality is influenced by factors on four different levels: context, user, system, and content. In image quality, the *content level* addresses format and resolution, but also the general type of content like landscape photos or clip-arts. The technical influence factors are abstracted on the *system level*. They cover influences of the devices, the displays, the transmission network to deliver the images, possibly causing waiting times or even image artifacts, but also the implementation of the application itself like progressive image download. Due to the low costs and fast turn-around of Crowdsourcing evaluation, these categories may be analysed with a statistically sufficiently large number of subjects.

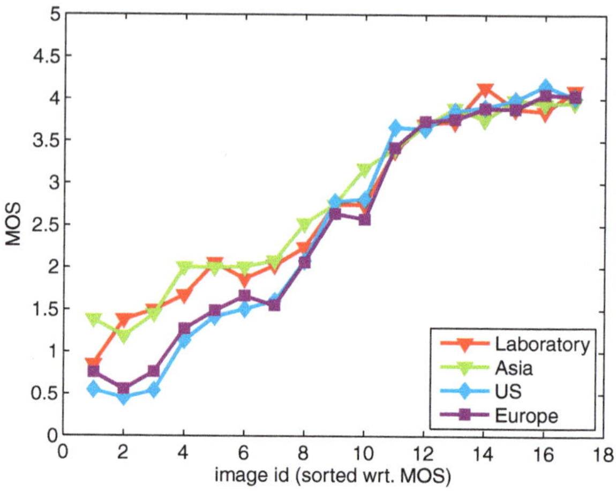

Fig. 21.3 Results for image quality tests conducted in a laboratory with local subjects and via crowdsourcing with international users from Asia, US, and Europe

The *user level* includes psychological factors like expectations and experiences of the user, but also memory and recency effects. The *context level* considers aspects like the environment in which the user is consuming the service, the socio-cultural background, or the purpose of using the service like entertainment or information retrieval. Crowdsourcing enables new ways to research all these factors and quantify their impact on QoE. For example, a QoE influence factor on the user level can be the users' expectations: those used to lower quality represented by low video resolution will rate differently than those typically consuming higher quality as represented by high video resolution. The expectation level may be closely related to the country of the subject and users from different regions may have different expectations about the provided content quality.

The impact of demographics on QoE is illustrated using an image quality test as example. Wang et al. [31] designed and conducted a laboratory test and from the resulting *LIVE* data set we selected 17 images and their corresponding MOSs from the JPEG compressed images within this dataset. Figure 21.3 compares the MOS values from the laboratory study with the results from the Crowdsourcing study from different regions. Figure 21.4 shows the squared MOS difference from the different regions to the LIVE dataset, normalised by the absolute MOS value from the laboratory. We can see that there are differences (up to 1 MOS value) across regions for MOS values below 2.5. For higher image qualities, there is no impact of the demographics. For this experiment, the results from US and Europe lead to very similar results, while the subjects in the laboratory and the Asian worker agree in their image quality ratings in that study. Note that the laboratory study was performed in the US in 2004 and this could explain the agreement of Asian and labratory subjects, as expectations may change over time regarding image quality. Similar results of the impact of demographics on MOS was also observed in other studies e.g. regarding web aesthetics [29].

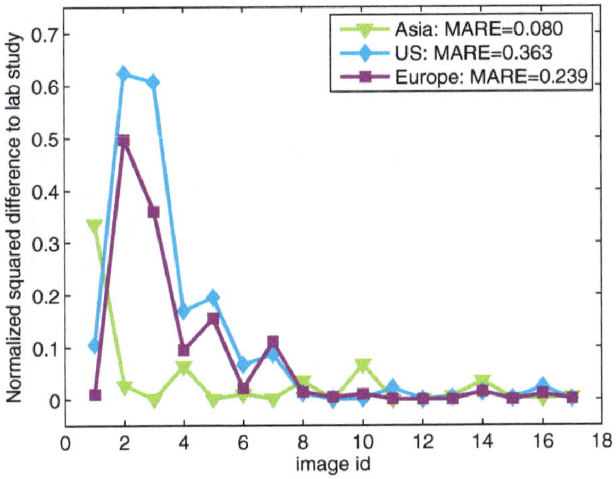

Fig. 21.4 Results for image quality tests obtained via Crowdsourcing with international users from Asia, US, and Europe, compared to the results from a laboratory study using the mean squared error, normalised by the absolute MOS value from the laboratory. Additionally, the mean absolute relative error (MARE) over all images is shown

In summary, Crowdsourcing enables extended studies on the various QoE influence factor, often not possible in a single laboratory, where the influence of these factors is often not noticeable due to the restricted pool of subjects. Still, the influence factors are not or only partly under control in a crowdsourcing campaign and it is necessary to monitor and track all these additional QoE influence factors.

21.5 Conclusions and Discussions

Crowdsourcing offers new possibilities for QoE evaluation by moving the evaluation task from the traditional laboratory environment into the Internet and enabling researchers to easily access a global pool of subjects for the evaluation task. The advantages of Crowdtesting are reduced time and costs for tests, a large and diverse panel of international users, and realistic user settings, allowing us to circumvent bottle-necks in traditional laboratory setups. However, conceptual, motivational, and technical challenges emerge due to the test environment. Therefore appropriate mechanisms like reliability checks or training phases must be included in the task design. In doing so, crowdsourcing offers the possibility to broaden QoE research and to study additional influences on content, user, system, and context level like the impact of incentives or demographics on QoE as discussed in this chapter. With advances in the research on incentives like gamification, crowdsourcing has the potential to be an even more powerful tool for quality testing. One of the major advantages of crowdsourcing studies is also that real-world problems and influences on QoE are intrinsically emphasized.

In principle, crowdsourcing could be used for the assessment of any stimuli and interactivity, using any type of subjective methodology. In reality, however, we are faced with several limitations on the possible scope of QoE crowdtesting. The main technical factors limiting the scope of QoE assessment are bandwidth constraints and support of the workers' devices to present the required stimuli. Although 2-D video and audio capabilities have become standard at most devices, 3-D video and audio capabilities or high dynamic range (HDR) displays cannot be readily assumed to be available. The support for other stimuli, for example, haptic or olfactory stimuli, is nearly non-existent in common computer hardware as used by the workers and thus these stimuli are currently not suitable for QoE crowdtesting. Besides these technical factors, QoE assessment methodologies requiring the interaction between different workers, e.g. for interactive video conferencing, are possible, but challenging in their execution. Taking these limitations into account, QoE crowdtesting is feasible for typical web applications like web browsing or file download, 2-D video, image and audio QoE assessment tasks, where the usable formats depend on the bandwidth requirement.

References

1. Alonso O, Rose DE, Stewart B (2008) Crowdsourcing for relevance evaluation. ACM SIGIR Forum 42(2):9–15. doi:10.1145/1480506.1480508. http://doi.acm.org/10.1145/1480506.1480508
2. Amazon Mechanical Turk (2013). http://mturk.com
3. Barkowsky M, Li J, Han T, Youn S, Ok J, Lee C, Hedberg C, Ananth IV, Wang K, Brunnström K et al (2013) Towards standardized 3dtv qoe assessment: cross-lab study on display technology and viewing environment parameters. In: IS&T/SPIE electronic imaging, International Society for Optics and Photonics, pp 864, 809–864, 809
4. Catellier A, Pinson M, Ingram W, Webster A (2012) Impact of mobile devices and usage location on perceived multimedia quality. In: 2012 fourth international workshop on quality of multimedia experience (QoMEX), IEEE, pp 39–44
5. Chen KT, Chang CJ, Wu CC, Chang YC, Lei CL (2010) Quadrant of euphoria: a crowdsourcing platform for QoE assessment. Network, IEEE 24(2):28–35. doi:10.1109/MNET.2010.5430141
6. De Simone F, Naccari M, Tagliasacchi M, Dufaux F, Tubaro S, Ebrahimi T (2009) Subjective assessment of H.264/AVC video sequences transmitted over a noisy channel. In: Proceedings of the first international workshop on quality of multimedia experience (QoMEX 2009), pp 204–209. doi:10.1109/QOMEX.2009.5246952
7. Downs JS, Holbrook MB, Sheng S, Cranor LF (2010) Are your participants gaming the system? Screening mechanical turk workers. In: Proceedings of the SIGCHI conference on human factors in computing systems, CHI '10, ACM, New York, NY, USA, pp 2399–2402 doi:10.1145/1753326.1753688. http://doi.acm.org/10.1145/1753326.1753688
8. Eickhoff C, Harris CG, de Vries AP, Srinivasan P (2012) Quality through flow and immersion: gamifying crowdsourced relevance assessments. In: Proceedings of ACM SIGIR conference on research and development in information retrieval 2012. ACM
9. Eickhoff C, de Vries A (2012) Increasing cheat robustness of crowdsourcing tasks. Inf Retrieval 16(2):121–137. doi:10.1007/s10791-011-9181-9
10. Facebook (2013). http://www.facebook.com
11. Gardlo B, Ries M, Hoßfeld T (2012) Impact of screening technique on crowdsourcing QoE assessments. In: 22nd international conference radioelektronika 2012, special session on quality in multimedia systems. Brno, Czech Republic
12. Hirth M, Hoßfeld T, Tran-Gia P (2011) Anatomy of a crowdsourcing platform—using the example of Microworkers.com. In: Workshop on future internet and next generation networks (FINGNet). Seoul, Korea. doi:10.1109/IMIS.2011.89

13. Hoßfeld T, Keimel C, Hirth M, Gardlo B, Habigt J, Diepold K, Tran-Gia P (2013) CrowdTesting: a novel methodology for subjective user studies and QoE evaluation. Technical report 486, University of Würzburg
14. Hoßfeld T, Seufert M, Hirth M, Zinner T, Tran-Gia P, Schatz R (2011) Quantification of youtube qoe via crowdsourcing. In: 2011 IEEE international symposium on multimedia (ISM), pp 494–499. doi:10.1109/ISM.2011.87
15. Hsueh PY, Melville P, Sindhwani V (2009) Data quality from crowdsourcing: a study of annotation selection criteria. In: Proceedings of the NAACL HLT 2009 workshop on active learning for natural language processing, HLT '09, Association for Computational Linguistics, Stroudsburg, PA, USA, pp 27–35 http://dl.acm.org/citation.cfm?id=1564131.1564137
16. ITU-R BT.500-13 (2012) Methodology for the subjective assessment of the quality of television pictures. International Telecommunication Union, Geneva, Switzerland
17. Keimel C, Habigt J, Diepold K (2012) Challenges in crowd-based video quality assessment. In: 2012 fourth international workshop on quality of multimedia experience (QoMEX), pp 13–18. doi:10.1109/QoMEX.2012.6263866
18. Keimel C, Habigt J, Horch C, Diepold K (2012) Qualitycrowd—a framework for crowd-based quality evaluation. In: Picture coding symposium (PCS), pp 245–248. doi:10.1109/PCS.2012.6213338
19. Keimel C, Habigt J, Horch C, Diepold K (2012) Video quality evaluation in the cloud. In: Packet video workshop (PV), 2012 19th, international, pp 155–160. doi:10.1109/PV.2012.6229729
20. Kittur A, Chi E, Suh B (2008) Crowdsourcing user studies with mechanical turk. In: Proceedings of the twenty-sixth annual SIGCHI conference on human factors in computing systems. ACM, pp 453–456
21. Kittur A, Chi EH, Suh B (2008) Crowdsourcing user studies with mechanical turk. In: Proceedings of the SIGCHI conference on human factors in computing systems, CHI '08, ACM, New York, NY, USA, pp 453–456. doi:10.1145/1357054.1357127. http://doi.acm.org/10.1145/1357054.1357127
22. Microworkers (2013). http://microworkers.com
23. Pinson MH, Janowski L, Pepion R, Huynh-Thu Q, Schmidmer C, Corriveau P, Younkin A, Le Callet P, Barkowsky M, Ingram W (2011) The influence of subjects and environment on audiovisual subjective tests: an international study. Selected Topics in Signal Processing, IEEE Journal of, 6(6):640–651. doi:10.1109/JSTSP.2012.2215306
24. Ross J, Irani L, Silberman M, Zaldivar A, Tomlinson B (2010) Who are the crowdworkers? Shifting demographics in mechanical turk. In: Proceedings of the 28th of the international conference extended abstracts on human factors in computing systems, ACM, pp 2863–2872. doi:10.1145/1753846.1753873
25. Sabou M, Bontcheva K, Scharl A (2012) Crowdsourcing research opportunities: lessons from natural language processing. In: Proceedings of the 12th international conference on knowledge management and knowledge technologies, ACM, p 17
26. Shaw AD, Horton JJ, Chen DL (2011) Designing incentives for inexpert human raters. In: Proceedings of the ACM 2011 conference on computer supported cooperative work, CSCW '11, ACM, New York, NY, USA, pp 275–284. doi:10.1145/1958824.1958865http://doi.acm.org/10.1145/1958824.1958865
27. Suri S, Goldstein D, Mason W (2011) Honesty in an online labor market. In: Human computation: papers from the 2011 AAAI, Workshop (WS-11-11)
28. Technische Universität München, Institute for Data Processing: Qualitycrowd (2013). http://www.ldv.ei.tum.de/videolab
29. Varela M, Mäki T, Skorin-Kapov L, Hoßfeld T (2013) Increasing payments in crowdsourcing: don't look a gift horse in the mouth. In: 4th international workshop on perceptual quality of systems (PQS 2013). Vienna, Austria
30. Von Ahn L, Dabbish L (2004) Labeling images with a computer game. In: Proceedings of the SIGCHI conference on human factors in computing systems, ACM, pp 319–326
31. Wang Z, Bovik A, Sheikh H, Simoncelli E (2004) Image quality assessment: from error visibility to structural similarity. IEEE Trans Image Process 13(4):600–612. doi:10.1109/TIP.2003.819861

Chapter 22
Web Browsing

Dominik Strohmeier, Sebastian Egger, Alexander Raake, Tobias Hoßfeld and Raimund Schatz

Abstract The Chapter provides an overview of Quality of Experience research for web-browsing, highlighting recent research trends. It indicates how Web-QoE assessment has evolved from the mapping of technically measured page-load times to quality estimates to the notion of perceived page-load time. Here, the consideration of the user's current task and respective role of individual element load times is discussed. The interactive nature of web-browsing is further analyzed in terms of temporal effects regarding the subsequent page access of users during typical browsing sessions. Finally, the chapter provides an outlook on future challenges related with the increasing complexity of web-services and respective page-loading processes.

D. Strohmeier (✉) · A. Raake
Assessment of IP-based Applications, Telekom Innovation Laboratories, TU Berlin, Berlin, Germany
e-mail: dominik.strohmeier@gmail.com

A. Raake
e-mail: alexander.raake@telekom.de

S. Egger · R. Schatz
Telecommunications Research Center Vienna (FTW), Vienna, Austria
e-mail: egger@ftw.at

R. Schatz
e-mail: schatz@ftw.at

T. Hoßfeld
Institute of Computer Science, Chair of Communication Networks, University of Würzburg, Würzburg, Germany
e-mail: hossfeld@informatik.uni-wuerzburg.de

S. Möller and A. Raake (eds.), *Quality of Experience*,
T-Labs Series in Telecommunication Services, DOI: 10.1007/978-3-319-02681-7_22,

22.1 Introduction

For many users, the Internet has turned into one of the most pervasive means of accessing multimedia services. For many of us, it plays an important role in our daily lives. Significant technological progress of recent years has transformed the Internet from a simple data sharing network into a sophisticated, but complex ecosystem [3, 12]. This evolution has also set new requirements on maintaining and improving the performance of the Internet, especially in terms of improved Quality of Experience.

Over the years, many sophisticated QoS-based approaches have been established for network performance assessment and optimization [1, 14]. However, the Internet's performance is still impaired by high latency or low responsiveness of web services, which represent critical problems for the end users. As shown recently, 67 % of web users encounter slowly loading web sites weekly, 25 % of the users even daily. For Internet Service Providers (ISPs), these figures show that user-perceived performance of web services still is a critical issue. Studies have shown that 49 % of the regular users will leave a web site and change to a competitor's service due to experienced performance issues [8, 20]. Consequently, optimization of Future Internet web services based on QoS-based approaches and purely technological paradigms does not sufficiently capture the requirements to meet or enhance the end-users' satisfaction with these services.

While existing QoS approaches for optimizing web services deal with performance aspects of the physical system, Web-QoE, defined as "Quality of Experience of interactive services that are based on the HTTP protocol and accessed via a browser" [9], focuses on the optimization of web services by understanding the end-users' perception of performance. The critical issue in this context is perceived waiting which occurs after requesting a web site until it has been fully loaded in the browser.

This chapter provides an overview of recent Web-QoE research. It discusses waiting times[1] as the key metric for assessing Quality of Experience for web-based services. Going from single page requests to a series of consecutive page views, it also outlines the importance of temporal considerations as well as the interactive nature of the service. Especially interactivity and the related tasks which users want to accomplish have shown to be a major QoE influencing factor beyond network-related performance parameters.

[1] Note that we use the term *time* when we address physically measured time as well as perceived time. However, in some cases the term *duration* may be more appropriate, for example for addressing the duration of a *stimulus* or in general *perceived duration*.

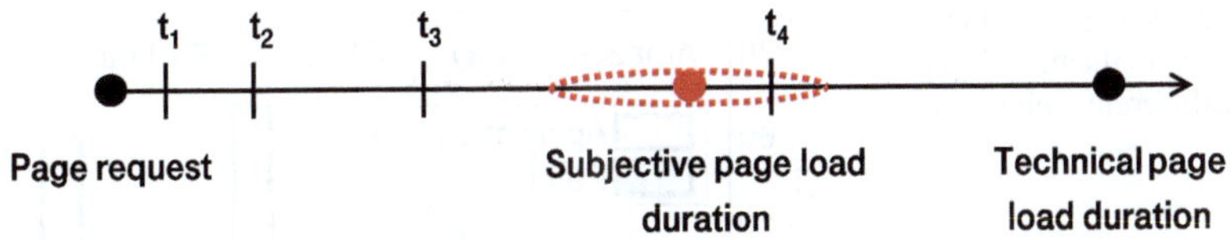

Fig. 22.1 Page view cycle

22.2 From Single Page Requests to Sessions and Flow Experience

Waiting times while requesting a web page have been regarded as the crucial QoE influencing factor on web-QoE. From a network perspective, these waiting times are the results of delays which occur at various points of the Internet. According to Butkiewicz et al. [3], especially the inclusion of third-party services (e.g. analytics, advertisements) into modern web services significantly increases the content and service complexity for a single website request—and leads to various delays. Ager et al. [1] and Poese et al. [14] show that content transmission over CDNs and multi-server structures has become an issue for network-based optimization of the Internet.

However, for an individual web user, it is irrelevant at which part of the complex end-to-end transmission a delay occurs. For him, all occurring delays sum up to a total waiting time from the page request to the fully loaded page visible on his screen. The perception of a single page request has recently been described using the page view cycle as shown in Fig. 22.1 [10]. Starting from requesting the page by an explicit click on a link for entering a URL, several conditions are reached as perceivable events for users to estimate progress of the page request—and as anchor points for estimating waiting times.

After requesting a page by a dedicated mouse click, key stroke, or even touch event, the browser window turns blank at t_1. The status of the progress bar changes at t_2 showing the progress of loading the requested website. Now, three important events occur whose importance we will discuss in this chapter. At t_3, the first element of the requested page becomes visible on the screen. Then, elements will appear progressively on the screen until, at t_4, all elements are visible on the screen. Still, elements may not be loaded on scrollable pages outside of the visible pane. Technically, this page load request is finalized after the technical page load time, marking the time until all HTTP requests for this page have been fulfilled. Discussing perceived waiting times as a key metric for understanding web-QoE presumes knowledge about the relationship of the page view cycle events, technical page load time, and perceived page load time is needed. Recently presented results show that significant differences between the technical page load duration and the subjectively perceived time until completion of a single page request exist [16]. Users were asked to request a series of different web pages. For each request, they marked the point in time at which they considered a page to be loaded. Figure 22.2 shows the technical and the perceived page load time for different page types (and three different pages within each

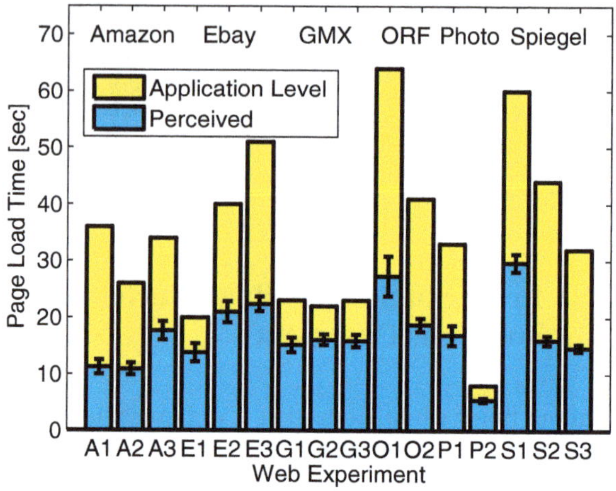

Fig. 22.2 Perceived subjective versus application-level PLT for different web pages

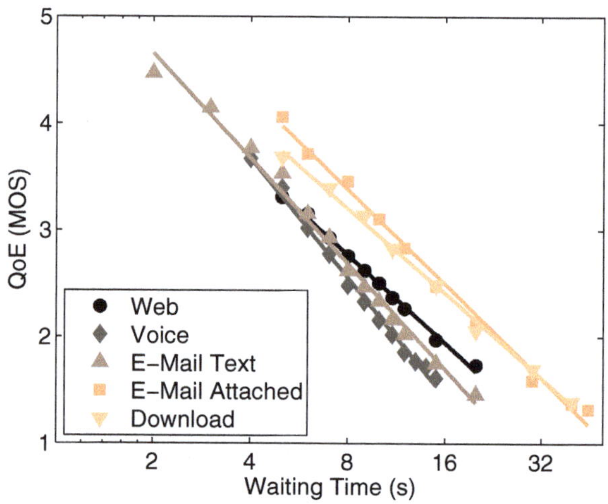

Fig. 22.3 QoE as a result of waiting time for certain applications with logarithmic mappings applied from [4] based on data from [13]

type, e.g. front page, search results and article detail page for Amazon). The results show large differences between technical and perceived completion time, with ratios ranging from 1.5 up to 3 (where 1 would be the exact match between subjective and application level PLT).

Although technical and perceived page load time differ significantly, web-QoE monitoring requires reliable metrics attached to controllable events during the page view cycle. The simplest solution which has been used so far is the application of the overall page load time (OPLD). In these applications, the controlled time events relate closest to the technical page load times. During different tasks like browsing through a picture album or performing online searches, the request for the next picture and search results are delayed for a certain amount of time. Figure 22.3 shows the results for different tasks performed.

Table 22.1 Logarithmic mapping between waiting time x and MOS $f(x)$ as illustrated in Fig. 22.3

Application	Mapping function
Web	$f(x) = -1.19\ln(x) + 5.23$
Voice	$f(x) = -1.61\ln(x) + 5.90$
E-mail text	$f(x) = -1.42\ln(x) + 5.64$
E-mail attached	$f(x) = -1.27\ln(x) + 6.03$
Download	$f(x) = -1.14\ln(x) + 5.57$

It can be seen that for different tasks the assumption is valid that the overall waiting times are the key metric for estimating QoE for these independent page requests. Differences between the tasks relate to different sensitivity or acceptable levels for waiting times. While other studies [17] observed an exponential relationship between download time and QoE based on the IQX hypothesis [6], a logarithmic relationship between waiting times and Mean Opinion Scores can be identified with the appropriate mapping functions given in Table 22.1. As basis for this finding, the Weber-Fechner law has been identified and was successfully applied for monitoring web-QoE of single page requests [5].

However, several web studies confirm that web browsing is an interactive activity. Even pages having plentiful information and links to other pages tend to be regularly viewed only for a brief period. Thus, users do not perceive web browsing as a sequence of single isolated page retrieval events but rather as an immersive flow experience. The notion of flow implies that the quality of the web browsing experience is determined by the timings of multiple page-view events that occur over a certain time frame during which the user interacts with a website and forms a quality judgment. Recent literature refers to these consecutive page views as sessions. The question to be answered is if the findings for single page requests still are valid for these sessions. Are subjective page load times (sPLT) as proposed by standardization bodies [10, 11] enough to capture Quality of Experience for web-based services? Can also other events during the page view cycle (Fig. 22.1) have impact on Quality of Experience for web browsing? Egger et al. [5] recently concluded that the choice of subjective page load time must be extended towards novel metrics to be able to reliably model web-QoE with respect to the highly interactive nature of web browsing.

22.3 The Importance of Tasks and User Interaction

While the previously discussed approach of requesting a page and judging web-QoE based on the overall waiting times must be regarded as a very passive way of accessing the page, browsing the web is usually a highly interactive process. During sessions, users access websites to perform specific tasks. Therefore, they request new web pages, digest information provided, proceed to the next page by clicking a link until finally they complete their dedicated task. Or users simply access news websites to browse the content quickly getting an update of what is happening without dedicated goals.

Already before looking into web-QoE and waiting times specifically, studies showed that a relationship between perceived delay and user tasks exist for task based web browsing as described by Galletta et al. [7]. In their study on waiting times for web services, they confirm perceived delays as the critical issue for the acceptance of web services. However, they conclude that negative impacts of delay are strongest when these delays are longer than users would expect them to be, or if they occur in unpredictable patterns. Also Bhatti et al. [2] summarize that users of an e-commerce service were strongly influenced by their expectations of the delay. They describe that among the influence factors user expectation, the type of task, and the method of page loading (all at once vs. incremental loading) are critical. Especially differences between free-browsing tasks and specific actions intended with a page request (like buying a product) showed to make a distinction between perception of waiting times.

Looking into web-QoE, these findings offer interesting approaches for improving the overall page load time approach. Recently, Strohmeier et al. [18, 19] introduced the per-element load times to web-QoE. Their approach is based on the studies of Bhatti et al. They assume that users perform dedicated tasks on websites—and therefore interact with specific elements per web site to accomplish this task. This can either happen on just a single page or expand over several pages to be requested. The motivation for the per-element load time approach can also be seen in better understanding the relationship between the different events in the page view cycle and their impact on determining subjective page load time (Fig. 22.1).

In a first study, Strohmeier et al. [18] applied a single page and varied the users' task on it (Fig. 22.4). The task was conducted on a news landing page. While in a first round of evaluations, users simply requested the website for different overall load times (free exploration), they were given a specific information assimilation task in a second round on the same pages. In addition to the overall load times, Strohmeier et al. [18] also varied the per-element load times for the elements which were needed to finalize the information assimilation task which they called "who ate what". These specific elements either occurred at the beginning, in the middle, or at the end of the page load process (element order)—and hence enabled users to fulfill their task after different waiting times.

The results of this study show that task as well as per-element load times have a significant effect on web-QoE beyond overall load times. For the free exploration task during which users only requested the web page, Strohmeier et al.'s findings are in line with the results of other web-QoE studies. Again, the logarithmic relationship between overall load times and Quality of Experience can be found. However, an interesting observation can be made for the task-based evaluation. In cases where the elements related to the information assimilation task appeared early during the page load process, Quality of Experience is not impacted anymore by load times longer than 8 or 12 s, respectively. Users seem to apply different perceived times for estimating Quality of Experience for these cases. In their results, they conclude that the inclusion of tasks and measurements of task achievement look promising to understand the differences between subjective and technical loading time.

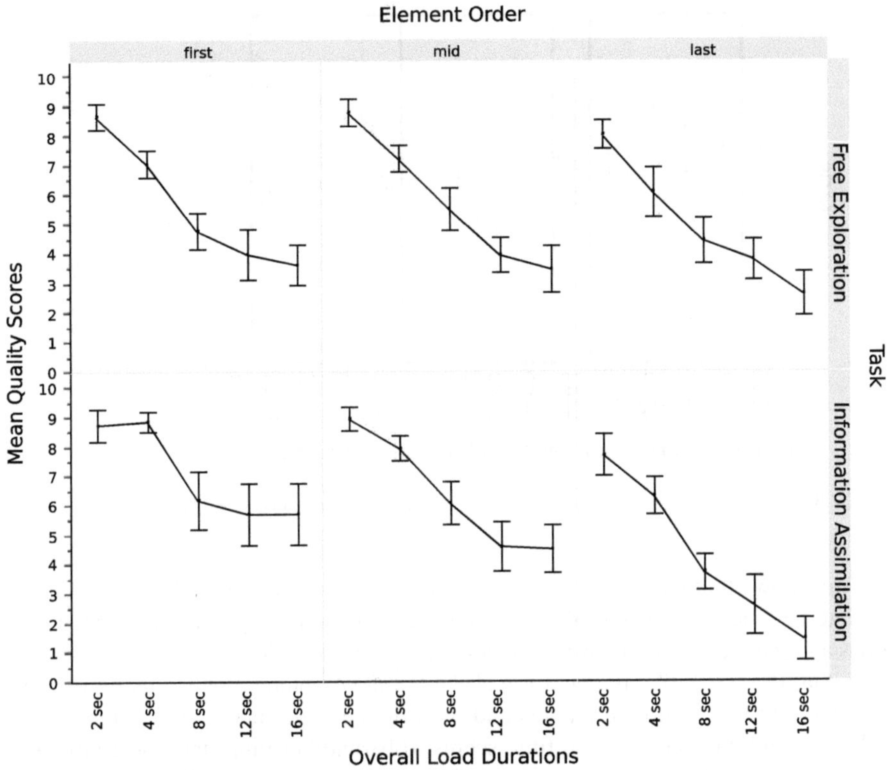

Fig. 22.4 Influence of task and element order on QoE

In a second study, Strohmeier et al. [19] extended the findings of their first study into the evaluation of sessions. In this study, the evaluation is based on four consecutive page requests on a news website which users perform to find the "team of the week". The task was accomplished by following a click path across the four pages. Within this second study, they varied overall load times, session duration, and per-element load times. Here, the clickable element to proceed in the task either appeared regularly (as fifth of 11 elements) or significantly delayed (as ninth out of 11 elements). The results in Fig. 22.5 confirm again that the overall load times as well as the per-element load times have an impact on web-QoE also within sessions. While, in general, the overall session times, i.e. the sum of single page load times per session, seems to play an important role, significant differences for constant overall load times can be found for varying element load times. In their study, Strohmeier et al. calculated correlations of MOS, overall session durations, and task completion times. Task completion time thereby was measured as the sum of measured times from page request to the click on the element per page across sessions. The results show that task completion time has significantly the highest correlation with the Mean Opinion Scores.

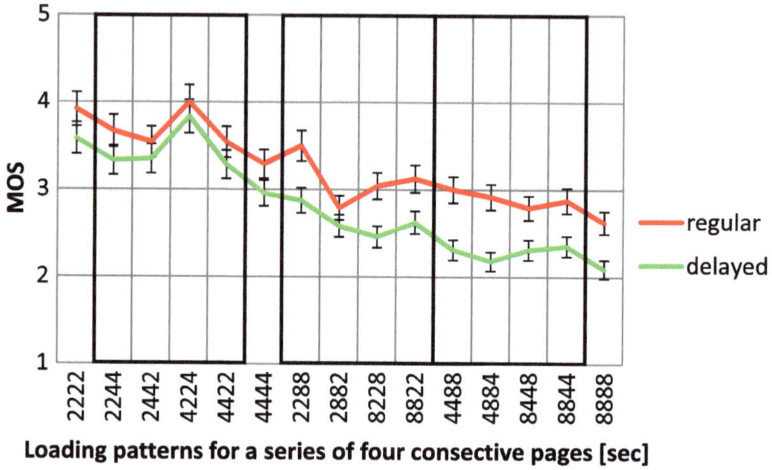

Fig. 22.5 Loading patterns for a series of four consecutive pages; task-based sessions

This shows that user tasks and related user interaction with web page elements seem to be crucial for web-QoE modeling and monitoring beyond overall page load times and the technically measurable delay times. Strohmeier et al.'s results [18, 19] show that more reliable models for web-QoE can be obtained by (a) including per-element load times into the models and (b) validate these models for different user tasks. Overall, task completion time seems to be another important QoE influence factor for web-QoE. Efficient execution of user tasks seems to contribute to web-QoE. This motivates a multidisciplinary evaluation of web-QoE by extending evaluation towards conventional usability assessment. There, extensive knowledge about user interaction with web services already exists.

22.4 Discussion and Conclusion

Quality of Experience of emerging web services has become an important topic for maintaining customer satisfaction as well as for optimizing existing technological infrastructure. However, recent work has shown that it is important to understand Web-QoE from a users' point of view. Until now, the focus has been set on understanding the impact of perceived waiting time on Quality of Experience and on the identification of which factors determine perceived waiting time. However, technically these waiting times can be the results of manifold delays in the end-to-end system. The current knowledge about Web-QoE allows several conclusions for future research and implications for using Web-QoE to be drawn.

Understanding key influencing factors: Existing models for measuring the performance of web services in terms of QoE are still addressing a too limited selection of metrics, and primarily technical page load time. Previous research and recommendations for Web-QoE [5, 11, 15] emphasize an understanding and modeling of Web-QoE based on overall page load times. Recent results, however, show that for accurate modeling of Web-QoE, more detailed diagnostic information about the page load process like per-element load times [18, 19] or temporal effects within sessions [9] is needed. This need is based on the interactive nature of web browsing and the identified impact of user tasks and task accomplishment on Web-QoE.

Understanding users' tasks: Beyond technical page load times, task completion times have been identified as a key influencing factor for Web-QoE. In combination with per-element load times, first results show that understanding tasks and the time until accomplishment lead to more valid Web-QoE models. Thereby, the increasing complexity and interactivity of web services leads to a limited applicability of purely network-based solutions for monitoring QoE. Challenges for translating the importance of knowledge about tasks into network-based monitoring approaches lies in additional requirements for estimating task performance at the client side. However, this challenge also comprises the need for valid Web-QoE models for services in which HTML5 or AJAX technologies are used. These recently enabled technologies have led to the introduction of purely web-based operating systems (e.g. Google's Chromebook, Mozilla's Firefox OS for mobiles) and the creation of desktop-like applications like Google Docs or Yahoo Mail as viable alternatives to traditional Microsoft Office or Mozilla Thunderbird, respectively. This advance in exploiting the technological possibilities of the Internet introduces asynchronous loading of web page elements from the network. This results in network traffic which is more and more decoupled from the users' experiences on their computer, tablet, or mobile devices. So eventually, there will be challenges to map Quality of Experience to measurable Quality of Service parameters.

Mapping Web-QoE and QoS: To successfully apply Quality of Experience for prospective optimization of web services, mapping functions for relating QoE influence factors on controllable or measurable parameters of Quality of Service will be needed. These measurable parameters need to rely on different monitoring points of the end-to-end delivery chain to allow accurate, yet generally applicable models for predicting end-user Web-QoE. Recently, studies showed that even simple matching of waiting times (as QoE factor) and variation of bandwidth (as one source of different delay times on QoS level) cannot be applied [4].

References

1. Ager B, Mühlbauer W, Smaragdakis G, Uhlig S (2011) Web content cartography. In: Proceedings of the 2011 ACM SIGCOMM internet measurement conference, IMC '11. ACM, New York, pp 585–600. doi:10.1145/2068816.2068870
2. Bhatti N, Bouch A, Kuchinsky A (2000) Integrating user-perceived quality into web server design. In: Proceedings of the 9th international World Wide Web conference on computer networks : the international journal of computer and telecommunications netowrking. North-Holland Publishing Co., Amsterdam, pp 1–16. http://dl.acm.org/citation.cfm?id=347319.346245

3. Butkiewicz M, Madhyastha HV, Sekar V (2011) Understanding website complexity: measurements, metrics, and implications. In: Proceedings of the 2011 ACM SIGCOMM conference on internet measurement conference, IMC '11. ACM, New York, pp 313–328. doi:10.1145/2068816.2068846
4. Egger S, Reichl P, Hoßfeld T, Schatz R (2012) 'Time is bandwidth'? Narrowing the gap between subjective time perception and quality of experience. In: IEEE ICC 2012—communication QoS, reliability and modeling symposium (ICC'12 CQRM), Ottawa
5. Egger S, Reichl P, Hoßfeld T, Schatz R (2012) Time is bandwidth? Narrowing the gap between subjective time perception and quality of experience. In: 2012 IEEE international conference on communications (ICC 2012). IEEE, Ottawa
6. Fiedler M, Hoßfeld T, Tran-Gia P (2010) A generic quantitative relationship between quality of experience and quality of service. Netw IEEE 24(2):36–41
7. Galletta DF, Henry RM, McCoy S, Polak P (2006) When the wait isn't so bad: the interacting effects of website delay, familiarity, and breadth. Inf Syst Res 17(1):20–37. doi:10.1287/isre.1050.0073
8. Gomez Inc (2011) When seconds count: national consumer survey on websites and mobile performance expectations. http://www.gomez.com/wp-content/downloads/GomezWebSpeedSurvey.pdf. Accessed 13 April 2012
9. Hoßfeld T, Schatz R, Biedermann S, Platzer A, Egger S, Fiedler M (2011) The memory effect and its implications on web QoE modeling. In: 23rd international teletraffic congress (ITC 2011), San Francisco
10. International Telecommunication Union (ITU), Study group 12, Question 13 (Q13.12) (2012) Work item quality of web browsing (G.QoE-Web). International Telecommunication Union, Geneva. http://www.itu.int/ITU-T/workprog/wp_item.aspx?isn=8208
11. ITU Recommendation, ITU-T G.1030 (2005) Estimating end-to-end performance in IP networks for data applications. ITU Telecom, Standardization Sector of ITU, Geneva
12. Kilkki K (2008) Quality of experience in communications ecosystem. J Univ Comput Sci 14(5):615–624
13. Niida S, Uemura S, Nakamura H (2010) Mobile services. IEEE Veh Technol Mag 5(3):61–67. doi:10.1109/MVT.2010.937850
14. Poese I, Frank B, Ager B, Smaragdakis G, Feldmann A (2010) Improving content delivery using provider-aided distance information. In: Proceedings of the 10th annual conference on internet measurement, IMC '10. ACM, New York, pp 22–34. doi:10.1145/1879141.1879145. http://doi.acm.org/10.1145/1879141.1879145
15. Reichl P, Egger S, Schatz R, D'Alconzo A (2010) The logarithmic nature of QoE and the role of the weber-fechner law in QoE assessment. In: Proceeding of the IEEE international communications conference (ICC), pp 1–5. IEEE, Cape Town. doi:10.1109/ICC.2010.5501894
16. Schatz R, Hoßfeld T, Janowski L, Egger S (2012) From packets to people: quality of experience as new measurement challenge. In: Biersack E, Callegari C, Matijasevic M (eds) Data traffic monitoring and analysis: from measurement, classification and anomaly detection to quality of experience. Springers computer communications and networks series, Springer, Berlin
17. Shaikh J, Fiedler M, Collange M (2010) Quality of experience from user and network perspectives. Ann Telecommun-annales des télécommunications 65(1–2):47–57
18. Strohmeier D, Jumisko-Pyykko S, Raake A (2012) Toward task-dependent evaluation of Web-QoE: free exploration vs. "who ate what?". In: IEEE globecom workshops 2012, Anaheim, pp 1309–1313. doi:10.1109/GLOCOMW.2012.6477771
19. Strohmeier D, Mikkola M, Raake A (2013) The importance of task completion times for modeling Web-QoE of consecutive web page requests. In: Proceeding of the Fifth international workshop on quality of multimedia experience (QoMEX 2013), Klagenfurt
20. Technology Review (2011) The need for speed. http://www.technologyreview.com/files/54902/GoogleSpeed_charts.pdf. Accessed 13 April 2012

Chapter 23
Mobile Human–Computer Interaction

Robert Schleicher, Tilo Westermann and Ralf Reichmuth

Abstract This chapter gives an overview on basic concepts and current research in mobile human–computer interaction (HCI) by showing where it extends the notion of interaction with stationary devices. Important differences next to basic hardware properties (size etc.) are that the corresponding devices offer instant access to the internet and are used in a variety of situations or contexts, where location information so far appears to be the primary way to assess this context. As a consequence, a couple of new research paradigms for field testing emerged, which are structured along the dimensions of scalability (number of users) and research outcomes. The majority of these studies appears to be rather exploratory, less explanatory at the moment. Possible reasons as well as future research directions are discussed.

23.1 Introduction

Mobile devices like smartphones or tablets are becoming increasingly popular, and may to some extent replace the desktop PC and laptop as the most prominent hardware for human–computer interaction (HCI) [1]. Thus they are also in the scope of research on HCI. Here, the approaches vary from considering them as just a rescaled variant of a pre-existent device type (i.e. a "very small" laptop or terminal) to seeing smartphones as the core element of *ubiquitous computing*, where people will constantly exchange digital information with their environment, exploiting numerous sensors and actuators. While the notion of very small laptops may immediately be

R. Schleicher (✉) · T. Westermann · R. Reichmuth
Quality and Usability Lab, Telekom Innovation Laboratories, TU Berlin, Berlin, Germany
e-mail: robert.schleicher@tu-berlin.de

T. Westermann
e-mail: tilo.westermann@telekom.de

R. Reichmuth
e-mail: ralf.reichmuth@telekom.de

S. Möller and A. Raake (eds.), *Quality of Experience*,
T-Labs Series in Telecommunication Services, DOI: 10.1007/978-3-319-02681-7_23,

refused by the reader as hopelessly outdated, it can still serve as a starting point for examining what is special to *mobile* HCI, and in many cases this conservative view can explain typical phenomena just as well as an approach that draws mostly from the notion of unprecedented ubiquitousness as envisioned by Mark Weiser (e.g. [32]). We will try to cover both aspects in this chapter and point out their potentials as well as shortcomings, especially with regard to related research paradigms and findings, where a similar distinction can be made between applying established research techniques to a new artifact, the mobile device, versus using this artifact to also develop new methods to study its use. This will be the second part of this chapter. The final section then outlines possible future developments and research directions.

23.1.1 Mobile Devices Versus *Mobile* Interaction

The fact that the devices are intended to be carried around already imposes several constraints on the corresponding hardware and basic functionalities, namely:

- Limited size and weight, with the most obvious consequence for interaction probably being the limited screen size next to the limitations in battery life.
- Increased demand for robustness, resulting in the desire to remove as many predetermined breaking points in the hardware as possible, e.g. less physical controls like buttons or rotating wheels.
- Varying access to networks and restricted bandwidth.

On a hardware level, these limitations are commonly tried to be compensated for by incorporating additional sensors in the device: For example, the desire to maximize screen size and avoid physical controls has led to the installation of touch-sensitive displays, so that the same area can be used for in- and output without requiring additional accessories like a mouse. While this interaction type is appreciated by many users and appears to be the prevailing interaction paradigm for mobile devices at the moment [5], it also brings new issues, e.g. the *fat finger problem* [34], referring to the fact that unlike with a mouse, the touched area is occluded for the eyes, and that a finger touch is much more imprecise than pointing with a mouse in terms of addressed pixels. As a consequence, the number of functionalities available on a single screen or within one mobile application ("app") is usually quite lower than on common PC software.

A more *interaction*-focused view would favor another explanation for this phenomenon: mobile phones are frequently used for occasional interactions like briefly checking mails or searching for a location on the go, where these tasks in part apparently serve to kill idle time [9]. Users spend an average of 2 h and 38 min per day on smartphones and tablets, 80 % of that time inside apps [20] with about 4 min per session on smartphones (8 min per session on tablets [11]). This fact taken together with demands of a competing main task like walking requires the user interface to be of manageable complexity, and probably manageable with a single hand [27]. In addition, context factors like lighting may also ask for simplified graphical user interfaces (GUI).

In a similar vein, the tendency to use mail programs on mobile devices to mostly check for and read new emails without writing longer replies can be understood as a new way of frequently passive *ambient mobile communication* which focuses on awareness (of what is going on) as a consequence of the increasing amount of information users are bombarded with [4]. Alternatively it could be explained with the awkwardness to use the virtual keyboard most mobile devices are equipped with for typing longer texts as still expressed by many users [25]. Another reason is that some users take advantage of the instant access mobile devices offer to emails etc. as a pre-check to subsequently decide whether it is worth to turn to the workstation, and directly refer to the lower physical as well as temporal effort as the driving force [2]. Here, the lower weight and size, and the always-on property of mobile devices may be the crucial factors, less a new way of human–computer interaction, especially as this behavioral pattern is also an important motivation for accessing the mobile internet at home [8, 9].

In this context, it is worth taking a brief look at the differences between smartphones and tablets again as described in Farago [11]: They are often summarized under the term *mobile*, but the differences suggest that these may also be treated differently from a research perspective: While on both apps are used throughout the whole day, tablets show a greater spike of usage during evening time from 7 p.m. to 10 p.m. This may indicate that tablets are more often used alongside leisure activities like watching TV. In addition, time spent across app categories differs between smartphones and tablets. On an abstract level, tablets seem to be best suited for consuming media and entertainment (Games, Entertainment, News) and smartphones rather for communication and task-oriented activities (Social Networking, Utilities, Health and Fitness, and Lifestyle). While *Games* is the category users spent the most time with on both devices, their usage is more prevalent on tablets with 67 % compared to smartphones with 39 % of total usage time [11], which also complies with the notion of relaxing on the couch. If an inherently stationary context like *at home* is so popular for mobile device usage, the question arises, what actually has to be considered *mobile*.

23.1.1.1 Mobile: The Context Factor Location

Mobile in a strict sense means that the device is *portable* and that it is intended for the use *underway*. To avoid dilemma such as whether waiting somewhere is really mobile, it is more common to speak of the *mobile context* in which the mobile device is used. Unfortunately, there is no consistent definition in the literature for the term *context* and its distribution into a set of context factors. For instance Schilit et al. [28] see context as "where you are, who you are with, and what resources are nearby". Furthermore, Schmidt et al. [29] divided the context of use into a set of different factors. First, human factors are classified into information on the user, the user's social environment and the user's tasks. Second, physical environment is classified into location, infrastructure and physical conditions.

However, for the designer of a specific mobile application this general model of the context of use may be difficult to implement. Two things have to be kept in mind when incorporating context-dependent information in applications: First, there are many factors given of which some are very complicated to detect like the user's social environment. Second, there must be a benefit for an application to include a context factor as it in most cases goes at the expense of application complexity and perceived data privacy, and the benefits of context factors may differ. Still, there are some factors that are of primal relevance in a lot of cases. We will focus on these. Next to time, the context factor *location* plays an important role. Humans have temporal and spatial regularities [14] which make this factor easier to detect and furthermore, the location is related to the user's activities. That is why this information is often used to design context aware applications. An important aspect of the location is whether the user is on the move or stationary, because this is associated with certain behavioral patterns. Verkasalo [31] found that people often use multimedia applications while they are on the move and use games rather at home, which also accounts for the major part of mobile web access there [8] up to the statement that the user's home appears the most common place for smartphone usage in general [31].

Relevant context configurations can be narrowed down more when looking at specific type of tasks, e.g. the use of mobile internet services. For instance Lee et al. [21] found that 0.2 % of all possible context configurations account for 24 % of all sessions, and in only about 18 % of all possible context configurations the participants used mobile internet services. This knowledge may be used to consider only certain contexts for an application and focus the classification on those. In line with this, some interesting contexts could already be determined automatically, i.e., requiring no user feedback. For instance, a small number of context-relevant activities like walking, jogging, going by bus or subway could be detected with help of an architecture using the smartphone sensors [16]. Knowing the first 12 h of a user's location pattern like being at home, work or elsewhere, this could predict the next 12 h states of a user with help of a machine learning algorithm [10]. Once the context is detected it may serve as the basis for designing context aware applications. In summary, the most relevant context information to achieve this goal so far appears to be spatial location, where the mere GPS data can meanwhile be enriched with additional information using *point-of-interest* OS library functions.[1]

23.2 Research Paradigms

The relevance of mobility and use context also affects the research paradigms applied to mobile HCI as classic laboratory research is to some extent limited here. Instead, the increasing demand to conduct field studies has widely been acknowledged amongst researchers, and established methods like online surveys, interviews, and

---

[1] http://developer.apple.com/library/ios/#documentation/CoreLocation/Reference/CoreLocation_Framework

observations are used to learn more about mobile phone usage *in situ*. In its most individualized form it means following mobile phone users like a shadow (thus also called *shadowing*) to observe their actual behavior (e.g. [19]). While this technique is also used in other domains with all its pros and cons like the need for substantial human resources, the danger to change behavior due to the presence of an additional person etc., one disadvantage may be particular to mobile HCI, namely the difficulty to see as an observer what is actually happening on the device screen. Brown and Laurier [5] try to overcome this by capturing the screen of the mobile device and equipping their subjects with additional wearable cameras. Here, using the device itself as a data collecting tool emerges as a new approach, be it for explicit user feedback (i.e. self-assessment) or implicit feedback via logging interaction and sensor data like session length, location, ambient sound etc. Inclusion of both data sources is preferable, as QoE ultimately concerns the user's perspective as described thoroughly in Chap. 2 of this book, and technical parameters thus need to be linked to self-assessment. Froehlich et al. [12], Liu and Wang [22], and Möller et al. [24] all present platforms that offer both, the option to collect sensor data as well as self-assessment for different platforms. Still, they all recruited the participants individually to either hand out mobile devices with the software pre-installed to the participants [22] or offered face-to-face assistance to install it on the participants' devices [24]. However, utilizing the mobile device for data collecting can also mean that a study is conducted completely anonymously by publishing the corresponding software in an app store and distributing it this way. As the number of participants usually increases with this approach, it is also called *Research in the Large* or *Large-Scale studies*. Good examples are González et al. [14] and Henze et al. [18], who released a game to examine the systematic shift or offset between touched area and target position on touch screens.

Extending the idea of releasing data collecting tools into the wild, it would in principle not even be necessary to publish a full-blown app, but only offer some libraries (APIs) other developers could include in their app to log data and thus collect usage data across apps. Böhmer et al. [6] in part follow this idea. In general however, this approach is rather pursued by commercial suppliers for app tracking as it requires substantial support on the backend side, e.g. server space for data storage etc.

Deploying data available from commercial providers is yet another way to conduct research on behavioral patterns associated with mobile phone usage as was pursued by González et al. [14] or Golder and Macy [13], who used twitter messages to evaluate mood changes throughout the daily, weekly, and seasonal changes. Which setup might be most appropriate relies on the specific research goal.

So far, the predominant research approach in mobile HCI appears to be explorative or data-driven, which could be summarized as "collect user data to determine usage pattern (to better understand/improve mobile HCI)", less confirmatory or theory-driven, where a clear research question with specific predictions, i.e. hypotheses derived from a theory, is the starting point. If done so, it is rather the application of an existing model of human behavior to mobile interaction. This might be due to the fact that to our knowledge there are no well-established theories specific to *mobile* HCI, rather to human–machine interaction in general—e.g. the understanding of touch and haptics as described in Chap. 18 of this book or the models on user experience given in Chap. 3. One could ask whether there need to be mobile HCI-specific models at all, but the discomfort many researchers express with simply transferring existing models to this domain, and thus their step back to start with explorative studies and the development of new research paradigms might indicate such a necessity.

As a consequence, many publications in that area so far devote much space to the description of their research method or framework they collect data with (especially if they aim at large datasets), and apparently less to the analysis of that very data, which is in most cases rather descriptive, i.e. *how* people use a device or a service, less explanatory, i.e. *why* people do so. For explanatory analyses, studies that rely on classic ethnographic/usability testing methods associated with smaller sample sizes appear to be at an advantage at the moment.

23.3 Research Outcomes

Although the generalizability of results based on a small sample is limited for behavior that appears to depend on so many individual influences as mobile phone usage does, Barkhuus and Polichar [2] for example report phenomena in their Californian sample of $n = 21$ we also encountered like users having installed several weather apps simultaneously on the phone to be able to compare predictions. Referring to Chalmers and Galani [7] and Bell and Dourish [3], they use these case-study like observations to exemplify how the notion of *seamless* ubiquitous computing may be better replaced by *seamful* interaction, where users utilize a variety of applications and accounts in a highly individual and sometimes messy manner. Still, subjects are apparently not frustrated of having to deal with a patchwork of individual services and do not express the wish to have one global system that satisfies all needs smoothly and unnoticed as envisioned by Weiser [32]. Instead, Barkhuus and Polichar [2] conclude "Our research suggests the power of the ability to mix, match and interconnect individual apps was in large part what has made the smart phone so successful as a ubicomp device. Enhancement of this functionality may be the important direction that distinguishes successful mobile phones in the future." (ib., p. 10). Such a statement may be of higher value for evaluating the quality of experience of mobile HCI than a mere listing of session lengths and app usage sequences determined by a large scale study. In a similar vein, Salazar et al. [27] conclude in their helpful literature review on usability guidelines for mobile phones that due to the use on the go, typos and abbreviations in user input should be tolerated by applications, a finding that could similarly have been derived from an ethnographic analysis of email signatures which regularly include statements like "sent via mobile phone—please excuse TYpos"—apparently users expect other users to accept such errors, and thus might wish the same from the device.

We tried to depict the aforementioned relation of study scalability and level of findings in Fig. 23.1, where we grouped the studies mentioned in this text according

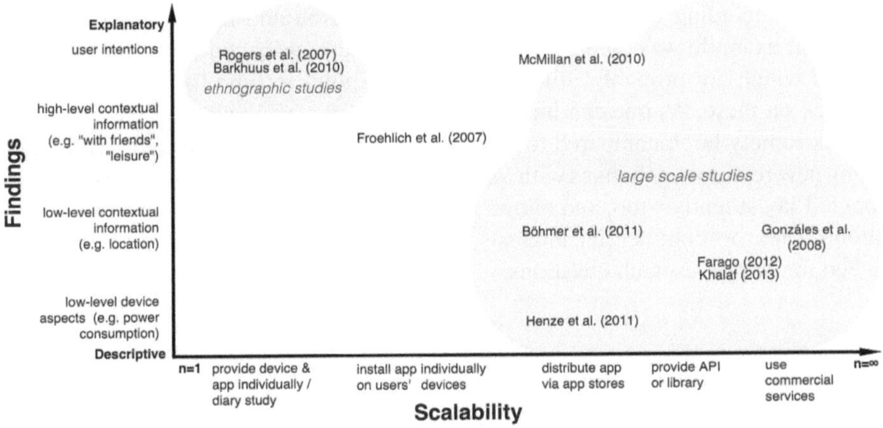

Fig. 23.1 Comparison of the scalability and level of findings in exemplary studies mentioned in this chapter extending a classification of Froehlich et al. [12]

to their sample size/research paradigm and their respective outcomes. Rogers et al. [26] used a combination of logging device data, observations and interviews with a limited set of users. Though costly in terms of time and effort involved, they found "a whole host of unexpected, context-based usability and user experience problems". While this setup may be not well-suited for large scale studies, insights were gained that were less likely to be found with a large scale approach (e.g. emerging usability problems when "placing the device in the palms of students on a cold spring day"). On the other hand, using data from commercial services (e.g. [11]) allows for more generalized but rather descriptive findings due to the large sample size. Please note that in principle also the extensive data logging of the American National Security Agency (NSA) as revealed by Edward Snowden [15] could be placed at this end of the x-Axis. We refrained from doing so as there is little information available regarding their level of findings as well as to what extent they are specifically targeting mobile communication.

Of course we are aware that the relative lack of explanatory results might be a transitional phenomenon, as one first needs to validate new methods before they can be used to derive new findings that can be trusted, and we will try to give examples of successful combination of both in the section on future research directions. However, we see one reason why research on mobile HCI may be prone to the danger of persevering on a descriptive level: the main innovation forces in mobile HCI are hardware and software (operating system) manufacturers, probably much more than in other areas of human–machine interaction, or at least with much shorter release circles. At that pace, simply describing the state of the art is already a challenge sometimes.

As academia and industry both share the same research goal, "to better understand mobile HCI", many of the improvements in new device/OS versions are likely to be based on research results that were achieved in proprietary labs and have not been published, including the development of new research and testing methods. One prominent example was Google's Gmail Labs[2] where users could opt in to try new features which are probably rolled out at a later point in time for all and provide feedback on these. As one can imagine, the size of the available user base is likely to be extremely large compared to large scale studies in academia. This method of testing new features or changes with a subset of the user base has been incorporated in Google Play recently,[3] too, and allows app developers to do beta-testing and staged rollouts. Next to commercial interests, this platform could also be an interesting testbed for future research questions.

23.4 Conclusions

We started by exemplifying that many phenomena observed in mobile human computer interaction can either be understood as a new variant of classic HCI topics, or as the manifestation of a completely new type of interaction. In our experience, many issues of daily mobile phone usage can still be explained with adhering to the former, conservative notion, e.g. adapted behavior due to response latency, interface restrictions etc., and thus would probably not even require the establishment of a new term. However, one area where this notion falls short is the instant access to information in various contexts, as mobile devices are used in a much larger variety of situations than desktop or laptop computers were. Of all the possible context factors, *location* appears to be the better-examined and exploited (in terms of available services) information, as opposed to temporal variation for example (see Golder and Macy [13] for an example of the latter). The importance of context factors also gave rise to a couple of new research paradigms leading to much larger sample sizes as compared to 'classic' field studies. However, one limitation we suspect is that the sheer amount of data is rather used for descriptive, less for explanatory or even *predictive* analysis. While this may be a transient phenomenon, one speculative explanation may be the close relation to commercial product development and its rapid progress.

23.4.1 Future Research Directions

The close vicinity of research in mobile HCI to commercial product development cannot be denied, but it should not lead to renunciation, because in many cases it were the commercial innovations that made new directions in research possible after all. Without the success of the iPhone, there would surely be much less large-scale studies. One could even argue that the long-claimed notion of user-centered design as an iterative process is now fully embraced and implemented in the mobile

[2] https://support.google.com/mail/answer/29418

[3] https://support.google.com/googleplay/android-developer/answer/3131213

sector with its continuous software updates and frequent hardware releases. The style guides of current mobile operating systems give extensive advice on how to exploit the possibility of mobile hard- and software to maximize user experience, e.g. with touch,[4] almost matching up to HCI text books.

With that being said, the challenge is probably rather to establish distinct research goals that are to some extent independent of progress in hard- and software development. Within the currently prevailing *Freemium* concept of mobile services, apps are almost considered as disposable articles by many users [23], and this has to be taken into consideration when devoting resources.

Research in the commercial sector is usually under intensified time pressure and thus focuses on results that can be monetized in the first place. The impact of mobile devices on human behavior beyond that is presumably less considered. With regard to future research methods, connecting large-scale logging with *explanatory* findings from individual users to allow for explanatory large-scale/logging studies might be a possible way to go. On the basis of McMillan et al. [23], we see great potential in combining large scale studies with personal interviews, derive specific hypotheses from those, and re-examine the available logging data in that regard. More often than not, additional questions arise from logging data, and in these cases, a personal statement may be desirable. Incorporating a functionality to obtain those in a way that is coherent with the app itself as seen in McMillan et al. [23], where survey items were presented as within-game tasks by which one could obtain additional points may limit the obtrusiveness and help answering these questions. One potential way of further minimizing the difficulties in directly approaching users is to make use of apps that are targeted at geographically limited areas. An example is described in Westermann and Möller [33], where an app is tailored to serve students of a university campus. Pursuing this approach has the benefit of being able to invite users to take part in a personal interview or additional lab study, as they are based nearby.

While it has some limitations, e.g. with respect to scalability or the target group, we see one main advantage in the fact that users are becoming increasingly skeptical on revealing detailed personal information without knowing about their further usage, a tendency that has been intensified with the recent disclosures of secret service data logging [15]. We explicitly mentioned it in the context of *Large Scale studies*, probably irritating to the reader as there are many fundamental differences, most obviously the proclaimed objective and the terms of participation. Still, this extreme example served to hint at a general issue that is relevant for mobile HCI research, namely to what extent the intrusion into the users' privacy is perceived as acceptable in the light of the outcomes. Chapter 2 of this book named the intention to assess random experiencing without affecting it as one of the key challenges for future QoE research. Mobile devices which are *always on, always with me* may be the ideal candidates for that purpose, and the more the process of data collection is automatized, the less the user appears to be aware of it. McMillan et al. [23] for example report that although explicitly stated on its initial starting page, later interviews revealed that no user was

[4] http://msdn.microsoft.com/en-us/library/windows/apps/hh465415.aspx

aware that playing their mobile game actually meant providing data for a research project. Thus, all involved parties have to carefully deliberate about whether it is for instance really necessary to have the device log that someone is at home in a specific study on interaction quality and how to convey this. Otherwise, research might be confronted with comparable suspicions of snooping.

At the same time, people are actively seeking after products and services to collect data on their own behavior to an unprecedented extent, for example in the *quantified self* movement [30]. Connecting these lay movements with academic research, where people can get detailed feedback on the later analysis results and the derived conclusions (at least in the form of accessing the corresponding papers), and where commercial or clandestine exploitation is not the key intention, might extend the *citizen science* idea [17] explicitly to mobile HCI. Finding research methods and setups that are in addition not vulnerable to the aforementioned misuse might be one of the greater challenges.

References

1. Andrew L, Aquino C (2013) Mobile future in focus. Available at http://www.comscore.com/Insights/Presentations_and_Whitepapers/2013/2013_Mobile_Future_in_Focus. Accessed 21 May 2013
2. Barkhuus L, Polichar VE (2010) Empowerment through seamfulness: smart phones in everyday life. Pers Ubiquit Comput 15(6):629–639
3. Bell G, Dourish P (2006) Yesterday's tomorrows: notes on ubiquitous computing's dominant vision. Pers Ubiquit Comput 11(2):133–143
4. Bentley F, Kaushik P, Narasimhan N, Dhiraj A (2006) Ambient mobile communications. In: Proceedings CHI 2006. ACM Press, pp 1–3
5. Brown B, Laurier E (2013) iPhone in vivo?: Video analysis of mobile device use. In: Proceedings of the SIGCHI conference on human factors in computing systems-CHI '13. ACM Press, New York, pp 1031–1040
6. Böhmer M, Hecht B, Schöning J, Krüger A, Bauer G (2011) Falling asleep with angry birds, facebook and kindle: a large scale study on mobile application usage. In: Proceedings of Mobile HCI, pp 47–56
7. Chalmers M, Galami A (2004) Seamful interweaving: heterogeneity in the theory and design of interactive systems. In: Proceedings of DIS 2004. ACM Press, pp 243–252
8. Church K, Ernest P, Oliver N (2011) Understanding mobile web and mobile search use in today's dynamic mobile landscape, pp 67–76
9. Cui Y, Roto V (2008) How people use the web on mobile devices. In: Proceedings of the 17th international conference on World Wide Web—WWW '08. ACM Press, New York, pp 905–914
10. Eagle N, Pentland AS (2009) Eigenbehaviors: identifying structure in routine. Behav Ecol Sociobiol 63(7):1057–1066
11. Farago P (Flurry) (2012) The truth about cats and dogs: smartphone vs. tablet usage differences. Available at http://blog.flurry.com/bid/90987/The-Truth-About-Cats-and-Dogs-Smartphone-vs-Tablet-Usage-Differences. Accessed 21 May 2013
12. Froehlich J, Chen MY, Consolvo S, Harrison B, Landay JA (2007) My experience?: a system for in situ tracing and capturing of user feed-back on mobile phones. In: MobiSys'07. ACM Press, pp 57–70
13. Golder SA, Macy MW (2011) Diurnal and seasonal mood vary with work, sleep, and daylength across diverse cultures. Science 333:1878–1881

14. González MC, Hidalgo CA, Barabási A-L (2008) Understanding individual human mobility patterns. Nature 453(7196):779–782
15. Greenwald G, MacAskill E (2013) NSA Prism program taps in to user data of Apple, Google and others. The Guardian, London
16. Han M, Vinh LT, Lee Y-K, Lee S (2012) Comprehensive context recognizer based on multimodal sensors in a smartphone. Sensors (Basel, Switzerland) 12(9):12588–12605
17. Hand E (2010) Volunteer army catches interstellar dust grains. Nature 466(August):685–687
18. Henze N, Rukzio E, Boll S (2011) 100,000,000 taps: analysis and improvement of touch performance in the large. In: Proceedings of Mobile HCI. pp 133–142
19. Jacucci G, Oulasvirta A, Ilmonen T, Evans J, Salovaara A (2007) CoMedia: mobile group media for active spectatorship. In: Proceedings CHI, pp 1273–1282
20. Khalaf S (Flurry) (2013) Flurry five-year report: It's an app world. The web just lives in It. Available at http://blog.flurry.com/bid/95723/Flurry-Five-Year-Report-It-s-an-App-World-The-Web-Just-Lives-in-It. Accessed 21 May 2013
21. Lee I, Kim J, Kim J (2005) Use contexts for the mobile internet: a longitudinal study monitoring actual use of mobile internet services. Int J Hum-Comput Interact 18(3):269–292
22. Liu N, Wang X (2010) Data logging plus e-diary?: towards an online evaluation approach of mobile service field, trial. In: MobileHCI'10. pp 287–290
23. McMillan D et al (2010) Further into the wild: running worldwide trials of mobile systems In: Floréen P, Krüger A, Spasojevic M (eds) Pervasive computing, 6030. Springer, Berlin, pp 210–227
24. Möller A, Kranz M, Schmid B, Roalter L, Diewald S (2013) Investigating self-reporting behavior in long-term studies. In: Proceedings of the SIGCHI conference on human factors in computing systems—CHI '13. ACM Press, New York, pp 2931–2940
25. Page T (2013) Usability of text input interfaces in smartphones. J Des Res 11(1):39–56
26. Rogers Y, Connelly K, Tedesco L, Hazlewood W, Kurtz A, Hall RE, Hursey J, Toscos T (2007) Why it's worth the hassle: the value of in-situ studies when designing ubicomp. In: Krumm J et al (eds) In: Proceedings of the 9th international conference on Ubiquitous computing. Springer, pp 336–353
27. Salazar LH, Lacerda T, Nunes JV, Gresse von Wangenheim C (2013) A systematic literature review on usability heuristics for mobile phones. Int J Mob Hum Comput Interact (IJMHCI) 5(2):50–61
28. Schilit B, Adams N, Want R (1994) Context-aware computing applications. In: IEEE workshopo on mobile computing systems and applications. pp 1–7
29. Schmidt A, Beigl M, Gellersen H-W (1999) There is more to context than location. Comput Graph 23(6):893–901
30. Swan M (2012) Sensor mania! the internet of things, wearable computing, objective metrics, and the quantified self 2.0. J Sens Actuat Netw 1(3):217–253
31. Verkasalo H (2008) Contextual patterns in mobile service usage. Pers Ubiquit Comput 13(5):331–342
32. Weiser M (1991) The computer for the 21st century. Sci Am 265(3):94–104
33. Westermann T, Möller S (2012) MoCCha: a mobile campus app for analyzing user behavior in the field. In: Proceedings NordiCHI. pp 799–800
34. Widgor D, Wixon D (2011) Brave NUI world: designing natural user interfaces for touch and gesture. Morgan Kaufmann Publishers, Burlington

Chapter 24
Sensory Experience: Quality of Experience Beyond Audio-Visual

Christian Timmerer, Markus Waltl, Benjamin Rainer and Niall Murray

Abstract This chapter introduces the concept of Sensory Experience which aims to define the Quality of Experience (QoE) going beyond audio-visual content. In particular, we show how to utilize sensory effects such as ambient light, scent, wind, or vibration as additional dimensions contributing to the quality of the user experience. Therefore, we utilize a standardized representation format for sensory effects that are attached to traditional multimedia resources such as audio, video, and image contents. Sensory effects are rendered on special devices (e.g., fans, lights, motion chair, scent emitter) in synchronization with the traditional multimedia resources and shall stimulate also other senses than hearing and seeing with the intention to increase the Quality of Experience (QoE), in this context referred to as Sensory Experience.

24.1 Introduction

Multimedia resources or multimedia content (i.e., combinations of text, graphics, images, audio, and video) has become omnipresent in our daily live. Each day we consume and also produce dozens of multimedia assets when reading electronic newspapers, listening to Internet radio, watching digital television (TV), and sharing them (including our own) within social networks. The quality of these assets range

C. Timmerer (✉) · M. Waltl · B. Rainer
Alpen-Adria-Universitaet Klagenfurt, Klagenfurt, Austria
e-mail: christian.timmerer@itec.aau.at

M. Waltl
e-mail: markus.waltl@itec.aau.at

B. Rainer
e-mail: benjamin.rainer@itec.aau.at

N. Murray
Athlone Institute of Technology, Dublin, Ireland
e-mail: nmurray@research.ait.ie

S. Möller and A. Raake (eds.), *Quality of Experience*,
T-Labs Series in Telecommunication Services, DOI: 10.1007/978-3-319-02681-7_24,

from professional high-quality content to user-generated content which sometimes but not necessarily is of low quality but of high sentimental value. The quality of the multimedia content as perceived by the end user is commonly referred to as Quality of Experience (QoE) (cf. Chaps. 2, 3) which is typically defined along specific influence factors (cf. Chap. 4) and features (cf. Chap. 5) influencing the QoE [13].

In most cases the QoE is defined for a certain modality (e.g., image) or simply combinations thereof (e.g., audio and visual) but mainly targeting two human senses, namely hearing and seeing. However, the consumption of multimedia content may stimulate also other senses such as olfaction, mechanoreception, equilibrioception, or thermoreception. Therefore, in this work traditional multimedia content (i.e., mainly audio-visual) is annotated with sensory information describing sensory effects (e.g., additional ambient light, wind, vibration, scent) which are synchronized with the traditional multimedia content and rendered on appropriate devices (e.g., ambient lights, fans, motion chairs, scent vaporizer). The ultimate goal of this approach is that the user will also perceive these additional sensory effects giving her/him the sensation of being part of the particular multimedia content and resulting in a worthwhile, informative user experience. In the context of this work, this kind of user experience is referred to as a truly immersive Sensory Experience.

The purpose of this chapter is to introduce the concept of Sensory Experience (cf. Sect. 24.2), its assessment in terms of the QoE in order to derive a comprehensive QoE model (cf. Sect. 24.3), and synchronization issues with traditional multimedia content, e.g., based on scent data (cf. Sect. 24.4). The chapter is concluded with Sect. 24.5 highlighting the major findings.

24.2 Sensory Experience

24.2.1 Concept and System Architecture

The concept and system architecture of receiving sensory effects in addition to audio/visual content is depicted in Fig. 24.1. The media and the corresponding Sensory Effect Metadata (SEM) may be obtained from a Digital Versatile Disc, Blu-ray Disc, or any kind of online service (e.g., download/play or streaming portal). The media processing engine acts as the mediation device and is responsible for playing the actual media resource and accompanying sensory effects in a synchronized way. That is done based on the users' setup in terms of both media and sensory effect rendering. Therefore, the media processing engine may adapt both the media and the SEM (and, consequently, the corresponding effects) according to the capabilities of the various rendering devices. The user environment (e.g., a living room) is extended with additional rendering devices enabling the stimulation of senses other than hearing and seeing. For example, a motion chair, fan/ventilator, heater/cooler, etc. may be used to address the somatosensory (human sensory) sub-system whereas a scent vaporizer device stimulates the olfactory sub-system. The visual sub-system

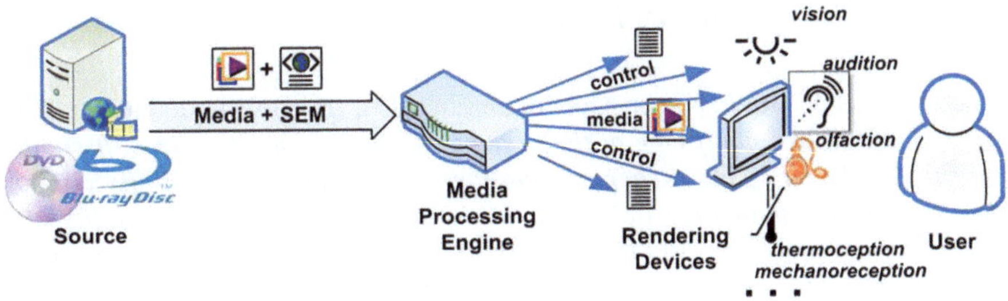

Fig. 24.1 Concept and system architecture of sensory experience [26]

may be further stimulated using (additional) ambient light devices in addition to the visual content.

24.2.2 Sensory Effect Description Language

The metadata format for describing such sensory effects is defined by ISO/MPEG in the context of "MPEG-V: Media Context and Control". In particular, Sensory Information (Part 3) [9] defines a Sensory Effect Description Language (SEDL), an XML Schema-based language, which enables one to describe sensory effects. The actual sensory effects are not part of SEDL but defined within the Sensory Effect Vocabulary (SEV) for extensibility and flexibility, allowing each application domain to define its own sensory effects, if applicable. A description conforming to SEDL is referred to as a SEM description and may be associated to any kind of multimedia content (e.g., movies, music, Web sites, games). For example, the SEM description may be attached to any (1) file-based data structure, i.e., ISO base media file format (e.g., mp4) as timed metadata track; (2) stream-based data structures, i.e., within an MPEG-2 Transport Stream in a similar way as the electronic program guide which is also XML-based or as a playload of the Real-time Transport Protocol; or (3) included into a Web page as an alternate representation of the current document.

In related works, the possible implementation within Universal Plug and Play (UPnP) is described in [19] and a broadcasting system including sensory information is proposed in [31]. The SEM description is used to steer sensory devices like fans, vibration chairs, lamps, etc. via an appropriate mediation device in order to enrich the experience of the user. That is, in addition to the audio-visual content of, e.g., a movie, the user will also perceive other effects such as the ones described above with the aim to improve the users' QoE.

24.2.3 Quality of Sensory Experience

In the Qualinet White Paper on QoE definition [13], the QoE is defined as "*the degree of delight or annoyance of the user of an application or service. It results from the fulfillment of his or her expectations with respect to the utility and/or enjoyment of the application or service in the light of the users personality and current state*". In addition to the QoE definition, [13] also highlights application areas, factors influencing the QoE, features of the QoE, and relationship between QoS and QoE. In particular, different application domains may adopt the generally agreed definition of QoE and provide a specialization thereof, taking into account its requirements formulated by means of influence factors and features of the QoE.

The Quality of Sensory Experience comprises the QoE of multimedia content annotated with sensory effect metadata and the synchronization thereof. Hence, the QoE is multidimensional and multi-sensorial. For the assessment of the quality of sensory experience, we focus on subjective quality assessments with the aim to derive a QoE model. Related research in this area is mainly foçusing on ambient light associated to audio-visual (TV) content. For example, [20] provides an initial study on how additional light is perceived by end users whereas [21] is more comprehensive and in the context of 3DTV. However, a comprehensive QoE model which potentially takes into account all human senses is currently not available and, thus, introduced in Sect. 24.3 based on multiple subjective quality assessments in that domain.

The synchronization of traditional multimedia content (audio and video) has long been an active research topic [22] and can be applied for sensory effects too. However, the delivery of multimedia to facilitate sensory experience (referred to as multiple sensorial media—MulSeMedia in [6]) brings another level of complexity to the synchronization field. Works with sensory related objects (e.g., olfaction, haptic, etc.) have reported unexpected results from synchronization perspective. For example, works reported by [1, 15] indicate that assessors are quite tolerant to certain levels of inter-media skew and to lingering effects associated with olfaction. Comparing the results of these works highlights interesting differences in perceived synchronization resulting from the information being presented by the media. In terms of haptic media, [8] report a system that supports haptic, olfaction and visual integration as part of a fruit harvesting game. The aim was to investigate the influence of inter-stream synchronization error between olfactory and haptic media on QoE. Eid et al. [4] also report a multiplexing framework to support synchronized delivery of sensory media (haptic in addition to audiovisual). Understanding the impact of inter-media skew on QoE is an active research topic and in Sect. 24.4 the focus lies with sensory experience works involving olfaction.

24.3 Assessing the Quality of Sensory Experience

24.3.1 Experimental Setup

In order to study the influence on the QoE when consuming multimedia content annotated with sensory effects, we report from various subjective quality assessments with slightly different contexts and goals as well as partially utilizing different methods. In all cases, we have adopted methods defined by ITU-T Rec. P.910 [11], P.911 [12], and BT.500-13 [10]. The setup for our experiments, i.e., location, participants, apparatus, and procedure for evaluation, are described in detail in [28] and only briefly described here. For all subjective tests, we invited around 20 (Sect. 24.3.2) and up to 32 (Sect. 24.3.3) participants, equally distributed among males and females, and not familiar with the subject which conforms to guidelines defined in [10–12]. The test sequences have been carefully selected in terms of content, genre, and qualities (when needed) and manually annotated with different sensory effects [29]. For all tests, the same setup has been used which was inspired by and partially based on [23].

24.3.2 Experimental Results

In our first experiment [28], we demonstrated that sensory effects provide a vital tool for enhancing the user experience depending on the actual genre. Therefore, we gathered test sequences of different genres, i.e., action (Rambo 4, Babylon A.D.), news (ZIB Flash), documentary (Earth), commercials (Wo ist Klaus), and sports (Formula 1), and annotated them with various sensory effect metadata, i.e., wind, vibration, and light effects. Note that light effects are actually not part of the SEM description but extracted automatically from the video content [26]. The sequences were chosen carefully to have all different types of effects within each sequence. For the actual method, we adopted the Degradation Category Rating (DCR) [12] and turned the five-level impairment scale into a new five-level enhancement scale. That is, the subjects rate on the enhancement of a stimulus annotated with sensory effects compared to a reference stimulus without sensory effects rather than on the impairment. The quality of the video content (independent of the sensory effects) was equal for both sequences of the same genre and did not contain any visual artifacts.

The detailed evaluation results are given in [28]. The DCR improvement score with a confidence interval of 95 % is depicted in Fig. 24.2. The x-axis shows the name of the sequences. As one can see, two sequences were presented twice but not directly one after the other in order to test the reliability of the participants. Additionally, the order of the sequences was randomized for each participant. The figure clearly shows the lower MOS for news compared to the higher MOS for action and documentary genres. In particular, the action, sports, and documentary genres benefit more from these additional effects. Interestingly, although the sequences Rambo 4 and Babylon A.D. are from the same genre, the results differ slightly. The commercial genre can

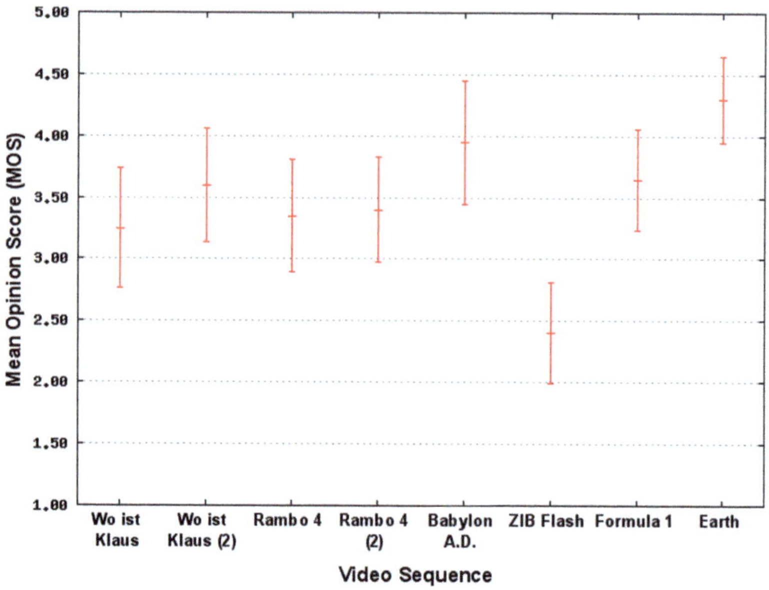

Fig. 24.2 DCR improvement score and confidence interval [28]

also profit from the additional effects but not at the same level as the documentary genre. Only the news genre may not profit from these effects. Furthermore, the figure also depicts that the two videos presented twice differ in the results but still have overlapping confidence intervals.

The aim of another experiment [27] was to investigate the relationship of the QoE to various video bit-rates of multimedia contents annotated with sensory effects. In particular, we were interested in the subjective quality gap between video resources annotated with and without sensory effects at different bit-rates. The overall setup of the second experiment was similar to the first one. The test stimuli comprise the two best performing video sequences from our first experiment. For each sequence, four versions with different bit-rates were prepared whereby only the video bit-rate was affected and the audio bit-rate remained constant for all versions of a given sequence. Additionally, each sequence has been annotated with sensory effects resulting in 16 different bit-streams to be evaluated. For the actual subjective assessment, we have adopted the Absolute Category Rating with Hidden Reference (ACR-HR) method using a five-point discrete scale from excellent to bad as defined in [11]. Like in the previous subsection, the detailed evaluation results are given in [27]. Thus, we will only concentrate on the MOS values depending on various bit-rates as depicted in Fig. 24.3 (sequence Earth, i.e., the documentary and results for the other sequence is similar). Interestingly, the sequences with sensory effects have always a higher MOS than their counterparts without sensory effects and almost steadily increase for higher PSNR/bit-rates. In general, the results confirm the observations from the previous experiment. Additionally, Fig. 24.3 also shows that the MOS of the lowest bit-rate

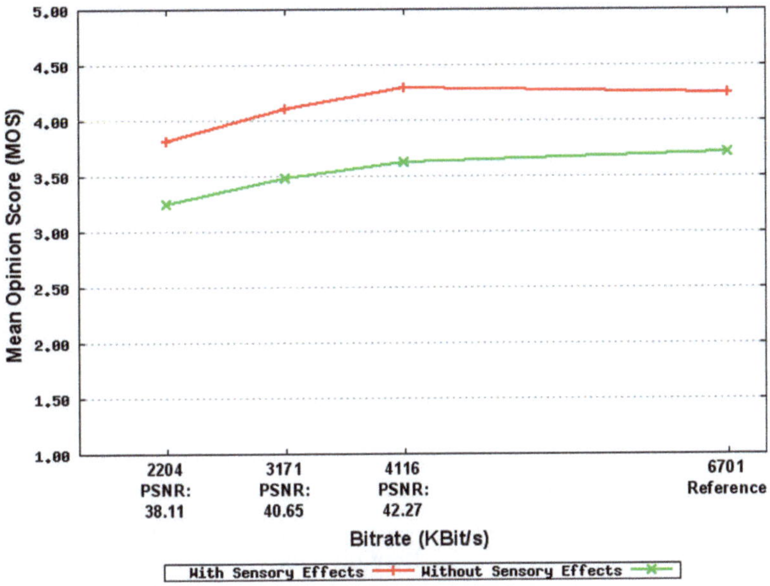

Fig. 24.3 MOS versus PSNR/bitrate for Earth sequence [27]

version with sensory effects is always higher than the MOS of all higher bit-rate variants without sensory effects. Furthermore, we calculated the average difference between the two curves using the Bjontegaard Delta (BD) method [2] with the result that the sequence enriched with sensory effects is 0.6 MOS points higher than without sensory effects (0.5 MOS points on average for both sequences).

24.3.3 QoE Model for Sensory Experience

The ultimate goal, however, is to come up with a QoE model for the sensory experience [25]. Therefore, we have conducted a subjective quality assessment which evaluates the influence of individual sensory effects and all combinations thereof. Our QoE model for sensory experience is defined complementary to existing approaches for predicting the QoE of audio-visual services. For example, these existing approaches aim to map QoS to QoE [32] or to predict the QoE [3, 30] with a main focus on audio-visual services and do not take into account additional assets such as sensory effects. Other QoE models such as that presented in [18] are based on perception, emotion, and sensation and mainly address adaptation and presentation issues without explicitly addressing sensory effects.

Figure 24.4 depicts the results for the video sequence 2012 from the action genre with all possible configurations. The results clearly indicate that without sensory effects the MOS for the QoE is around 40 and adding sensory effects will increase

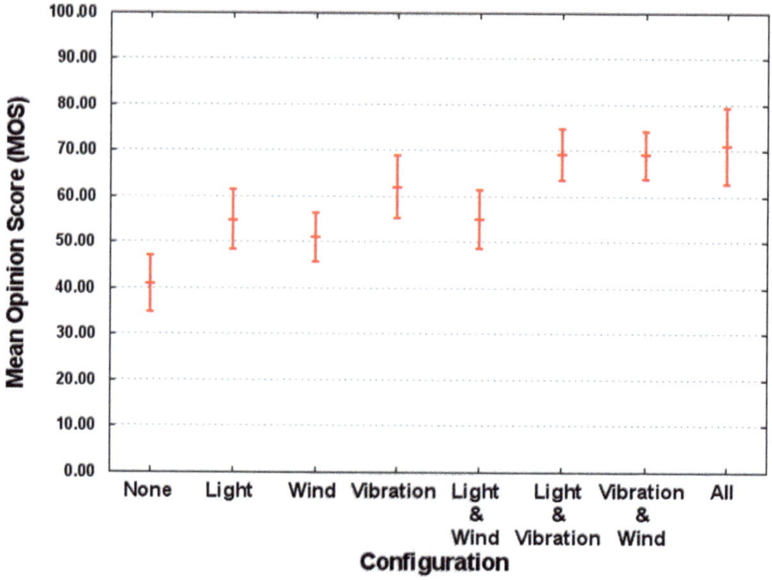

Fig. 24.4 MOS and confidence intervals (95 %) for the sequence 2012 [25]

the MOS. For example, adding light effects to the video sequence increases the MOS from around 40 up to about 55. The MOS slightly decreases when exchanging the light effect with a wind effect. The highest impact of a single sensory effect on the QoE is achieved by using the vibration effect which increases the MOS up to approximately 62. One can see that using only vibration effects has a bigger impact on the QoE than using only light or wind effects. Surprisingly, the combination of light and wind effects does not result in a higher rating than the vibration effect on its own. Moreover, it shows a similar rating as with only light or wind effects, at least for this sequence. Any other effect combined with vibration increases the QoE. Finally, the highest MOS is achieved by combining all three sensory effects.

As indicated above, the available models that try to map QoS to QoE or to estimate QoE from different parameters do not take into account additional assets such as sensory effects. The results from the study presented led us to the hypothesis that there exists a linear relationship between the number of effects and the actual QoE. Thus, we introduce a linear QoE model for sensory experience. The aim of this model is to enable an estimation of the QoE of the multimedia content with sensory effects (QoE_w) from the QoE of the multimedia content without sensory effects (QoE_{wo}). Equation 24.1 shows our QoE model for sensory experience.

$$QoE_w = QoE_{wo} \times \left(\delta + \sum w_i b_i\right) \tag{24.1}$$

In our QoE model, w_i represents the weighting factor for a sensory effect of type i, e.g., in our setup $i \in \{light(l), wind(w), vibration(v)\}$. Please note that further sensory effect types (e.g., scent) may be incorporated easily. The binary variables $b_i (b_i \in \{0, 1\})$ are used to identify whether effect i is present for a given setup. Finally, δ is used for fine-tuning. The QoE_{wo} may be assessed through any existing model such as those given in [18] or by an appropriate QoS to QoE mapping [5]. The results of the conducted study request for a model that deals with all types of sensory effects separately. Therefore, we introduce the model illustrated by Eq. 24.1 with weighting factors and binary variables for each type of sensory effect. An instantiation and validation of the proposed QoE model is provided in [25]. With haptics (sense of touch) already addressed in Chap. 18, the next section of this chapter reports the effect of inter-media skew with respect to olfaction enhanced multimedia and the impact of such on the QoE.

24.4 Subjective Evaluation of Olfactory and Visual Media Synchronization

Synchronization of media enhanced with olfaction is particularly complex due to its slow moving, lingering nature and variable perception based on factors such as age, sex and nationality [15] among others. This is demonstrated in the lack of reported works involving olfaction, even compared with other complex sensory media such as haptic. In fact, the principle focus of works associated with olfaction has been the development of olfactory displays [17, 24]. These works focus on the hardware that enables controlled emission of minute amounts of scent. It is arguable that these works are dealing with olfactory data from an intra-stream perspective [15]. Works reporting the user perception of the inter-stream synchronization of olfactory data with other media are now available in the literature, e.g., audio-visual [1], haptic [4, 8], and visual [15]. A consistency exists across all of these approaches in that the impact on the user's QoE is analyzed in the presence of varying degrees of inter-media skew between the media. The methodology of integrating artificial skews between media and examining the user perception was initially documented in [22]. The results below, discussed in detail in [14–16], recount the findings of an empirical study analyzing the effect of inter-media skew between olfaction and visual media on QoE. The same videos and scents were used as in [1, 7], but the contextual audio was replaced with the sound of a blowing fan, hence the focus was to examine the relationship between olfactory and visual media.

24.4.1 Experimental Setup

In order to study the impact of inter-media skew on QoE, a number of subjective quality assessments were preformed. The aim of the experiments was to determine (a) assessor ability to detect skew; (b) assessor perception of skew; and (c) impact, if any, of skew on QoE. The methodology employed in these works is based on ITU-T P.910 [11] with alterations for some of the statements. The impairment scale was replaced with a Likert Scale for QoE comparative analysis. The experimental setup (video sequences and scents, lab details, the emitter, questionnaires and rating scales) is described in detail in [14, 15]. The skews introduced between the olfaction and video were introduced in step sizes of 5 s.

24.4.2 Experimental Results

Details on assessor detection and perception of olfactory-visual inter-media skew can be found in [14–16]. The impact of skew on QoE was determined via assessors answering questions on their (1) sense of enjoyment; (2) sense of relevance; and (3) sense of reality having experienced an olfaction-enhanced video clip. Assessors answered the questionnaire on the test clip which may or may not have had an inter-media skew, by comparing it with a reference clip, which presented synchronized olfaction and visual media. This is consistent with the Degradation Category Rating (DCR) [12]. The MOS with 95 % confidence interval are presented in Figs. 24.5, 24.6 and 24.7 for impact of inter-media skew for each of the aforementioned criteria. The figures here report general results without consideration for age, sex, or nationality of assessors (see previous works for relevant MOS scores based on age [14], gender [14], and nationality [14, 16]. The x-axis shows the skew size in seconds between the olfaction and the video. Olfaction presented before video is represented in terms of the negative skew times and olfaction after video is presented in terms of positive skew times. Synchronized presentation is represented by a skew size of 0 s. The y-axis represents the MOS scores of the groups of participants for each of three statements relevant to QoE.

24.4.2.1 Impact of Skew on Enjoyment

Figure 24.5 shows the MOS reflecting assessors level of enjoyment of olfaction-enhanced multimedia in the presence of varying degrees of inter-media skew. When synchronized presentation takes place, assessors agreed that they enjoyed watching the video clip. It is clear that assessors rated their enjoyment of the experience higher with olfaction presented after video as opposed to olfaction presented before video. As per [14, 16], younger females enjoyed the experience least in the presence of large skews and most when presented in sync, whilst for older male groups, the presence skew, relatively speaking, had the least impact. In contrasting the views of males and females for enjoyment, the female group reported greater sense enjoyment of olfaction before video than their male equivalent for olfaction after video as reported in [16].

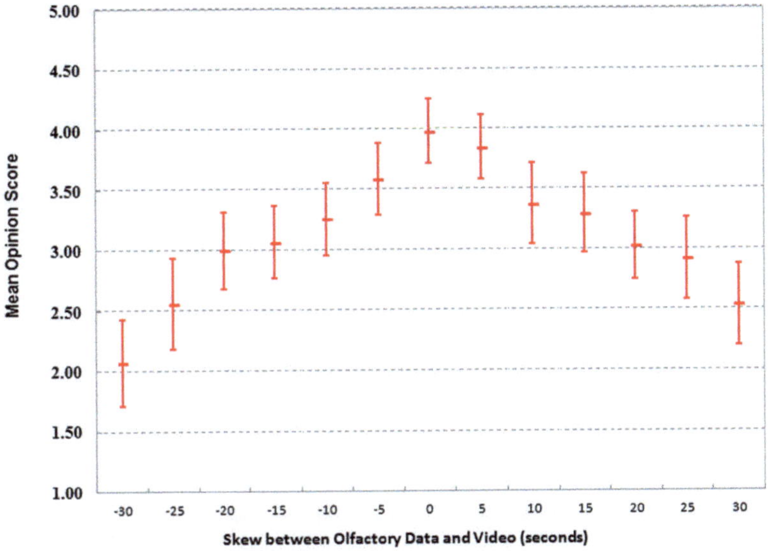

Fig. 24.5 Analysis of sense of enjoyment per skew with confidence interval based on 95 % confidence level [15]

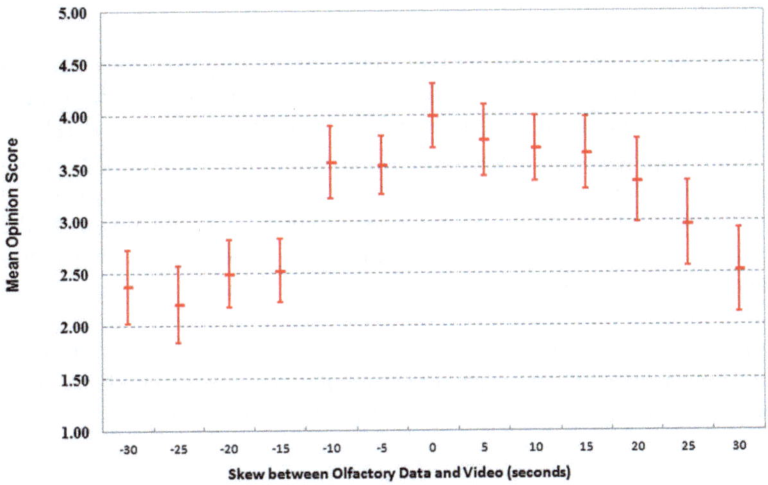

Fig. 24.6 Analysis of sense of relevance per skew with confidence interval based on 95 % confidence level [15]

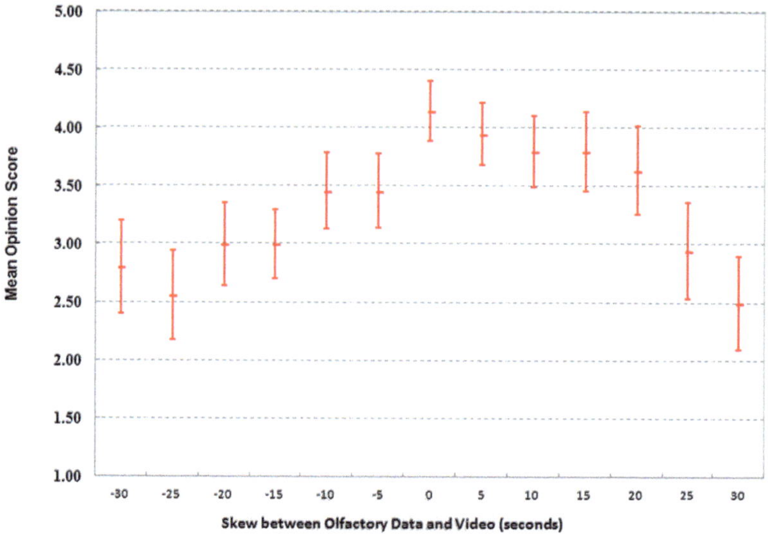

Fig. 24.7 Analysis of sense of reality per skew with confidence interval based on 95 % confidence level [15]

24.4.2.2 Impact of Skew on Relevance

Investigating the impact of inter-media skew considering relevance of the scent, Fig. 24.6 shows that olfaction after video provides a higher sense of relevance when compared with olfaction presented in advance of video. This trend of relevance is consistent for both male and female groups [14]. In general, the female group reported higher scores of relevance both in the presence and absence of skew compared with males. Considering nationality [16], the Asian group reported the scent to be the least relevant to the video content, with the African group indicating that they considered the scent to be most relevant during their experiences in both the presence and absence of skew.

The most interesting finding from Fig. 24.6 is the slow reduction in perceived sense of relevance for olfaction presented after video as compared with olfaction presented before video. In fact, assessors rated skews of −5 and +20 s similarly in terms of relevance.

24.4.2.3 Impact of Skew on Reality

Figure 24.7 shows MOS scores reflecting assessors sense of reality in the presence of varying degrees of inter-media skew. It is clear from Fig. 24.7 that skew adversely affects assessor sense of reality. Particularly interesting from analysis of the MOS is the slow reduction in heightened sense of reality for olfaction after scent. In terms of contrasting the responses between males and females, the female group was particularly affected by skew, much more so than their male equivalents. From [14], for skews of −10 to +10 s, the male groups perceived the olfaction as equally contributing to an enhanced sense of reality. The highest MOS rating was reported for the male group at +5 s (as opposed to 0 s skew). The female MOS scores indicate less heightened sense of reality at skew levels when olfaction is presented before video, greater sense of reality at no skew and with skews for olfaction presented after video. When comparing the ratings of the female group when scent was presented ahead or after the video, the sense of reality was higher at skews of +20 s then it was for skews of −5 s. A similar trend, although not as significant, exists for males rating the sense of reality for −10 and +15 s [15]. Considering both their age and gender, the older female groups reported to be the most sensitive to skew, i.e., they reported higher sense of reality when synchronized presentation or small skews were present and had the largest fall off in the presence of large skews. The youngest female group reported a similar trend but not as sensitive to the sense of reality loss in the presence of large skews. Interestingly, skew had the least impact on the older male group.

24.5 Conclusions

In this chapter, we have extended the concept of Quality of Experience towards Sensory Experience by attaching sensory information as an additional data track to traditional multimedia content (audio-visual). It allows for rendering of sensory effects synchronized with traditional multimedia content with the aim to increase the QoE.

We have shown how to assess the QoE of this emerging application domain thanks to publicly available datasets, evaluation setups, and available off-the-shelf hardware to render these sensory effects which enables objective comparison of results among research laboratories. Furthermore, we have derived a QoE model which allows to quantify the Sensory Experience based on the QoE of the multimedia content without sensory effects.

Additionally, we report on the findings of an empirical study analyzing the effect of inter-media skew between olfaction and visual media on QoE which shows that, in general, a higher QoE is perceived with olfaction presented after video as opposed to olfaction presented before video.

Based on the differences reported here and in [1], it is clear that the presence of contextual audio affects the impact of skew on QoE, hence a study to further understand the impact audio has on the perception of scent is necessary. Indeed, there is significant scope to develop models akin to Sect. 24.3.3 to represent the human perception of olfaction integrated with other media.

References

1. Ademoye O, Ghinea G (2009) Synchronization of olfaction-enhanced multimedia. IEEE Trans Multimedia 11(3):561–565. doi:10.1109/TMM.2009.2012927
2. Bjontegaard G (2001) Calculation of average PSNR differences between RD curves. ITU-T VCEG meeting VCEG-M33, Austin, USA
3. Cherif W, Ksentini A, Negru D, Sidibe M (2011) A_PSQA: efficient real-time video streaming QoE tool in a future media internet context. In: IEEE international conference on multimedia and expo (ICME 2001), pp 1–6. doi:10.1109/ICME.2011.6011993
4. Eid M, Cha J, El-Saddik A (2011) Admux: an adaptive multiplexer for haptic audio-visual-data communication. IEEE Trans Instrum Meas 60(1):21–31. doi:10.1109/TIM.2010.2065530
5. Fiedler M, Hoßfeld T, Tran-Gia P (2010) A generic quantitative relationship between quality of experience and quality of service. IEEE Network (special issue on improving QoE for network services)
6. Ghinea G, Ademoye O (2010) A user perspective of olfaction-enhanced MulSeMedia. In: Proceedings of the international conference on management of emergent digital ecosystems (MEDES '10). ACM, New York, pp 277–280. doi:10.1145/1936254.1936308. http://doi.acm.org/10.1145/1936254.1936308
7. Ghinea G, Ademoye O (2010) Perceived synchronization of olfactory multimedia. IEEE Trans Syst Man Cybern Part A Syst Hum 40(4):657–663. doi:10.1109/TSMCA.2010.2041224
8. Hoshino S, Ishibashi Y, Fukushima N, Sugawara S (2011) QoE assessment in olfactory and haptic media transmission: influence of inter-stream synchronization error. In: IEEE international workshop technical committee on communications quality and reliability (CQR 2011), pp 1–6. doi:10.1109/CQR.2011.5996082
9. ISO/IEC 23005–3 (2010) Information technology—media context and control—sensory information. ISO/IEC JTC 1/SC 29/WG 11/N11425, Geneva, Switzerland
10. ITU-R Rec. BT.500-13 (2012) Methodology for the subjective assessment of the quality of television pictures
11. ITU-T Rec. P.910 (2008) Subjective video quality assessment methods for multimedia applications
12. ITU-T Rec. P.911 (2008) Subjective audiovisual quality assessment methods for multimedia applications
13. Möller S, Le Callet P, Perkis A (eds) Qualinet white paper on definitions of quality of experience—output version of the Dagstuhl seminar 12181: European network on quality of experience in multimedia systems and services (COST Action IC 1003), Lausanne, 1.1 edn
14. Murray N, Qiao Y, Lee B, Karunakar AK, Muntean GM (2013) Age and gender influence on perceived olfactory and visual media synchronization. In: 2013 IEEE international conference on multimedia and expo (ICME '13). IEEE, p 16. doi:10.1109/ICME.2013.6607467
15. Murray N, Qiao Y, Lee B, Karunakar AK, Muntean GM (2013) Subjective evaluation of olfactory and visual media synchronization. In: Proceedings of the 4th ACM multimedia systems conference (MMSys '13). ACM, New York, pp 162–171. doi:10.1145/2483977.2483999. http://doi.acm.org/10.1145/2483977.2483999
16. Murray N, Qiao Y, Lee B, Karunakar AK, Muntean GM (2014) User profile based perceived olfactory and visual media synchronization. ACM Trans Multimedia Comput Commun Appl 1(10)
17. Nakamoto T, Yoshikawa K (2006) Movie with scents generated by olfactory display using solenoid valves. In: Virtual reality conference 2006, pp 291–292. doi:10.1109/VR.2006.102
18. Pereira F (2005) A triple user characterization model for video adaptation and quality of experience evaluation. In: IEEE 7th workshop on multimedia signal processing 2005, pp 1–4. doi:10.1109/MMSP.2005.248674
19. Pyo S, Joo S, Choi B, Kim M, Kim J (2008) A metadata schema design on representation of sensory effect information for sensible media and its service framework using UPnP. In: 10th international conference on advanced communication technology (ICACT 2008), vol 2, pp 1129–1134. doi:10.1109/ICACT.2008.4493965

20. de Ruyter B, Aarts E (2004) Ambient intelligence: visualizing the future. In: Proceedings of the working conference on advanced visual interfaces (AVI ’04). ACM Press, New York, pp 203–208. doi:10.1145/989863.989897http://dx.doi.org/10.1145/989863.989897
21. Seuntiens P, Vogels I, van Keersop A (2007) Visual experience of 3D-TV with pixelated ambilight. In: Proceeding of the 10th annual international workshop on presence, pp 339–344
22. Steinmetz R (1996) Human perception of jitter and media synchronization. IEEE J Sel Areas Commun 14(1):61–72. doi:10.1109/49.481694
23. Storms RL, Zyda MJ (2000) Interactions in perceived quality of auditory-visual displays. Presence: Teleoper Virtual Environ 9(6):557–580. doi:10.1162/105474600300040385. http://dx.doi.org/10.1162/105474600300040385
24. Sugimoto S, Noguchi D, Bannnai Y, Okada K (2010) Ink jet olfactory display enabling instantaneous switches of scents. In: Proceedings of the international conference on Multimedia (MM ’10). ACM, New York, pp 301–310. doi:10.1145/1873951.1873994. http://doi.acm.org/10.1145/1873951.1873994
25. Timmerer C, Rainer B, Waltl M (2013) A utility model for sensory experience. In: Proceedings of the fifth international workshop on quality of multimedia experience (QoMEX 2013). IEEE, Los Alamitos, CA
26. Waltl M, Timmerer C, Hellwagner H (2009) A test-bed for quality of multimedia experience evaluation of sensory effects. In: Ebrahim T, El-Maleh K, Dane G, Karam L (eds) Proceedings of the first international workshop on quality of multimedia experience (QoMEX 2009). IEEE, Los Alamitos, CA, pp 145–150. doi:10.1109/QOMEX.2009.5246962. http://www.qomex2009.org
27. Waltl M, Timmerer C, Hellwagner H (2010) Improving the quality of multimedia experience through sensory effects. In: Perkis A, Mller S, Svensson P, Reibman A (eds) Proceedings of the 2nd international workshop on quality of multimedia experience (QoMEX’10). IEEE, Trondheim, Norway, pp 124–129. http://www.qomex2010.org
28. Waltl M, Timmerer C, Hellwagner H (2010) Increasing the user experience of multimedia presentations with sensory effects. In: Leonardi R, Migliorati P, Cavallaro A (eds) Proceedings of the 11th international workshop on image analysis for multimedia interactive services (WIAMIS’10). IEEE, Desenzano del Garda, pp 1–4
29. Waltl M, Timmerer C, Rainer B, Hellwagner H (2012) Sensory effect dataset and test setups. In: 2012 Fourth international workshop on quality of multimedia experience (QoMEX 2012), pp 115–120. doi:10.1109/QoMEX.2012.6263841
30. Wang T, Pervez A, Zou H (2010) VQM-based QoS/QoE mapping for streaming video. In: 3rd IEEE international conference on broadband network and multimedia technology (IC-BNMT 2010), pp 807–812. doi:10.1109/ICBNMT.2010.5705202
31. Yoon K, Choi B, Lee ES, Lim TB (2010) 4-d broadcasting with mpeg-v. In: IEEE international workshop on multimedia signal processing (MMSP 2010), pp 257–262. doi:10.1109/MMSP.2010.5662029
32. Zinner T, Hohlfeld O, Abboud O, Hoßfeld T (2010) Impact of frame rate and resolution on objective QoE metrics. In: International workshop on quality of multimedia experience 2010, Trondheim

Chapter 25
Gaming

Justus Beyer and Sebastian Möller

Abstract Playing is the first activity newborn humans immerse themselves in besides fulfilling basic needs. There is no ultimate goal to be achieved: Playing is a process that is only kept alive by the player's experience of it. This chapter provides an overview over existing concepts and current research on the experience of playing video games. It does so by taking the perspective of a quality engineer, who identifies influencing factors, quantifies them in terms of performance metrics, and analyzes their impact on perceived quality features. To support the development of empirical test methods as well as instrumental prediction models for video gaming QoE, the concepts are grouped in a taxonomy. The chapter is concluded by a discussion of the empirical application of the framework in experiments, a brief look at an existing QoE prediction model, and an outlook at promising future research directions.

25.1 Introduction

With video gaming becoming more and more popular, the effort to produce high quality titles has risen dramatically. While it required one developer to create Tetris in 1984, modern game productions consume the budget of a Hollywood movie [1], making the processes and influencing factors of a player's quality perception not only of scientific, but also of commercial interest. However, the user-perceived QoE of games has not seen the thorough investigation other multimedia services have in the past.

J. Beyer (✉) · S. Möller
Quality and Usability Lab, Telekom Innovation Laboratories, TU Berlin, Berlin, Germany
e-mail: justus.beyer@qu.tu-berlin.de

S. Möller
e-mail: sebastian.moeller@telekom.de

S. Möller and A. Raake (eds.), *Quality of Experience*,
T-Labs Series in Telecommunication Services, DOI: 10.1007/978-3-319-02681-7_25,

A game has been defined as "a rule based system with a variable and quantifiable outcome, where different outcomes are assigned different values, the player exerts effort in order to influence the outcome, the player feels emotionally attached to the outcome, and the consequences of the activity are negotiable" [22]. Based on the platform they are implemented on, video games have for long been broadly classified into computer games, which are played on general purpose PC hardware, console games (Xbox, PlayStation, Wii, etc.), mobile games, which run on devices such as smartphones, tablets, or special gaming hardware such as the PlayStation Portable, and online games, which are often browser-based and require a constant Internet connection. As many recent computer and mobile games also contain features of online games, these sets are not disjunct: Both single- and multi-player games on computer, console and mobile platforms make use of Internet connections to coordinate interactions or exchange information such as leader boards, high scores, or updates. A special case are so-called "cloud games", where the control execution, game logic and rendering of a computer or console game is physically executed on a remote server farm (cloud), and just the display and input interpretation take place on the player's device.

A reason for the limited understanding of gaming QoE is that video gaming is a human-machine interaction activity and not merely a pure media delivery activity. Thus, standard methods for assessing the impact of transmission parameters are not sufficient. Moreover, in addition to the game content itself, the backend platform the game is built upon, the user interface (device and software), any transmission channels involved, as well as the characteristics of the user may have a significant impact on user-perceived QoE.

In contrast to the goal of completing tasks with minimal effort in task-oriented human-machine interaction, the primary aim of games is to provide an entertaining activity, where challenges are put in front of the user on purpose and difficulty is optimized to meet the players capabilities. That difference prevents the easy application of standard methods for determining usability (including effectiveness, efficiency, but also hedonic quality aspects), which are used in productivity-oriented human-computer interaction. Furthermore the outcomes of a game themselves are not necessarily the most rewarding aspect, but the process of overcoming the challenges and achieving the desired outcomes is [23]. Productivity-oriented applications, on the other hand, are designed to minimize challenges while achieving the desired outcome, which is their most rewarding aspect.

25.2 Taxonomy of QoE Aspects

To open up the space of quality aspects for video gaming, this chapter follows the considerations which were laid out in Möller et al. [27] and will take a detailed look on which aspects of quality might hold for this situation, and which others might not. The basis of this discussion is the definition of QoE as set up in Chap. 2 of this book. Following the approach of that chapter, as well as the taxonomy of multimodal human-machine interaction developed in [25], we will differentiate between influence factors (on QoE), interaction performance aspects, and quality features. Those of the factors and features which are relevant for gaming, are put into a logical relationship to highlight their dependencies. Wherever possible, we cite metrics, which can be used for a quantitative assessment.

25.2.1 Influence Factors

As introduced in Chap. 4 we will categorize factors influencing gaming QoE into human factors, which we call user factors here, system factors, and context factors.

25.2.1.1 User Factors

All factors which are specific to a certain player or to a type of player shall fall into this category. In the following we will briefly discuss experience, playing style, intrinsic motivation, and a set of static and dynamic user factors as attributes to describe QoE-relevant influences of a player.

- Experience: Both in scientific and in popular literature a classification in "hardcore gamer" and "casual gamer" is widely used, distinguishing the classes based on the average time of playing per time period. Despite the division's broad adoption, no common threshold exists to delimit the two groups. Another popular distinction exists between "newbie" and "pro gamer" based on the experience with a particular game or game genre. These characteristics are related to the gamer's skill, change dynamically, and can be measured with demographic questionnaires.
- Playing style: Bartle [2] differentiates between "achiever", "explorer", "socializer" and "killer" in the context of the Multi User Dungeon game genre, a predecessor of the modern genre of role-playing games. The degree to which a player belongs to these classes (in percent) can be determined using a self-reporting questionnaire [13]. This differentiation has been criticized [8] for overlaps between the four classes and its missing empirical validation.
- Intrinsic motivation: Starting from Bartle's classes, Yee [43] developed a classification of intrinsic motivation, which is necessary for playing, along three axes: achievement, social, and immersion. Each axis is composed of 3 sub-components, and a 39-item questionnaire is available for classifying users along these axes (e.g. a player achieving a high value on the axis "socializing" has a higher motivation for interacting with other players than another who has a lower value on this axis).
- Static and dynamic user factors: As in multimodal interaction [25] static characteristics include age, gender, native language, etc., whereas dynamic ones include the current emotional status, boredom, distraction, curiosity, and intended relaxation, see also [11]. To elicit the static factors, screening questionnaires can be

used at the beginning of an experiment. For dynamic characteristics psychological metrics exist.

A further discussion of human influence factors can be found in Chap. 4.

25.2.1.2 System Factors

The factors which we will discuss in this section contain both the game, which is being played by the user, but also the whole setup and user-perceivable design of the hard- and software. In case of a handheld offline gaming device, this would refer to all QoE-relevant properties of the apparatus and the software within. For online and cloud games, however, it would also include properties of the server, the involved transmission channel, and the error-handling for problems arising from the distributed nature of the system (e.g. delays, loss of packets/information, establishing a synchronized state between all involved parts).

- Game genre: Wolf [42] defines 42 different genres, but many games belong to several of these genres. For marketing purposes a differentiation into 13 "super genres" is common, e.g. action, fight, flight, shooter, strategy, sport, etc. [29]. The game genre may strongly influence the effect a technical platform has on user-perceived QoE, e.g. the sensitivity to parameters like delay is more influential to some genres than to others.
- Game structure: Fullerton [12] differentiates e.g. single player against the game, several players against the game, several players against each other, cooperative game, team game, etc.
- Game mechanics and rules: These largely influence and determine game outcomes and are individual to each game. To our knowledge, no easy classification exists.
- Technical system set-up: This includes client and server characteristics, any transmission systems involved (characterized by parameters like their bandwidth, packet loss, delay, jitter, packet reordering, etc.), interface and client software characteristics (which may include potential counter-measures for insufficient or varying bandwidth, buffering, etc.; these can be very influential as Pommer [32] shows), and device characteristics (interaction capabilities and modalities, feedback capability, ergonomics, etc.).
- Design characteristics: These describe the design of a system, which can be experienced by the user, and is commonly specified by design experts and developed together with the specification of the game. No simple classification of game designs is known to us.

Most technical system factors are specified in a qualitative and/or quantitative way in corresponding specification documents by the game developer or by the provider of the technical platform, whereas design characteristics can better be specified by design experts or experienced salesmen.

25.2.1.3 Context Factors

While in some rare cases, such as special car or airplane simulators, the player's environment is formed, lit, and even moved in a way to resemble the corresponding original, the ordinary case is that it lies outside the reach of the games designers. Besides these physical environment factors the success of online multiplayer games (e.g. Massively Multiplayer Online Role-playing Games such as World of Warcraft) or party games (e.g. various karaoke games such as SingStar) have demonstrated the significance of the social context.

- Physical environment factors: Those factors include room characteristics (space, acoustics, lighting) and usage situation (in-house, on the move, etc.).
- Social context: Relationships to other players who are involved in the game, potential parallel activities of the player, privacy and security issues, which might be particularly relevant in multiplayer games.
- Extrinsic motivation: May be financial or social reward or alike, depending on the user group (see above).
- Service factors: Includes access restrictions, availability of the system, resulting costs (sometimes expressed as cost per gaming time), etc.

Context factors are specified by the developers of the game and the technical platform, as well as by service providers.

25.2.2 Interaction Performance Aspects

As in the taxonomy for task-oriented multimodal interaction [25], we define interaction performance aspects separately for the user, and for the system, but further differentiate the latter into performance aspects of the user interface (device and software), performance aspects of the backend platform, and performance aspects of the game logic. Between these blocks, communication channels may exist: A physical channel between the user and the user interface; IP-based channels between the user interface and the backend platform (e.g. in cloud gaming) or not (e.g. when playing on a fixed platform), and between the platform and the game (e.g. in multi-player games where the game of one user is influenced by another user and this information is exchanged on a game-level).

25.2.2.1 System Performance

- User interface performance: Includes the input and output performance of the user interface.
- Backend platform performance: Can be subdivided e.g. into the performance of the processing of commands from the user interface, and the performance of generating corresponding output.

- Game performance: Mostly influenced by the control the user has over the interaction in the game, the game rules, and the game reaction. Can be expressed e.g. in terms of game errors, or alike.
- Communication channel performance: Includes all aspects of any involved transmission channels, i.e. the effectiveness and efficiency of forwarding user controls to the game, and the performance of forwarding game output to the user (especially relevant with cloud gaming).

25.2.2.2 User Performance

As with other multimodal interactive systems, user performance can be differentiated into perceptual effort, cognitive workload, and physical response (action) effort; The latter may largely be influenced by the device used for the interaction. As stated above, it is actually the task of a game to put a certain load on the user. Thus, keeping the load low does typically not result in good gaming experience.

- Perceptual effort: Effort required to decode the system messages, understand and interpret their meaning [44].
- Cognitive Workload: Commonly specifying the costs of task performance (e.g. necessary information processing capacity and resources). As we do not have a "task" here, we consider the effort to achieve a desired "outcome" as the gaming task. Subjective and objective methods for assessing cognitive workload are given in [41], but they have rarely been used in the context of gaming to our knowledge.
- Physical response effort: Physical effort required to interact with the game. No special metrics have been defined for this aspect to our knowledge.

25.2.3 Quality Features

As can be seen in the taxonomy of Fig. 25.1, we have separated quality features into five groups, which will be addressed in the following sections.

25.2.3.1 Interaction Quality

When looking into the quality of computer games, a common feature is the playability of a game. However, there seems to be no consensus on the definition of the term. As an example, Sánchez et al. [35] define playability as "the set of properties to describe the players experience with a particular game system, that the principal goal is fun/entertainment to the player in a satisfactory and credible way, playing alone or with other players. Playability reflects the players pleasure, experience, sensations and feelings when he/she is playing the videogame". Similarly, Foraker Labs define it as "the degree to which a game is fun to play and usable, with an emphasis on the

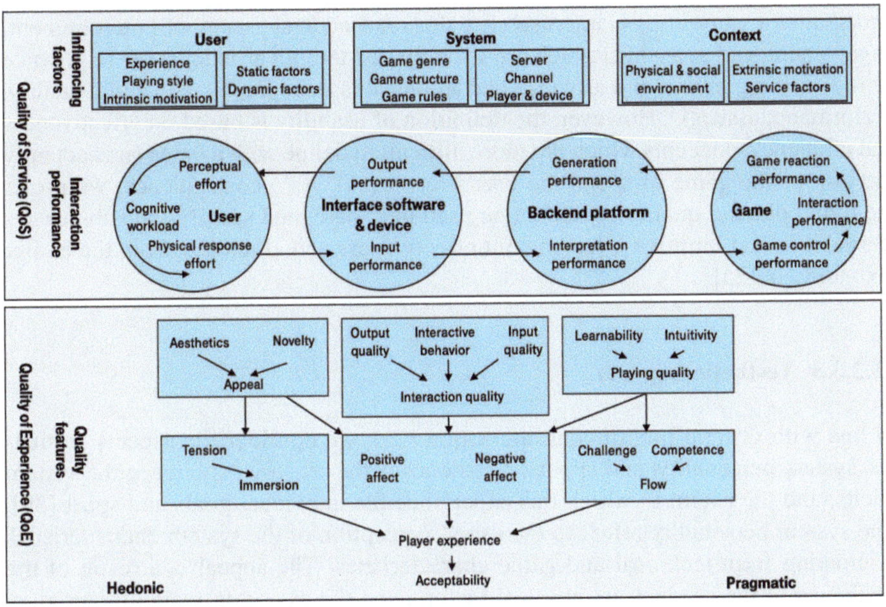

Fig. 25.1 Taxonomy of gaming QoE aspects. *Upper panel* influence factors and interaction performance aspects; *lower panel* quality features

interaction style and plot-quality of the game; the quality of gameplay. Playability is affected by the quality of the storyline, responsiveness, pace, usability, customizability, control, intensity of interaction, intricacy, and strategy, as well as the degree of realism and the quality of graphics and sound" [10]. In contrast to this, Engl [9] defines playability as the degree, to which all functional and structural elements of a game (hardware and software) enable a positive user experience for the gamer. This definition considers playability as a prerequisite of positive player experience (like usability can be considered as a prerequisite for user satisfaction, see definitions of usability [26]), or as a technical and structural basis for this, but not the player experience itself. In the following, we would like to adopt this narrower definition and would like to call this the "interaction quality" of a game. This meaning falls in line with the general definition of this term for multimodal systems and includes input quality (gamer to system), output quality (system to gamer, e.g. in terms of graphics quality, video quality, sound quality), as well as the interactive behavior (in task-oriented interaction we called this "cooperativity", but as a game storyline is not designed to be cooperative to the user, we prefer the general term interactive behavior here).

25.2.3.2 Playing Quality

Playing quality can be considered as a kind of game usability. This is defined by Pinelle et al. [30] as "the degree to which a player is able to learn, control, and understand a game. [...] Game usability does not address issues of entertainment, engagement, and storyline, which are strongly tied to both artistic issues (e.g. voice acting, writing, music, and artwork) and technical issues (graphic and audio quality, performance issues)." However, the definition of usability is based on effectiveness and efficiency, concepts which are more difficult to define with a game (it is actually the task of the game to spend the user's resources). As a consequence, we prefer the term "playing quality" over "game usability" here, and specify the sub-aspects learnability and intuitivity, leaving out effectiveness and efficiency from the earlier taxonomy in [25].

25.2.3.3 Aesthetic Aspects

In line with general multimodal interaction [25], we consider the aspects aesthetics, system personality, and appeal. Aesthetics is the sensory experience the system elicits, and the extent to which this experience fits individual goals and spirit [39]. The system personality refers to the users' perception of the system characteristics originating from technical and game characteristics. The appeal is a result of the aesthetics of the product, its physical factors, and the extend to which the product inherits interesting, novel and surprising features [16, 38]. Some of these aspects can be measured via questionnaires like [15], or via psycho-physiological measures [24].

25.2.3.4 Player Experience

Player experience is a broad concept, which covers a large set of sub-aspects. As mentioned before, we consider the "degree of delight or annoyance of the user" as a key aspect of QoE, which should be reflected in player experience as well. To our knowledge, Poels et al. [31] defined the most comprehensive taxonomy of player experience, which we would like to adopt in the following. According to their definition, player experience consists of the sub-aspects challenge, control, flow, tension, immersion, competence, positive affect, and negative affect. These seven sub-aspects can be measured with the purposely-built Game Experience Questionnaire (GEQ), see Ijsselsteijn et al. [18]. In the following paragraphs, we would like to highlight some of these sub-aspects.

- Flow, challenge, control: According to Csikszentmihalyi [7], flow is an equilibrium between boredom and fear, between requirements and abilities, and it is a dynamic experience of complete dissolution of an acting person in his/her activity. The activity itself constantly poses new challenges, so there is no time for boredom or sorrows. Intrinsic motivation is important for flow, as well as control over the game [5]. Hassenzahl relates flow to user experience: "Briefly, flow is a positive experience caused by an optimal balance of challenges and skills in a goal-oriented environment. In other words, flow is the positive UX [User Experience] derived from fulfilling the need for competence (i.e., mastery); it is a particular experience stemming from the fulfillment of a particular be-goal" [14]. In general, everybody can experience flow, but there seem to be factors which reduce flow in games, like age, reaction time, abilities, exposure to computers (digital natives vs. newbies), see e.g. [17]. Flow can be measured e.g. with the "Flow-Kurzskala", a scale with 16 sub-items, see Rheinberg et al. [34].

- Immersion: Is the degree of feeling to be in another (virtual) reality; in turn, the perception of the own (real) reality reduces. For games, immersion has been classified into three phases as "engagement", "engrossment", and "total immersion" [4, 20] which build upon each other. Jennett et al. [21] showed that immersion can be measured e.g. by eye-tracking, but also using the Immersive Experience Questionnaire (IEQ).
- Positive and negative affect: Positive affect can come in many different forms, and it is usually the goal of all gaming activity. According to Murphy [28], fun is "the positive feelings that occur before, during, and after a compelling flow experience. [...] It is not perfect, but it is concrete. The list of positive feelings associated with this definition of fun is quite long and includes: delight, engagement, enjoyment, cheer, pleasure, entertainment, satisfaction, happiness, fiero, control, and mastery of material"; negative ones might be frustration and boredom. Applied to computer games, Lazzaro and Keeker [23] investigated emotions and classified them into 4 types of fun: Hard fun (linked e.g. to computer games; typical is a constant change between frustration and fiero), easy fun (linked e.g. to curiosity, mostly covered by immersion), serious fun (linked e.g. to relaxation from stress), and people fun (linked to social interaction). The fun types may be linked to the playing style user types from Bartle, e.g. an achiever mostly searches for hard fun, an explorer for easy fun, a socializer for people fun, and a killer for hard and people fun [36].

25.2.3.5 Acceptability

Following the general definition, acceptability describes how readily a user will actually use the system. Acceptability may be represented by a purely economic measure, relating the number of potential users to quantity of the target group.

25.2.4 Use Cases

The presented taxonomy can be used in a wide range of applications. For example, the provider of an online gaming platform might be interested in how the bandwidth of the provided IP channel, or how the delay, with which controls from the user device are forwarded to his platform affect player experience. The taxonomy indicates that the effect may depend on the characteristics of the user (casual vs. power gamers), of the game (shooter vs. strategic exploration game), or of the device used for control and display (video codec and resolution). In order to identify and quantify these effects, the service provider would have to carry out subjective experiments where he controls most of the influencing factors of the taxonomy, and measures the interesting independent variables (e.g. with the Game Experience Questionnaire). In case it fits his interests, he might also focus on sub-aspects of the taxonomy (e.g. positive/negative affect, or flow), which he can address with additional metrics. In this case, the taxonomy helps to select the appropriate metrics and the right experimental set-up.

Another use case is a service provider who wants to monitor gaming behavior in order to find out about player experience. This service provider would need to set up models, which link measurable parameters of the game and the platform it is implemented on to user behavior during the game, and subsequently to player experience. For this purpose, it may be desirable to first collect performance metrics, then link them to the upper aspects of the QoE aspect layer (such as input quality, output quality, learnability, etc.), and subsequently link them to player experience in a second step.

25.3 Empirical Application and Results

This section will demonstrate the use of the presented taxonomy in planning and interpretation of studies. It will do so by looking at two experiments, in which the influence of different influence factors on various quality features were evaluated for an online role-playing game and cloud gaming versions of an adventure, and a shooter game (Fig. 25.2).

Schmidt investigated the effects of varying input delay and packet loss on a transmission channel on various quality features and overall Quality of Experience of an online role-playing game [37]. Unlike solely network-based studies such as [6] or [3], Schmidt modified the "device" influencing factor and deliberately reduced its input performance by using a software module on the test participant's computer, which delayed the processing of keystrokes by specifiable time. For the test runs, 18 test participants were invited, who had never before played the game used in the experiment (Guild Wars 2). To elicit user factor data, each test participant had to fill out a screening questionnaire at the beginning of an experiment, asking them about personal data such as age, gender, and profession, but also questions about their experience (number of playing hours per week), game genre preferences, and their affinity towards video gaming in general. After an initial training phase, the testers had to play different scenarios of about 15 min length each. Subsequently, the Game Experience Questionnaire (GEQ) was used after each test scenario to measure the quality features challenge, competence, flow, immersion, tension, negative affect, and positive affect. Besides the GEQ the persons had to rate the perceived input delay (interactive behavior), jerkiness while performing actions (in this case also interactive behavior), and overall QoE (player experience).

The results of the study show an impact of the tested independent variables and influence factors input delay and packet loss not only on player experience, but on all tested dependent variables (challenge, competence, etc.). An input delay

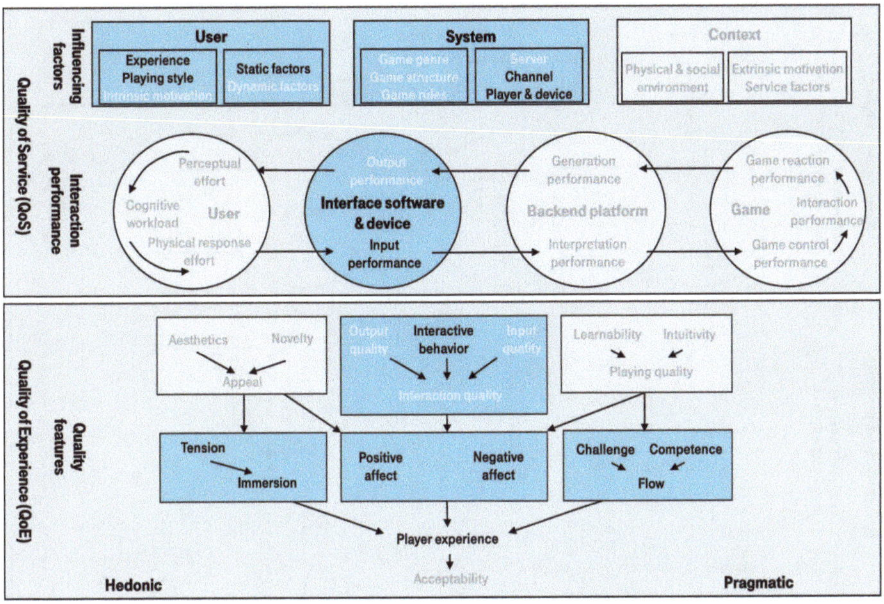

Fig. 25.2 Taxonomy with aspects highlighted, which were subject in Schmidt's study

of 300 ms was noticed and recognized by all test participants (i.e. it had an effect on the Interactive behavior quality feature) and influenced the overall quality rating significantly. This connection was stronger for male than for female players, indicating the relevance of user factors as influence factors. As for the other quality features input delay had an influence, but it was not significant. Packet-loss inducing jerkiness in the game was also noticed by the test participants and affected user-perceived QoE significantly, whereas its effect on the GEQ dimensions remained insignificant.

As stated above, another application of the taxonomy is the evaluation of gaming services (Fig. 25.3). Pommer conducted a study [32], in which he investigated the link between the influencing factors network channel bandwidth and delay, and user-perceived quality (player experience) for two different games, which were running on a cloud gaming platform [32]. The first game was an adventure game built around solving puzzles, whereas the second represented the first person shooter genre and also featured a different game structure and different game rules. Involuntarily, the interface (client) software turned into another independent variable, as a software update changed the behavior of the system. To elicit data from the user factors, an extended screening questionnaire had to be filled at the beginning of the experiment. It contained questions to determine previous experience with gaming, the test participants playing style, common static factors (name, age, gender, etc.), and also items to estimate the person's intrinsic motivation to use commercial cloud gaming services. After each test run, a custom questionnaire had to be filled, asking for various

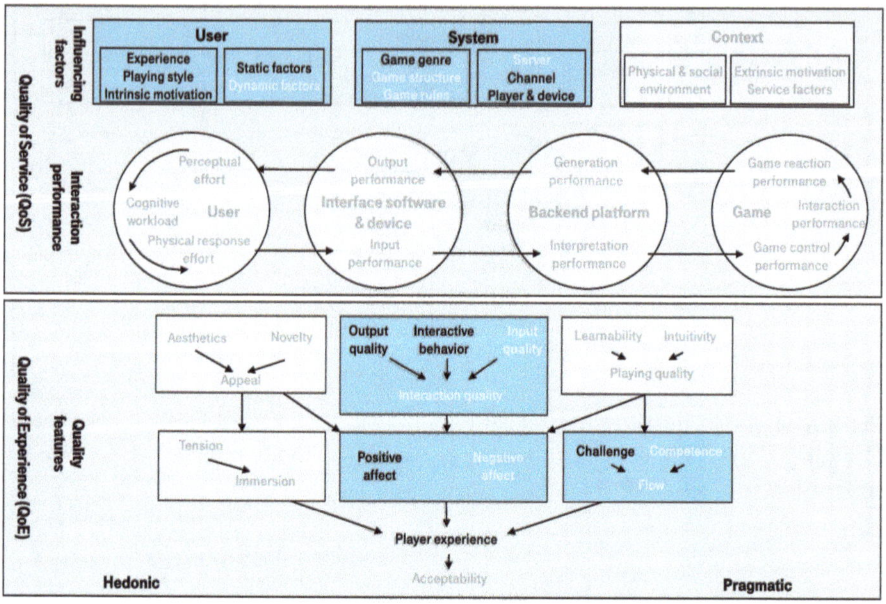

Fig. 25.3 Taxonomy with aspects highlighted, which were subject in the Pommer's study

aspects of interactive behavior and output quality, challenge and positive affect quality features. The collected data showed that bandwidth and delay degradations have an impact on user-perceived quality. An increase in transmission delay led to significantly lower player experience for the adventure game. In both games experienced gamers noticed the delay more than casual gamers. The effect of variations of the bandwidth yielded mixed results on the cloud gaming system. As Pommer was able to show, the specific implementation of the algorithms, which adjust the system's data traffic to the performance of the transmission channel was unable to adapt properly to the imposed limitations, causing an oscillating channel use in some test cases with noticeable image distortions.

25.4 Quantitative Model

Only limited work has been done to model the link between influence factors and quality features. One example, where such a model has been developed is the work by Wang et al. The Mobile Gaming User Experience (MGUE) model [40] is a parametric model, that uses system parameters to compute a MOS rating (referred to as the GMOS, Game Mean Opinion Score) for mobile cloud gaming scenarios. The model's input parameters can be grouped in 4 categories: Source Video Factors, Cloud Server Factors, Wireless Network Factors, and Client Factors. These groups can be mapped to this taxonomy's server, channel, and player and device influence factors. The structure of the computation is derived from the E-model [19] and proposes a number of impairment factors which degrade the experience, assuming that the undegraded experience is perfect regardless of the game that is being played. To accommodate the varying susceptibility of different game genres towards degradations such as latency or packet loss, they introduce a set of tuning factors. However, constants for these are only provided for three specific games.

25.5 Discussion

In the taxonomy, we have tried to classify influence factors, interaction performance aspects, and quality features on three layers. This choice was motivated by our earlier taxonomy on quality aspects of multimodal human-machine interaction, but has here been refined and adapted to reflect the particularities of gaming. For each of the factors and aspects, we have tried to specify how they can be quantified, in order to make Gaming QoE a measurable object of investigation. Whereas such metrics are available for some factors and aspects (e.g. screening questionnaire for gamer classes), there are no metrics for all of the concepts yet. An interesting approach might be physiological methods, to establish a direct relationship between aspects of the taxonomy and biological measures.

For some concepts, there might be a dispute as to where in the taxonomy they belong. Whereas we consider a significant step between different layers of the taxonomy (influence factors, interaction performance aspects, quality features), there might be arguable differences as to which of the quality aspects exercise an influence on which others.

Overall, we did not display all possible relationships between factors/aspects, and in particular we did not depict the relationships across layers. Such relationships across layers might be of particular interest to users of the taxonomy. Ultimately, our aim is to develop models, which are able to predict certain features of player experience from influencing factors of the system, user and context. One such example, the MGUE model, which predicts a MOS from parameters describing IP transmission channel and video coding effects on mobile cloud games, was presented in this chapter. We would like to further develop such models by understanding the related perceptual features, so that they might provide more robust and more diagnostic predictions. This approach also bears the chance to build a model that is valid for more than just a few player types, game genres and scenarios.

References

1. Androvich M (2008) GTA IV: Most expensive game ever developed? Gamesindustry International. http://www.gamesindustry.biz/articles/gta-iv-most-expensive-game-ever-developed
2. Bartle R (1996) Hearts, clubs, diamonds, spades—players who suit muds. http://www.mud.co.uk/richard/hcds.htm

3. Beigbeder T, Coughlan R, Lusher C, Plunkett J, Agu E, Claypool M (2004) The effects of loss and latency on user performance in unreal tournament 2003. In: Proceedings of 3rd ACM SIGCOMM workshop on network and system support for games (NetGames '04). ACM, New York, USA, pp 144–151
4. Brown E, Cairns P (2004) A grounded investigation of game immersion. In: Proceedings of CHI '04 extended abstracts on human factors in computing systems. pp 1297–1300
5. Chen J (2007) Flow in games. Commun ACM 50(4):31–34
6. Claypool M, Claypool K (2006) Latency and player actions in online games. Commun ACM 49(11):40–45
7. Csikszentmihalyi M (2010) Das flow-Erlebnis: Jenseits von Angst und Langeweile: im Tun aufgehen [The Flow Experience: beyond fear and boredom: opening in doing], 1985/1975, 10th edn. Klett-Cotta, Stuttgart
8. Dixon D (2011) Player types and gamification. In: Proceedings of CHI 2011, workshop gamification: using game design elements in non-game contexts. http://gamification-research.org/wp-content/uploads/2011/04/11-Dixon.pdf
9. Engl S (2010) Mobile gaming—Eine empirische Studie zum Spielverhalten und Nutzungserlebnis in mobilen Kontexten. Magister thesis, Universität Regensburg
10. Foraker Labs: Glossary—playability (2012) http://www.usabilityfirst.com/glossary/playability
11. Fritz J (2011) Mit Computerspielern ins Spiel kommen [Get into play with computer gamers]. Schriftenreihe Medienforsch. der LfM NRW 68, Vistas, Berlin
12. Fullerton T (2008) Game design workshop: a playcentric approach to creating innovative games. Morgan Kaufman
13. Bartle RA (2012) GamerDNA, Bartle test of gamer psychology. http://www.gamerdna.com/quizzes/bartle-test-of-gamer-psychology/
14. Hassenzahl M (2008) User experience (UX): towards an experiential perspective on product quality. In: Proceedings of 20th French-speaking conference on human-computer interaction
15. Hassenzahl H, Burmester M, Koller F (2003) AttrakDiff: Ein Fragebogen zur Messung wahrgenommener hedonischer und pragmatischer Qualität. In: Ziegler J, Szwillus G (eds) Mensch and computer 2003. Teubner, Stuttgart, pp 187–196
16. Hassenzahl M, Platz A, Burmester M, Lehner K (2000) Hedonic and ergonomic quality aspects determine a software's appeal. In: Proceedings of CHI 2000, Den Haag, pp 201–208
17. Hugentobler von Zürich U (2011) Messen von flow mit EEG in Computerspielen [Measuring flow with EEG in computer games]. PhD thesis, University of Zürich
18. Ijsselsteijn W, De Kort Y, Poels K, Bellotti F, Jurgelionis A (2007) Characterising and measuring user experiences in digital games. In: Proceedings of international conference on advances in computer entertainment technology
19. Bergstra JA, Middelburg CA (2003) The E-Model, a computational model for use in transmission planning. International Telecommunication Union, Geneva. http://citeseerx.ist.psu.edu/viewdoc/summary?doi=10.1.1.103.3547
20. Jennett CI (2010) Is game immersion just another form of selective attention? An empirical investigation of real world dissociation in computer game immersion. PhD thesis, University College London
21. Jennett C, Cox AL, Cairns P, Dhoparee S, Epps A, Tijs T, Walton A (2008) Measuring and defining the experience of immersion in games.Int J Hum-Comput Stud 66(9):641–661
22. Juul J (2005) Half-real: video games between real rules and fictional worlds. The MIT Press, Cambridge
23. Lazzaro N, Keeker K (2004) Whats my method? A game show on games. In: Proceedings of CHI 2004, Vienna
24. Mandryk RL, Inkpen K, Calvert TW (2006) Using psycho-physiological techniques to measure user experience with entertainment technologies. Behav Inf Technol 25(2):141–158
25. Möller S, Engelbrecht K-P, Kühnel C, Wechsung I, Weiss B (2009) A taxonomy of quality of service and quality of experience of multimodal human-machine interaction. In: Proceedings of 1st international workshop on quality of multimedia experience (QoMEX09), 29–31 July 2009, San Diego, CA

26. Möller S (2010) Quality engineering. Qualität kommunikationstechnischer Systeme. Springer, Heidelberg
27. Möller S, Schmidt S, Beyer J (2013) An overview of concepts and evaluation methods for computer gaming QoE. Accepted for: 5th international workshop on quality of multimedia experience 2013 (QoMEX13), Klagenfurt, 3–5 July
28. Murphy C (2010) Why games work and the science of learning. Alion Science and Technology, Virginia
29. NPD Group Inc (2008) NDP software category definitions. https://www5.npd.com/tech/pdf/swcategories.pdf
30. Pinelle D, Wong N, Stach T (2008) Heuristic evaluation for games: usability principles for video game design. In: Proceedings of the ACM conference on human factors in computing systems (CHI 2008). pp 1453–1462
31. Poels K, de Kort Y, Ijsselsteijn W (2007) It is always a lot of fun! Exploring dimensions of digital game experience using focus group methodology. In: Proceedings of future play 2007, Toronto, Canada, pp 83–89
32. Pommer D (2013) Quality of experience of gaming: cloud gaming—Einfluss von Verzögerung und Übertragungsrate [Quality of experience of gaming: cloud gaming—influence of delay and transmission rate]. Study Project thesis, TU Berlin
33. European network on quality of experience in multimedia systems and services (COST Action IC 1003). In: Le Callet P, Möller S, Perkis A (eds) Qualinet white paper on definitions of quality of experience, Lausanne, Version 1.1, 3 June 2012
34. Rheinberg F, Vollmeyer R., Engeser S (2003) Die Erfassung des Flow-Erlebens. In: Stiensmeier-Pelster J, Rheinberg F (eds) Diagnostik von motivation und Selbstkonzept (Tests und Trends N.F.) 2, Hogrefe, Göttingen, pp 261–279
35. Sánchez G, Gutiérrez FL, Cabrera MJ, Padilla Zea N (2008) Design of adaptive videogame interfaces: a practical case of use in special education. In: Proceedings of 7th international conference on computer-aided design of user interfaces (CADUI 2008), pp 57–63
36. Schaffer N (2009) Verifying an integrated model of usability in games. PhD thesis, Rensselaer Polytechnic Institute, Troy, NY
37. Schmidt S (2013) Messung der Quality of Experience von Computerspielen [Measurement of the quality of experience of computer games]. Study thesis, TU Berlin
38. Stelmaszweska H, Fields B, Blandford A (2004) Conceptualising user hedonic experience. In: Reed DJ, Baxter G, Blythe M (eds) Proceedings of ECCE-12, living and working with technology, EACE, York, pp 83–89
39. Vilnai-Yavetz I, Rafaeli A, Schneider Yaacov C (2005) Instrumentality, aesthetics, and symbolism of office design. Environ Behav 37(4):533–551
40. Wang S, Dey S (2012) Cloud mobile gaming: modeling and measuring user experience in mobile wireless networks. ACM SIGMOBILE Mob Comput Commun Rev 16(1):10–21
41. Wickens CD (1992) Engineering psychology and human performance. HarperCollins, New York
42. Wolf MJP (2001) Genre and the video game. In: Wolf MJP (ed) The medium of the video game. University of Texas Press, Austin, pp 113–134
43. Yee N (2005) Motivations of play in MMORPGS—results from a factor analytic approach. http://www.nickyee.com/daedalus/motivations.pdf
44. Zimbardo PG (1995) Psychologie. Springer, Berlin

Chapter 26
Recognition Tasks

Lucjan Janowski, Mikołaj Leszczuk, Mohamed-Chaker Larabi and Anna Ukhanova

Abstract This chapter proposes a definition of Quality of Experience (QoE) in the case of task based applications. The definition is followed by the describing of the current work in the field of the QoE methodology in the specific case of a security system. Different metrics predicting QoE proposed in the literature are discussed.

26.1 Introduction

Users of video to perform tasks (e.g. CCTV/IP surveillance public safety, telemedical services, firefighters, production lines' monitoring) require sufficient image/video quality to recognize the information needed for their application. Therefore, the fundamental measure of video quality in these applications is the success rate of these recognition tasks, which is referred to as visual intelligibility or acuity. This is because in the recognition tasks subjective user satisfaction depends only, or almost only, on the possibility of achieving a given functionality (event detection, object detection/recognition/identification). Additionally, the quality of video used

L. Janowski (✉) · M. Leszczuk
AGH University of Science and Technology, Kraków, Poland
e-mail: janowski@kt.agh.edu.pl

M. Leszczuk
e-mail: leszczuk@kt.agh.edu.pl

M.-C. Larabi
XLIM, Université de Poitiers, Poitiers, France
e-mail: chaker.larabi@univ-poitiers.fr

A. Ukhanova
Technical University of Denmark, Kongens Lyngby, Denmark
e-mail: annuk@fotonik.dtu.dk

S. Möller and A. Raake (eds.), *Quality of Experience*,
T-Labs Series in Telecommunication Services, DOI: 10.1007/978-3-319-02681-7_26,

by a human observer for recognition tasks is considerably different from objective video quality used in image/computer processing (Computer Vision).

One of the notable use cases where video quality is crucial are security applications. Video services with blurred images may have far more severe consequences for video security practitioners than just quality degradation. Therefore, the Quality of Experience (QoE) concept has to be tailored for security tasks (by security, one means here all applications linked with video IP surveillance, CCTV, aerial surveillance, preventing terrorists/criminal acts, etc.).

One of the major causes of reduction of visual intelligibility is loss of data, through various forms of compression. Additionally, the characteristics of the scene being captured have a direct effect on visual intelligibility and on the performance of a compression operation—specifically, the size of the target of interest, the lighting conditions, and the temporal complexity of the scene. Moreover, acquisition conditions are very difficult to control since there are no or just few recommendations regarding the setup for such applications. A few countries have focused on this problem in order to provide common rules. However, it is often observed that crime scenes are captured by means belonging to small shops, individuals, etc., where the quality/performance of the used devices is poor. Investigators have to deal and work on such data to solve the targeted case.

The effects and interactions of compression and scene characteristics can be studied only by performing a series of application specific tests. An additional challenge is how to test existing or develop new objective measurements that will predict the results of the subjective tests of visual intelligibility. To develop accurate objective measurements (models) for security applications quality, subjective experiments must be performed. For this purpose, in case of video signal, the ITU-T[1] P.910 Recommendation "Subjective video quality assessment methods for multimedia applications" [1] addresses the methodology for performing subjective tests in a rigorous manner. However, one needs to take some precautions while using this recommendation because the initial targeted application is, as discussed previously, far from security application targets.

Numerous security applications require a special framework appropriate to the function—i.e. its use for security tasks. Once the framework is in place, methods should be developed to measure the usefulness of the reduced signal quality. The precisely computed usefulness can be used to optimize not only the signal quality but the whole security system. It is especially important in surveillance systems, which often aggregate a large number of cameras the streams of which have to be saved for possible future investigation. For example in Chicago at least 10,000 surveillance cameras are connected to a common storing system [2].

[1] International Telecommunication Union—Telecommunication Standardization Sector.

26.2 Definitions of Quality of Experience in Task-Based Applications

In order to define QoE in task-based applications we start from defining task-based applications by a system using a signal to perform a specific task. A task application can be a medical image annotation, an object/situation recognition, a product control to name a few. The task it self can be as simple as detecting a motion in the scene or as complex as persons/behaviors/interactions tracking. An interesting example is a security system. Such systems are created in order to increase security. Such a definition does not describe a specific task which is specified on demand of future needs. Nevertheless, a security system is always used to perform a task. A task is the identification or interpretation of the signal.

The QoE in its task-based definition should reflect the task centric QoE, fitting at the same time the general definition described in Chap. 2. The QoE definition starts from stating what is an event, experience and quality. We believe that the concept of quality, as defined in [3] and Chap. 2, is valid for security applications as well. It involves comparison and judgment processes. In case of task-based applications, comparison with other cases helps to judge what is the probability that the task was run correctly.

QoE in case of a task involves: the probability of being correct, including the amount of time required for the completion of the task (here called "modified quality") and the interpretation of the event represented by the signal (here called "modified experience"). Therefore, the QoE defined for the task-based applications changes the "degree of delight or annoyance" used in the general definition described in Chap. 2 to usability and assurance of correct task performance. The following definition based on Chap. 2 and [3] is proposed here:

Quality of Experience in Task-based Applications: Is the degree of satisfaction generated by the performed task. Satisfaction depends on a combination of the degree of assurance about the correctness of the performed task, the expertise of the user, and the processing time.

Task-based applications often involve larger systems, in which case QoE should consider the whole system and the degree of usability of a particular application in solving the task.

The proposed definition includes many different aspects such as: inexperienced system operator, incorrect task for specific signal, usability of a particular system under concern. All those and many other specific factors influence the obtained QoE in task-based applications.

Besides the human operated systems, automatic detection algorithms are more and more often used. In those cases, quality is also a crucial parameter of the system. Of course, in such a case QoE does not have much sense since experience cannot be achieved by a machine. Here, instead of quality the interesting concept is performance. The authors propose to use, in case of automatic systems, the term Recognition Performance (RP). We propose the following definition:

Recognition Performance for an Automatic System: Defines the potential of the signal to be used for the successful achievement of the recognition task.

Recognition in case of automatic systems is often described by receiver operating characteristic (ROC) or other methods involving true positive and negative values. Nevertheless, such a description is valid for an algorithm, and in this case we are focusing on the input signal quality, not the recognition itself.

26.3 Psychophysical Experiments

To develop accurate objective measurements and models for QoE in the case of task-based applications, subjective experiments must be performed. Some standards exist in case of video quality assessment. The ITU has recommendations that address the methodology for performing subjective tests [1, 4]. These methods are targeted at the entertainment application of video and were developed to assess a person's perceptual opinion of quality. They are not entirely appropriate for task-based applications, in which video is used to detect/recognize/identify objects, people or events.

Assessment principles for the maximization of task-based video quality are a relatively new field. Problems of quality measurements for task-based video are partially addressed in a few preliminary standards and a recommendation [5, 6] that mainly introduces basic definitions, methods of testing and psychophysical experiments. ITU-T Rec. P.912 describes multiple choice, single answer, and timed task subjective test methods, as well as the distinction between real-time and viewer-controlled viewing, and the concept of scenario groups to be used for these types of tests. Scenario groups are groups of very similar scenes with only small, controlled differences between them, which enable testing recognition ability while eliminating or greatly reducing the potential effect of scene memorization. While these concepts have been introduced specifically for task-based video applications in ITU-T Rec. P.912, more research is necessary to validate the methods and refine the data analysis.

Section 7.3 of ITU-T Rec. P.912 ("Subjects") says that, "Subjects who are experts in the application field of the target recognition video should be used. The number of subjects should follow the recommendations of ITU-T P.910 [1]." There do also exist some potentially applicable, similar ideas incorporated from industry. For example, large television companies hire expert subjects to monitor their quality [7].

Unfortunately, expert subjects (police officers, doctors, etc.) are costly and difficult to hire compared to non-expert subjects (colleagues, friends, students, pensioners). Nevertheless, to the best of the authors' knowledge, in fact this issue of necessity of expert subjects has not been confirmed in any specific academic research. There is also no evidence that television companies have applied any serious effort to determine standards for psychophysical quality experiments. What is more, recent research results even show that clearly, in terms of the quality of psychophysical experiment (subjective test) results, it is more important to motivate the subjects than to acquire experts. They can either be paid or (for public safety scenarios) they can be police officers or practitioners of other public safety agencies [8].

The presented standards tell little about the data analysis which is different from traditional quality tests. In case of recognition task there is a danger that a subject will remember the correct answer from the previously seen sequence. Therefore, a single source sequence should be shown only once. Showing each source only once calls for a different experiment design and data analysis.

One of many problems which have to be addressed is how to limit irrelevant subjects. The most popular way to validate subjects is the correlation. It is simple and intuitively correct. We compute correlation between individual scores and the scores obtained by all other subjects. It is used for example by Video Quality Expert Group (VQEG) in [9]. Another technique described in BT.500 is based on an outlier detection. If a significant number of answers for particular subject are counted as outliers the subject is removed. In case of task based subjective experiment different sequences are scored by different users and computing correlation or outlier ratio becomes more difficult. An alternative proposed in [10], where a detailed analysis of the problem is presented, is based on measuring how often a subject performed a task worst than other subjects. Tasks are compared even if the sequences are not perfectly the same since a task difficulty is taken into consideration.

The other problem the authors identified is "experiment hacking." In case of object recognition we could clearly see that subjects recognize an object by eliminating possible answers rather than really seeing it. It is especially dangerous since a specification suggesting that a particular system makes it possible to detect a gun can not be based on assumption—"it cannot be a phone so I will guess gun."

Last but not the least there are no standards on how an automatic system should be tested. In this case it is not intended to define a psychophysical experiment. In opposition, it would be important to define a standard describing the correct testing methodology for automatic recognition systems. This will be as important as the definition of [1] for human subjects. Such a standard should take into account specific features of automatic system testing, like a much larger number of different source and distortion conditions which can be analyzed. Also a clear description of the type of sources and delivery systems which were used by the evaluation process is the key information for the correct system implementation.

When dealing with security applications, observers are often or always experts from police department, military services and so on. It is then better to speak about Quality of Expertise rather than QoE. This can be justified by the fact that an expert will use his expertise in order to solve a given case. They are hence required to perform standard tasks of detection, recognition and identification (DRI). Of course, the ability to perform these tasks is closely linked to the quality of the recorded image/video. Depending on the addressed case, experts are able to provide DRI scores combining both quality of the target and its understandability.

In the framework of a French funded project (*QuIAVU*) aiming at quality maximization of legal evidence images, Larabi et al. [11] proposed, together with experts from French police departments, an evaluation procedure dedicated to video-surveillance. Four cases have been identified: vehicles, license plates, persons, and faces. For each case, a discrete scoring table has been defined going from 0 (no detection), 1 (limited detection), ..., 5 (average recognition), ...to 9 (full identification) where criteria have been given for each step. For instance, a score of 4 for vehicles corresponds to: *Possible recognition—Defects on the vehicle* (e.g. *lighting, missing items), distinguishable features* (e.g. *logos, stickers,...*) while the same score for license plates corresponds to : *Possible recognition—partial readability of a group of characters without formal recognition.*

A recent complete work dealing with QoE has been presented by Tsifouti et al. [12] about defining acceptable bitrates for human face identification in video-surveillance, in collaboration with the UK Home Office. The described investigation was composed of four steps: (1) Collection of representative video footage, (2) Characterization and grouping of video scenes based on four content attributes: Brightness, Distance, Busyness and Angle, (3) Identification of key scenes with regards to H.264/AVC compression and (4) Testing of five video-surveillance recording systems commonly used on London buses by experts grouped in three categories : Bus analysts, Metropolitan Police Service (MPS) police officers and MPS surveillance officers. Each category is composed of experimented agents using video-surveillance images in their daily work. In this work, experts were asked to respond with a *yes* or *no* to the question: *Is the compressed version(s) as useful as the reference in terms of facial information*? Exploitation of results allowed to draw several conclusions about QoE for this specific condition. For instance, it was recommended to use a 60 % of observers yes responses on London buses, which is higher then the absolute threshold of 50 %. Also, during daytime, when there is variable illumination, it was recommended to set the bitrate of approximately 1,500 kbps (derived from the worst-case performance) and during nighttime, when the bus illumination is on, to reset the bitrate to around 700 kbps (constant bus illumination).

From both studies, one can notice that it is somehow difficult or even impossible to define a psychophysical evaluation standard for task-based recognition purposes, taking into consideration all aspects of the field. Obviously, it is very different from multimedia applications where even with technical variations, cultural differences between countries, and so on, the notion of quality remains quite similar leading to a kind of easiness in the definition of standards. In the case of task-based recognition, the procedure is highly dependent on the targeted application/problem and it is rather unthinkable to be exhaustive while defining common rules. For video-surveillance, the standardization of facilities will undoubtedly help in the definition of common best practices and standards for psychophysical evaluation.

26.4 Modeling Approaches for Objective Metrics

Some subjective recognition metrics, described below and applicable for security applications, have been proposed over the last years. They usually combine aspects of Quality of Recognition (QoR) and QoE. These metrics, for most of them, have not been focused on security practitioners as subjects, but rather on naïve participants. The metrics are not context specific, and they do not apply video surveillance-oriented standardized discrimination levels.

One of the metrics being definitively worth mentioning is Ghinea's Quality of Perception (QoP) [13, 14]. The QoP metric does not entirely fit video surveillance needs. It targets mainly video deterioration caused by frame rate measured in frame per second (fps), whereas fps not necessarily affects the quality of IP surveillance/CCTV [15]. The metric has been established for rather low, legacy resolutions, and tested on rather small groups of subjects (10 instead of the standardized 24 valid, correlating subjects). Furthermore, a video recognition quality metric for a clear objective of video surveillance requires tests in a fully controlled environment [4], with standardized discrimination levels (avoiding ambiguous questions) and with minimized impact of subliminal cues [6].

Another metric being worth mentioning is QoP's offshoot, Strohmeier's Open Profiling of Quality (OPQ) [16]. This metric puts more stress on video quality than on recognition/discrimination levels. Its application context, being focused on 3D, is also different than video surveillance which requires rather 2D. Like the previous metric, this one also does not apply standardized discrimination levels, allowing subjects to use their own vocabulary. The approach is qualitative rather than quantitative, whereas the latter is preferred by public safety practitioners for e.g. public procurement. The OPQ model is somehow content/subject-oriented, while for video surveillance a more generalized metric framework is needed.

OPQ partly utilizes free sorting, as used in [17] but also applied in the method called Interpretation Based Quality (IBQ) [18, 19], adapted from [20, 21]. Unfortunately, these approaches allow mapping relative, rather than absolute, quality.

In [22], Leszczuk et al. attempted to develop quality thresholds in license plate recognition tasks, based on video, streamed in constrained networking conditions. The measures that have been developed for this kind of task-based video provide specifications and recommendations that will assist users of task-based video to determine the technology that will successfully allow them to perform the required function.

Since the number of surveillance cameras is still growing, it is extremely likely that automatic systems will be used to carry out the tasks. Research presented by Janowski et al. in [23] includes the analysis of automatic recognition algorithms.

In [24], Dumke explores using visual acuity as a video quality metric for public safety applications. An experiment has been conducted to track the relationship between visual acuity and the ability to perform a forced-choice object recognition task with digital video of varying quality. Visual acuity is measured according to the smallest letters reliably recognized on a reduced LogMAR chart (commonly used to measure an individual's visual acuity).

The work [25] by Leszczuk introduces a typical usage of task-based video: surveillance video for accurate license plate recognition. The author presents the field of task-based video quality assessment, from subjective psychophysical experiments to objective quality models. He defines a subjective assessment procedure including several source video sequences encoded at various bitrates. The task assigned to observers was to recognize car license plate numbers and assess them using a variant of the famous Absolute Category Rating (ACR) test procedure. A threshold detection parameter has been defined by tolerating no more than one error on the character set of the license plate. Finally the logarithmic model learned from the experiments allows to predict the detection probability function of the bitrate. The continuation of this research is given in [26] by the same author, presenting a quality optimization approach driven by recognition rates.

In [27], Maalouf et al. propose an offline monitoring procedure for legal evidence images. The main target is to provide for a given scene (crime, theft, etc.) the best quality match of the region of interest. The proposed monitoring tool allows the selection of an object/region of interest (vehicle, license plate, face or person). This object is tracked over the whole scene thanks to a robust tracking algorithm based on foveal wavelet and mean shift. In parallel, the quality of this object of interest is assessed using a no-reference metric based on the sharpness feature. This choice comes from the expertise learned from investigators when assessing QoE. Finally, an intra super-resolution service allows increasing the resolution of the targeted object on the best quality matches. From a legal point of view the application of this super-resolution is admitted since it does not bring any side information. This work combines quality and expertise in the same framework allowing to guarantee the QoE for security experts.

26.5 Standardization

There exist only a very limited set of standards for psychophysical quality experiments in task-based video applications. The nature of these standards depends on the task being performed.

The Video Quality in Public Safety (VQiPS) Working Group, established in 2009 and supported by the U.S. Department of Homeland Security Office for Interoperability and Compatibility, has been developing a user guide for public safety video applications. The aim of the guide is to provide the potential consumers of public safety video equipment with specifications that best fit their particular application. The process of developing the guide will have a further beneficial effect of identifying areas in which adequate research has not yet been conducted, so that such gaps may be filled. A challenge for this particular work is ensuring that it is understandable to public safety practitioners, who may have little knowledge of video technology [28].

Internationally, the number of people and organizations interested in this area continues to grow, and a task-based video project under the Video Quality Experts Group (VQEG) [29] was created. The new project, The Quality Assessment for Recognition Tasks (QART), addresses precisely the problem of lack of quality standards for video monitoring [30]. The initiative is co-chaired by the Public Safety Communications Research (PSCR) program, U.S.A., and AGH University of Science and Technology in Krakow, Poland. Other members include research teams from Belgium, Denmark, France, Germany, and South Korea. The purpose of QART is exactly the same as the other VQEG projects—to advance the field of quality assessment for task-based video through collaboration in the development of test methods, performance specifications and standards for task-based video, as well as predictive models based on network and other relevant parameters. The QART project is performing a series of subjective tests to study the effects and interactions of compression and scene characteristics. An additional goal is to test existing or develop new objective measurements that will predict the results of the subjective tests of visual intelligibility.

Another important effort is being held within the International Standards Organization (ISO) under the technical committee called "*ISO/TC 223-Societal Security*" to make national security organizations work together [31]. ISO/TC 223 aims at developing standards with the goal to help to increase the security, i.e. protection of the society from and response to incidents, emergencies, and disasters caused by intentional and unintentional human acts, natural hazards, and technical failures.

26.6 Future Work

Subjective evaluation of security applications in controlled laboratory conditions may be difficult to achieve, even if the number of sequences to be evaluated may be reduced significantly if objective measurements show a high correlation. A possible solution to this problem may be found in the recent advances on crowd-sourcing [32]. Crowdsourcing is currently considered as a rapid way of obtaining estimations of video quality. While their judgment performance is not as high as those obtained in standardized lab conditions, they may prove particularly useful in the scenario when objective metrics and subjective data are available. Such research is useful for security applications.

Other plans/next steps for standardizing test methods and experimental designs include verification of issues like: subliminal cues, Computer-Generated Imagery (CGI) source video sequences, possible alternative test designs to avoid repetitions/memorization as well as automated eye charts. The agreed tasks include verifying requirements, refining methods/designs and, finally, making subjective experiments both more accurate and feasible. Work on comparing task-based versus task-free experiments (e.g. results from eye-tracking [33] experiments in free exploration versus task-based tests) is planned as well.

As usual quality metrics are needed especially for the end users which would like to know what is the limit of their security system and how expensive it would be to make it better. On the other hand, it is rather difficult to provide metrics without a solid and well proved methodology of subjective tests not forgetting about automatic systems' evaluation. The work on metrics and evaluation methodology should advance in parallel.

Finally, security applications are, and still will be, user-centric and experts, whether they are civilian or investigators, should solve cases or perform DRI tasks. Learning from their expertise will certainly help calibrating monitoring tools in order to guarantee QoE continuously.

Acknowledgments The research leading to these results has received funding from the European Communities Seventh Framework Programme (FP7/2007-2013) under grant agreement no. 218086 (INDECT).

References

1. ITU-T Recommendation P.910 (1999) Subjective video quality assessment methods for multimedia applications. International Telecommunication Union, Geneva
2. ACLU (2011) Chicago's video surveillance cameras. Technical report, ACLU of Illinois
3. Möller S, Le Callet P, Perkis A (eds) (2012) Qualinet white paper on definitions of quality of experience: output version of the Dagstuhl seminar 12181, 1.1 edn. In: European network on quality of experience in multimedia systems and services (COST Action IC 1003), Lausanne
4. ITU-T Recommendation BT.500-13 (2012) Methodology for the subjective assessment of the quality of television pictures. International Telecommunication Union, Geneva
5. Ford CG, McFarland MA, Stange IW (2009) Subjective video quality assessment methods for recognition tasks. In: Proceedings of the SPIE, vol 7240, pp 72, 400Z–72, 400Z–11
6. ITU-T Recommendation P.912 (2008) Subjective video quality assessment methods for recognition tasks. International Telecommunication Union, Geneva
7. Spangler T (2009) Golden eyes. Multichannel News
8. Leszczuk M, Koń A, Dumke J, Janowski L (2012) Redefining ITU-T P.912 recommendation requirements for subjects of quality assessments in recognition tasks. In: Dziech A, Czyewski A (eds) Multimedia communications, services and security, communications in computer and information science, vol 287. Springer, Heidelberg, pp 188–199
9. VQEG (2010) Report on the validation of video quality models for high definition video content, version 2.0 edn. http://www.vqeg.org
10. Janowski L (2012) Task-based subject validation: reliability metrics. In: Fourth international workshop on quality of multimedia experience (QoMEX), pp 182–187
11. Larabi MC, Nicholson D (2011) Monitoring image quality for security applications. In: IS&T/SPIE electronic imaging: image quality system performance. Burlingame, CA
12. Tsifouti A, Triantaphillidou S, Bilissi E, Larabi MC (2013) Acceptable nitrates for human face identification from CCTV imagery. In: IS&T/SPIE electronic imaging: image quality system performance. Burlingame, CA
13. Ghinea G, Chen SY (2008) Measuring quality of perception in distributed multimedia: verbalizers vs. imagers. Comput Hum Behav 24(4):1317–1329
14. Ghinea G, Thomas JP (1998) QoS impact on user perception and understanding of multimedia video clips. In: Proceedings of the sixth ACM international conference on multimedia, MULTIMEDIA 98. ACM, New York, USA, pp 49–54. doi:10.1145/290747.290754. http://doi.acm.org

15. Janowski L, Romaniak P (2010) QoE as a function of frame rate and resolution changes. In: Zeadally S, Cerqueira E, Curado M, Leszczuk M (eds) Future multimedia networking. Lecture notes in computer science, vol 6157. Springer, Heidelberg, pp 34–45
16. Strohmeier D, Jumisko-Pyykko S, Kunze K (2010) Open profiling of quality: a mixed method approach to understanding multimodal quality perception. Adv Multiedia 3:1–3:17. doi:10.1155/2010/658980. http://dx.doi.org
17. Duplaga M, Leszczuk M, Papir Z, Przelaskowski A (2008) Evaluation of quality retaining diagnostic credibility for surgery video recordings. In: Proceedings of the 10th international conference on visual information systems: web-based visual information search and management, VISUAL'08, Springer, Heidelberg, pp 227–230
18. Nyman G, Radun J, Leisti T, Oja J, Ojanen H, Olives JL, Vuori T, Hakkinen J (2006) What do users really perceive—probing the subjective image quality experience. In: Proceedings of the SPIE international symposium on electronic imaging 2006: imaging quality and system performance III, vol 6059, pp 1–7
19. Radun J, Leisti T, Hakkinen J, Ojanen H, Olives Radun J, Leisti T, Hakkinen J, Ojanen H, Olives JL, Vuori T, Nyman G (2008) Content and quality: interpretation-based estimation of image quality. ACM Trans Appl Percept 4(2):1–2:15. doi:10.1145/1278760.1278762. http://doi.acm.org
20. Faye P, Bremaud D, Daubin MD, Courcoux P, Giboreau A, Nicod H (2004) Perceptive free sorting and verbalisation tasks with naive subjects: an alternative to descriptive mappings. Food Qual Prefer 15(7–8):781–791. doi:10.1016/j.foodqual.2004.04.009. http://www.sciencedirect.com/science/article/pii/S0950329304000540. (Fifth Rose Marie Pangborn Sensory Science Symposium)
21. Picard D, Dacremont C, Valentin D, Giboreau A (2003) Perceptual dimensions of tactile textures. Acta Psychol 114(2):165–184. doi:10.1016/j.actpsy.2003.08.001. http://www.sciencedirect.com/science/article/pii/S0001691803000751
22. Leszczuk M, Janowski L, Romaniak P, Glowacz A, Mirek R (2011) Quality assessment for a licence plate recognition task based on a video streamed in limited networking conditions. In: Dziech A, Czyewski A (eds) Multimedia communications, services and security, communications in computer and information science, vol 149. Springer, Heidelberg, pp 10–18
23. Janowski L, Kozlowski P, Baran R, Romaniak P, Glowacz A, Rusc T (2012) Quality assessment for a visual and automatic license plate recognition. Multimedia Tools Appl 1–18. doi:10.1007/s11042-012-1199-5
24. Dumke J (2013) Visual acuity and task-based video quality in public safety applications. In: Proceedings of the SPIE 8653, image quality and system performance X, 865306 pp 865, 306–865, 306–307. doi:10.1117/12.2004882. http://dx.doi.org/10.1117/12.2004882
25. Leszczuk M (2011) Assessing task-based video quality—a journey from subjective psychophysical experiments to objective quality models. In: Dziech A, Czyewski A (eds) Multimedia communications, services and security, communications in computer and information science, vol 149. Springer, Heidelberg, pp 91–99
26. Leszczuk M (2012) Optimising task-based video quality. Multimedia Tools Appl 1–18. doi:10.1007/s11042-012-1161-6
27. Maalouf A, Larabi MC, Nicholson D (2012) Offline quality monitoring for legal evidence images in video-surveillance applications. Multimedia Tools Appl 1–30. doi:10.1007/s11042-17012-17126817-9
28. Leszczuk MI, Stange I, Ford C (2011) Determining image quality requirements for recognition tasks in generalized public safety video applications: definitions, testing, standardization, and current trends. In: IEEE International symposium on the broadband multimedia systems and broadcasting (BMSB), pp 1–5. doi:10.1109/BMSB.2011.5954938
29. VQEG (2013) The video quality experts group. http://www.vqeg.org
30. Leszczuk M, Dumke J (2012) The quality assessment for recognition tasks (QART), VQEG. http://www.its.bldrdoc.gov/vqeg/project-pages/qart/qart.aspx
31. ISO-22311:2012 (2012) Societal security videosurveillance format for interoperability. Technical report, ISO 2012

32. Keimel C, Habigt J, Horch C, Diepold K (2012) Video quality evaluation in the cloud. In: Proceddings of 19th international packet video workshop (PV) 2012, pp 155–160
33. Kunka B, Kostek B (2009) Non-intrusive infrared-free eye tracking method. In: Signal processing algorithms, architectures, arrangements, and applications conference proceedings (SPA) 2009, pp 105–109

Chapter 27
Perception of Quality Changes in Wireless Networks

Blazej Lewcio and Sebastian Möller

Abstract Over the top services so far discussed in this book, such as speech and video telephony, gain the attention of mobile users, who nowadays do not expect the connectivity only, but also demand for Quality of Experience (QoE). In this sense, the concept of an "always best connected" mobile user faces new challenges. Nomadic use of services and heterogeneity of technology make it impossible to constantly provide the highest transmission quality. As a matter of fact, intelligent management of quality is required and to be successfully in doing so, QoE in heterogeneous networks must be explored and user perception of changing quality must be understood. This chapter addresses user perception of quality while using speech and video telephony in heterogeneous wireless networks. It is in particular focused on user perception of quality changes due to switching between networks, codecs, and encoding bit rates during an ongoing transmission. This knowledge is inevitable for a perception-based design of service and mobility management in modern networks.

27.1 QoE in Wireless Networks

Heterogeneous technology platforms enable to access speech and video telephony services seamlessly by nomadic users. Nonetheless, an ideal communication path does not exist and many fundamental trade-offs must be considered in provisioning the service quality [8]. Network coverage versus throughput and reliability of a connection is one aspect of the decision [21, 28]. Efficiency versus robustness of signalcompression is another [6, 17, 40]. In this sense, to make the best of

B. Lewcio (✉) · S. Möller
Quality and Usability Lab, Telekom Innovation Laboratories, TU Berlin, Berlin, Germany
e-mail: blazej.lewcio@telekom.de

S. Möller
e-mail: sebastian.moeller@telekom.de

S. Möller and A. Raake (eds.), *Quality of Experience*,
T-Labs Series in Telecommunication Services, DOI: 10.1007/978-3-319-02681-7_27,

available resources when a telephony service is used by a nomadic user, an intelligent adaptation of the ongoing transmission is required.

The simultaneous availability of diverse wireless technologies in certain geographical areas creates overlaid wireless architectures [30]. In this network hierarchy, the lowest level is characterized by a small network coverage and high-speed connection, and the highest level represents the opposite, large geographical coverage, and low connection throughput. Having such a system, management of the connectivity according to user location and utilized service type is required [9]. The attachment point to the Internet must be frequently changed, which is referred to as a network handover, further differentiating this term in a homogeneous and heterogeneous handover, depending if the handover is performed between access points that belong to the same or to different wireless technologies [34]. If the ongoing connection of a mobile station can be maintained during and after a handover, seamless mobility is assured.

There are several protocols that enable end-devices to change the attachment point to the Internet. Mobile IP [22] or SIP [24] are only two of them that operate in different transmission layers. Both of the approaches introduce certain trade-offs, such as transmission overhead and reliability [13, 25]. Due to the latter feature—higher reliability of the networking-based solution—Mobile IP was applied in the research presented in this chapter. Mobile IPv4 defines mechanisms that enable a mobile terminal to change its point of attachment to the Internet whilst remaining reachable through a permanent address, so called home address, at the same time preserving all the active connections while travelling to a new network; the interested reader may refer to [22] for a detailed description of the protocol.

Notably, most of the service and mobility management efforts neglect user perception and focus on numeric measurement of system performance instead. Network researchers benchmark the quality of a mobile system using parameters, such as packet loss or throughput [29], as well as connection degradations due to a network handover itself [16, 20, 23, 33]. In this sense, network mobility is managed based on parametric triggers, such as the prediction of user mobility patterns according to Signal-to-Noise Ratio (SNR) changes [3], or to network layer metrics such as variation of inter packet delay [4]. In addition, more advanced efforts towards context-aware mobility management already exist, which combine the knowledge of a mobile station with information from the network to support the decision process [31].

Although there are many mobility-related challenges in the networking layer, also a broader context of mobility must be considered. This comprises possibilities of quality adaptation in the telephony application itself. Application-layer elements, such as speech and video codecs, enable to control the efficiency or robustness of signal compression [39], which applies for example modern means of scalable multi-rate coding and of error resiliency [2, 26, 37]. These mechanisms are used to optimize the coding process and to balance between compression efficiency and error robustness, either manually, or automatically, according to quantitative metrics such as the number of decoding errors that can be detected [5, 27, 42].

Fortunately, in the last decade an increasing number of researchers made a step towards service and mobility management that is based on the knowledge of Quality of Experience. For example, in [32] speech quality is estimated to decide when a handover between WiFi and HSPA networks should be scheduled. In [19], the effect of packet loss in heterogeneous networks is mitigated by a changeover between two narrowband speech codecs, PCMA and GSM, according to parametric quality prediction. In [1, 12, 41], the encoding bit rate of speech and video codecs is adjusted to maximize the Quality of Experience of mobile users.

However, even though there are many studies that aim at perception-based management of quality in wireless networks, they neglect the fact how the process of quality adaptation, itself, is perceived by the users. Moreover, even though previous studies of user perception of time-varying quality already exist [7, 35, 38], this knowledge is not tightly integrated in the management of speech and video telephony in heterogeneous wireless networks. In these networks, however, time-variation is an intrinsic characteristic, and neglecting QoE when the service is dynamically adapted can have tremendous consequences. For example, intended quality improvements, such as switching from narrowband to wideband speech transmission, can actively degrade the overall Quality of Experience, if it is performed too late in a call [18, 36]. Another prominent example is frequent adaptation of transmission that might not improve the overall Quality of Experience [7], and thus can be abandoned, at the same time reducing the effort of service management and improving the scalability of a telecommunication system.

Therefore, the goal of this chapter is to explore Quality of Experience of speech and video telephony in wireless networks. The study is focused on user perception of quality changes, which enables to derive perception-based mobility guidelines. However, please note that a technical implementation of a mobility management system as such is not in scope of this chapter. Therefore, in Sect. 27.2 a methodology to measure user perception of quality in wireless networks is discussed, which lays the foundation for Sects. 27.3 and 27.4 that address user perception of speech and video telephony quality in wireless networks, respectively. At the end of this chapter, in Sect. 27.5, the main findings are summarized and discussed.

27.2 Measurement of User Perception in Wireless Networks

In order to get analytic insights into user experience of telephony quality in wireless networks, user perception of entire phone calls must be considered. Standard listening-only tests, such as ITU-T Rec. P.800 [11], which make use of speech samples with a length of approx. 4–8 s are not suitable for this purpose (cf. Chap. 10). On the other hand, conversational tests—despite being comparable to normal telephone usage and thus being ecologically valid—place a content-related focus on the user's attention (cf. Chap. 11). In such a situation, users are generally less analytic in their judgments, and it might happen that subtle perceptual differences get blurred. As a compromise, so-called simulated conversation tests (or call quality tests), which are outlined in detail in Chap. 10, were used. This protocol specifies that five approx. 6–12 s long samples that correspond to utterances of one call participant are presented with pauses in between the stimuli to the participant of the test. In the pauses, the invited person has to fulfill a content related task, such as to orally answer a content-related question. At the end of such a call simulation, the overall Quality of Experience is judged. The results of the simulated conversation tests were related to the findings from short-sample listening-only tests according to ITU-T Rec. P.800 and P.911, for speech and video telephony respectively. This way, it became possible to establish relationships between the perceptual effects resulting from instantaneous link changes and the overall perception of time-varying transmission quality at the end of a call. As an example, it was investigated how a switch of a speech codec is perceived, and what the effect of this single change on the quality judged at the end of the call is.

In this chapter, results from simulated conversation tests as well as short-sample listening-only tests will be presented to analyze the mentioned effects. The quality of the individual samples that were used in each of the simulated conversations as well as of some additional material was judged in the short sample test separately. The test files were sequentially presented to the test participants who judged the quality of each sample on the 5-point MOS scale. The short samples were used to quantify the influence of heterogeneous technology on user perception, in particular addressing how the phenomenon of a technology switch during an ongoing transmission is perceived. This way the aforementioned link between the perception of single effects and the perception of the overall call quality could be created.

The test material was processed in a dedicated research testbed [15]. The testbed provides access to heterogeneous wireless networks, such as WiFi and HSDPA, and enables to roam between those networks during ongoing transmission. The network handover support is implemented through the Mobile IP protocol. When a handover is performed, the "make-before-break" policy is applied. This means that connectivity to both involved networks is assured before the network interface is switched. Moreover, the testbed is equipped with a telephony client that has been extended with several research features. The most notable feature is the support of switching between diverse speech and video codecs during an ongoing connection. For this purpose a dedicated switching technique that enables to reduce the side effects of a switch was developed. The solution is based on a Session Initiation Protocol (SIP) handshake that is used to trigger the codec changeover procedure in the application, where additional algorithms enable to gracefully replace the active codec so that data loss and play-out interruptions are reduced. The interested reader may want to refer to [14] for detailed information about the experimental setup and the conducted tests.

27.3 Time-Varying Quality of Wireless Speech Telephony

The first call quality test with 13 participants was designed to identify the main perceptual effects that may occur in heterogeneous wireless networks. Network handover between WiFi and HSDPA, changeover between narrowband ITU-T Rec. G.711 and

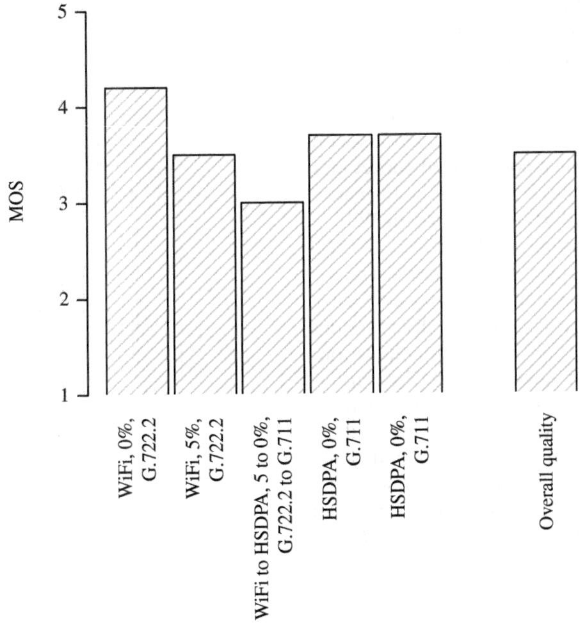

Fig. 27.1 Visualization example of a call quality profile when a user is leaving WiFi coverage area and a handover to HSDPA is performed. Quality judgements of consecutive samples of a call and of the overall call quality. Phantom values for demonstration only

wideband ITU-T G. 722.2 codec, and emulated packet loss with random distribution were the addressed mobility phenomena.

A visualization of an example of a call quality profile is depicted in Fig. 27.1. This profile is intended to represent a situation when a user initializes a wideband phone call in an unimpaired WiFi network, but due to his movement away from the access point, he leaves the network coverage area, which leads to link-layer impairments and packet loss. Consequently, a handover to a HSDPA network, in which a narrowband codec is enforced, is triggered to maintain the call. The example profile consists of 5 samples that are sequentially played out and that introduce different quality levels. The first sample introduces high speech quality provided by the wideband G.722.2 codec in a WiFi network. The second sample is degraded by 5 % of packet loss. The third sample is affected by a network and codec switch from the impaired WiFi and G.722.2 to an unimpaired HSDPA network and G.711. The unimpaired narrowband transmission is provided until the end of the call in the last two samples. In overall, the quality judged at the end of this call is below that of pure narrowband speech and no benefit of the initial wideband transmission is gained.

This way several quality profiles were constructed and the ratings of the overall quality collected [14]. The quality judgements that have been collected during the conducted tests (cf. Fig. 27.2) revealed that pure wideband transmission was rated best (#2) and substantially better than pure narrowband quality (#1). However, this relationship was changed when degradations due to packet loss emulation occurred. If packet loss was increasing within a call, the speech quality provided by the narrowband codec was rated considerably higher than that when the wideband codec was

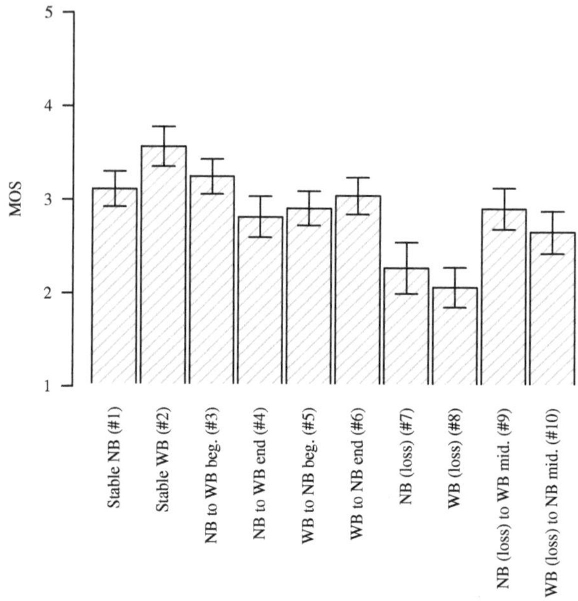

Fig. 27.2 User judgements of the overall call quality extracted from [14]. Narrowband (*NB*) transmission in HSDPA using ITU-T Rec. G.711 codec at 64 kbit/s with the recommended packet loss concealment, and wideband (*WB*) transmission in WiFi using the ITU-T Rec. G.722.2 codec at 23.05 kbit/s. Transmission eventually affected by increasing in discrete steps (10 % per sample) emulation of random packet loss up to 20 %, and by switching between WB and NB at the beginning (mid. of 2nd sample), in the middle (mid. of 3rd sample), or at the end (mid. of 4th sample) of a call. *MOS*, and 95 % CI

used (#7, 8), which was related to a higher packet loss robustness of the narrowband codec [10]. Moreover, in any case when the transmission was affected by packet loss, the quality was judged worst in the entire test. This observation reveals that packet loss was the most dominant factor of call quality degradation. As a result, if a connection is degraded, packet loss robustness may be a more important characteristic of a codec than the supported bandwidth of the speech signal.

If the increasing packet loss in a call was eliminated by performing a handover to a loss-free network in the middle of an emulated call, which was simultaneously accompanied by a switch of the speech signal bandwidth, the perceived quality was always substantially improved (#9, 10). This finding held when the network and codec were changed in both directions (either from wideband to narrowband or from narrowband to wideband). However, the improvement was lower if the narrowband codec was used at the end of a call. Therefore, if packet loss occurs in a call, early switching to a loss free network is advantageous in any case, even if it is required to reduce the bandwidth of the speech signal. However, the quality judged after the call was always lower than that of pure narrowband transmission, which is the first indication that, if a network and a codec switch can be foreseen, and at the same time a stable narrowband link is available, delivery of low, but stable narrowband quality improves the Quality of Experience. There is no benefit of provisioning impaired wideband speech at the beginning of a call.

When no packets were lost, raising the speech bandwidth from narrowband to wideband was only advantageous if a sufficient duration of a call remained in order to take profit of the improved quality (#3). Although the relationship between the negative impact of this kind of a switch itself and the time that is necessary to profit from the wideband quality requires a dedicated study, it can be derived that raising the speech bandwidth from narrowband to wideband is profitable if it happens at the beginning of a call only, as the duration of a real call is unpredictable. Moreover, if the switch to wideband was performed late in a call, the overall quality experience was degraded below that of pure narrowband (#4), which is related to degradation of short-term perception through a codec switch that is explored later in this section. Thus, switching from narrowband to wideband is advantageous only if it occurs early in a call and if the wideband part is long enough. As a result, direct raising of the speech bandwidth from narrowband to wideband is a sensible instrument for quality adaptation.

In turn, decreasing the speech bandwidth from wideband to narrowband always degraded the call quality below the rating of pure narrowband, and the quality judged at the end of the simulated call was the lower, the earlier in a call it was switched (#5, 6). Once again there was no significant profit taken of the wideband period at the beginning of a call. As a general consequence, once narrowband occurred in a phone call, the end quality judgement was close to that of pure narrowband transmission. Simultaneously, the direction and temporal position of a switch had only a marginal impact on the overall quality perception, which means that if packet loss does not affect the quality, switching between wideband and narrowband speech in most of the cases is not recommended. This is yet another argument that if a switch can be foreseen and a stable network that provides stable narrowband speech quality is available, this network should be used.

Further tests in [14] revealed that frequent network and codec changeovers within a loss-free call turned out to be rated worse than a single change between wideband and narrowband. This is yet another argument that frequent and rapid changes of speech quality are not appreciated by the users and the changeover between narrowband and wideband codecs turned out to be the second main effect that influenced the quality perception in this study.

The perception of switching itself was addressed in a dedicated short sample test with 24 participants. The evaluation of codec changeover conditions in this test showed that any switch between wideband and narrowband was always rated substantially lower than pure narrowband transmission (cf. Fig. 27.3a). The switching direction was not of much relevance for the short-term perception of quality, although the user experience was slightly higher if wideband was provided at the end of a sample. As a result, it was proven that a rapid change between wideband and narrowband speech has a negative impact on the short-term perception of quality, and that instant quality degradation is immediately perceived, but users require more time to recover after a switching event and to take profits of wideband speech. These observations also explain the judgements of call quality from the previous tests when a codec changeover was included.

If one codec was used for the entire transmission, the network handover itself turned out to have only a minor impact on the Quality of Experience (cf. Fig. 27.3b). This effect is related to the variation of the inter-packet delay (IPD) that may lead to packet discard in the jitter buffer of a telephony client due to two reasons. First, the

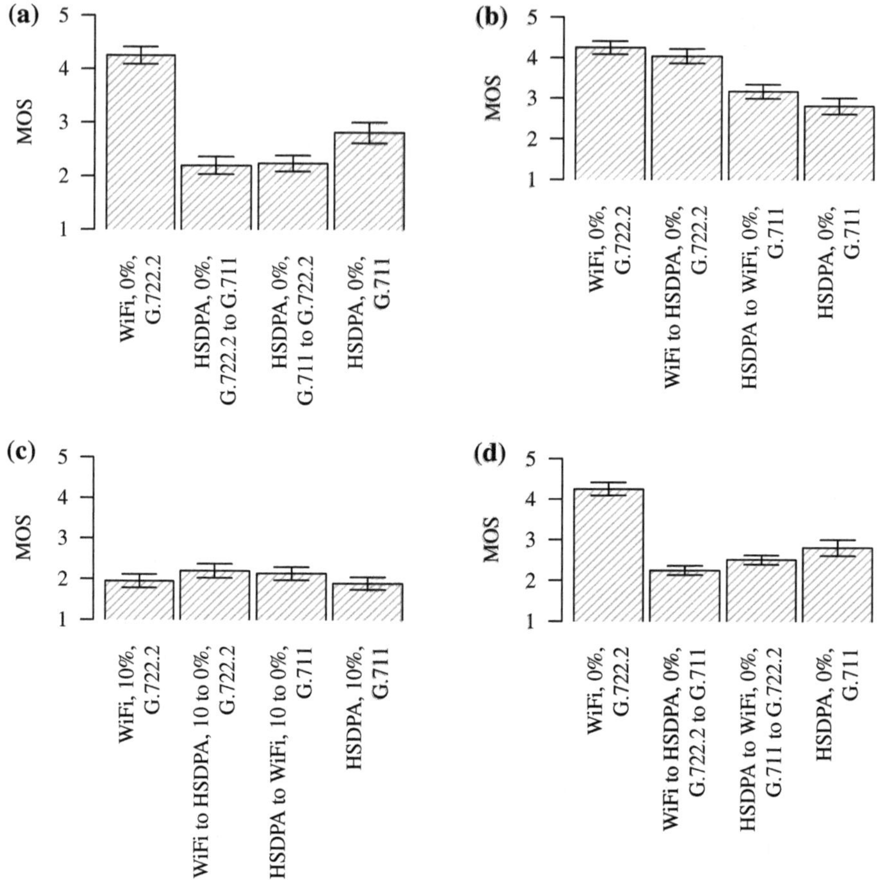

Fig. 27.3 User judgements of quality of short speech samples extracted from [14]. *MOS*, and 95 % CI. **a** Codec changeover. **b** Network handover. **c** Network handover, loss 10 %. **d** Network and codec switch

cellular HSDPA technology offers less stable IPD than a local WiFi hotspot. Second, a handover between heterogeneous networks results in an unavoidable change of network delay. This may lead to an IPD jump, which is particularly observed when the connection is switched from a less to a more delay-affected network, in this case represented by WiFi and HSDPA, respectively. Therefore, a handover from WiFi to HSDPA slightly degraded, and a handover in the opposite direction slightly improved the quality perception as compared to constant transmission in the initial network. Apparently, switching to a higher or lower performance network is instantly perceived along with the switching direction, even though the perceived effect is marginal. This, however, makes the network handover a useful tool to cope with packet loss when the same speech codec can be used in the new network. If the connection was initially affected by packet loss of 10 %, a handover to a loss-free network always improved the overall experience. Even though the end quality was rated similarly in both cases, the improvement was considerable if the wideband codec was used only, because substantially higher quality was experienced after the handover due to lower robustness of the wideband codec. To this end, this result also confirms that an abrupt improvement of speech quality while preserving the speech signal bandwidth is perceived according to the improvement direction, as compared to a codec changeover from narrowband to wideband that degraded the perceived quality. As a result, the power of quality adaptation by network handover can be exploited if a substantial quality gain is expected after the switch. This is the case when strong connection degradations are experienced or if the speech codec is not robust against packet loss.

If both, network and codec switch, were combined within a sample, the user judgements were similar to that of conditions that were affected by codec changeover only. Apparently, the phenomenon of switching between wideband and narrowband speech dominates the user experience. In any case, the end quality of the combined conditions was rated below that of pure narrowband which confirms that when no other degradations are experienced no benefits of switching the network along with the codec can be achieved. This effect was occluded by the high packet loss rate of 20 % (not depicted, please refer to [14]), where a handover to a loss-free network along with a codec changeover always improved the end quality perception. This observation confirms that if the link-quality exhibits a certain degree of packet loss, the negative effect of a codec changeover can be occluded and a quality gain can be achieved, but more research is required to detail the threshold values.

27.4 Time-Varying Quality of Wireless Video Telephony

Similarly to the study that was presented in the previous section, user perception of system dynamics that might occur during a video call in heterogeneous wireless networks was addressed. User perception of network handovers between WiFi and HSDPA, changeovers between MPEG-4 and H.264 coding, switching the video encoding bit rate between 256 and 1,536 kbit/s, and degradations due to packet loss up to 5 % were in focus of the study. Also the design of the user experiments was guided by a similar methodology to that already presented in the previous section. User perception of call quality was tested based on approx. 90 s long call simulations according to the recommendation from ETSI TR 102 506 v.1.2.1. The perception of switching itself was analysed in a short-sample test according to ITU-T Rec. P. 911. The test material that presented pre-recorded (approx. 9 s long) utterances of one call participant was displayed in the VGA (640 × 480 pixels) format in the middle of a 10.1 inch WSVGA (1,024 × 600 pixels) display of a Dell Inspiron 1,012 notebook.

The collected results of the video call quality test with 20 participants (cf. Fig. 27.4a) confirmed some known facts and expectations. Efficient video coding, in this case H.264 at 1,536 kbit/s, enabled to maximize the call quality experience under stable networking conditions. However, any degradation of the reference

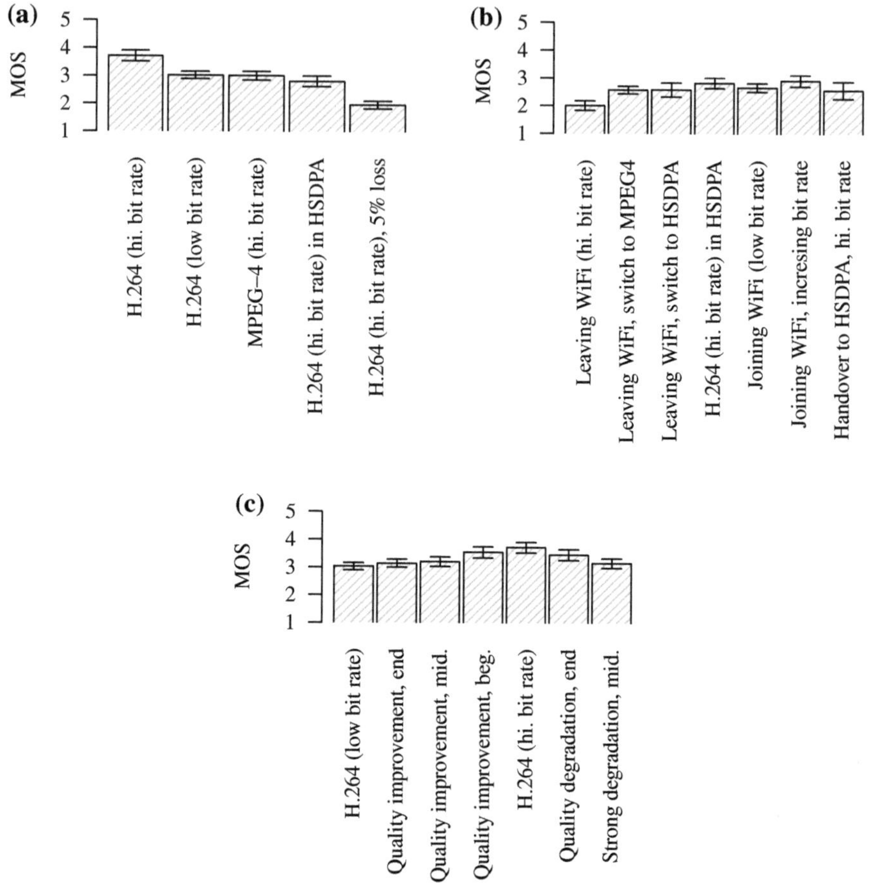

Fig. 27.4 User judgements of the overall quality of emulated video calls extracted from [14]. MOS_{av}, and 95 % CI. **a** Stable transmission. **b** Mobility. **c** Temporal position of quality switching

streaming configuration in any of the available dimensions, such as use of the HSDPA network, of the less-efficient MPEG-4 codec, or a reduction of the encoding bit rate to 256 kbit/s caused a substantial loss of overall call quality.

When the quality of a video call was degraded through emulation of leaving a WiFi area and increasing packet loss rate (0, 3, 3, 5, 5 % per call segment, respectively), the Quality of Experience was substantially degraded (cf. Fig. 27.4b). However, when the effect of an increasing packet loss was countered by a handover to a loss-free HSDPA network or by a changeover to the more robust (in the setup MPEG-4) codec in the middle of a call, the QoE was always substantially improved. However, in any case, the overall quality was judged lower than that of constant transmission in HSDPA. Therefore, when packet loss and the need of quality adaptation can be foreseen, it is advantageous to proactively schedule the entire transmission in a HSDPA network, as no perceptual profit of the initially unimpaired WiFi was taken, and the overhead of mobility management can be reduced. This finding was additionally confirmed by the quality judgement of the quality profile that included a network handover from WiFi to HSDPA in the middle of a call. The quality of this profile was rated below that of constant transmission in HSDPA as well, which also reveals that a handover in case of video telephony is a more critical operation from the perceptual point of view than in the case of pure speech telephony.

In turn, when joining a WiFi area was emulated through a decreasing packet loss rate during a video call (5, 5, 3, 0, 0 % per call segment, respectively) and the encoding bit rate was increased from 256 to 1,536 kbit/s in the middle of a call to boost the user experience, the Quality of Experience was improved, but the perceived gain was not substantial. This once again confirmed that a quality degradation in a call is remarkably stronger perceived than a quality improvement.

The timing of a switching event within a video call that is not affected by packet loss has also its implications on user perception (cf. Fig. 27.4c). When the encoding bit rate of the H.264 codec was increased from 256 to 1,536 kbit/s at the beginning, in the middle, or at the end of an emulated call, the overall QoE was always improved. The earlier in a video call the adaptation was performed, the stronger was the perceived effect, but the improvement was substantial only, if the bit rate was increased at the beginning of a call. In any other case, the quality judgements were similar to each other.

In turn, when the video encoding bit rate was decreased at the end of a video call, the observed effect was once again stronger than that of a quality improvement at the same position in a call. The observed effect was not substantial, although the quality change happened close to the moment of the overall quality rating. This observation implies that not the position when the quality is changed, but the duration of low quality transmission is the deciding aspect of quality experience in a video call. This finding was different from the case of switching between narrowband and wideband speech quality in the previous section, where degradation of quality at the end of a call degraded the user experience substantially.

The perception of switching itself was once again addressed in a dedicated short sample test with 20 participants. When the network was switched from WiFi to HSDPA in the middle of a sample, the video quality was always degraded. This is the cost of resource allocation procedures in the cellular HSDPA network, which may lead to direct packet loss or inter-packet delay variation and indirect data discard in the application. As a result, the overall user experience was always worse than that in both of the networks (cf. Fig. 27.5a). The quality degradation was perceived stronger when the video bit rate was high. Therefore, when stable transmission in either HSDPA or WiFi can be guaranteed during a video call, switching between these networks is not recommended. If the necessity of a handover can be predicted, proactive scheduling of the entire transmission in a lower quality network allows to maximize the user experience and the number of service management operations is simultaneously reduced. In any other case, robust video compression and low encoding bit rate help to mask visible artifacts that are perceived due to a handover.

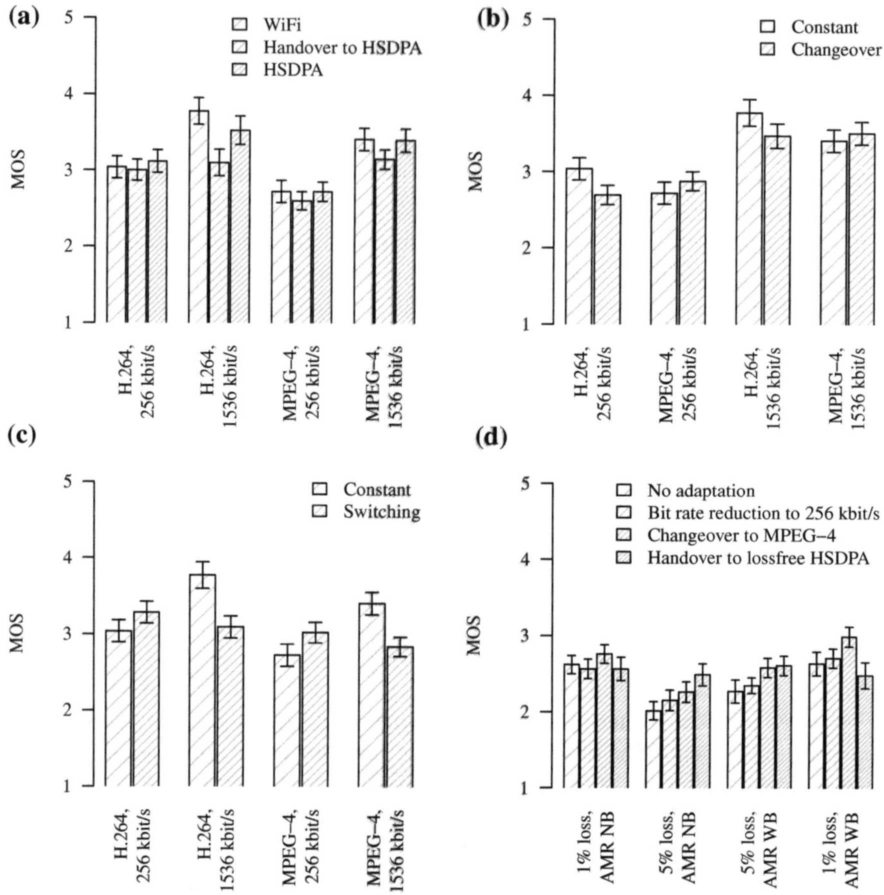

Fig. 27.5 User judgements of quality of short video samples extracted from [14]. MOS_{av}, and 95 % CI. **a** Network. **b** Video codec. **c** Bit rate. **d** Countering loss

In turn, when the video codec was switched during an ongoing session, the end quality of such a sample was always judged between the ratings that were collected for MPEG-4 and H.264, and the quality perception was changed accordingly (cf. Fig. 27.5b). A changeover from MPEG-4 to H.264 always improved, and a switch in the opposite direction always degraded the Quality of Experience. Doing so, the quality improvement was less perceived than the quality degradation. In the latter case, the overall quality was rated close to that provided by the lower quality codec. Therefore, if a switch to a low quality codec can be foreseen (e.g. to increase coding robustness due to handover necessity), constant use of the low quality codec enables to reduce the number of quality management operations at no substantial loss of perceived quality.

When the encoding bit rate was switched during an ongoing transmission, the user experience was changed according to the logical expectations (cf. Fig. 27.5c). When the video bit rate was increased, the overall quality impression was improved. However, the improvement was only substantial when MPEG-4 was used, which is related to the fact that this codec is less optimized for low transmission bit rates than H.264. When the encoding bit rate was reduced, the quality perception was always substantially degraded, and the overall quality of such a sample was close to that observed for constant use of the lower bit rate. Once again it should be noted that degradation of quality has a remarkably stronger impact on user perception than a quality improvement, and hence oscillation of the delivered quality should be avoided.

Finally, all of the above switching techniques were applied to counter the effect of packet loss in the middle of a sample (cf. Fig. 27.5d). As a baseline scenario an audiovisual stream that was encoded with H.264 at 1,536 kbit/s and with either AMR wideband or with AMR narrowband speech codec, and transmitted in WiFi that was affected by either 1 or 5 % of packet loss. The transmission was adapted by a handover to a loss-free HSDPA network, a changeover to a more robust (in the setup) MPEG-4 codec, and by a reduction of the encoding video bit rate to 256 kbit/s in the middle of a sample.

The results confirmed that in the context of video telephony a changeover to a more robust video codec turned out to be the adaptation technique that always improved the Quality of Experience. However, if packet loss was high, network handover was the most recommended technique of transmission adaptation, which is of particular relevance when a mobile user is leaving the WiFi coverage. Therefore, it is recommended to change the video codec first, and to switch the network when a certain threshold of degradation is exceeded. Moreover, it was also confirmed that the control of the video encoding bit rate is not an effective technique of transmission adaptation when link-layer impairments occur.

This test also confirmed that if instability of a connection can be foreseen, constant use of a less efficient but robust video codec, such as in this case MPEG-4, or permanent transmission in a wide-coverage network, such as HSDPA, helps to proactively deliver higher Quality of Experience than any other of the analysed quality adaptation methods.

27.5 Conclusions

Taking into consideration the analysis presented in this chapter, it is crucial to include user perception of quality when speech or video telephony in wireless networks is managed. As a matter of fact, proper selection of the wireless network, codec, and encoding bit rate is essential to provide high quality of a call. But, when the transmission is adapted during an ongoing call, it has to be considered that a reduction of quality is experienced quicker and stronger than a quality improvement. The latter requires considerably more time to generate perceptual profits. Therefore, if a drastic reduction of quality can be foreseen, proactive utilization of a low-quality but robust configuration reduces the service management effort at no significant costs of the perceived quality. Otherwise, the use of quality adaptation techniques during a call is in most of the cases advantageous, but the quality improvement is substantial only when 1) an essential boost of quality can be achieved, 2) the adaptation is used early in a call, or 3) when the process is applied to counterbalance the effect of packet loss.

Some of the addressed quality adaptation methods require particular knowledge of how they are perceived. For example, switching between narrowband and wideband speech codecs degrades the perceived quality below that of pure narrowband in most of the cases, even if improvement of quality would be expected when the encoding bandwidth is switched to wideband. Therefore, a switch from narrowband to wideband speech is not recommended late in a call, instead pure narrowband quality should be proactively provided. Another operation that is worth of particular attention is the network handover that may have marginal influence on the quality of speech telephony, but it might result in a significant quality degradation during a video call. Finally, it was also proven that the reduction of the video encoding bit rate is not an effective technique of quality adaptation when link-layer impairments occur and alternative measures, such as codec or network switching, should be preferred. As a consequence, when the presented aspects of quality perception are considered, not only the Quality of Experience in wireless networks can be improved, but also the scalability of a communication system can be enhanced by avoiding ineffective transmission management operations.

References

1. Agboma F, Liotta A (2008) QoE-aware QoS management. In: Proceedings of international conference on advances in mobile computing and multimedia (MoMM). Linz, Austria, pp 111–116
2. Bessette B, Salami R, Lefebvre R, Jelinek M, Rotola-Pukkila J, Vainio J, Mikkola H, Jarvinen K (2002) The adaptive multirate wideband speech codec (AMR-WB). IEEE Trans Speech Audio Process 10(8):620–636
3. Chien S, Liu H, Low A, Maciocco C, Ho Y (2008) Smart predictive trigger for effective handover in wireless networks. In: Proceedings of IEEE international conference on communications (ICC). Beijing, China, pp 2175–2181
4. Cunningham G, Perry P, Murphy L (2004) Soft, vertical handover of streamed video. In: Proceedings of IET international conference on 3G mobile communication technologies. London, United Kingdom, pp 432–436
5. Girod B, Farber N (1999) Feedback-based error control for mobile video transmission. Proc IEEE 87(10):1707–1723
6. Girod B, Färber N (2000) Wireless video. In: Reibman A, Sun M-T (eds) Compressed video over networks, Marcel Dekker, New York
7. Gros L, Chateau N (2001) Instantaneous and overall judgements for time-varying speech quality: assessments and relationships. Acta Acustica united with Acustica 87(3):367–377
8. Gustafsson E, Jonsson A (2003) Always best connected. IEEE Wirel Commun 10(1):49–55
9. Huber JF (2004) Mobile next-generation networks. IEEE Multimedia 11(1):72–83
10. ITU-T Recommendation G.113—Amendment 1 (2006) New appendix IV-provisional planning values for the wideband equipment impairment factor Ie, wb. International Telecommunication Union, Geneve

11. ITU-T Recommendation P.800 (1996) Methods of subjective determination of transmission quality. International Telecommunication Union, Geneve
12. Khan A, Sun L, Ifeachor E (2012) QoE prediction model and its application in video quality adaptation over UMTS networks. IEEE Trans Multimedia 14(2):431–442
13. Kwon T, Gerla M, Das S (2002) Mobility management for VoIP service: mobile IP vs. SIP. IEEE Wirel Commun 9(5):66–75
14. Lewcio B (2013) Management of speech and video telephony quality in heterogeneous wireless networks, Berlin, Germany. http://link.springer.com/book/10.1007%2F978-3-319-02102-7
15. Lewcio B, Möller S (2011) A testbed for QoE-based multimedia streaming optimization in heterogeneous wireless networks. In: Proceedings of IEEE international conference on signal processing and communication systems (ICSPCS). Honolulu, HI, USA, pp 1–9
16. Ma L, Yu F, Leung V, Randhawa T (2004) A new method to support UMTS/WLAN vertical handover using SCTP. IEEE Wirel Commun 11(4):44–51
17. Möller S, Raake A, Kitawaki N, Takahashi A, Wältermann M (2006) Impairment factor framework for wideband speech codecs. IEEE Trans Audio, Speech, Lang Process 14(6):1969–1976
18. Möller S, Wältermann M, Lewcio B, Kirschnick N, Vidales P (2009) Speech quality while roaming in next generation networks. In: Proceedings of IEEE international conference on communications (ICC). Dresden, Germany, pp 1–5
19. Ng SL, Hoh S, Singh D (2005) Effectiveness of adaptive codec switching VOIP application over heterogeneous networks. In: Proceedings of international conference on mobile technology, applications and systems (MobiSys). San Juan, Puerto Rico, pp 1–7
20. Pahlavan K, Krishnamurthy P, Hatami A, Ylianttila M, Makela J, Pichna R, Vallstron J (2000) Handoff in hybrid mobile data networks. IEEE Pers Commun 7(2):34–47
21. Pan Y, Sun, Y, Hsu C, Chen M (2009) A user-decided service model and resource management in a cooperative WiMAX/HSDPA network. In: Proceedings of IEEE international conference on communications (ICC). Dresden, Germany, pp 1–6
22. Perkins C (2002) IP mobility support for IPv4. RFC 3344 (proposed standard), Internet Engineering Task Force
23. Pyun JY (2008) Context-aware streaming video system for vertical handover over wireless overlay network. IEEE Trans Consum Electron 54(1):71–79
24. Rosenberg J, Schulzrinne H, Camarillo G, Johnston A, Peterson J, Sparks R, Handley M, Schooler E (2002) SIP: session initiation protocol. RFC 3261 (proposed standard), Internet Engineering Task Force
25. Schulzrinne H, Wedlund E (2000) Application-layer mobility using SIP. Mobile Comput Commun Rev 1(2):47–57
26. Schwarz H, Marpe D, Wiegand T (2007) Overview of the scalable video coding extension of the H. 264/AVC standard. IEEE Trans Circ Syst Video Technol 17(9):1103–1120
27. Seo J, Woo S, Bae K (2001) Study on the application of an AMR speech codec to VoIP. In: Proceedings of IEEE international conference on acoustics, speech and signal processing (ICASSP). Salt Lake City, UT, USA, pp 1373–1376
28. Si P, Ji H, Yu F (2010) Optimal network selection in heterogeneous wireless multimedia networks. Wirel Netw 16(5):1277–1288
29. Stallings W (2000) Data and computer communications, 7th edn. Prentice Hall, Upper Saddle River, NJ
30. Stemm M, Katz R (1998) Vertical handoffs in wireless overlay networks. Mobile Netw Appl 3(4):335–350
31. Taniuchi K, Ohba Y, Fajardo V, Das S, Tauil M, Cheng Y, Dutta A, Baker D, Yajnik M, Famolari D (2009) IEEE 802.21: media independent handover: features, applicability and realization. IEEE Commun Mag 47(1):112–120
32. Varela M, Laulajainen J (2011) QoE-driven mobility management—integrating the users' quality perception into network-level decision making. In: Proceedings of IEEE international workshop on quality of multimedia experience (QoMEX). Mechelen, Belgium, pp 19–24
33. Vatn J (1999) Long random wait times for getting a care-of address are a danger to mobile multimedia. In: Proceedings of IEEE international workshop on mobile multimedia communications (MoMuC). San Diego, CA, USA, pp 142–144

34. Vidales P (2005) Seamless mobility in 4G systems. Tech. Rep. UCAM-CL-TR-656, University of Cambridge, Computer Laboratory
35. Voran SD (2005) A basic experiment on time-varying speech quality. In: Proceedings of international conference on measurement of speech and audio quality in networks (MESAQUIN). Prague, Czech Republic
36. Voran SD (2010) Subjective ratings of instantaneous and gradual transitions from narrowband to wideband active speech. In: Proceedings of IEEE international conference on acoustics, speech and signal processing (ICASSP). Dallas, TX, USA, pp 4674–4677
37. Wang Y, Zhu Q (1998) Error control and concealment for video communication: a review. Proc IEEE 86(5):974–997
38. Weiss B, Möller S, Raake A, Berger J, Ullmann R (2009) Modeling call quality for time-varying transmission characteristics using simulated conversational structures. Acta Acustica united with Acustica 95(6):1140–1151
39. Wenger S (2003) H. 264/AVC over IP. IEEE Trans Circuits Syst Video Technol 13(7):645–656
40. Wiegand T, Schwarz H, Joch A, Kossentini F, Sullivan GJ (2003) Rate-constrained coder control and comparison of video coding standards. IEEE Trans Circuits Syst Video Technol 13(7):688–703
41. Zhang H, Zhao J, Yang O (2008) Adaptive rate control for VoIP in wireless Ad Hoc networks. In: Proceedings of IEEE international conference on communications (ICC). Beijing, China, pp 3166–3170
42. Zhang R, Regunathan S, Rose K (2000) Video coding with optimal inter/intra-mode switching for packet loss resilience. IEEE J Sel Areas Commun 18(6):966–976

Chapter 28
QoE-Based Network and Application Management

Raimund Schatz, Markus Fiedler and Lea Skorin-Kapov

Abstract This chapter presents an overview of a set of recently proposed QoE-based management approaches that all try so resolve a central dilemma: maximizing user satisfaction while at the same time maximizing resource efficiency and economy. To this end, it first builds bridges between recent approaches towards QoE-based Network Management and standardized Network Management functions. This is contrasted by a discussion of recent approaches towards QoE-based Application Management. Further, it is shown how both Network Management and Application Management can work together in concert. Finally, open issues regarding a better integration of management and QoE are outlined.

28.1 Introduction

Proactive management of applications and networks has the potential to resolve the central dilemma of delivering applications to end users at maximum quality, while at the same time minimizing the costs of the other stakeholders involved in the delivery, including network, service and cloud providers. The so-far typical Internet control paradigms "best effort", "one size fits all" and "prevent performance problems by overprovisioning" have led to inadequate and uneconomical ways of providing sufficient levels of QoE. Indeed, users and providers may have different (and potentially

R. Schatz (✉)
Telecommunications Research Center Vienna (FTW), Vienna, Austria
e-mail: schatz@ftw.at

M. Fiedler
Blekinge Institute of Technology (BTH), Karlskrona, Sweden
e-mail: mfi@bth.se

L. Skorin-Kapov
Faculty of Electrical Engineering and Computing, University of Zagreb, Zagreb, Croatia
e-mail: lea.skorin-kapov@fer.hr

S. Möller and A. Raake (eds.), *Quality of Experience*,
T-Labs Series in Telecommunication Services, DOI: 10.1007/978-3-319-02681-7_28,

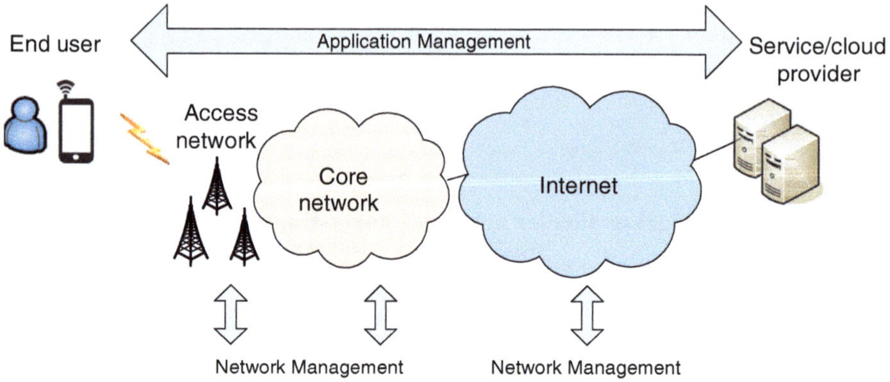

Fig. 28.1 Network management (NM) and application management (AM) constitute complementary approaches that utilize different monitoring and control points

conflicting) views, experiences and understandings of a service [48]. In this context, QoE is supposed to enable a broader, more holistic understanding of the impact of networked communication and content delivery systems on the end-user and thus to complement management perspectives on quality and performance that have traditionally excluded the user perspective.

This chapter presents an overview of a set of recently proposed QoE-based management approaches that are specifically related to *Network Management* (NM) and *Application Management* (AM). While NM is based on monitoring and exerting control on access, core network and Internet level, AM seeks to adapt quality and performance on end-user and application host/cloud level. The different, complementary perspectives applied by AM and NM are illustrated by Fig. 28.1. NM focuses on monitoring and control onto the network entities in order to keep the network up-and-running. Thus, it is not surprising that "Over-The-Top" (OTT) services running on top of Internet, such as YouTube, Skype and Netflix have implemented their own AM, i.e. QoE control schemes on application level such as forward-error coding or adaptation of video resolution, which aim at decreasing the risks of spatial (blocking etc.) or temporal (stalling etc.) artifacts, respectively. Naturally, this type of control that adapts the application to the conditions found in the network is situated much closer to the user than the network-level control. Thus, AM can act as a "mediator" between network and user interface, optimizing QoE under the given circumstances.

What is common to both categories of QoE-based management approaches (AM, NM) is that they are based on the results of various QoE research fields: QoE assessment, modeling, measurement and monitoring. Consequently, this chapter builds on the previous chapters in this book, illustrating how QoE management serves as a major crystallization point and catalyst for advancing this area of research.

The remainder of this chapter is structured as follows: Sect. 28.2 introduces a set of recent approaches towards QoE-based Network Management and relates them to the FCAPS classification. Likewise, Sect. 28.3 presents a set of recent approaches towards QoE-based Application Management. Section 28.4 then shows how both types of approaches are put together in a combined fashion using QoE management in walled-garden IPTV settings and YouTube video streaming as examples. Finally, Sect. 28.5 wraps up the chapter and points at some aspects that need further attention.

28.2 QoE-Based Network Management

Given the broad range of issues that may be considered under the umbrella term of Network Management (NM), we consider it beneficial to identify those areas that may in particular be exploited to optimize service quality as perceived by end users. In that sense, we draw links between QoE-driven NM approaches and the ISO-standardized FCAPS framework, which serves to classify NM objectives across five different levels, as elaborated on in the first subsection. We then present an overview of recent approaches to QoE-based NM, with focus on QoE-driven resource management and multi-operator scenarios.

28.2.1 The FCAPS Classification

The ISO-standardized minimal set of functional areas of NM are defined as FCAPS (Fault, Configuration, Accounting, Performance and Security Management) [31] and commonly referred to within NM [12, 21, 23, 48]. With regards to QoE, the following areas are of specific importance:

- Fault Management is aiming at isolating and fixing network failures as quickly as possible in order to minimise the time that the users are disconnected from network service(s). Thus, it provides a central lead in assuring QoE by limiting the impact of network problems on user annoyance.
- Performance Management potentially has the most obvious connection to QoE and user delight, although its monitoring and control facilities are rather limited [20, 21]. The performance aspect of NM focuses on monitoring network-related parameters such as byte counts and link loads (which may include the generation of alarm messages once pre-configured thresholds are crossed, and by subsequently allocating more resources (which is commonly referred to as "throwing bandwidth at the problems").

The relative high importance of Fault Management within NM as compared to Performance Management is motivated by the observation that users react much more to uncontrolled quality degradations (e.g., due to packet losses because of congestion) than to controlled degradations (e.g., congestion avoidance through throughput reduction) [15, 25]. Performance management that focuses on provisioning of QoE is recognised as a key topic for future NM [48]. From a user perspective, the other NM functional areas related to Configuration (monitoring and managing the system configuration), Accounting (focusing on billing and charging), and Security (managing network authentication, authorization, and auditing) may be considered as generally having a less prominent and more indirect link to QoE improvements. However, for certain service scenarios (e.g., e-commerce, e-banking, e-health), the latter functions may prove to be of high importance.

28.2.2 QoE-Driven Network Resource Management

With regards to QoE-driven network resource management approaches, a distinction has been made between user-centric and network-centric approaches, whereby the former explicitly take into account end-user QoE-related feedback, while the latter implicitly treat QoE while conducting QoE optimization based on network-collected data [57]. While resource allocation decisions are inherently made in the network, feedback collected from the client device or triggered by the end user can provide valuable input to the decision making process. Furthermore, certain information which may be relevant in making optimal resource allocation decisions may only be available in the network (e.g., operator policy, subscriber data, service priority, network resource availability). Resource management can actually take place in two different parts of the network: access and core network. Regarding the FCAPS classification, there are clear links with Performance Management (in terms of QoE-driven control of network resource allocation) and Fault Management (i.e., the collection of relevant data influencing QoE can serve to both identify and manage faults in both the network and at the client device).

Access network. While the access network can be wireline or wireless, it is in the domain of wireless networks that we find resources to be both more constrained and more variable over time, due to issues such as time-varying transmission channel conditions, user mobility, etc. [22], see also Chap. 27. As a result, the majority of research dealing with QoE-driven resource allocation targets this domain (as will be the focus of our review), with clear impacts on end user perceived service quality [11, 49].

Utility functions have been used to correlate user perceived value with QoS metrics such as delay, loss, error probability, and throughput [19, 40, 65]. In [65], the authors use utility functions to maximize utility across multiple users accessing different video contents in a wireless network by calculating the optimal radio resource allocation per user. They propose an enhanced objective function to avoid noticeable quality fluctuations (shown to have a negative impact on user perceived service quality). The maximization of aggregate utility across all users in a cell is also addressed in [9], where the mapping of service response time and user data rate (in the case of Web browsing) to MOS serves as input for a proposed radio resource allocation algorithm applicable in beyond 3G networks. Further solutions address QoE-driven traffic management in network access points by way of admission control, prioritized scheduling, and bandwidth management, relying on traffic differentiation and the customer subscription scheme [55].

A challenge with utility-based resource allocation mechanisms lies in the fact that certain applications have resource demands that may change over time (for example, the relationship between user perceived value and allocated bandwidth may change based on application state, such as a new media component added or removed from an ongoing session, or user behaviour such as pausing a video stream). In certain cases, dynamic feedback provided by the client can be used to drive network scheduling mechanisms, such as in the case of YouTube (to be discussed further in Sect. 28.4).

As opposed to feedback automatically generated by the client, the approach in [10] proposes mechanisms for end users to dynamically and asynchronously express their subjectively perceived (dis)satisfaction with respect to the instantaneous experience of their service quality. Based on direct user actions indicating preferences regarding service performance and corresponding cost, user's service utility functions are adapted, consequently driving the utility maximization problem being solved at the wireless base station.

While the majority of existing research addressing QoE-driven radio resource allocation focuses on downlink transmission, the need for optimized uplink resource allocation has been recognized in light of end users increasingly upstreaming multimedia content. A distributed QoE optimization approach is proposed in [14], supporting both optimized allocation of uplink resources and media adaptation decisions at the source client (e.g., video rate adaptation and decision on which video layers to transmit).

Core network. In the context of converged core network evolution, the 3rd Generation Partnership Project (3GPP) has specified the Policy and Charging Control (PCC) architecture, supporting differentiated service quality based on the mapping of service flows to different bearers [4]. The decisions regarding bearer assignment may be driven by service requirements specified and negotiated at the application-level and passed on to underlying network mechanisms. In [58], the authors propose mechanisms for the E2E negotiation and calculation of both optimal and suboptimal multimedia service configurations and corresponding network resource allocations, given service utility functions and user preferences. Such calculations may further serve as input to PCC mechanisms responsible for performing domain-wide QoE-driven resource allocation decisions [34].

Related approaches have proposed the inclusion of a QoE estimation/control server as a novel application server in the 3GPP architecture, responsible for collecting relevant data (e.g., related to network performance, client device performance, subscriber profiles, service requirements, or operator policy), estimating QoE, and invoking QoE control mechanisms [19]. Examples of such mechanisms include prioritized network resource usage, modified service bandwidth limits, or notifications sent to subscribers informing them of potential actions to take to improve QoE.

28.2.3 Towards QoE Management in Multi-Operator Settings

Considering QoE from an end-to-end (E2E) perspective, it is clear that communications may span multiple types of networks (fixed or wireless) belonging in turn to multiple operators. While QoS assurance in independent transport networks has been well studied, challenges remain on how to secure E2E QoS and QoE across multiple network domains, relying on inter-domain signalling and inter-provider agreements [62].

Network convergence and quality assurance in multi-operator networks are fundamental issues addressed in the scope of Next Generation Network (NGN) standards. A high-level framework addressing E2E QoE assurance has been proposed in [70], relying on the assumption that client devices are capable of reporting QoE/QoS performance to network QoE management components along the E2E path. Given that in practice, different networks will generally manage and optimize QoE locally, in which case E2E QoE will depend on the traversed networks, QoE management in the network may be integrated or complemented with application-level QoE control mechanisms [63].

Seamless communications is a specific multi-operator setting that actually tries to exploit quality diversity by automatically choosing the best-fitting network to a set of decision criteria, typically involving quality, cost and security [30] and thus addressing the FCAPS dimensions Accounting and Security Management. While seamless communications were initially QoS-oriented, attention turned to QoE as a driving paradigm for making optimal network switching decisions [13, 29]. Switching decisions can be made in both proactive (in order to optimize starting conditions and load distribution) and reactive (to performance degradations and link losses) ways.

In addition, the commercialization of QoS in heterogenous networks with multiple operators (i.e., inter-operator/inter-domain QoS as a good) has recently received a strong impetus. For example, in the ETICS project [1] the user-centric understanding of demand, i.e., willingness-to-pay and QoE for network services, has been piggybacked on course-granular inter-domain end-to-end QoS Service Level Agreements (SLAs) used for efficiency reasons aggregating the required QoS guarantees for several users or whole domains. In this context, recently initiated studies are addressing the notion of evolved QoE-driven Service Level Agreements[1], incorporating measures of user-perceived service quality (and stemming from knowledge regarding correlations between QoS and QoE) [3]. Further considering business opportunities, the exchange of monitoring data collected at different points along the service delivery chain among different players involved (application/service providers, network operators, etc.) may provide valuable insight into the causes of QoE degradations and potential for QoE control, both from a network and an application perspective.

28.3 QoE-Based Application Management

The management approaches described in the previous section have focused primarily on controlling quality on access and core network level. In contrast, QoE-based application management targets the application server at the head-end as well as the

[1] For a more extensive discussion on user-centric SLAs please refer to Chap. 7 in this book.

client terminal as the main control points. This section discusses QoE-based application management with a focus on non-interactive media streaming for services like online video and IPTV. With respect to such (typically passively consumed) video streaming services, a clear distinction can be made between more traditional streaming techniques based on push-based paradigms and server-side decisions as opposed to newer pull-based paradigms involving intelligent clients and HTTP adaptive streaming [52]. In addition to media-related metrics (e.g., frame rate, encoding, content type), in the former case, QoE management solutions for UDP/RTP media streaming are driven by intrinsic network metrics such as packet loss ratio and transfer delay, while the latter case generally focuses on HTTP/TCP-related metrics such as re-buffering rate and duration [7].

28.3.1 UDP/RTP-Based Multimedia Streaming

Several studies have addressed QoE-driven adaptation schemes for video delivery via UDP/RTP over different types of networks, aiming at alleviating the impact of packet loss and media distortion on the user experience. The adaptation of video sender bitrate to meet end user QoE requirements (derived based on application and network parameters, and taking into account content type) is addressed in [39]. In their subsequent work [38], the authors apply a newly proposed video quality prediction model for the purpose of QoE control via sender bitrate adaptation targeting UMTS networks. Feedback regarding network QoS information is collected via transmitted RTCP reports. In a similar fashion, but with a focus on voice scenarios, [35] propose a QoE-driven VoIP adaptation scheme based on different network conditions and available bandwidth. In a more generic approach targeting multimedia access networks, [41] proposes an autonomic QoE management architecture that monitors network problems, determines QoE optimization actions (using an approach based on neural networks), and executes necessary actions (e.g. activating Forward Error Correction packets or selecting the delivery bit rate).

What is common to the above QoE-centric Application Management approaches is that they focus on bitrate adaptation, with most of the intelligence residing at the server side. However, with the growing popularity of TCP (and HTTP) based media streaming, the research focus has shifted accordingly towards more client-centric approaches, as discussed in the next subsection.

28.3.2 HTTP Adaptive Streaming

Adaptive streaming over HTTP [52, 64] is becoming an increasingly popular way of delivering videos over IP networks using the TCP protocol. It is typically implemented as a combination of streaming servers and intelligent clients that make adaptation decisions based on local observations. Nonetheless, providing high QoE remains a challenge particularly in mobile networks featuring bandwidth fluctuations and outages that ultimately cause buffer starvation and frequent picture quality changes. These issues necessitate the development of intelligent QoE-aware adaptation mechanisms (i.e., a quality scheduler) on the application level.

In this context, [54] benchmarked the quality adaptation strategies of several commercially available solutions. Their results confirm the large QoE impact of the quality scheduler, highlighting the inherent tradeoffs between high average quality, stable quality, protection against buffer underruns and bandwidth utilization as well as the need for more sophisticated solutions. Further, evaluating commercial bitrate-adaptive players in the context of competing for shared resources, the authors in [36] constitute that they lack to satisfy fairness, efficiency, and stability goals. To this end, they developed a suite of techniques for improved chunk scheduling and bitrate selection that can systematically guide the tradeoffs between reaching the aforementioned goals.

While DASH (Dynamic Adaptive Streaming over HTTP) is typically used in the context of single-layer codecs (H.264/AVC), recent studies have addressed streaming adaptation algorithms for scalable video coding based on H.264/SVC [51, 56]. In [56], the authors propose an adaptation algorithm which they present as outperforming other DASH mechanisms in terms of video quality, low switching frequency and usage of the available resources in a realistic mobile network scenario. A general analysis of the impact and trade-offs of SVC-based quality adaptation algorithms is given in [5], with a focus on Peer-to-Peer (P2P) Video on Demand (VoD) provisioning systems that feature dynamic optimization of what the authors term 'session quality' (rebufferings, playback delay, etc.).

While client-side bitrate adaptation is the de-facto approach today, the authors in [46] argue that CDN (Content Delivery Network) performance variability is difficult to detect when relying simply on such approaches. Consequently, they present a coordinated Internet video control plane that can use a global view of client and network conditions to dynamically optimize video delivery via control over two parameters: suitable choice of bitrate, and choice of CDN/server. The goal is to provide a high quality viewing experience despite an unreliable delivery infrastructure, supporting bitrate adaptation at both the start and during a session. Their analysis shows that such a control plane can potentially improve the rebuffering ratio by up to 100 % in the average case and by more than one order of magnitude under stress.

Hoßfeld et al. [28] discusses technical challenges emerging from shifting services to the cloud as well as how this shift impacts QoE and QoE management, with a focus on multimedia cloud applications such as video streaming. Discussing the different ways how to address these challenges, the authors show how different players in the ecosystem (including network, service, and cloud providers) have to interact and exchange information in order to realize QoE-based management for cloud-based multimedia services. This QoE management proposed in [28] clearly goes beyond pure Application Management, a topic addressed in the next section.

28.4 Bringing Application and Network Management Together

As a synthesis of the previous two sections, we are now going to discuss how network and application management can work together in a complementary fashion. This is illustrated in the context of two different scenarios, with the first one being more telco operator-centric and the second one being more Internet/OTT-centric.

28.4.1 QoE Management for Managed Services: Walled-Garden IPTV

As defined by ITU standards, IPTV refers to the delivery of multimedia services (e.g., television, video, audio, graphics, data) over managed IP networks that provide required levels of QoS/QoE, security, interactivity, and reliability [32]. The phrase *walled-garden* IPTV has been used to refer to proprietary operator solutions offering full control of the service delivery chain, from acquiring and managing content, to delivery via broadband networks to set-top boxes in customer homes. Given full control, operators are able to employ both NM and AM approaches to provide a certain level of quality assurance to end users. An example QoE management approach applicable in such a traditional IPTV environment is presented in [47]. The authors propose a QoE estimation process (per IPTV channel) based on measured network QoS parameters, zapping time (channel switching time), audio/video quality, and media synchronization. The resulting estimations are used to invoke various NM or AM QoE optimization actions, such as modification of traffic flow prioritization, selection of other routing paths, or media transcoding at the server side.

Considering architectural solutions for IPTV, proprietary, walled solutions have been noted as being faced with issues related to interoperability, multi-vendor environments, and third-party provisioning [42]. Different solutions have involved the integration of IPTV services within NGN environments, for example based on a fully NGN-integrated quality-assured IPTV provisioning model [67]. Standardization efforts that have been made by organizations such as the ITU and ETSI/TISPAN have proposed different architectural options, focusing on those based on the NGN architecture [2, 33]. Considering the concrete case of NGN-based IPTV, service control functions corresponding to a *service layer* (e.g., session control and management, media control and processing) are inherently linked to resource and admission control functions in the *network layer*. Given that the network resource allocations requested are based on media requirements that are negotiated and established at the service layer, AM outcomes (e.g., choice of different content or encoding schemes) provide input for making NM decisions (e.g., resource reservation). On the other hand, data collected along different monitoring points in the network can be used to make AM decisions. Consequently, with the QoE-oriented service control and application functionalities intertwined with the transport layer QoS control mechanisms, it becomes evident that in the context of NGNs, application and network management schemes are conceptualized to work together in assuring end-user QoE. Given such functionalities, QoE management approaches such as the one presented in [47] could be considered, but in a standardized, multi-service, open environment rather than in a proprietary IPTV network.

28.4.2 QoE Management for OTT Video: The Case of YouTube

The previous scenario has outlined how the combined, complementary use of NM and AM is being addressed in an operator-controlled IPTV setting. In contrast, this complementary use can also be driven by the need to manage the QoE of a concrete resource-intensive video service delivered over the Internet: YouTube.

YouTube accounts for more than 30 % of the overall Internet's traffic [17], with over 4 billion videos viewed every day in 2012 [69]. This outstanding success also creates serious challenges for network operators and service providers, who need to engineer their systems to correctly handle the resulting huge volume of OTT video traffic and the large number of users in efficient ways. For these reasons YouTube has become a primary target not only for the networking community at large [6, 8, 16, 66], but also for QoE research, resulting in a growing amount of work on YouTube QoE management, e.g. [59, 61, 68].

From a technical perspective, YouTube is an online video platform that utilizes non-adaptive HTTP streaming to deliver multimedia content to clients via an inherently unreliable best-effort Internet in the form of a progressive download[2] [16]. Due to this technology choice, the smooth playback of the video (i.e., fast startup, no rebufferings) rather than visual image quality is the main QoE management challenge [27, 50]. In this respect, YouTube already features some performance improvement measures that have direct QoE impact: on the application level, YouTube streaming utilizes custom application flow control techniques referred to as 'block sending' as well as dual-threshold buffer management (cf. [8, 18]). The main purposes are throughput smoothing via rate control (however, not without side-effects due to interactions with the already present TCP flow-control [8]) and the prevention of stalling effects caused by buffer starvation. On the CDN-level, YouTube employs a three tier caching infrastructure distributed over four continents with two goals: enhanced streaming performance by selecting a nearby cache as well as load balancing among cache clusters [6].

Albeit these measures were introduced for the purpose of improving the overall performance of the service (including other aspects such as fairness, efficiency and robustness), they do not represent full-fledged proactive QoE management, thus leaving room for further optimization [26]. This issue has been addressed by recent work on QoE-based AM and NM for YouTube that concentrates on two different network environments: (1) a local wireless mesh network access network environment that

[2] This refers to the implementation of YouTube as of end of 2012.

foresees central resource management; and (2) a global Internet environment where resource management can only happen decentrally.

As regards the former, approaches for local mesh networks have been addressing various network resource management options for QoE management based on application-level client feedback (generated by a custom application observing buffer levels at the client side): QoS differentation via traffic shaping [59], routing [61], and physical reconfiguration of nodes [60]. However, these options are not directly applicable to the global Internet environment with its inherent requirements for scaleability and decentralization. Thus as regards the Internet scenario, [26] suggests a controlled exploitation of selected tradeoffs in order to manage and improve YouTube QoE by means of combined AM and NM. For example, recent user studies on YouTube have found that increasing initial buffering delay before playback has less negative QoE impact than increasing the amount of stalling during playback [24]. Thus, if the QoS properties of the network transmission path as well as the properties of the video clip being requested are known, one can compute the optimal initial delay that minimizes the likelihood of stalling without annoying the user with unduly startup waiting times [26].

The key challenge that remains is that exploitation of such QoE-related tradeoffs requires a level of information exchange between network and application that cannot be passed to the network stack with today's APIs. Furthermore, the network stack must be able to react to these requirements dynamically. To this end, new APIs like the GAPI [43] and forwarding concepts such as Forwarding on Gates [44, 45] are currently being investigated to enable network-application interaction on a large scale.

Both examples in this section have shown that QoE-based network and application management should not be understood as separate, mutually exclusive paths towards QoE improvement. Indeed, as also suggested by [37, 71], the QoE management becomes most effective when NM and AM are allowed to work together in terms of a combined complementary approach.

28.5 Conclusion

This chapter has identified relationships between Network Management (NM), Application Management (AM), and QoE. While AM has a direct connection to QoE through the application's presentation layer and user interface, in practice there have been developed rather few ties between NM and QoE so far. This may make it hard for network managers to precisely locate the reason for specific user annoyance, or to create specific conditions for user delight.The latter is not surprising, as the control points within NM are much farther away from the users' points of perception than the control points within AM. Today, AM typically acts as mediator between network(s) and user(s) and aims at leveling off non-optimal network behaviour.We observed both pro-active and re-active management approaches that try to follow the different dynamics in the networked system in order to level out the QoE to the desired level(s), eventually determined by the user.

When applying the (within NM well-known) FCAPS classification to both QoE-based NM and AM, it becomes obvious that most of the proposed QoE-based management approaches fall into the domain of Performance Management. The presented examples are dominated by resource and access control, which even touches upon Accounting Management in particular if billing plans correlate with perceived utility [10, 53, 57]. However, within NM, Fault Management is seen at the number-one duty, followed by Configuration, Accounting and Security management, while Performance Management is often considered to be freestyle. We observe Fault Management functionality related to resource (re-)allocation and re-active routing of traffic, amongst others in the context of mobility and seamless communications. Furthermore, we observe that many contributions are rather patchy (i.e., they address just parts of the networked system) or found on high levels of abstraction (i.e., rather far away from practical implementability), and that NM and AM are typically not coordinated. Indeed, AM performed by "Over-The-Top" (OTT) services (such as YouTube, Skype, etc.) is not necessarily in line with network operator preferences. On the other hand, service differentiation on Internet level might violate the network neutrality principle if users are not notified about such measures by the corresponding operator.

Tying QoE, AM and NM closely together puts forward the need for aligned views and mindsets. For instance, the understanding of a fault can be completely different for a network provider (broken link) or for a user (missed goal in live soccer streaming due to a single freeze in the wrong moment). Besides clarifying and synchronizing the meaning of different concepts and notions (like "quality" or "performance"), their importance for the different communities need to be assessed and aligned in order to make the vision of truly user- and QoE-centric Network and Application Management a reality.

References

1. EU FP7 Project ETICS (2010–2013) Economics and technologies for inter-carrier services. https://www.ict-etics.eu
2. (2011) NGN integrated subsystem architecture. ETSI TS 182 028, v3.5.1
3. EU Celtic-Plus Project QuEEN (2011–2015) Quality of experience estimators in networks. http://www.celticplus.eu/Projects/Celtic-projects/Call8/QUEEN/queen-default.asp
4. (2013) Policy and charging control architecture. 3GPP TS 23.203, release 12. http://www.3gpp.org/ftp/Specs/html-info/23203.htm
5. Abboud O, Zinner T, Pussep K, Al-Sabea S, Steinmetz R (2011) On the impact of quality adaptation in SVC-based P2P video-on-demand systems. In: Proceedings of the second annual ACM conference on multimedia systems, MMSys '11, ACM, New York, USA, pp 223–232. doi:10.1145/1943552.1943582
6. Adhikari VK, Jain S, Chen Y, Zhang ZL (2012) Vivisecting YouTube: an active measurement study. In: Proceedings of the IEEE INFOCOM 2012 mini-conference, Orlando, Florida.
7. Alberti C et al. (2013) Automated QoE evaluation of dynamic adaptive streaming over HTTP. In: Fifth international workshop on quality of multimedia experience, QoMEX

8. Alcock S, Nelson R (2011) Application flow control in YouTube video streams. ACM SIGCOMM Comput Commun Rev 41(2):24–30
9. Ameigeiras P, Ramos-Munoz JJ, Navarro-Ortiz J, Mogensen P, Lopez-Soler JM (2010) QoE oriented cross-layer design of a resource allocation algorithm in beyond 3g systems. Comput Commun 33(5):571–582
10. Aristomenopoulos G, Kastrinogiannis T, Kaldanis V, Karantonis G, Papavassiliou S (2010) A novel framework for dynamic utility-based QoE provisioning in wireless networks. In: IEEE global telecommunications conference (GLOBECOM 2010). IEEE pp 1–6
11. Baraković S, Skorin-Kapov L (2013) Survey and challenges of QoE management issues in wireless networks. J Comput Netw Commun 2013:1–28
12. Boutaba R, Polyrakis A (2001) Projecting FCAPS to active networks. In: Proceedings of the enterprise networking, applications and service conference, pp 97–104
13. Dillon E, Power G, Ramos MO, Rodriguez MC, Argente JR, Fiedler M, Tonesi D (2009) PERIMETER: a quality of experience framework. In: Proceedings of the future internet symposium (FIS), Berlin, Germany
14. El Essaili A, Zhou L, Schroeder D, Steinbach E, Kellerer W (2011) QoE-driven live and on-demand LTE uplink video transmission. In: 2011 IEEE 13th international workshop on multimedia signal processing (MMSP). IEEE, pp 1–6
15. Fiedler M, Hoßfeld T (2010) Quality of experience-related differential equations and provisioning-delivery hysteresis. In: 21st ITC specialist seminar on multimedia applications-Traffic, performance and QoE, Miyazaki, Japan
16. Finamore A, Mellia M, Munafo M, Torres R, Rao S (2011) YouTube everywhere: impact of device and infrastructure synergies on user experience. In: Proceedings of the 2011 ACM SIGCOMM internet measurement conference (IMC'11), Berlin, Germany
17. Gehlen V, Finamore A, Mellia M, Munafò MM (2012) Uncovering the big players of the web. In: Proceedings of the 4th international conference on traffic monitoring and analysis, TMA'12, Springer, Berlin, Heidelberg, pp 15–28
18. Ghobadi M, Cheng Y, Jain A, Matthis M (2012) Trickle: rate limiting Youtube video streaming. In: Proceedings of the 2012 USENIX annual technical conference, Boston, USA
19. Gomez G, Lorca J, Garcia R, Perez Q (2013) Towards a QoE-driven resource control in LTE and LTE-A networks. J Comput Netw Commun 2013:1–15
20. Gorod A, Gove R, Sauser B, Boardman J (2007) System of systems management: a network management approach. In: Proceedings of the IEEE international conference on system of systems engineering (SoSE'07), pp 1–5
21. Goyal P, Mikkilineni R (2009) FCAPS in the business services fabric model. In: Proceedings of the 18th international workshop on enabling technologies (WETICE'09), pp 45–51
22. Han Z, Liu KR (2008) Resource allocation for wireless networks: basics, techniques, and applications. Cambridge university press, Cambridge
23. Hong DK, Hong C, Hyoung Y (2002) An integrated network management framework for internet access service using ATM over ADSL technology. In: Proceedings of the 5th IEEE international conference on high speed networks and multimedia, communications, pp 24–31
24. Hoßfeld T, Egger S, Schatz R, Fiedler M, Masuch K, Lorentzen C (2012) Initial delay vs. interruptions: between the devil and the deep blue sea. In: 2012 fourth international workshop on quality of multimedia experience (QoMEX), p 16
25. Hoßfeld T, Fiedler M, Zinner T (2011) The QoE provisioning-delivery-hysteresis and its importance for service provisioning in the future internet. In: Proceedings of the 7th EURO-NGI conference on next generation internet networks (NGI 2011), Kaiserslautern, Germany
26. Hoßfeld T, Liers F, Schatz R, Staehle B, Staehle D, Volkert T, Wamser F (2012) Quality of experience management for YouTube: clouds, FoG and the AquareYoum. PIK–praxis der informationsverarbeitung und kommunikation 35(3). doi:10.1515/pik-2012-0024. http://www.degruyter.com/view/j/piko.2012.35.issue-3/pik-2012-0024/pik-2012-0024.xml
27. Hoßfeld T, Schatz R, Seufert M, Hirth M, Zinner T, Tran-Gia P (2011) Quantification of YouTube QoE via crowdsourcing. In: IEEE international workshop on multimedia quality of experience—modeling, evaluation, and directions (MQoE 2011), Dana Point, USA

28. Hoßfeld T, Schatz R, Varela M, Timmerer C (2012) Challenges of QoE management for cloud applications. IEEE Commun Mag 50(4):28–36
29. Ickin S, Vogeleer KD, Fiedler M, Erman D (2010) On the choice of performance metrics for user-centric seamless communication. In: Proceedings of the third Euro-NF workshop on socio-economic issues of networks of the future, Ghent, Belgium
30. Isaksson L, Fiedler M (2007) Seamless connectivity in WLAN and cellular networks with multi criteria decision making. In: Proceedings of the NGI 2007, Trondheim, Norway
31. ISO/IEC (1998) Information technology—open systems interconnection—systems management overview
32. ITU-T Recommendation Y.1901 (2009) Requirements for the support of IPTV services. International Telecommunication Union, Geneva
33. ITU-T Recommendation Y.1910 (2008) IPTV—functional architecture. International Telecommunication Union, Geneva
34. Ivesic K, Matijasevic M, Skorin-Kapov L (2011) Simulation based evaluation of dynamic resource allocation for adaptive multimedia services. In: 7th international conference on network and service management (CNSM), pp 1–4
35. Jammeh E, Mkwawa I, Khan A, Goudarzi M, Sun L, Ifeachor E (2012) Quality of experience (QoE) driven adaptation scheme for voice/video over IP. Telecommun Syst 49(1):99–111. doi:10.1007/s11235-010-9356-5. http://dx.doi.org/10.1007/s11235-010-9356-5
36. Jiang J, Sekar V, Zhang H (2012) Improving fairness, efficiency, and stability in HTTP-based adaptive video streaming with FESTIVE. In: Proceedings of the 8th international Conference on emerging networking experiments and technologies, CoNEXT '12. ACM, New York, USA p 97108. doi:10.1145/2413176.2413189. http://doi.acm.org/10.1145/2413176.2413189
37. Tutschku K, Fiedler M (eds) (2011) Euro-NF deliverable D.SEA.10.2. Second update of the Euro-NF vision regarding the network of the future
38. Khan A, Sun L, Ifeachor E (2012) QoE prediction model and its application in video quality adaptation over UMTS networks. IEEE Trans Multimedia 14(2):431–442
39. Khan A, Sun L, Jammeh E, Ifeachor E (2010) Quality of experience-driven adaptation scheme for video applications over wireless networks. IET Commun 4(11):1337–1347
40. Khan S, Duhovnikov S, Steinbach E, Kellerer W (2007) MOS-based multiuser multiapplication cross-layer optimization for mobile multimedia communication. Advances in multimedia
41. Latré S et al (2009) An autonomic architecture for optimizing QoE in multimedia access networks. Comput Netw 53(10):1587–1602
42. Lee CS (2007) IPTV over next generation networks in ITU-T. In: IEEE 2nd IEEE/IFIP international workshop on broadband convergence networks, 2007. BcN'07. IEEE, pp 1–18
43. Liers F, Volkert T, Martin D, Backhaus H, Wippel H, Veith E, Siddiqui AA, Khondoker R (2011) GAPI: a G-Lab application-to-network interface. In: EuroView2011, Würzburg, Germany
44. Liers F, Volkert T, Mitschele-Thiel A (2010) Demonstrating forwarding on gates with first applications. In: EuroView2010, Würzburg, Germany
45. Liers F, Volkert T, Mitschele-Thiel A (2011) Scalable network support for application requirements with forwarding on gates. In: EuroView2011, Würzburg, Germany
46. Liu X, Dobrian F, Milner H, Jiang J, Sekar V, Stoica I, Zhang H (2012) A case for a coordinated internet video control plane. SIGCOMM Comput Commun Rev 42(4):359370. doi:10.1145/2377677.2377752. http://doi.acm.org/10.1145/2377677.2377752
47. Lloret J, Garcia M, Atenas M, Canovas A (2011) A QoE management system to improve the IPTV network. Int J Commun Syst 24(1):118–138
48. Lu X, Zhou W, Song J (2010) Key issues of future network management. In: Proceedings of the international conference on computer application and system modeling (ICCASM 2010), pp V11–649-V11-653
49. Martini MG, Chen CW, Chen Z, Dagiuklas T, Sun L, Zhu X (2012) Guest editorial: QoE-aware wireless multimedia systems. IEEE J Sel Areas Commun 30(7):1153–1156
50. Mok RKP, Chan EWW, Chang RKC (2011) Measuring the quality of experience of http video streaming. In: IEEE/IFIP IM (Pre-conf Session), Dubland, Ireland

51. Muller C, Renzi D, Lederer S, Battista S, Timmerer C (2012) Using scalable video coding for dynamic adaptive streaming over HTTP in mobile environments. In: 2012 IEEE proceedings of the 20th European signal processing conference (EUSIPCO). IEEE, pp 2208–2212
52. Oyman O, Singh S (2012) Quality of experience for HTTP adaptive streaming services. IEEE Commun Mag 50(4):20–27
53. Reichl P (2010) From charging for quality of service to charging for quality of experience. Ann Telecommun-Ann Des Télécommun 65(3–4):189–199
54. Riiser H, Bergsaker HS, Vigmostad P, Halvorsen P, Griwodz C (2012) A comparison of quality scheduling in commercial adaptive HTTP streaming solutions on a 3G network. In: Proceedings of the 4th workshop on mobile video, MoVid '12. ACM, New York, USA, p 2530. doi:10.1145/2151677.2151684. http://doi.acm.org/10.1145/2151677.2151684
55. Seppänen J, Varela M (2013) QoE-driven network management for real-time over-the-top multimedia services. In: IEEE wireless communications and networking conference 2013, Shanghai, China
56. Sieber C, Hoßfeld T, Zinner T, Tran-Gia P, Timmerer C (2013) Implementation and user-centric comparison of a novel adaptation logic for DASH with SVC. In: IFIP/IEEE international workshop on quality of experience centric management (QCMan), Ghent, Belgium
57. Skorin-Kapov L, Ivesic K, Aristomenopoulos G, Papavassiliou S (2013) Approaches for utility-based QoE-driven optimization of network resource allocation for multimedia services. In: Biersack E, Callegari C, Matijasevic M (eds) Data traffic monitoring and analysis, vol 7754. Lecture Notes in Computer Science Springer, Berlin Heidelberg, pp 337–358
58. Skorin-Kapov L, Matijasevic M (2009) Modeling of a QoS matching and optimization function for multimedia services in the NGN. Proceedings of the 12th IFIP/IEEE international conference on management of multimedia and mobile networks and services. MMNS, Venice, Italy, pp 55–68
59. Staehle B, Hirth M, Pries R, Wamser F, Staehle D (2011) Aquarema in action: improving the YouTube QoE in wireless mesh networks. In: 2011 Baltic congress on future internet communications (BCFIC Riga). IEEE, pp 33–40
60. Staehle B, Wamser F, Deschner S, Blenk A, Staehle D, Hahm O, Schmittberger N, Günes M (2011) Application-aware self-optimization of wireless mesh networks with AquareYoum and DES-SERT. In: Euroview 2011, Würzburg, Germany
61. Staehle B, Wamser F, Hirth M, Stezenbach D, Staehle D (2011) AquareYoum: application and quality of experience-aware resource management for YouTube in wireless mesh networks. PIK—praxis der informationsverarbeitung und kommunikation
62. Stankiewicz R, Jajszczyk A (2011) A survey of QoE assurance in converged networks. Comput Netw 55:1459–1473. http://dx.doi.org/10.1016/j.comnet.2011.02.004. http://dx.doi.org/10.1016/j.comnet.2011.02.004
63. Sterle J, Volk M, Sedlar U, Bester J, Kos A (2011) Application-based NGN QoE controller. IEEE Commun Mag 49(1):92–101
64. Stockhammer T (2011) Dynamic adaptive streaming over HTTP— standards and design principles. In: Proceedings of the second annual ACM conference on multimedia systems, MMSys '11ACM, New York, USA, pp 133–144
65. Thakolsri S, Kellerer W, Steinbach E (2011) QoE-based cross-layer optimization of wireless video with unperceivable temporal video quality fluctuation. In: 2011 IEEE international conference on communications (ICC), pp 1–6. doi:10.1109/icc.2011.5963296
66. Torres R, Finamore A, Kim JR, Mellia M, Munafo MM, Rao S (2011) Dissecting video server selection strategies in the youtube cdn. In: Proceedings of the 31st international conference on distributed computing systems (ICDCS'11), Minneapolis, Minnesota, USA
67. Volk M, Guna J, Kos A, Bester J (2008) IPTV systems, standards and architectures: part II–quality-assured provisioning of IPTV services within the NGN environment. IEEE Commun Mag 46(5):118–126
68. Wamser F, Staehle D, Prokopec J, Maeder A, Tran-Gia P (2012) Utilizing buffered YouTube playtime for QoE-oriented scheduling in OFDMA networks. In: Proceedings of the 24th international teletraffic congress, pp 113–120

69. Youtube press room—statistics. http://www.youtube.com/t/press_statistics (last accessed Oct 2012). http://www.youtube.com/t/press_statistics
70. Zhang J, Ansari N (2011) On assuring end-to-end QoE in next generation networks: challenges and a possible solution. IEEE Commun Mag 49(7):185–191
71. Zinner T, Klein D, Hoßfeld T (2012) User-centric network-application interaction for live HD video streaming. In: 4th international conference on mobile networks and management (MONAMI 2012), Hamburg, Germany

Index

S. Möller and A. Raake (eds.), *Quality of Experience*,
T-Labs Series in Telecommunication Services, DOI: 10.1007/978-3-319-02681-7,

MIX
Papier aus verantwortungsvollen Quellen
Paper from responsible sources
FSC® C105338

If you have any concerns about our products,
you can contact us on
ProductSafety@springernature.com

In case Publisher is established outside the EU,
the EU authorized representative is:
Springer Nature Customer Service Center GmbH
Europaplatz 3, 69115 Heidelberg, Germany

Printed by Libri Plureos GmbH
in Hamburg, Germany